조봉옥 선생님께

건강하시고 행복하시기를.
기도합니다.

2018. 8. 28.

고왕인 드림.

Published by Cedar Publishing Co. (Registered in the Republic of Korea)

Printed in Seoul, South Korea.

First printing July, 2010.

ISBN-978-89-7152-061-1 93300

The Creative Peace

Wong In Koh

Dedicated to my Parents,

Mr. and Mrs. Sung Hoon and Joo Shim (Kang) Koh,

Who Implanted the Seed of this Book

Believing in the Prayer of

"Thy Will Be Done on Earth,

As It Is in Heaven."

of the Jesus of Nazareth

CONTENTS

PART Ⅱ. The Science of Humanity

Chapter 4. The Science of Humanity /85

Chapter 6. The Science of Cruxology (I): Social Continuum /185

Chapter 7. The Science of Cruxology (II): Menesysal Technology /229

PART Ⅲ. Integritium and the Creative Peace

Chapter 9. Status Quo Diagnosis /307

THREE MINUTES READING

This book tells about

The Creative Peace Achieved by Integritium

The author introduces an entirely different set of terms, ideas and approaches in how to deal with confronted global problems. The essence of our problem is avoiding imperium while maintaining human-centric systemicity. Imperium existed as a part of immature integritium, which is about to ripen to admit diversified cultures and politics. The optimum solution to all problems would be achieving a political economy of global integritium, one that would be generated by the creative peace. The pursuit of this solution should be commenced from the scientific approach toward reviving humanity. In this study we attain a methodology of formulating the science of humanity.

Based on the scientific findings, it is generally thought that the initiality of humanity originates from its physicality. Brain science has found profound evidences in support of this proposition. However, natural sciences tend to ignore the phenomenon of the duality of contrary monism, i.e. between physicalism and spiritualism. Physicalism is solely supported by epistemological evidences. On the other hand, spiritualism stands on the foundation of ontological understanding which has yet to be verified by quantified scientific corroboration.

In accordance with the science of complexity, **anthropomorphosis** is defined as **the process of human 'transcent'** associated with strange attractor. Strange attractor of humanity stems from human physicality that is intercommunicative with the substantial universe. Anthropomorphosis is a process of **manifestation of the intrinsic attributes of humanity**. Those attributes are **the result of the positioning of humanity, which is in between Plexion**, the actual infinity of the real universe, **and pragma**, the growing substantial environment of existential humanity.

The Creative Peace is a Dynamic Continuum.

Peace is a dynamic continuum, a continuous body of flow within the vast human environment. Continuum is newly defined in the science of humanity of our research, adopting existing notions of natural science while creating novel concepts within the realm of social science.

Creativity is the Keyword of "the Creative Peace"(創意的 平和).

"Creativity" (創意) is the keyword of this book. The phenomenon of creativity is presented as the goal and solution of peace. Creativity is a matter of communality (共同體性) as well as individuality (個體性). "The Creative Peace" has eventually attained a new and essential concept liberated from the past notion of "creative peace" which merely denoted "the peace pursued through arts and culture."

The creative peace is reinterpreted to be "**the Continuum of Activated Humanity in Stabilized Condition**." All the phases of human history are the output of creativity of the human race. With an adequate scientific viewpoint of the past, history presents entirely fresh insight into humanity (人間性) itself in terms of creativity. "The Science of Creativity" is presented in this research as the science which organizes overall phenomenon of creativity.

The creative peace is what human beings have been pursuing all the time in the name of the kingdom of the millennium. In the present, humanity is confronting such an important historical threshold that we must find a way to overcome and change this threshold into an unprecedented opportunity to enter into in-depth transformation.

"Persontology" (格體論) *integrates the Ideas of General Sciences.*

Anthropomorphosis, the process of the formation of humanity, is essential in the science of creativity (創意科學). Creativity is the reflection of complex humanity, which is personality (人格性). Life (生命) is the outcome of nature with a necessary and sufficient level of complexity accomplished to sustain negentropy in entropic circumference. Likewise, the person (人格) is the question and the answer of humanity, so that people may recognize the complex humanity in a simplex form of their own fractal mode.

The ontology based on the universal phenomenon of person, which is called "Persontology", is the way in which a society and the universe are framed, and the way that a person understands the outer world. "**The Science of Integrism**" (統幾科學) provides theoretical fundamentals to **introduce persontological views into rationale system**.

"The Science of Cruxology" is the Methodology of the Creative Peace.

The solution to achieve the creative peace is based on the analysis of human personality which is reinterpreted in terms of "menesys," short for "MEgalo-NEtwork-SYStem." The concept and methodology of "the Creative Peace" introduced herein could be adopted as an effective tool for the global orientation of peace in the next half century.

The methodology of the creative peace is named "**The Science of Cruxology**" (幾틀科學, Ghi-teul Science), various portions of which have **already been used before recognition in various civilizations of human history**. Cruxology is the framework of the most natural way of human initiatives and responses to spatiotemporal environments. The science of cruxology is **the conceptual foundation for the technology of menesys**. Menesysal management contains the solution of the creative peace.

I dedicate this book to those people and countries that would like to exert their efforts to build up the legacy of "**Integritium**," a next step of planetary development in political economy.

PROLOGUE

"The world is a unity."

Please do not jump to conclusions regarding this sentence; it may be simply stating but within it are innumerable ideas.

What is a human? The science of humanity has not yet been set up to enough satisfaction to induce a full agreement among all the kinds of mutually offset existential bases of humanity. People want to know how to live in happiness and satisfaction. The answer lies in the fact of that how to mutually harmonize and integrate being steadfastly founded on the knowledge of what we are as humans.

I will try to reveal in a most contemporary way what we have to do and what we must become. Why did I select this theme? Because it is the only subject that has captured me since my juvenile days. It was in the vast mist spread over my realm of thought. Like a spirit, it had no name or identity.

Long ago when I was 15 years old, I had imagined writing a poem of a shaman boy sitting under a shiny sun leaning on a hay stack immersed in deep thought in the presence of the Oneness Spirit. The Spirit talked to him gently and lovingly in a voiceless voice with wordless words. It was a man-to-man encounter of perfect peace. It was like a dream.

I have always worried subconsciously whether I would be quitting everything when I had taken on a definite form of life. I wanted to be a wholly mixed existence being nothing but doing everything. Later in my mid forties, I named this an undiversified embryonic being. The vocabulary "undiversified embryonic community" which I used in my writing comes from such experience. Intuitively illustrating my stance of view, the source of the thought comes from nihilism and its consequent anarchism.

As I see, nihilism, anarchism, agnosticism and most other negative thoughts are not destructive to humanity, as long as those thoughts have strong ties with ontological view points. I had a strong source of ontological energy. I had lived into vivid biblical Christianity since my birth. I have tried to live in ceaseless antagonism all my life.

I have been living with bottomless nihility since the pupil of my right eye was pierced by a small medical injection needle. The needle was shot by one of my class mates by accident in an anatomy class making a fun of rocket shooting.

At that time, I was inspecting a cute brown rabbit I had brought from home. I was thirteen when it happened in the first semester of my junior year of middle school. I have now lived for half a century using my left eye for material things and the other left for intangible things.

My understanding on the human cognitive process is overhauled. Every morning, I open my eyes discovering my physical cognitive system gives me a slightly offset four layered image. My damaged right eye had an ophthalmological operation to remove a cataract. It was about fifty years ago my doctor tried to make three intentional scratches on the cataract to induce easier adsorption. It was not completely successful.

It was from then on I stated experiencing peculiar images of the world. My theory of cognition is too practical to discuss only in a philosophical manner. I had a strong desire to dig to the core of my existence, deeper than psychological level.

As I have studied human beings psychologically, I discovered that human beings started basically from nihilism and anarchism in origin. The intangible numbers, zero and infinitude come from nihilistic attitudes from which a man of thought cannot escape from even in the days long before the invention of numerical innovations.

Here, I would comment that humanity, in other words being human psychologically, started from the encounter with infinity, which later diversified and structured in specific ways in accordance with specific personal history or individual hysteresis of experiences. A notable difference between nullity and infinity is the psychological weight imposed on a person when the person recognizes subconsciously unidentifiable externality. This concept will be connected to the orthogonality of phase space of humanity.

Nihilism is the mother of anarchism. In my definition, pure anarchism is the willingness to initialize. Humans intend to initialize through restoring the foundation and completely cleaning the surface to the bottom; as spiritually done by highly virtuous monks in ancient days, passing down to philosophical anarchists in medieval and recent times, and eventually deteriorated and became depraved by destruction by modern anti-governmental anarchists.

Since my "pupil accident," it became my major job to understand the relation between the internality and externality of myself. It also uncorked my understanding on the connection of subjectivity and objectivity, on the distinction of consciousness and subconsciousness, and gradually on discerning everything, in most practical ways mobilizing all the nerves in my later life. The indentations of lives make people different from each other. The specifications (特化) of persons are realized because such indentations grasp lives in various differentiated forms.

I have precious experiences with the Christian Bible in my heart. However my interpretation of the Word, different from the conventional ways, encompasses the whole span of human thought; from western to eastern, from ancient to contemporary, from Hinduism to Confucianism, from major orthodox theologies and heretical thoughts, from divine to anthropic, and etc. I emphasize that none of the historical developments and even meaningless internal psychological incidences are to be neglected.

They are reflecting certain aspects of the whole signal from the eternity. I am not cutting out that which does not belong to the main stream of humanity. Reductionism, as I see it, is mostly attributable to a psychological tendency. All things have the right to be duly interpreted. It does not mean that nothing needs to be cured.

I tend to have refused to become a part of any existing pattern and at least to unalterably make myself certain patterns with no significance. I have tried to enjoy the joys of the growing points of human life alike those of plants.

Despite being very creative, I have not done much with that creativity I was given. It shows that my critical concerns remained with the spiritual aspects of this world rather than the materialized aspects.

What does being spiritual mean in the real world? I would say it has to do with most material things and "not going further than needed." It thus makes humanity concentrate solely on what should be attained. Spirituality has another implication that it is in factual world identical with pure spontaneity. Thus a personal spirituality can be defined roughly by his voluntariness.

Spirituality always meets with materiality in this world. In this context, in a very subjective understanding of mine, spirituality is the concrete source of creativity. I will leave a comment here that I differentiate mentality from spirituality. Western thoughts usually do not give distinctive conceptual differences between these two notions. But there is difference. Mentality is similar to virtual reality produced by brain-neurophysiological activities.

The science of complexity, which I met in my late forties, was very identical in its structure of thinking with my way of thoughts, even though it has yet to be generally acknowledged among people in the world. The science of complexity penetrates all of human history and even deep into the dimensions of spirituality. I have saved my argument on this matter but now would like to express that the science of complexity is the science of humanity. At the least, I will say the science of complexity has stemmed from humanity.

The science of complexity is a good tool for the illustration of evolutionism following the line of thought, "adaptation - natural selection - evolution." However, I believe that this science must be developed more complexly than most contemporary thinkers would accept.

CAS (complex adaptive system), usually dealt in present complexity science, focuses more on subjective aspects. The directives(嚮導) of intentionality(志向性), for instance, the notion of which, I believe, should be introduced in brain science, should essentially be conceptualized to extend the horizon of this science. This can be explained through the balanced theorization between a subject and its circumference. It means we need to avoid epistemological thinking methods when we think.

The epistemological view shows a way of constructing a person. Building up a person is based on a serial process of experience. Experience is personalized by personal allowance. Allowance or tolerance of an individual is governed by the character, structure, and intensity of his belief. Belief is a wandering terminology in present days and thus not easy to define. Belief has to do with intentionality, which totally represents a person, but should be more than that. Belief is the source of the projection of consciousness. Belief is not mystic but very practical.

In the Buddhist's view, "seeing nullity"(見空, the view of vacuity) is different from "nullity of seeing."(空見, the view of nullity) "A man seeing nullity will be saved but man with the view of nullity cannot be saved." When I was experiencing different kinds of psychological depression in my high teens, I had gone through understanding the meaning of the revelation of 'seeing nullity'.

The crevice and nihil in and of mysticism was overcome by understanding the paradigm of the right mysticism. I am sure that mysticism is not a foe of science and intellectuality if properly understood. I believe it has much to do with the human physicality of on-brain and off-brain activities. It is true that various advanced religions mutually agree to the others in their intrinsic central rationale, although they strictly disagree in central dogma.

Belief is also the central rationale of all the paradigms of humanity. "Now faith (=belief) is the substance of things hoped for, the evidence of things not seen." (Hebrews 11:1) The "substance" means in Koine, i.e. in the ancient Alexandrian Greek dialect, *huipostasis*, the real body or the true entity.

Unseen evidence tells of before the circumferential happening. That evidential substance is in reality in us before materialization (in a manner of information, in a form of superposed wave patterns, or in a pattern of persontological plexus of affairs). Some of the experimental results of brain science support certain reversions of temporal processes in human response to the environment.

In such a context, belief is a sort of inversion in cause and effect. The human creativity is generated in the course of interaction between internal subjectivity and external environment. And exchanges of distinctively different values in the same rationale system are made through nullification (無化) and infinification (無限化).

As I have observed my inner thought process, the nullification and infinification has an undividably firm league with emotional processes of humanity at peak levels. The nullification and infinification is named "N&I" in my dictionary. I found that belief is at the center of creativity. I believe a well-balanced science has the same effect in our material as with an unbiased religion. I am basically a scientist and an engineer who has chosen a methodology of discernment. The harmonization with spiritual truth is another task.

I certainly believe humanity has a hopeful future as long as it finds a proper way. All the paradigms have their own particular positions in a dimensional coordinate of humanity. All the races and nations have their own such positions. Their historical agonies and even the antagonisms among them also possess their own marked places to be located.

My personal pursuit and expedition for the creative peace came from my early experience of a war which swept over the Korean Peninsula in 1950s. Despite my family being comparatively wealthier than other famished families, the pain that struck me was deep. I was five years old when I felt that pain, which I still feel when I see newly troubled nations appear in the present world.

Sometimes I burst out in tears and my heart is torn during my everyday dawn meditations. I have a longing to design a global mechanism which will help establish the creative peace for all the nations.

An empire is what many contemporary intellectuals see the future of the world. I do not believe so. Even the strongest country of this day should be considering an enhanced way of managing political economics. We are not at the top of the sinuous curve doomed to drop down; we are still going up. I believe in market economy. I also believe that adjustments will stay valid only when a variety of seeds and domains remain in them.

"Continuum economy" is the picture in my mind. The positive mechanisms and utility of imperialism and elitism should also be duly recognized within proper constraints. But the practices of power need to be altered.

The future of the human race is in outer space. It is thought even more feasible now that the global impact of financial collapses have been overcome. Space is not a dream anymore and has already become a concrete aim. We need to find possible means to concentrate all the available tangible and intangible resources of the planet.

But the choice of methods of concentration and mobilization always make the difference. This book will deal with general methodology of managing the future of humanity getting along with establishing the science of humanity.

And also pursued is how we can construct a soundly evolving and dynamically stable status; "the creative peace," in conservation of the diversity of humanity and the source of directives and energy of "anthropomorphosis."

PREFACE

This book is the fruit of the writer's life-long search for an integrated solution for the future of humanity. The frame of the study covers a relatively broad range of disciplines. Some topics if necessary are touched in depth. It is recognized that certain topics must be dealt with more detail later on. Karl Popper expressed that living science is an eternal process imperative for a creative hypothesis to be supportive, and researchers are supposed to frame hypotheses and sort out unacceptable notions through experiments and other empirical methods. The basic rationality behind the discussions of this writing is somewhat hypothetical, but can be perceived to be practically useful when recognized in an integrated picture.

Humankind has developed several major streams of thoughts, such as Judeo-Christian heritage and Islamism from the Middle East, Confucianism and Taoism from East Asia, Hinduism and Buddhism from South Asia and so forth. A fact to be noted is that, in this century of space entrance, people want to establish a mutual coincidence of global rationalities and consequent communication channeling among different thought systems.

It is a common understanding that while beliefs and thoughts stay different among societies, cultures, civilizations and historical backgrounds. The genetic coincidences of the mental and physical humanity are undeniable as is the essential mode of life in spite of all the racial differences. What is the phenomenon of humanity?

The Old Testament is an important source of human thoughts. Genesis starts with "*Bereshieth Bara Elohim ~,*" which means "In the beginning God created ~." Divine Creation is the source of human creativity, because human beings resemble God.[1)] The "beginning" is sometimes interpreted as to refer to the head of all creation, the "*arche*" in Greek. The first three verses of Genesis which explains up to the creation of the first light have important implications to the birth of the universe. As they seem to be a metaphor for the Big Bang. As some people might think, this somewhat coincides with the concept of emergence in the science of complexity.

In Genesis 1:2 following just after the great premise of Creation mentioned in verse 1:1, the writer will humbly interpret that the Divine Heart has been revealed. There are a number of commentaries made on 1:2. It can be carefully suggested that the verse shows the Divine Mind of Creation. The world, interpreted in English as the earth, and *ha erez* in Hebrew, is not God Himself but Divine Mind. *Ha Erez* does not include the heavens.

> The earth was without form and void (*tohuwabohu*),
> and darkness (*hosek*) was upon the face of the deep (*tehom*);
> and the Spirit (*ruah*) of God was moving over (*merahephet*)
> the face of the waters. (Genesis 1:2)

Tohuwabohu is known to be a twin word that usually appears together. "Formless and void" is the scriptural foundation of "creatio ex nihilo," creation from nothingness, one of the most important doctrines in Christianity. (The interesting commentary of Nahmanides cited by G.L. Schroeder needs some discussion in depth. According to the interpretation, *bohu* is a status filled with the building blocks of matter. It is agreeable when the building blocks are interpreted vacuity that generates everything.)

The earth (*ha erez*), equivalent to the world in Hebrew, was nothingness (vacuity out of the wholeness); then, how could 'darkness' and 'the deep' exist in or come out of nothingness? *Tehom*, the bottomless abyss, is filled with water; then, 'the water of the deep' and 'the waters' could not be substantial.

"*Ruah*" is traditionally interpreted "the Spirit of God", the Third Person of the Trinity who is unsubstantial, only spiritual.[2)] The abyss is the infinite desire as a metaphor; which can only fit God's Desire only. (Ezekiel 28:2) The Spirit of God moves and vibrates[3)] over the face of the deep in the darkness of *Deus absconditus*, the self-concealing God. (Isaiah 45:15)

Hebraism is said by H. Trocme to take dynamics from asymmetry and unbalance.[4)] Messianism, the self-propelled dynamism of human history, stemmed from the asymmetry of humanity. Messianism, either individual or communal, is a model of human endeavor to overcome obstacles in history. Messianism is in some sense is the growth point of humanity. Transcendental excessiveness and concentration of energy and substantiality of a life appear at the growth point. It is the internal logic of a life form.

Another term "integriti-sm" which generalizes such messianic mechanism in humanity needs to be given. In general such dynamic asymmetry is hidden in static symmetry. Integritism is a creative approach in achieving individual and communal integrity in a person. "Integrator," a reformer, was held in higher regard than the emperor in ancient Rome. Integration, or, if acceptable, integritizing, is the source of living energy and unifying rationale for a society. A messiah, in this sense, is an integrator that renews the society as can be seen in a shoot with a growing point from a stump of a tree. In Genesis 1:1, *bereshieth*, *arche*, head, or the beginning is the initiator and integrator of the world. In this context, Genesis 1:2 was suggested to be considered as being a paradigm to understanding Divine Providence of driving history.[5)]

Now let us look at the Neo-Confucian views on a person. A person is the subjective actor of one's own destiny. The static body (體, Ti, or Che) of a person is generally structured in symmetry. But the dynamic use (用, Yong) is realized in asymmetry. The body and its use, Che and Yong, is a good combination that illustrates the phenomenon of humanity. The kinetic energy originating from the unbalanced dynamics in a personal body apparently presents forwarding action.[6)] Che is the source of stabilization while Yong is the source of movement. Human mentality possesses both of these to process the history forward. When Che-factor prevails, human history circulates and shrinks into a stabilized mode. When Yong-factor takes the initiative, history is oriented to striving to expand the existing body and/or create a new body. However, Che and Yong behave in a manner of being together. Yong is the movement of Che while Che is changed by Yong.

Human beings by nature always have been living in mental transition zones. These transition zones are what Yong has produced. Transition generally sustains its stability because of Che. Che furnishes balance and belief on its rational conformity. Human beings assume ways of conformity in the boundary zones of mixing opposite and contradictory thinking methods; myths and reality, faith and fact, divine and human, ontology and epistemology, and materialism(唯物論) and mentalism(唯心論). Man is in the middle, floating between the dipolar extremes. They define the real world in between.

For instance, ontology and epistemology stand opposite to each other. However, the two philosophical attitudes have been collaborating in history for the benefit of human civilizations. Each party respectively has several branches of thinking methods within itself. But in a broad sense, there have been mutual shifts and movements of boundaries and zones toward either side several times in history, which generated refreshing societal energy and sometimes induced remarkable breakthroughs in our civilization.

To create something, we need to start from a stabilized position which contains a good symmetrical balance of past knowledge, but consistently have to keep asymmetry in ourselves to drive forward. It is how a human being behaves. Even enterprises work in this way. Accounting and organization are usually managed on symmetrical principles, while marketing and research & development are done on asymmetrical principles.

We know the above comments are somewhat too simplified. These instances focus on the necessity of integration between symmetry and asymmetry. Integration is a matter of conformity to be developed between opposite or conflicting thoughts and attitudes. The conformity is in common sense supposed to be based on a unified rationale.

The creative peace is a peace with asymmetry. Usually peace is maintained when the condition of symmetrical balance is attained. However, the pursuit of "peace with asymmetry (which is peace with creativity)" that is defined by "stable structure and the dynamic energy of humanity" is one of this book's main points. Thus we must call the creative peace a young and energetic peace, which is capable of creating, producing, and endeavoring. But the energetic creative peace does not allow for violence and decadence.

Symmetry is a principle which enables the process of systemization, while asymmetry is a principle for the process of networking. Asymmetry is generally realized through the development of networking. Networking does not originate from an intentional design. On the other hand systemization needs a well planned and balanced design.

In a strict sense, a system cannot determine randomity that exceeds its own quantity of information. This means no system can understand its own program by itself. Randomity is recognized in our thinking by being furnished by nexusity. In an evolutionary view, the compatibility of networking and systemization would together have made it possible for their well combined pathways to the present.

However, in our contemporary society, systemization has been heavily processed through the teleological development of human society. Thus a network is usually developed until the systematic characteristic of the network is achieved to certain intensity that the efficiency of the network community is high enough for its survival. Systems are usually designed in order to concentrate all capabilities and resources of its constituents to accomplish totally high efficiency.

A system easily experiences failure when drastic changes of either internal properties or circumferential conditions occur. It happens most frequently from failure of symmetry previously attained by balancing.

Systems in many cases tend to take on network behavior in order to avoid catastrophes and to maintain long-term maximum efficiency. Perhaps the phenomenon of human cognition by the brain would be a proper example for explaining the interweaving of the two contradicting mechanisms of system and network.

The human brain is probably the most effective system that has ever appeared in the organic world. It would be a good question to ask why the human brain usually exceeds the body in efficiency. There is no way yet to quantitatively measure the efficiency in the operation of human body and brain. But it is understood that in the final round the brain seems likely to have won the game against the body. This is achieved, ironically through maintaining a usual manner of mutual collaboration.

Cyborgs, as substitutes for body parts, may be necessary in the near future at least for the excavation of extreme environments. The brain looks like it expels inefficient parts of the natural body while introducing artificial ones in its place.

In recent years, brain science has dug deeper into the matters of cognition, emotion, consciousness, and personality. In most cases, researchers tend to adopt the idea of evolution for analyzing investigated data. The century we live in is unfortunately no where anyone can escape from the influence of evolutionism.

However, as far as humanity is concerned, it seems inevitable for us to ruminate over the essence of ideas, paradigms, and even metaphors applied to the building up of our main frame of mentality. Our mentality should be, to a large extent, interactive with science and technology, which has revealed its dominant power over all civilizations. The mentality, in higher hierarchy over the materialistic mechanism of the brain, is a link between past civilizations and future integrative developments in humanity. It sounds odd to understand that materiality produces mentality; either side is able to establish its own complete systematic chain of logic using with persuasive ideas.

Factual evidence speaks for itself. The case of Copernicus – the long journey until the final acceptance by the Vatican – ensures us of the victory of science. However, at the moment, this science has become responsible for human mentality, including paving a way for social science and cultural science (usually called the "humanities"; however, it looks adequate to expand the usage of "humanity" over all the manifestations of the mentality of humankind).

Some scientists say in the next half century mathematics will be transformed into biological mathematics. It is known that biology has been affected and changed by physics. Psychology has found her inseparable partner in neuroscience. New fields such as "brain philosophy" have been established since the start of the 1990s.

What we realize now is that we need integration. We are at the point where a reliable integrity of human ideals must be achieved to drive our civilization further on "the creative peace." This study has tried to establish such ideas in a coarse manner (as seen to most refined intellectuals), in order to integrate a wide range of subjects, encompassing West and East, past and present, science and culture, and many other different ideals, as we have encountered.

An earthworm, wriggling ever so slowly towards a spring of water, leaves traces of its body on the wet surface of mud. This can become a sign to the thirsty man searching for water, after which multitudes of people may follow and be saved. Like an earthworm we may be doing nothing more than wriggling towards another life-giving discovery.

2008.1.7. Wong In, Koh
Author

---

1) *Imago dei* is thought to have originally provided the birth of the terminology of Anthropomorphosis. "*Ad imaginem ad similitudinem*" (Genesis 1:26) Anthropomorphosis is warned of its danger to have Christianity mix up with pantheism. Ancient Greeks had invented human forms of gods and goddesses. These cases could have decreased the possible outcomes of scientific interpretation for theoretical generalization. But the key point of this problem is how an *anthropo* (human) can enjoy the association with infinite Divinity. Humanity is the matter of infinity, which is the essence of ancient Divinity.

2) In order to discuss "*ruah*", the Trinity of God has to be commented in theology. Trinitarianism is of "One God in Three Persons." However, person is a notion yet to be developed further in philosophy and science. (*Persona* and *prosophon*, the sources of the vocabulary, will be later discussed.)

3) Von Rad, a commentator, annotated that "The much disputed 'merahepet' is not to be translated by 'brood' ~, the verb appears to have the meaning of 'vibrate', 'tremble', 'move', 'stir'", and "'Ruah' is better translated 'storm of God', i.e., terrible storm ~."

4) Andre Trocme, *Jesus et la revolution non violente*, 1961.

5) A similar paradigm is in the calling of the Hebrew people to the perfection of Divine Holiness starting from the social structure of symmetric talion, the law such as "eye for an eye."

6) Che-Yong or Ti-Yong in Neo-Confucianism developed and prevailed in East Asia, China of Song Dynasty (since 11th century), Korea of Koryo Dynasty (since 13th century), and Japan (since 16th century). The concepts were used in a more liberal interpretation of personal taste. The concept of Ti-Yong here is not used in a strict manner as among Neo-Confucians.

INTRODUCTION

The content of this book is inseparable from the personal history of the writer. The motives and developmental thought processes of this book are all coils of bundle of his life-long search. The contents are sincerely scholastic and scientific as balance was striven for in avoidance of any bias and deviations. It is recommended that the prologue and epilogue be read first, in which is presented entanglement of the writer's destiny which has led him to this writing.

The purpose of this book is to express what the creative peace is and to find deduce an approach of attaining the creative peace locally and globally. The topic is subject to the support of vast and profound investigation of data and information. However, it was intended to deal with the topic largely at a hypothetical level to show at a glance, the whole scheme on which the thought and methodology stand, despite being at a premature stage of verifying the feasibility of the topic in an examined and quantifiable mode. It is preferentially deemed necessary to alter the way of thinking by changing basic standpoints in approaching the glocal (global and local) and permunal (personal and communal) problem of peace.

The definition of the creative peace is "the continuum of the activated state of humanity in stabilized condition." The three keywords within the definition of the creative peace are humanity, integrity, and creativity. Humanity is redefined to be the spatiotemporal output of the whole human race encompassing all the phases of human phenomenon. It does not have to be as holistic as it is usually thought to be.

Integrity is achieved to a sufficient degree in a complex phase space of humanity composed of three domains: individuality, communality and external environment. Human integrity is achieved mainly in the communality domain which is actually a transition zone between individuality and environment. But when social communality is excessively agitated by the distortion of the esthetic condition, human integrity finds itself in the realm of individuality.

It is true that individuality at its own level also has to reveal integrity of its own. The integrity of individuality is mainly originated from the physical conditions of an individual, who makes his own intellectual, emotional and/or volitional abilities highly concentrated. Sometimes, individuals disconnect most of their connection with community and revive right standard through their internal reformation. This Ethical phenomenon can be seen as a sort of fluid flow. Similarly, the phenomenon associated with human creativity can be seen as deemed analogous with fluid mechanics and phase behavior, if the theoretical system is well defined.

Many geniuses practice their outstanding creative activities in unstable psychological conditions at individual level. However, there is a need for their creations and innovations to be processed and accepted by their community in accordance with communal standards and necessities, at least within communal tolerance. Creativity, an activated and elevated state of humanity, is generated only when and where a significant and sufficient degree of integrity in the domain of communality (from peaceful through tolerable stabilized condition) is attained. Creativity, as has been observed in the special case of geniuses, is frequently generated successfully at low degrees of integrity in the domain of individuality.

This writing is based on the thought of how to produce the best output from mutually competing systems, including the most frequently applied arrangement of dipolar, tripolar and other multiply compatible systems. Bifurcation seems inevitable in human-originated society and in brain-produced thought systems. It is thought that "the science of creativity" should be developed in order to support the overall approach to the creative peace. The science of creativity should be able to treat all phases of the creativity phenomenon based on why and how human beings respond to their environment, which in fact is the dynamic plexus of accidental occurrences. Therefore it is necessary that the science of creativity be supported by "the science of cruxology" and "the science of integrism." Cruxology and integrism is explained later on briefly in this introduction and deeply in Part II.

This study is fundamentally based upon the thought that a human, as a person(人格), is presented in a mixed frame of systematic and network mechanisms. The essences of these mechanisms are respectively named systemicity and nexusity. The mixed dynamic frame of these distinctive mechanisms is called an integrism. The individuality of a human being is embedded in human physique, as is the brain. It is deemed that individuality is subject to the good combination of the closed mechanism and the openness of the human brain. The closed mechanism is observed within the self-recognition process while openness is presented(observed) while communicating with the external world and in maintaining "dynamic modularity(動變 分格性)" of the brain. The concept of integrism is thought to lead us to a new understanding of the individual for a more profound analysis of human response to community and environment.

Integrism(統合能, or 統幾) is a measure of the integrated degree of integers, which represent people in individual case and various forms of communities at different communal levels, within their local environment. The study on integrism is called the science of integrism. Integrism is comparable to organisms in a terminological context, yet is differentiated by dynamicity. The term "organism" denotes the form of a body in static state, while integrism of an integer denotes the mode of dynamic plexus, which consists of interwoven serial occurrences of accidental events of an integer. The accidentality of an integer from all the occurrences of succeeding events is the outcome of the total externality of the concerned integer. Thus this difference in their existential definitions generates ontological differences between organism and integrism. Integrism is also a different notion from paradigm in the sense that paradigm is solely ideal and conceptual. Integrism, however, is a respondent and communicative frame of humanity embedded in material physique and the neuronal brain.

The science of integrism(統幾科學) provides the basis for the science of cruxology(幾틀科學). Crux is an initiality or initial value of human initiative which contains personal orientation in both emotional and logical view. Human initiative exposed as crux decides the subsequent proceeding of human behavior. Crux is an initiality we usually tend to express within the science of complexity. Crux is the seed of cruxological development. Cruxology is the way in which human beings understand, interpret, organize their own knowledge system; make future plans, and implement as their response to the external world. Cruxology is a very similar notion to wisdom. Wisdom is more or less intuitive while cruxology is aiming at constructing scientific frame of wisdom developed from scientific and technological findings. Thus the study on cruxology is called the science of cruxology.

The science of integrism also gives foundation to the science of creativity(創意科學). Creativity ripens when integrity of relevant sets of knowledge converges at the point of vitalizing. The concept of creativity used in this study is broader than used in the original sense. It is thought that creativity emerges similarly with transcendental mechanisms which exceed the rationality and cognizability of the prior system. Creation or equivalent creative activity occurs in the transition process from the previous system to the present system. The vitalization of creativity appears to be classified into three groups, which are to be later discussed in detail, in accordance with the demands of individual or society. The science of creativity is concerned with maintaining proper level of humanity activation so that the elevated energy level of humanity may be kept sound and stable within a certain range.

If we were to oversimplify the content of this writing for an easier understanding, it could be illustrated such by a focusing on the phenomenon of person. Person is usually regarded as a static entity. However a person is hereafter considered as an integrated dynamic spatiotemporal existence, which has been generated from a number of precedent accidental occurrences with high enough relevancy and which will affect a number of posterior accidental occurrences with high enough relevancy. A human is regarded as a person when a human fulfills the sufficient requirements as an integer in a community by integrating the effects of network and system of his own will. Why do we emphasize network and systematic mechanisms together in a dipolar pair? The reason is that humanity stands in between systemicity and nexusity. Both essences are originated and developed from the need of survival. Human survival is not only a matter of material flesh but also of relational existence. Relational dynamic existence is at the center of humanity. Thus, it is encouraged that to readers understand that this writing is far and situated safe from the debates between evolutionism and creationism. It is the writer's foresight that such debates will pass away, and instead converge upon deeper reasonable logical foundation as sufficient scientific findings and wider understanding accumulates.

The way of thought adopted in this research is not recommended to be classified within social engineering. The approach is fundamentally scientific. It is a peace study which could be named the science of humanity (人性科學). It contains three distinctive realms of humanity expressed in ancient and contemporary versions: heaven (天), human (人) and land (地) expressed in the ancient version; spirituality (or, if acceptable, mentality), communality and environment in the contemporary version.

Here, environment contains the materialized portion of intersubjective human products, which is frequently called civilization, as well as natural environment. The match between heaven and spirituality (mentality) is made, because spirituality is reinterpreted to mean heaven in the phase space of human noosphere (if we try to use the closest terminology).

As commonly conceived of the Absolute Other (*Das Ganz Andereheit*), the total ontological frame creates the notion of centeredness in human cognition such as the Divinity and the self. This frame of thought constitutes theoretical basis for ontological centripetality, which proposes that all the substantial entities exist in a unified and centralized manner. (There is a contradictory opponent way of thought which is usually called epistemological philosophy. It can be expressed as the ontological centrifugality of humanity.)

The aforementioned three sources of human value interact mutually to eventually constitute the overall activities of humanity. Such an idea is called hereafter "Triradixism (三本主義)," which means three-rooted idea. The standpoints of this essay are that scientific explanation has to be based on materialism, which is centrifugal approach in ontological view, and thus that the mental and spiritual aspects of humanity should be dealt by centripetal approaches within the ontological view. Ontological centripetality is the key to understand the self-centeredness of humanity and human history. When centripetality and centrifugality of ontology appear together in contradiction, either side of mentalism and materialism does not take a leading role over the other side.

Thought and imagination are palpus of humanity. Humanity is always in deep and calm palpitation before the presence of infinities, or "the infinity" if acceptable to readers. The phenomenon of humanity commences from the encounter of the astonishing infinity. The intellectual encounter with the infinity did not happen only to Cantor and Gödel, who tried to conceive actual infinity in their intellectual capacity. The conception of actual infinity into human intellectual, which is based on the finite physical condition, is most likely impossible. I believe that these two great theoretical mathematicians, Cantor and Gödel, were neither satisfactorily gainful nor successful in approaching the real infinity intellectually, as can be seen in unhappy finale of their mental and psychological lives.

However, common people can cope with the feeling of infinity in an emotional aspect and process it into digestible manners in their brain activities. It is named "human infinities-matrix."

The Infinities-matrix is the structure of infinities which are suscepted and engraved mentally. The Infinities-matrix is important because of its overall influence to human integrism and cruxology. It is thought that the humanity of Homo sapiens was started from the trial-and-error simulation process of unidentified sets of experienced infinities by their brain function.

Now, going back to the main subject, the buildup of the creative peace in the real world can be manipulated by framing the aforementioned three sciences and applying those tools under implementation. The three sciences concerning integrism, cruxology and creativity constitute the most demanded knowledge system of the present generation in order to solve the current problems of the human race.

The creative peace is the arbitrating management of the dynamic humanity between the ideal of human subjectivity and the environmental reality. The endless efforts of humanity on the way toward the infinitely extended ideal from the present reality supposed to expose human transcendence over their limits. The identity of human beings is grounded on the fact that they do not give up in changing and overcoming their boring state of present situation, no matter the circumstances.

In pursuing the creative peace, we need to clarify our vision first to show what we seek to surmount and endeavor after as presented in Part I. Thus in Chapter 1, the vision is described in a common and realistic sense by defining the creative peace, creativity and the essence of peace.

In Chapter 2, the "Alpha Path," which is a major property of "the Creative Peace," is discussed at this current point of transition into the new millennium. In Chapter 3, discussion becomes more detailed and in-depth to show the keys of attaining the ideal of the creative peace. Its main theme is humanity encountered by and responding to the actual infinity, without which we cannot understand in our abstraction in semiotic mode.

In Part II, four main ideas are presented in an illustrated scheme of thoughts that the writer has developed. Chapter 4 has contents of what humanity is and has to be; i.e. humanology. The science of humanity encompasses the science of integrism, the science of cruxology, and the science of creativity.

In Chapter 5, the concept of integrism is introduced to show the criteria of assessing the state of humanity in certain circumstances.

Chapter 6 explains a way of thinking with the cruxological approach, in which cruxology is a general mechanism of the internal phenomena of humanity. Cruxology is a measure of how humanity produces variable and complex form of social entities.

In Chapter 7, we have explained a technological concept of menesys that is extremely variable and complicated in its manifestation. Menesys is the basis for the management of humanity in all personal, communal and global phases. Thus we can say that integrism is a criterion, cruxology a mechanism, creativity an essence, and menesys a management.

Chapter 8 is about creativity. Creativity is at the center of the external phenomena of humanity in a socio-psychological view. Creativity is an essence of humanity.

Part III deals with *ad hoc* issues. In Chapter 9, we will analyze the present situation of global conditions in terms of political economy. In Chapter 10, some of the essential measures that humankind should take in order to achieve further advancement are presented.

If a brief picture of this books material is sought to be attained, please read the 3-minute summary at first in order to get acquainted with the mutual connection of the new vocabulary. Afterwards, read in the following order: Chapter 10, Chapter 9, introduction, Chapter 1 and then Chapter 3. It would be also helpful if the summaries of Part I, II, and III are skimmed through briefly.

Theoretical readers would be recommended to read steadily chapters 1, 3, 4, 5, 6, 7, 8 successively.

PART I.

The Real World and Humanity

Chapter 1. A Realistic Vision of the Creative Peace

Creativity is the keyword of the next 200 years. Why is it applied to only about 200 years and not less or more? It is intuited in historical observation that human race goes through cycles of roughly 200 years in maintaining a healthy period of civilization and in keeping the high tide within spirit of a well managed dynasty. After the 200 year period, the human race experiences a new tide of alteration in their existential mechanism. Creativity is the next immediate agenda that must be tackled in order to overcome the inner conflict that the human race has introduced into reality out of the unbalanced contradiction of their ontology.

At the end of 20th century and at the dawn of the third millennium, our planet recorded a double shock in the economic history. Both competitive branches of economic thoughts were hit by hard blows of iron hammer. The crises of the socialist collapse and the financial crash of global capitalism are attributable to the fact that human race does not have the eyes to predict long-term outcomes from various sources. It is the duty of our generation to establish the criteria of conditions needed to achieve a desirable status of global political economy.

The creative peace is presented in this study to satisfy the wish of humanity in their endless struggle toward the delighted life of enough food, comfortable wear and safe shelter in a loving their beloved community. The creative peace is the technique of political economy which allows us to keep the toolkits which achieve and sustain such common happiness; those toolkits are global dynamics, hyper-nations, nations, and other subordinate levels. The creative peace was aforementioned in introduction to be "**the continuum of the activated state of humanity in stabilized condition.**"

In this chapter, the creative peace in reality will be defined. The chapter will try to establish a methodology to reduce the negative impact of collisions and conflicts to provide foundation for a new commencement and for the maintenance of the balanced unsymmetrical regime of the creative peace in a long-term sense.

1-1. Definition of Creativity

Activation of Humanity
Three Kinds of Creativity

About half a century ago, it was only in the field of creativity education that scientific approaches could be systematically found regarding issues on creativity. In cultural programs, creativeness has always been a major concern during all periods. However, a consensually agreed-upon definition of creativity has not been able to be reached.

The Present research trend in creativity seems to be more devoted to individual cases of the creative originalities of geniuses. We can find voluminous works of research by Mihaly Csikszentmihalyi and other groups, which concentrate on the individual originality of creativity.

In this book, both of communal and individual creativities have been taken into account for a general understanding of the phenomenon of human creativity. It is our strong belief that humanity cannot stand firmly if creativity is lacking.

Activation of Humanity

Humanity can be considered to be the accumulation of creativity. Economy is the fruit of the cumulative value of creativity and labor, both expressed by acting in the labor market and freezing in capital value. This viewpoint on economy will be deeply illustrated in Chapter 8 to give a more enhanced theory on value.

Creativity is generally activated at dead-end situations. If a person or a community is creative without any threat to life, their creativeness may have been activated due to internal attributes which have been formed through peak or summit experiences. When these internal attributes have come together to construct a structure, we call this process "summit build-up."

The skeleton of the summit build-up is the result of the neural treatment of the human brain and neuro-system against an excessive impulse affected by so called infinities (or internal attributes affected by infinities-transfer). A summit build-up is thus called an "infinities-matrix." The intangible infinities-matrix which is responsible for the emergence and construction of human mentality is the solid eventual source of anthropomorphosis, the expanded reproduction of humanity.

It is true that a person is creative when he is concentrating to find unique solutions which have never been seen before. The concerned concentration does not resemble the mental concentration required of memorization, or the repetitive actions of training for a sports game. It has somewhat similar characteristics to the relaxed selection of baby babbling in evolutionary debate. It is what I would name a nexusity effect in peripheral condition.

The phenomenon of creativity is as complex as the phenomenon of humanity. Creativity is at the center of humanity. The Creativity of human being was originally identified as a reflection of Divine Creation in Christianity. However, once the concept of Divinity has been put under the examination of disbelief since the days of the Enlightenment, liveliness of vivid creativity has been hidden under the cover of all-explicable intellectuality.

Explicability of modern intellectual thoughts is based on a rigid structure of rationality. Do I suggest here that modern intellectuals are not creative? I would comment that modern creativity, being dragged by overly systematic rationality, is apt to miss the summit part of vivid creativity. I believe that readers will be able to comprehensively grasp what the summit creativity is after reading this book.

Post-modern ideas seems to be effectively dismantled in the oppression of solid modernity except restoring the reframing the creative humanity in a whole. But, we find some possible exit in scientific thought like the science of complexity, as long as it is interpreted in a balanced way. Creativity is the main factor of interpreting humanity with the science of complexity, especially as to the quasi-phase behaviors of the change in individual personality and in communal society.

Creativity is an elevated part of humanity. Thus it has a different mechanism from the usual way of human reaction. We can say that humanity can be activated far higher than routine life if creativeness becomes our major aim of life. Sometimes and perhaps most frequently, creativity is attained at the cost of losing stability of life, as most people experience. The creative peace is presented as the tool for humankind to use to overcome the pitfalls of adopting creativity heavily into humanity at the cost of losing stability.

If we fail to introduce an effective way of utilizing human creativity, humanity will be unable to manage the rapid up-rise and bursting expansion of technological innovations. The resulting probable singularity problem could hit contemporary civilization. We believe that present humanity has enough mental and cultural self-control on the matter of discontinuity caused by probable singularity. Such an effective way of managing human creativity will be organized by materializing the creative peace.

We would like to define that creativity is basically the activation of humanity. Humanity is based on the cognitive mechanism which consists of more than concentration, memory, learning, reasoning, systems of knowledge, superficial variation and combination, and systemized intellectuality. It has been gradually recognized that humanity must be more superior to cyborgs entities in terms of the infinitely transcendental character which may lead the human-nature complex to the eternity, as long as humanity never gives up its identity. (We have to carefully comment here that the infinite number of steps of transcendence does not accomplish the real infinity.)

Three Kinds of Creativity

Where does this quality of transcent come from? It is my sincere desire that this terminology "transcent" to be widely used in human intellectual world. It is neither a descent nor an ascent. It is a step higher than those two brethren terminology, opening a novel horizon for humanity.

It is like the rationale of humanity hidden in the logic of 非想非非想處定(the abode of consciousness neither of perception nor of non-perception) in Buddhist thought. It is (or could be interpreted as) the place of humanity (in topological locality) where all conflicts in humanity will have been overcome.

Fundamentally, creativity is the mechanism of finding solutions against conflicts spread over our lives. Conflicts have certain hierarchical structures, depending on the causes and frames of static and dynamic relation of the universal humanity.

In this context, creativity is defined in three stages of different magnitudes of complexity. Detailed explanations will be presented in Chapter 8 in the science of creativity.

The first kind of creativity is generated from the simple combination of ideas. The first kind can be called contingent creativity as it usually happens accidentally, although it may be a case of well intended finding (searching). Combinations of pieces of different ideas seem to be achieved usually by synchronized plastic deformation of imagination. It means neuro-plasticity is the main drive of the first kind of creativity.

The second kind of creativity streams out of the integrated structure of rationales within an internal process. This second kind has to do with systematic ideas in humanity. It is an unconscious fruit of existing systems of intrinsic attributes. We know that languages are offspring of various emotional identities of concerned people. Creativity bound to rationale and language is mostly contributable to intellectuality.

The motive for attaining the second kind of creativity is not necessarily derived solely on the language level. The motive for attaining the second kind is spread over and derived from the entire phenotype domain of humanity, which encompasses logic, language, music, dance, and any other systematic tools for communication.

The third kind of creativity is achieved by the escape out of the systematic process of internal logic. Thus the third kind arises usually at the periphery of a system. The periphery of a system behaves like a transition zone in phase behavior and can thus be considered as an extended boundary zone for a system.

We know that novel sprouts of new civilization essentially require such an escape from existing past logics and ancient styles of emotional expression. The third kind spreads also over the entire scope of human life, from individual to global, and from everyday life to extended periods of history. The most important steps of advancement at any level have been achieved at the third kind of creativity.

Are these three kinds of creativity the activation of humanity? We accept it to be so. They are a stepwise activation of humanity to achieve the transcendence of a human being in the course of answering the deepest query of the actual infinity.

It is often discussed that the plasticity of the evolutionary mechanism, in the sense that the changes are made in humanity are attributable to the partial plastic alterations in the DNA scheme. This concept also seems to be applicable at the level of the neuro-brain mechanism.

The creativity process is heavily affected by plasticity in the first kind of creativity. But a novel concept of elasticity has to be adopted into this process, as the degree of complexity in the process of creativity has been increased to achieve the second and the third kind of creativity. Elasticity describes the ability of a substance to assume its original shape when the deforming force is removed. I think that elasticity in the neural system is realized intangibly by a tendency of replication, as characteristically presented in the imaginary process of the human neural mechanism.

It is recently being recognized that social mechanism is deeply associated with bio-elasticity. Bio-elasticity is sustained by the habits, instincts, and other causes engraved at the DNA level. The social structure is greatly dependent on the systemicity of humanity generated from the bio-elasticity and neuro-elasticity.

Plasticity is the segmental changeability of humanity. Sometimes, plasticity occurs on a large scale as can be found in historical collapses of civilizations. In most cases, plasticity causes partial alterations in order to sustain elasticity of a whole or higher level entity. Elasticity is dedicated to the conservation of systematic character and structure in bio-neuro-social environment.

Thus the activation of humanity derived from the superficial changes of life style through the essential alteration of life structure is deemed to be a matter of the ratio between elasticity and plasticity in the cognitive mechanism of a human individual. It would be consequently a matter of the ratio between systemicity and nexusity which governs the life of an entity, either individual or communal.

In this context, the mechanism around the activation of humanity which is concerned with the hierarchical creativity process cannot be easily and effectively handled if the continuity of humanity is not to be treated carefully.

1-2. Definition of Peace2

Peace is a Dynamic Continuum.
Stabilizing Interactivity
Economic Utility of Peace Continuum

It is a great question of the human race on how we have come to reach the conclusive desire for the perpetual peace. The search for the perpetual peace looks like is carved into humanity as the source of all human phenomena. The wailing of a baby at the moment of its birth, good and bad feeling, mood and emotion in our everyday life, an old person departing this life at the instant death gazing at his lifelong memory of revolving lantern: all these phenomena could be interpreted by the essential influence of the peace of eternity.

The peace of eternity is the source of the thought on peace in humanity. Perpetual peace is not exclusively only to religion. Peace without eternity is frequently with us very shortly in the name of happiness in our rest, leisure and sound sleep.

Happiness is such a complex concept that sometimes it looks streaming too short with mere temporary stability. However, when we call peace status, it means that stays long and stable enough at undamaged usualness that cost us at easy level.

Peace is a Dynamic Continuum.

Peace is not an abstract notion but a reality. Peace is a substantial body of pacified humanity. Thus we understand that peace is one of the real states of human existence. We can feel, taste, touch, define and achieve peace if strongly intended to be sought after.

Peace is not the opposite of war, nor the complete elimination of violence. It is our desperate desire to escape from the danger and damage of violence. However violence does not come out of our inner quality only. Unfortunately we find that violence will stay with us in any form to certain extent, nevertheless the impact on us would be significantly decreased.

We think of peace in a different way from our general desires. What we are interested in is a state of real peace, neither abstractive peace nor an apparent form of peace. Peace is a dynamic continuum itself, which consists of a positive relation among entities. If the problem of peace is approached in the method of conflict elimination, our viewpoint will not be capable of describing the peace phenomenon. Positive relation between entities means that entities are supposed to be beneficial to one another and gain profit through the relation.

Global and local peace should be observed as an organic entity. When we consider ideal peace such as Pax Christi or Grand Peace(太平), there are intrinsic attribute of the designated peace in our imagination. There is an order and structure of justice to build the peace on. Without the internal structure of rationale and the process of justification, a state of peace cannot be for long periods of time.

The Continuum of peace is a nexus of warm peaceful relation among the constituents of a society. We need to understand that peace continuum has certain levels of manifestation. Apparent continuum of peace is what we can easily perceive by the existence of warfare and violent affairs. Such apparent continuum of peaceful relations in diplomatic and military domains is originated from underneath political economy.

Economics, one century ago, was called political economics. Politics was not separable from economy. However, politics was later detached from economics as a separate discipline. Economics in the last half century turned out to be numerical and a matter of computation for several decades.

But at the same time it was gradually realized that political economics only was not sufficient enough to express human affairs to forecast significantly. The next level of excavation was social psychology. Thus political economy has extended its integrated scope to become socio-psychological politico-economics.

In the last two decades, we have seen a brilliant outburst of knowledge in neuro physiology, which has introduced novel insights to old and rigid political economy. We have yet to wait an emergence of new integration of this field with political economics. New thoughts on political economy would be effective to solve the mechanism of peace continuum.

This research has tried to open a way of integration to express peace continuum, despite being done with limited effort. Our endeavor encompasses scopes from neuro-physiology, and socio-psychology, to political economics; we could call it NPSP political economics. The basic research was implemented to establish rationale of the value system tracing from the internal mechanism of humanity to structural study of the external mechanism of humanity.

Peace is a dynamic continuum that represents ceaseless interaction between social entities. It can be described rather as a continuous body of flow with limited fluidity. Therefore, there is a greater need of a regulatory intrinsic order of peace in that ever changing mechanism of flow. When the flow slows down, peace becomes rotten, as we see in the case of holding absolute power which ends up with unconditionally getting rotten.

Rotten peace is far worse than violence. The continuum of peace alone is not the solution of human desire. 'Justice and love together,' 'peace, freedom and order altogether,' and 'peace, humanity and creativity in a set' would present comprehensive understanding of the desirable peace in our realistic world.

Stabilizing Interactivity

In this research, peace is considered to be a stabilized condition in human life. Stability is usually a matter of the range of tolerance in dynamic condition. Humanity dealing with peaceful state has a constraint of range in terms of how human life oscillates within stable variance in its intensive and extensive properties.

In physics, intensive property is independent to mass, while extensive property is dependent to mass. In our discussion we will use intensive and extensive property in a specific circumstance matched to its respective intrinsic and environmental condition. Intensive and intrinsic property decides the internal tolerability of an entity, while extensive and environmental condition provides the scale of external impact imposed on an entity.

Why is peace defined as a stabilized condition? What does the stabilized condition in human life mean? When asking specifically, what is it stabilized in? As mentioned above, peace is concerned with the intensity of the external impact. It may also denote the aspect of stabilizing the intensity of internal human cognition. We can see that stabilization of humanity should be observed in balance by the viewpoints of both internal and external spheres of humanity.

However, we prefer to define that stability is the outcome of interactive relationships within a society. We are interested in finding indicative variables affecting the construction of the state of peace, not necessarily precise quantitative measures. Precise measures will have to discriminate the respective effect of internal and external impact on one another.

Usually community and society are mutually synonymous apart from differences in scale. However it is necessary to once again consider that the sociology of humankind be approached again following after the relatively primitive concept of community which is easy to understand the basic textures: the connectivity of thought and spectrum from individual, family, extended family, local community, functional society, complex society, multinational enterprises, nation, hyper-nations, and to even the global society.

If interactivity among individual entities of various hierarchies in a specific society is expressed in a form of oscillatory wave, the wave profile would generally have a characteristic form and scale. The interactivity represents the density of relation among entities. The density of interaction would be expressed as the body temperature of a biological entity.

Stabilized dynamic condition for a certain period of time is achieved around strange attractor in the science of complexity. Strange attractor in common sense the terms, can be illustrated by the metaphor of a vortex in a river stream that maintains the same position for a long period of time. Such a vortex is stabilized long in an ever changing dynamic situation.

It is well recognized that metabolism could be a case of dynamic stabilization. It could be better illustrated as stabilizing interactivity, if we take homeothermic metabolism into consideration. Homeostasis as a stabilizing mechanism is largely related to the exchange of material and energy into or out of the metabolic system.

Homeostasis is relatively more stable and less diversified than homeochaos because of its dependence to physicality. But as we understand, the on and off process of DNA is to some extent governed by the orchestra of IF-THEN genes, which is influenced by our emotional, behavioral and volitional activities. It has to do with the conversion of social actions to biological signals, performed by the human brain.

(Homeochaos is defined in our study differently from K. Kaneko's study, as will be explained in Ch. 6. We used homeochaos for the case free from substantial constraints such as embedment to materiality. We believe that it is related to whether diverging or converging occurs in its elasticity. Elasticity of the human brain has convergence character when successive impacts of the same quale are given to human intentionality.)

It is my observation that the range of homeomenesys matching to the social level exists in between the ranges of homeochaos and homeostasis. Homeomenesys is much alike the pattern of brain-neuronal action because of the intrinsic resemblance of social action to neuronal activities and also of the direct influence of neurophysiological activities to the social level.

In comparison to the micro case of metabolism, the hidden mechanism of evolution guides the overall macro orientation of stabilizing interactivity at a higher level, whether it is adaptation (in the systemic view) or exaptation (in the nexal view) to new dimensions for the development of the next evolutionary step.

Homeostasis on the micro scale and evolution on the macro scale are major mechanisms of stabilization of social interactivity, apart from the steering factor of human abstractive mentality in between. This abstractive mentality is responsible for the development of intersubjective sphere of human kind. Abstractive mentality is the effectuator of introducing new stressless dimensions, or equivalently stress–tamed imaginations.

Social interaction generates intersubjectivity of a society. Intersubjectivity has a direct functional relation with the characteristic scale of social interaction. Some societies experience deeper, wider and more intensive mutual interaction between constituents while others do not. The scale of interaction characterizes the concerned society.

It seems true that the representative scale of interactive relationship is governed, or at least affected, by the variable characteristic size of the intersubjectivity of the specific society. The intersubjectivity is much too hierarchical and structural to be defined simply. The reason we consider characteristic size for an intersubjective domain is that a specifically concerned society evidently has specific partial dominance in the whole expression of an intersubjective domain of global humanity.

However, the hierarchical structure of intersubjectivity of a society has to do with the occurrence of meme. Meme is generated essentially from the structural intrinsic condition of somatic gene in response to environmental impacts. Meme is sticky to gene-constraints and prone to be restrained by the social environment as well as natural conditions like geographical specifics. Intersubjectivity is in major governed by the structure of meme collection as we can observe in sensing organs of the human physique and in perceptive processes of social interaction.

Discussing the stabilization in dynamic condition, we said above that the range of tolerance or tolerability is a measure of expressing dynamic stability. When we consider only the tolerability itself, repressive emotion becomes meaningful. If we are interested in the range of tolerance, we can put the entire scope of emotion from repressive to emanative emotion altogether into the variable condition of defining stabilization.

Emanative emotion helps humanity introduce creativity into a stabilized condition. Creativity is so essential in developing the creative peace. It is noteworthy that extraordinary expression of positive (emanative) emotion such as excessive delight and passion could be regarded as unstable, although they might become sources or sometimes fruits of creativity. The Internal instability of humanity, good or bad, which eventually causes unavoidable fluctuation in the external condition, used to be observed in social life but not necessarily ending up as an uncontrollable situation.

Controllable stability serves human creativity generally. However we know that creative answers against hard challenges are, unexpectedly and frequently, of the best quality among creative reactions. It indicates that there is a certain threshold value (related to survival instinct) of emotional reservoir to compose complex ideas into an organic system and to have it burst out.

Ceaseless fluctuation of the living condition breeds collapse in many cases. In a stabilized state, a human being is pacified. Pacification with internal awakening provides the best environment for human mentality to activate itself. But we must remember that keen attention by itself (which is mostly equivalent to momentary internal awakening and decisiveness) does not last long unless it is sustained by tough perseverance. It is generally agreed upon that perseverance has to do with the volitional mechanism.

Intellect, emotion, and volition have intrinsic mutual mechanisms which allow them to coexist and collaborate. It is due to the cognitive process of the neural system operated by abstracting from the relevant sets of reliable and meaningful information. When a purport of living is satisfied by verified somatic and mental procedures, such procedures are settled down in our lives. It is noteworthy that the human mental identity operates metabolically as well as somatically. Mental metabolism needs to be studied further to establish sound theory. Mental metabolism has to do with the disposal of undigested information. Such disposal could be performed by plastic deformation of information within neuronal network or sometimes by total oblivion.

However, at this moment, we would like to comment that a large part of human mentality has considerably alienated territory from the somatic source of cognition and thus has this generated its own system departed from our general life based on physicality. It looks to have attained a somewhat schizoid character. The schizoid division occurs between the ontological condition and the epistemological condition. Ontological condition of humanity is the projective capability of general human being stretching over the wholeness of the universal world, while epistemological condition of humanity means cognitive intercommunication with piecewise reality in which they happen to concentrate from one moment to another.

I believe these two compatible attitudes in living are formed through the conceptional and developmental process of the human physique, which includes the brain. Great portion of human culture and intellectuality are thought to be originated, strictly speaking, from the autistic behavior of the human brain. The Aforementioned schizoid division hurts human mentality, causing autisticity. Mental identity sometimes experiences abrupt change as seen in the development of secondary sexual characteristics but usually experiences gradual changes as observed in the growth of intellectuality through one's life span. Even the secondary sexual property contributes to the stabilization of interactivity in a society if it is interpreted in a Yin-Yang philosophy.

What we can say at this point of our discussion is that ontological and epistemological views, centrifugal and centripetal reactions, systematic and nexal activities, and somatic and mental interactions serve to stabilize the interactivity of human society.

Communal and social reactions behave like the compound image of shades of individual personal reactions. Generally speaking, compound shade is the composition of a bell-shape distribution of the characters of an uncountable number of individual reactions. It is peculiar that humanity has an unavoidable trend towards monopoly or oligopoly.

The accomplishment of compound shade in human society is a very much alike to the situation we experience in the strategic inflection status. In business, the strategic inflection point usually indicates the point of expansions of a single commodity or an individual business entity which achieves a nearly exclusive, defined position in a market or in a competitive territory by occupying magnitude differences against the rest of its competitors. It suggests that such dominant occupiers would replace the communal character. Stabilizing interactivity has to do with the factors of dominance or governance.

Economic Utility of the Peace Continuum

Here we must comment again that peace is a very factual state of living. Absolute peace that humans imagine or dream of is perpetual and of non-violence. However, what plain citizens seek as a peace status in the realistic world is one that is relatively sustainable and tolerable. Factual peace is a matter of meals. It is essentially similar in the international situation. Thus what really matter is the economic utility of peace, or that of the peace continuum.

Having to be sustainable and tolerable, a biological entity is required to be subject to metabolism. The cognitive mechanism associated with such biological metabolism has, in history, extended its territory of mental subjectivity to the domain of sociality. Here we realize that how to eat decides how to be sociable in the distribution of earning and in the collaborating for the acquisition of meal. All human history is full of reforming stories of collaboration and negotiation for food, goods and intangible valuables.

Reformation is a term indicating adaptation to a given situation. The stance of this book is that reformation in the domain of sociality should be made consistently. It must be the ideal of democracy and further the subconscious visions stored with in humanity. Such consistent reformation of sociality should appear in economy and in economic institutions.

Our present concern is how to develop sound methodology in order to keep the domain of sociality stable and vivid. The domain of sociality is supposed to match correspondently with that of intersubjectivity, which has been attenuated into a controllable range while being vividly and consistently active, even though it does not maintain a continuant motive of creativity.

Then a society would be able to sustain the power to live mutually interconnect for certain period time, experiencing a gradual downward slope. If a society can be restored to another summit experience of creativeness, their life spans will be prolonged because of the subsequent push.

The definition of peace should be more than mere balance of power and removal of violence. Oppression frequently observed in fierce competition is actually considered as a violent factor. Peace should also come to include a peace that lives together with violence and injustice in reality.

But in intellectual talks on the peace subject of peace, non-violence issues frequently take center stage. Religion and intellect have taken perpetual and complete pacification as the Polaris of peace. We believe that the study of global peace and international politics will be able to produce a practical definition of peace and a reliable methodology to maintain the administration of worldwide peace. We are at a turning point of history. The Economic utility of peace ought to be considered as a main concern in the concept of peace.

In this context, peace is no longer about maintaining the current power balance and reduces costs by merely settling the tension. Positive peace, currently well recognized in peace studies promotes active exchanges of knowledge, information and goods and creates a continual synergy effect in knowledge and production. The following illustration shows the economic utility of passive peace - indicating the balance of power-in relation to positive peace.

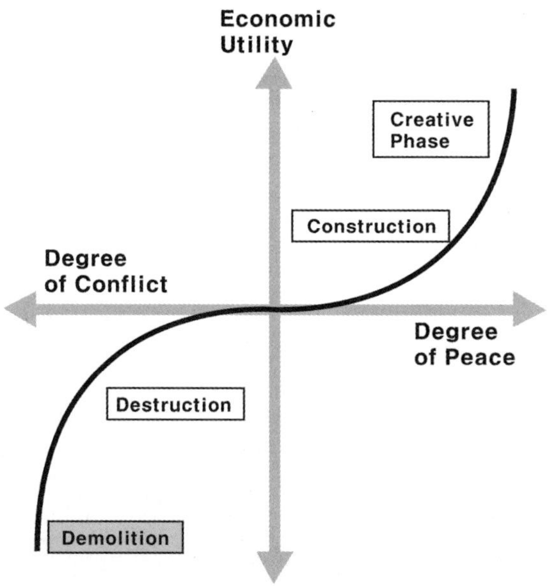

Figure 1-2-1. Economical Utility of Peaceful Status

As seen above Figure 1-2-1, dimension II shows the most efficient way to achieve the balance of power. Due to the inertia of potential opposition and the geometric increase in costs, the negotiation and acquisition of information is inevitable as the state of confrontation is prolonged. Whereas, in dimension I, we can see that a geometric rise in knowledge and produce is induced through positive peace, promoting active interchange among constituents, and thus resulting in the elimination of political-economic conflicts and barriers.

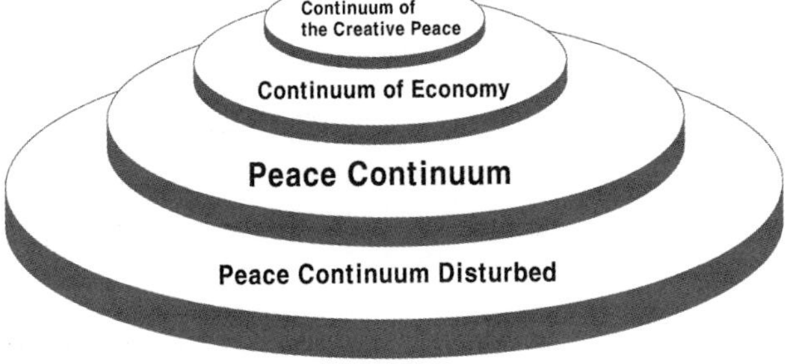

Figure 1-2-2. Continuum of Economy in Peace Continuum

The Major and fundamental portion of peace continuum is the continuum of economy as seen in Figure 1-2-2. When economic values are allowed to interflow in the continuum, we have to acknowledge that the continuum is effectively interconnected with peace, although political discordance might interrupt the process.

Economic value as market price is the only known quantifiable measure of value estimation. Therefore it is natural to conceive that a realistic vision of the creative peace should be expressed based on the continuum of economy. The continuum of economy must be regarded as larger than market economy, since the economy of the human race has variety in characteristic.

1-3. Definition of the Creative Peace

Continuum in NPSP Political Economics
Establishment of Fundamentals of the Creative Peace Continuum
Maintenance Criteria of the Continuum of the Creative Peace

We have explained what creativity and peace is respectively in the two preceding sections. Both notions are matter of complexity. We are making progress in compounding the two terminologies into a single concept.

The creative peace is defined to be "a continuum of the activated state of humanity in stabilized condition." On this definition, we are immediately facing with four questions. The first three are regarding what is meant by three constituent notions of the above definition: activation, humanity, and stabilized condition. These are dealt with in certain combined forms to explain creativity and peace.

The fourth and the last question is regarding how, in what sense "a continuum" could define the creative peace. Continuum is a terminology which has not yet been defined clearly even in cutting-edge movements of major academic disciplines. We have already used this terminology in previous sections.

Mathematics and physics have introduced the idea of continuum separately. In physics, even plain people have been acquainted with the spatiotemporal continuum of the universe. In mathematics, the Continuum hypothesis of Cantor has been widely recognized amongst intellectuals. The Continuum hypothesis deals with the continuity of real numbers.

Most representative academic disciplines have not agreed even on the logical foundation regarding the subject of continuum. However in our study we are going to attempt to define it as simple as possible through common sense.

Continuum in NPSP Political Economics

In political economics, continuum has not yet become a matter of theoretical consideration. However, this research strongly recommends that it be brought into discussion, as the concept of continuum of NPSP political economy has great potential to open up new avenues of academic discipline and practical application. NPSP stands for "neuro-physiological socio-psychological."

Regarding persuasion, Pascal had once advised that one should not use indefinite terms but use proven or self-evident propositions. Thus, we are going to attempt to establish the structure of the definition of the creative peace sturdily grounded on as much material evidences and rationality as we can manage at this stage.

The essential term "humanity" is utilized to express the quality of being human, which is expanded to include all positive outputs of human beings, either tangible or intangible. Being human will be considered more in-depth later on in our text. Positive outputs imply that there is also an opposite side: negative outputs- which are inhumane - that hurt humanity.

Humanity is the pronoun of complexity attached to the nature. Such complexity is originated from the composite components in human intellect, emotion and volition. These composite components cross over to the other sometimes in mutual agreements and upon other occasions in mutual disagreements.

In particular, the tendency of bipolarity and its subsequent balancing, which usually appear in the human interpretation on the nature, looks to be the principal generator of human complexity. Balancing in some sense is also a sort of complication to overcome, as conflicts caused by dualism appear. We know that the bifurcation process in many social and political occasions produce conflicts and confrontations.

However the real problem exists in the fact that human beings cannot give up dualism, balancing, symmetry and consequential fractal modes introduced thereon. Monism, dualism, and polymorphism are not allowed to be discarded from humanity. Our objective should be focused on how we can treat and live with these problematic tendencies in beautiful harmony.

"Activation" indicates the activated process and also purports initiation and creation of integrated conditions to make activation, which is introduced into reality, possible. Such integrated conditions are attained by consuming much energy under deep and wide influences of comparatively long durations.

Hence, we can compare that "activated humanity" is metaphorically to the status of highly elevated electrons orbiting around nucleus at a higher energy level. The status of highest energy potential comes from the interaction of the most contrary opposite factors present together in a same system. We would say "in a same 'menesys'" instead of "in a same 'system,'" as this research is meant to represent more complicated condition than merely a system.

Menesys is more complicated than a system because of the essential affection of networking character of the ubiquitous nature imposed on the entire natural phenomenon. It is simply understood when we consider that there is not a completely isolated part in the universe and in human society, even in the cases of black hole and heavily autistic people.

The reality in which we live in is so complicated since it is not a so-called system with simple, recognizable textures. We have recognized in this long research that a system is only an ideological concept which takes an end of bipolarity being compatible with network or nexus on another end in natural phenomenon.

People call an unrecognizable complexity a chaos. Chaos is chaotic solely in human interpretation performed by the intellectual brain. The Human brain categorizes its entire experience into definite partitions to be ready for a decision making process in order to avoid perilous path and find satisfaction.

A "stabilized condition" can be described as humanity following the tracks of gentle oscillating modes within a tolerable range of established humanity. This stabilization of humanity refuses both diffusive and attenuating behaviors. When extremely diffusive or attenuated behaviors arise, humanity in effect does not exist.

In actual history, the traces of long pursuits for perpetual peace are found in most civilizations. Perpetual peace has been usually the ideal goal of religions. Perpetual peace associated with eternity after death has been engraved into the thoughts of humans. It does not take long to enhance the idea of perpetual peace in real battling world.

We do not know if we can acknowledge the Viking's restless battles in the Valhalla palace of Odin a special pattern of peace. Could violence be rendering a sort of sound humanity? Is it correct to recognize and consider corruption and decadence to be healthy ingredients of humanity?

However, the human race is wisely vicious or viciously wise. Whatever they are, it wasn't before long that they insisted that wars were needed for peace and that at least, as expressed in the most modern and modified version, competitions were inevitably necessary.

The creative peace would be an outstanding case among various peace regimes which have been suggested in history. The creative peace concerns a full picture from the initial seed and to its final fruits/ through the whole process of fruition. It ought to be substantial for people to enjoy.

Thus it deals with reality; the realities of essence and appearance, the realities of nature and humanity, the realities of internality and externality, the realities of tangible and intangible things, and so forth.

Peace can usually be defined by two mutually complementary aspects: relation and order. Relation is a way of great variance consisting of dynamic activities or communications, while order is of a more static structure between entities, either tangible or intangible.

People tend to recognize changing relationships easier in a dynamic mode than in the state of order. Orders are stemmed generally from the embryonic stage of substantial entities, because an entity has to be in systematic order to form an organ or an organization. Without substantiality, humanity is apt to lose the standard of order in the imaginary sphere of mentality. Virtuality of imagination cannot provide a firm grip of reality.

Now we need to recollect that realistic peace is not static but dynamic. In a practical point of view, peace is a matter of management such as maintaining the considerably long stable status of strange attractor in the science of complexity. Peace should be realized under relational and dynamic thoughts.

In the mid 19th century, Claude Bernard, a French physiologist, thought that the free and independent status of life meant the homeostasis of *milieu interieur* (interior environment). Such a metaphor of physique leads us to the assumption that an individual body is basically an isolated circumstance of a minimal sea water environment.

However, this isolation is subject to a stable, dynamic equilibrium based on introducing and expelling of materials and energy. A physiological body is required to maintain material balance (metabolism) to keep it alive during its life cycle through the aging process.

The growth process and successive symptom of senility of a living thing suggests that the existence of continuous change, although it is limited within a certain range, in its intrinsic attributes and nexus of relations among mechanisms consist of the whole set of the living process. This change is considered by evolutionists to contribute also to genetic evolution.

This internal process is understood to be nonlinear due to the external environment of the body being a nonlinear source of material and energy. The nonlinearity of the internal process is not necessarily attributable only to externality but also to the consequence of variable internality.

Here, externality is the integrated influence of external conditions while, on the contrary, internality means a composite understanding of internal conditions, such as, for instance, the implication of the concept of market price in an economy and the bowel feeling of the composite emotional senses of a body.

A market and a physical body are both relatively isolated systems with unique boundaries respectively. But it has to be noted that external conditions influence the internalities of the systems over the boundaries.

In a market, the boundary condition is usually defined by supply and consumption. Sometimes the boundary condition can be extended to political factors and even to solar spots if needed in interpretation.

However, overall socio-psychological factors have not been considered yet. For a physical body as a system, it has comparatively definite boundary conditions because of its skin surface of keratinosite (skin cells). But a body is in general also open to the effect of intersubjectivity of a society.

For an individual entity, not only in biological life forms but also in other categorical existences, the intercommunication between internality and externality is the prevailing factor for its survival and mode of life. The intercommunication consists of the interflows of information, energy, and material.

These three interflows mutually influence one another. The creative peace would be recognized as a specific portion of dynamic balance of such nonlinear stability of global, regional, and/or local human society.

It is worthy to note that original peace can be attained through the conditioning of solely external factors while the creative peace must be supported by the conditioning of internality of concerned groups or individual constituents of the group to be achieved. The difficulties of establishing and maintaining the creative peace lies in that creativity is not merely a calm factor but rather a fluctuation mode made within a certain range among a number of concerned parties.

If we recognize the creative peace as a dynamic continuum, it can be assumed that the above interflows in a peace continuum can be analyzed scientifically. NPSP political economics is not just a dream but a good theoretical toolkit to approach global problems.

Establishment of Fundamentals of the Creative Peace Continuum

It has been widely discussed that without positive peace talks it would be impossible to break the vicious cycle of wars and power balancing. The concept of "positive peace" was originally suggested by J. Galtung. Positive peace is defined to be a peace status without physical violence, structural violence and cultural violence.

Positive peace is historically defined as a process, in which a just international order is made, production increases, and thereby, once the basic needs of life are satisfied, a culture flourishes to maturity to realize its potential for prosperity (Koh, W.I., 1995). This concept of justice was derived from the ancient oriental thoughts (such as the achievement of social justice from religious or statecraft commandments) and also from the observation of long historical coincidences which are usually found in up heaving process of restoration from periods of turbulent ruin toward subsequent reigns of peace.

Positive peace in this research also has a declining process as well as the aforementioned climbing process. Even in the declining historical cycle, physical, structural and cultural violence have been sufficiently reduced enough not to cause excessive dismantling of the working peace regime. In our study we are much more interested in the intrinsic attribute and dynamic mechanism of a historical entity, whether the object is an individual, community, nation, region or universe. Such tendency is explained in the in-depth conversation on the theory of creativity.

The great advantage of positive peace is that the interchange of commodity, information and culture are secured in an entire set of entities. Positive peace is the stand for the creative peace. These two concepts should be used in an intermingled way. But we have to understand that the creative peace is a superior dynamic regime of peace to positive peace. Humanity has merely scratched the surface of the creative peace over the course of history and has not achieved it in sufficient scale and depth.

The dynamic regime of the creative peace is the essence of creative society. The continuum of the creative peace will be established only under a comprehensive understanding of the entire mechanism. This entire mechanism is concerned with the infinite diversity of humanity.

This becomes a matter of real complexity when it is brought to in implementation. We must find some method to solve this complexity problem. For this purpose several variables were designed to clarify and functionally inter-relate internal processes of social changes. They are integrism, crux, and so on, which will be introduced further along in this research.

In the recent development of information theory, a trend is being formed that suggests that essential existence is information. If an ultimate extension of a thought is made, then all the affairs and entities are to be interpreted into the infinite information plexus. How are we to deal with this progress of thinking which is presently being emerged even into the realm of sciences?

Quantum theory looks to have solidly gained its position within the common sense of most intellectuals of the world. Humanistic philosophy needs to deal with this infinite information plexus, the theoretical basis of which is quantum theory. Infinite information plexus is actually the plexus of life, and consequently the plexus of existential affairs and events.

In this research we are going to name it "Plexion." Plexion is originated from the Greek words, "plexi-" and "-onto." Plexion will be illustrated in Chapter 3, under its comprehensive relationship with other existential mechanism of human being.

Plexion has to do with the Absolute Other (Das Ganz Andereheit, ff. R. Otto, DGA hereafter) which is the whole universe deducted by its single observer. The intersubjectivity portion of the observer is retained by DGA. I believe that DGA can help the methodology of science avoid mysticism and pantheism. The confrontational issue of concurrent humanity is an opening from the infinite universe for which human intellect is about to prick up its ears. It is the subject of infinity. That is the scientific discernment on the DGA and Plexion.

This study could be the start of a continuing pursuit of research in accordance with aforementioned orientation. It is necessary to comment at this point that the creative peace is not merely a metaphor of the homeostasis of milieu interieur of a living entity. The creative peace will be identified as a mental mechanism of the nature. Humanity has to find the transcendental way up to the soaring height in the course of accomplishing the creative peace.

The major purpose is to find the most reasonable method to establish the creative peace, in a most persuasive form to the public which would be workable in real world and applicable in any various aspects and hierarchies of the phenomena of humanity.

It is recommended that the creative peace should be regarded in historical view as the target of very near future of humanity. The way to practically pursue the creative peace is the topic of major concern in this work.

Maintenance Criteria of the Continuum of the Creative Peace

This study is extensively one of social engineering. But the fundamental of it is science. The criteria for the maintenance of the creative peace in a manner of continuum can be managed technologically by technocrats if the mechanism is thoroughly understood.

The technology of maintenance should be based on the quantified scheme to forecast with somewhat reasonable precision. Reasonable precision implies that we would not be led as tray from reality by the forecast. We require simulation models and calibration methods in order to obtain accurate forecasts into reality.

Our present study is suggested to be the commencing work of a great movement of research of the creative peace. If our work becomes considered as a preliminary proposal for a global project, while we would be glad and humbled to accept humbly the honor, we would rather focus on what is important: how to attain the eventual success of the creative peace.

Maintenance of the global continuum is subject to the excavation of new dimensions to decrease the stress of global political economy. **Stress management** would be the first solution to the problem has in mind. The stress management is simply dealt with in Chapter 2 under the subject of alpha path and is further studied in Chapter 5 to 7.

The second managerial point is that quality and the order of frequency of global creativity should be our major concern. We will call it **creativity management**. Creativity management appears in the science of creativity in Chapter 8.

The third is maintaining proper conditions for the creation of adequate temporary policies in the global scale. This will be introduced as **material balance management**. The ability to cope with this complex part of the research is not completely nor satisfactorily recommended due to the lack of research. The little amount of research that we can comment on at the moment is revealed in Chapter 9 and 10.

In the original sense, being creative looks to be a little bit deviated in understanding from the correct picture of creativity. Peace has become another victim of misunderstanding due to the excessive desire of most people to do project on it. How can we then reach the most probable way in which the creative peace could be attained and maintained?

Creativity, when interpreted, shall bring a breakthrough to humankind. We are going to explain creativity as a novel concept of dynamic complexity in the phenomena of humanity.

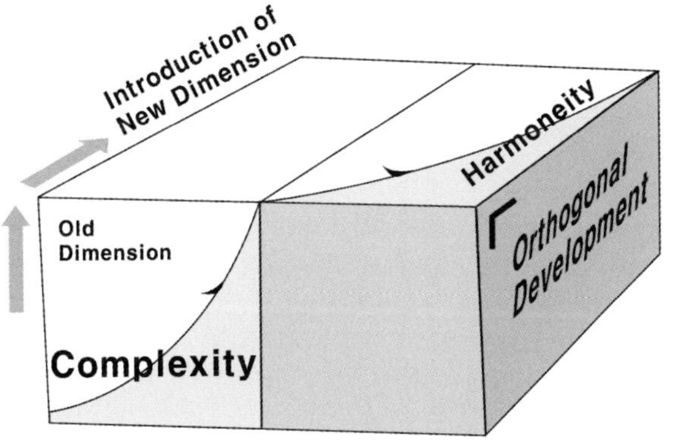

Figure 1-3-1. The Global Creative Peace Enhancing to the Universal Perpetual Peace

In Figure 1-3-1, the path of the global creative peace in our future shows how human race will be able to proceed toward the universal perpetual peace by introducing novel dimensions into the complex structure of human history. The path should follow the principle of orthogonal development, which will guarantee the harmoneity of human civilization.

Chapter 2. Alpha Path: Introduction of Novel Dimensions

The creative peace is the omega path toward the perpetual peace for which humanity has longed for thousands of years. The omega path may be regarded as a parody of the omega point with which we are all familiar. But it is not just a parody anymore for those people who dare to achieve perpetual peace on the Earth.

Humanity, if not enervated excessively further than it being recoverable, must search for ways to add new dimensions of self-adjustment toward durable peace. The omega path also implies in its name some negative aspects that should we fail to discover and pass over in this 'last' path, humanity could be severely misled or lost forever.

2-1. Singularity, or Anomaly?

The first concern, raising the issue of global peace, is the basic conditions to laying the groundwork. In introducing new patterns, it is generally important to examine existing foundations on which new patterns will be established. The discordances, if any, between the geographies of old foundations and new emerging patterns could cause new turbulence. Peace status is an integrated set of numbers of such complex conditions. Thus it is essential to investigate the general conditions of historical development.

In this chapter, cautiously dealt is a query whether the encountered future will be shocking and inevitably devastating, or somewhat astonishing and just alarming. For instance, R. Kurzweil has provided plenty of information about the singularity problem of present civilizations in his "Singularity is near (2005)." Originally, singularity is a notion introduced from theories on black holes. The singularity in our research is seen in the light of humanity.

The definition of humanity is variable depending on the purpose of research. At this moment let us easily define it as a general character of the human race and its offspring. Anyway, the rate and intensity of change are important factors which usually determine the situation to be of singularity, or of just anomaly. Singularity in social science is a drastic notion which accompanies definite discontinuity as the situation develops further. But anomalies in a more narrow concept, which is adopted here, are comparatively sudden changes but continuous in their vicinities and thus leave room for restoration.

2-2. Anomalies in Changing Humanity.

Continuity of Humanity
Homimmunage and Personage
Taking on Stress-free Dimensions, Abandoning Stressful Dimensions

Continuity of Humanity

There are several major factors that are the current governing foundational conditions on this issue of anomalies in changing humanity; science and technology, humanity and culture, and politics and economy. Changes in science and technology affect the change of humanity and culture, and the change in humanity and culture in turn influences the change of politics and economy. These three factors look mutually related in inter-synergism.

In fact the idea of singularity is a matter of the choice of humanity. The natural reality is that humanity is unbelievably conservative as far as its physicality and its physical derivatives are concerned. (The natural derivatives of human physicality are articulatory intrinsic attributes and their consequent communal characteristics.)

We may think of the certain possibility that extreme pursuit of eugenics can lead us to unfavorable situations in sustaining continuity of humanity. It might be hazardous for humanity to accept the unrestrained passion in a sense that humanity would lose hold of the safe position in the edge of chaos and fall into an undesirable fixed position of lowest fitness landscape. Does the highest efficiency of a perfect system always promise the best output in the long run? It is deemed negative.

Our standpoint is that humanity in a whole figure has been successful in curing its past hazards even though certain minimal parts of it might have been continuously torn out and lost. However, being human, in other words, being a person means natural healing and the ability to succeed its humanity. Within self-healing, we find the secret and mystery of humanity. Probably the verification of this idea could be the main theme of this research for peace in a sense that humanity has definite aspects of elasticity deeply rooted into social commensalism.

We should want to exclude singularity from the picture of our future, not because we want to escape psychologically from its danger, but because inter-synergism exists in human affairs to avoid perilous discontinuity. In Kurtweil's opinion dealt previously, the chances of singularity causing unrecoverable discontinuity of humanity seem most probable.

Does discontinuity of being human threaten the human race to losing their self-healing power before the collapse of no-return has settled definitely? What are the boundary affairs of becoming significantly discontinued from the past? Does that sharp shift of boundaries start from transformation into cyborgs? What is the essence of being human?

Now, it is necessary to understand what humanity really means in this fast changing situation. Inhumanity has been spreading for centuries in the intersubjective realm of human kind. However, it is most likely that ceaseless work has been the cure of humanity to prevent excessive inhumanity from spreading into overall civilization. The mystery of humanity is its immunity against the destruction of humanity and mankind.

The immunity, provided that we define it adequately, is resistant to inhumanity which induces the eventual destruction of humanity and mankind. It has been observed in overall historical events where sound humanity fighting against intemperate violence and excessive decadence saved the human society in a long cycle view. Continuity is very significant in human history, and so is immunity inherent in humanity as the cause of continuity.

Homimmunage and Personage

"Homimmunity(守人性)," a terminology designed to denote humane immunity, seems recommendable in showing the ability of self-inspection and the mechanism of self-protection, of humanity and of humanity-concerned entities and environments. Extending a little bit, "homimmunage"(守人態) would be used for the complex state of the hominized or humanized condition of a person. As to physical immunity, the boundary of the immune body of a person does not coincide with the boundary of the personal physical body. For instance, various parasitic worms are considered inclusive to the immune body while the personal body would resist them.

Exterior invaders are targets of the immune system when penetrated through the skin but usually not regarded so when materials pass through the digestive organs. The embryology of the body cavity of Echinoidea gives an implication of how delicately but definitely the internality and externality can be differentiated in considering human homimmunage.

"Personage"(人格態) would be another terminology introduced in this work to describe a socio-psychological entity with a centripetal "self" who has an independent interpretability of the environment and autobiographic memory. Personage is different from homimmunage in terms of existential boundaries as aforementioned and in its functional character to be frontal-advancing and positive while homimmunage is tail-protecting and conservative. Thus homimmunage and personage are considered mutually contrastive concepts.

However, human beings are ruminants of thinking. Findings in brain science describe that the human brain works ceaselessly in subconsciousness, consuming a lot of energy in order to form a rational movie of one's personal history. Rumination is the source of being and becoming human, through introspection.

René Descartes saying "*Cogito, ergo sum*" can be permitted in a somewhat deviated sense from what he originally meant. Human thinking is based on emotional foundation and becomes the source of ceaseless imagination. Human build up them through imagining ceaselessly, day and night centered in their rational thinking connected to their episodic memory.

Le Phènoméne of Pierre Teilhard de Chardin tells how human ruminate individually and communally. It is thought that personage is positioned around intentional concentration in conscious states, while homimmunage is positioned around the examination process of information received in subconscious states. So, homimmunage is heavily concerned with the function of rumination.

But personage should take some parts of the strengthening process of constructed stories and reasons. Personage and homimmunage together are constituents of "personhood(人性)," the outcome of a person. Hence, we can say that personage is relatively active in conscious and physically awaken diurnal environments, and homimmunage in subconscious and physically hypnotic nocturnal environments.

It is my frank feeling that, if Descartes had connected his suggestive theory on the pineal gland in this sinuous manner of periodic alternation, he would have had greater room in theorization. It is because the conception of the time factor of human physicality is attained by periodic environmental changes due to solar radiation which organizes human consciousness. Evolution on the Earth is inescapable from the effects of the solar environment.

Anyway, both personage and homimmunage are to be dealt together in establishing rationales for humanity and their derivatives. If the hypothesis developed further from the last paragraph, inhumanity existing over vast scales of meaningless information is sieved by homimmunage, the traces of which are found around the virtues of temperance and self-control.

Mainly, self-control and temperance are conservative virtues of maintaining communality.

The eventual source of homimmunage is existing communal relationships which defines present individual personality. Concerning the anomalies of humanity, which is the subject of this section, homimmunage has effects of diminution on the change, which means the transition, or equivalently the metamorphosis, of humanity. Homimmunage might work like a kind of constraint factor against changes, which we would call anthropomorphosis.

It is my observation that personage is represented by the virtues of capacity and connectivity of an entity, either in individual or communal situations. Capacity implies how extensively and how intensively a personal entity possesses its contents. Connectivity is meaningful basically when it is expressed in a manner of direction and intensity (that is vector) between two concerned entities.

Humanity is defined by the internal character of a personal entity, but the personal entity, in order to maintain humaneness, has yet to maintain connectivity with other humans in the society it belongs. Personage has a positive function which accelerates changes that reduce stress and for survival. For survival, everybody should concentrate fully to take necessary responsibility which is usually stressful.

Personage is a main factor to bringing metamorphosis in humanity, and thus in most cases positively influences anomalous phenomena to increase. However, our communal relationships are supposed to introduce external constraints to decrease anomalous effects.

Extending the idea a little further, we can comment that the multi-millennium history of human beings has already been engraved in the DNA level, and thus personage and homimmunage should have certain characteristics which have been developed and settled to sustain humanity in a continuous mode among entities.

Stabilized conditions are relevant to the concepts of personage and homimmunage. Personage provides prerequisite conditions for the endless extension of humanity toward plexion. Such extension initiates the transcent of humanity by activating transcendental intentionality. Homimmunage contains personage and the impact of pragma on personage. Pragma is the outer range of personal extension that admits the influence of externality. The ingested external flux of affairs or material, for instance, brings about impacts of pragma on personage.

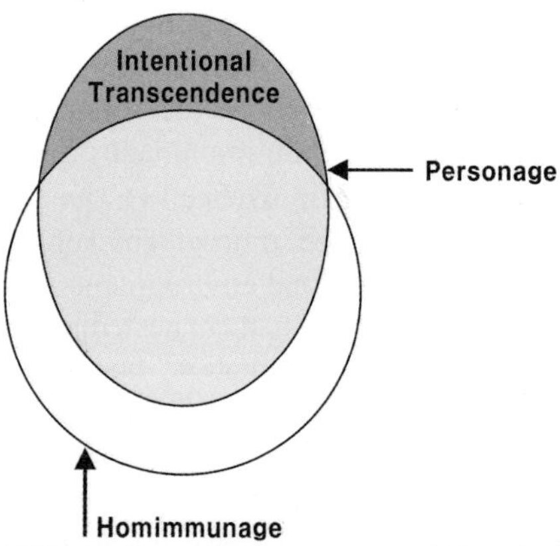

Figure 2-2-1. A Visualized Illustration of Personage and Homimmunage

Figure 2-2-1 shows a visualized illustration of the relation between personage and homimmunage. Both personage and homimmunage must work together in complementary manners in order to construct stabilized conditions of humanity. Stabilization is the basis for sound humanity to build up activation processes.

Taking on Stress-free Dimensions, Abandoning Stressful Dimensions

There are general trends in human history that have provided protection for it from unrestrained divergence. Presented here are two characteristic curves for the comparison of these concepts.

The first characteristic curve shows the divergent geometric increase of the effect of a single factor. The second characteristic curve shows the change due to the orthogonal interference of the effect of another factor from external conditions. Orthogonal interference is thought to have the highest probability of interfering with the original status because it has the least resistance against the first prevailing factor.

The second factor suppresses the divergence and creates a new stream of the Aristotle curve to converge to a stabilized condition. This characteristic conversion would be the answer to the more or less spontaneous stability of humanity regardless of the variety of global human conditions.

Figure 2-2-2 shows why the anomaly-hypothesis is more likely to happen than singularity. Anomalies are manifested when drastic changes caused by singularity, potentially developing to infinity, are tempered with by the other dynamic factors grown from the stress free dimensions, which could be understood as orthogonal to original dynamics.

This is very similar to the quantum effect of observation in a way that decoherence happens when we concentrate on specific physical and tangible affairs. The interceptions of subsequent orthogonal elements, presented as an inflexion on Figure 2-1(b), are intuitively considered to be a new source to initialize the next steps of situational development.

The influence of stress-free orthogonal elements or the quasi-orthogonal element(s) with the least stress manifold(s), or domain(s) are introduced from adjacency as soon as equilibrium is reached and the internal energy of the first system loses its driving force.

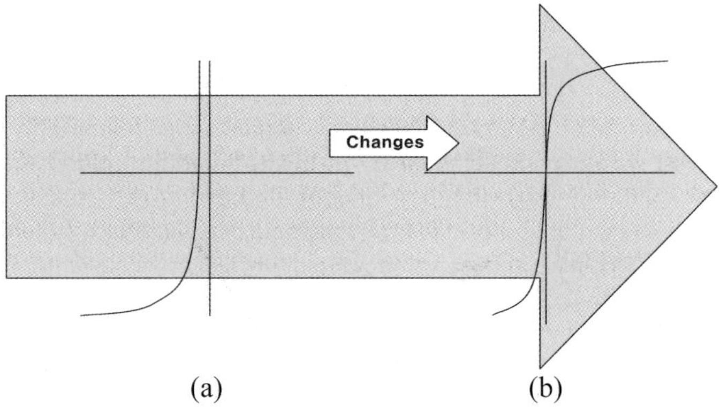

Figure 2-2-2. Divergence to Infinity (Singularity) Changes to S-curve by the Introduction of Inflexion. (Inflexion comes from hierarchical changes from boundary responses, or effects from dimensional changes reduced from chaos situations with orthogonal relations.)

Matters of boundary conditions are present in overall circumstances in nature and humanity. However, boundary problems are usually not scholastically considered since most of disciplines take isolated systems for their concern in theorization. Usually the latter part of the S-curve is thought to be the attenuation of internal effects from subjective sources of energy. But it is deemed to be rather caused by the combination of peripheral effects from boundaries and the gradual weakening of energy from the original source. The inflexion effects present in overall phases of natural affairs including humanity prevent humanity from plunging into catastrophic singularity.

It is worthy to note that unusual anomalies are presently before us in temporally mapped changes of humanity. The question asked at the start of this section was "singularity or anomaly?" Singularity is not that we are facing. Now we have confirmed that the answer is an anomaly, a consequent question to be raised would be whether we are at the stage of "collapse, or regeneracy?"

2-3. Collapse or Regeneracy?

Two Sources of Humanity Failure

If the problem of current civilization is thought to be of anomalous characteristics, excluding singularity, then our solution will be found around the collapse-or-regeneracy problem. There are three problems to overcome in order to regenerate humanity while avoiding collapse. The collapse of humanity seems to originate from either excessive independency of human individuals or excessive dependency in communal relationship. These are the two sources of failures.

Excessive individual independence causes isolation from other entities and the environment. The outlet of individual independence is connected to self-indulgence, or the development of a solely self-centered mind system. Self-indulgence causes diffusion of humanity which in various situations consequently results in "the dissolution of society", which is the first exit to the failure of humanity.

On the contrary, the development of a solely self-centered mind system unquestionably induces a society to be solidified. "Solidification of society" represented by despotism is certainly a failure in humanity, and is the second exit. Individual independence, when strengthened without deep consideration and only dependent on the rate of return in commercial criteria, for instance, unlimitedly aided by HIT (Human Interface Technology) and Augmented Virtual Reality, should cause the dissolution and solidification of our society in unrecoverable measures.

The excessive dependency in communal relationship provides the third exit to the failure of humanity. When a community is well trained and controlled by norms, in current expression, by communally defined patterns, they are apt to lose humanity as found in unduly strict Confucius countries. This, we would call, "the attenuation of humanity." Such social phenomenon appears not only as a matter of social norms but also in power relationships, as can be seen in the network of social evils when have penetrated into every corner of the world's power regimes leaving no room for reformation without breaking the whole system.

Regeneracy, Then How?

The long-lasting impacts of the Renaissance and the philosophy of enlightenment, when favorably evaluated, have lead common contemporary intellectuals to entertain the thought of "humanistic spontaneity sustaining both communal moderation and individual liberty."

"Spontaneity" might not represent the present situation well enough - for instance, technoetic aesthetics now dream of new ideas such as bio-telematics, sentient net, hyper-cortex, telenoia, and neuro-cosmology (Roy Ascott, 2002, Technoetic arts).

But it is not desirable to be too radical when the majority of the global population is staying behind in poverty and deterrence. It is most probable that a large part of the extremes of current humanity will later be rated as to be included in "isolated, partitioned, and discarded sects." Isolated humanity is apt to be discarded.

Extremities always exist. Our concern needs to be concentrated on restoring extreme cases to a healthy range and levels of humanity. Regeneracy of humanity is reinstated when wedging is done successfully. Wedging is not costly but should be done wisely infiltrating all the phases of events and substances.

To deal with the collapse-or-regeneracy question, three cases of collapse and three cases of regeneracy can be examined. Diagnosis on three collapse cases is made; Columbus as shown in the film of 1492, Jesuit Reduxion as presented in the film of Mission, and Pizarro, the Conquistador of Inca.

Christopher Columbus failed in establishing the first settlement in Haiti after his discovery of the West Indies Islands. It is the story of a movie named "1492," the year of his discovery. If the content of discussion is based only within the synopsis of the film, Columbus was not an administrator but an adventurer who dared to venture looking for discoveries.

Remnants of the first voyagers were left among harsh Indian tribes. They were all found slain. On his second trip to the new land, Columbus started building a stable colony with a number of people who had accompanied him.

The second colony experienced a rebellion and a hurricane disaster that destroyed all his efforts poured into the new world. The first Christian convert of the Indian headmen departed him in despair at the end of the story because Columbus slaughtered an Indian tribe due to the homicide of Spaniards. This is the sad finale of "1492".

It was deemed necessary to make a comparison with a case of virtuous interception of the founder of the Ming Dynasty in Chinese history, Zhu Yuanzhang. Zhu was an orphan and a Buddhist monk before he became a rebel leader. He was a person who deeply understood the agony and yearning of the common people. (It is my desire for someone to prove the reliance of this episode.)

In one of his major confrontation against the enemy, he was supposed to have his troops march many miles to attack. A patrol party reported to him that there were no obstacles except for a female ringdove brooding under a bush on the road to their proposed battlefield. Zhu showed a definite difference in leadership from other military leaders when he gave strict orders to keep their present position until the dove's brooding was complete.

Days of the most strategic and tactical value passed. However, what Zhu gained was not the military failure that most of his soldiers expected but the peaceful surrender of his enemies who happened to be deeply impressed by Zhu's decision.

The wedging effect of Zhu's virtuous decision can be interpreted in a way of successful propaganda warfare. (But it also must be noted that Zhu later eliminated most of his powerful and influential subordinates saying that the peace of the Chino-continent had to be kept in order to justify his actions.)

The youngest son of Columbus, Ernando, one generation later, brought plenty of gold and silver to Seville and decorated a tabernacle there. There exists a coffin with the corpse of Columbus lying inside supported by statues of four kings. We can find Ernando's castle and the traces of Columbus's brother attack by pirates in Santo Domingo of the Dominican Republic.

It is true that Columbus was an envisioned idealist in many respects. But it is noteworthy that the precedent failure of Columbus might have given clear instructions to the following Spaniard Conquistadors to grip the things with strong iron fingers although it was their very first time having encountered the Inca civilization.

The case of Columbus is good to illustrate the need of connection, a strong tie, between West Indians and newcomers in order to avoid conflicts. How could we connect with different culture and personalities to initiate stabilized peace in this world of conflict?

Pizarro was a born military leader. Slaughter was not a taboo to him and his army. Alexander the great was also a born conqueror. He ordered the destruction of Jerusalem because the Jewish people did not agree to erect his statue in their sanctuary.

The intellectual intervention of Simeon ha-Zadeque, the high priest of that year, proved the value of wisdom. He suggested Alexander that all the Jewish boys born in the year (probably A.D. 332) be named after Alexander so that Jerusalem may be saved.

Why did most negotiations have to fail? Why did some cases end with beautiful solutions? It looks like Columbus and Pizarro were not successful in reserving humanity in their historical inheritance.

Why do we find good successful examples of introducing new social models into completely different societies in cases such as Columba-cille (Iona Island, Scotland) and Jesuit Reduxion (Paraguay, Latin America)?

Jesuit missionaries established several Reduxion of Guarani Indians; they thence cultivated the population to which is known to be around 400,000, and managed those communities very well with only around 40 missionary monks. The story of the film "Mission" is based on their history. It is a well known fact that all the Reduxion were demolished by the Portuguese army.

In the film the tragedy was largely due to the ecclesial politics concerning the whole Jesuit movement in Europe at that time. Anyhow it was evidently not an easy task to maintain idealistic religious Utopia in territory where greedy colonization was in progress.

Columba-cille, who was Columba of the Church and known as the Great Columba, was a much more influential person than most kings of the day. The fruit of his work in Iona Island of northern Britain was accomplished by his deep piety and spiritual charisma under his strong political leadership.

He always maintained a good network with kings and aristocrats. Sometimes he was very active in helping a prince to the throne. He dared risking himself by being voluntarily exposed while converting an antagonistic ruler. The tragedy of Reduxion and the happy ending of Iona had definite differentiation in networking. Networking in this case was the core to survival.

Regeneracy is identical to survival. Survival functions as the stabilized foundation of regeneracy. Here is the story of a president who thought that way, without his close surveillance and advice, his nation and he would not have survived. He tried to utilize a new communication technology to gain greater probability of survival.

He was President Abraham Lincoln in the Civil War, who systematically intercepted telegrams sent from the battlefield to his secretary of Defense. His T-gram method was so effective that his infiltration and gathering of combat information may have influenced the army from the level of tactics to strategies in military decisions. For instance, a few of his decisions made through information collecting such as "territorial occupation, or exterminatory war were very important." In ideological issues, his long consideration was said to have been given to "federalism or liberation" problems.

President Abraham Lincoln was creative enough to wedge himself into the warfare mechanism of his contemporary age adopting the new telegram technology. He was frank to his ultimate needs and had comparatively less stress in wedging in new approaches to the factual world. Actually what he had done was a wedging operation not only with technology but also with the in-depth build-up of a more effective system-plus-network idea.

President Lincoln was a great figure who maintained the continuity of the society and achieved regeneracy while avoiding collapse through creative and substantial wedging decisions. There is a high possibility that the achievement of the creative peace is supposed to be of a similar outlook.

Now one can ask why all three cases were selected from conflict models. It is to show how to gain access to creative peace. Creativity and peace are both integrative ideas even being a real complexity concerning human factor being generated.

When we recognize creative peace as a complexity problem we get to an idea related to the "initiality concept" of how to start things, the "transition concept" of how emergence takes place, "regulation and modulation concept" of how to control in complex environment, and more profound dynamic models which include socio-psychological phenomena.

It is good to summarize again at this point that we have three situations that must be overcome for a human society; first, the dissolution; second, the solidification; third, the attenuation.

The dissolution and solidification of humanity in a society have to do with the character of personage whether it is prone to concentration toward individual independence, or communal dependence. Likewise, the attenuation of humanity in a society has to do with the homimmunage effect on the diminution of social change. How will we quantify these? If we could invent some way to describe it, it would be very beneficial for policy makers.

2-4. Alpha Path: Introduction of Novel Dimensions

Alpha Path and Omega Path
The Character of Confronted Omega Path

Omega path has us recollect the omega point of Pierre Teilhard de Chardin. The omega point has a notion of the designated forthcoming goal of humankind. He wrote in 1940 making an annotation on omega point that there was something superior to the omega point ahead of us. It means the omega point must be a higher notion than explained.

Here we will talk about omega path, which has not much to do with the theological implication of his omega point. But I believe we have to think about the somewhat complicated general process of the transcendence and transcent of humanity, later in Chapter 4, framing the science of humanity.

Alpha Path and Omega Path

A natural family generally begins in a loving relationship with the least stress between espousers. A civilization starts from its least resistant situation to achieve solidarity among constituents; people, disciplines, ways of thinking, and so on.

This least resistant condition in an individual or communal relationship paves a way to alpha path. Alpha paths are processes of bringing in new flows of stressless influence from new adjacent dimensions. It could be expressed as an introduction to a new stress-free dimension. Such stressless dimensions in relation to the existing phase space are called orthogonal to the old or ancient space.

Alpha path opens its way easily in a less dominant territory of a particular major constituent in which peripheral mix usually takes place with the other minor constituents. When we call a geographical territory a "common periphery," it would mean that the periphery is in between several core constituents which possess their own dominant territories respectively. (However, Alpha path is the most effective and characteristic when new dimensions are introduced as seen in the discovery of the new continent, establishment of the Internet interface, and forthcoming movement into space.)

Common periphery does not have positive predominant system that enforces its constituents to generate a new generation in accordance to its fixed systematic rule. For instance, the Netherlands was a common periphery several centuries ago. Palestine-Phoenicia is another example of common periphery a few millennia ago.

On the contrary, the well known concept of the melting pot indicates a core section of mixed constituents that positively maintains strong governance and active interaction in order to create new outcomes. Ancient Egypt, ancient imperial China and the United States of present could be identified as melting pots.

Melting pots and common periphery may both serve to initiate an emergence of a new stream of alpha path. Common periphery is deemed to easily cause new alpha path because it does not have enforced pre-orientation.

Alpha path is a beginning process of an entity as an integer to self-organize relevant affairs and constituents. A living organism should start with strong ties supported by organic mechanisms in its physiological body. Basic organic mechanisms, in a well known hypothesis as in earliest evolution by Carl Woese, opened its way to systematic complexity toward vertebrates from a very simple nexusity of symbiosis shown in an evolving process to eukaryote.

An active society usually maintains its high level of voluntariness and overall control in the society. Networking voluntariness (one of typical nexusity) is the starting spot for the formation of a community which will evolve to a more complex society with matured systemicity.

Now, what alpha path shows contrast to omega path is that the complexity of an old regime at the stage of omega path would reach the maximum than not. Generally speaking, the greater the complexity of an entity, the higher the stress is.

We live in a special age. The human race is about to finish a long of omega path of humanity which has been lasting in the past several centuries. This road originated from the significant events unfolded about 2500 years ago when humanity climbed up to the remarkable first summit of intellect.

At that time, great intellects had intuited all the aspects of humanity in mutually different ways of insight. Their comments were respectively about the actual contents of what they "felt" in their own distinctive manners of understanding the world. It means they had the fresh reality of objects they met in their minds. The reality was neither exactly physical, nor strictly mental.

It was a kind of mixed up emergence of self-organization of humanity. Each of these intellects tried to leave the impressions of the integrated scopes of his view; they were Socrates, Buddha, Confucius, Zoroaster, etc. The initiation of the synthesis of humanity has been started in the name of religion. The representative religions in human history started to think the views had better be rather recognized as mutually differentiated styles of thinking. Humanity, in order to get on such high summits of editing their wisdoms, had to activate rationales of the highest level to encompass all the phases of human life.

There is an interest aspect in humanity. Tolstoy, a great writer, said "Unhappy families are all in different shapes, but happy families appear all similar." Good character tends to come together but excessively bad character is found scattered.

If we take the topographical shape of entire humanity, as realistic evidence it will most likely seem that humanity is constituted virtues-centric. For instance, humble brightness is assessed superior than being aggressively predominant, which shows that the tendency of complementary contrast exists in the phase space of human virtues.

Character of Confronted Omega Path

The omega path we currently face is almost at the outlet of the tunnel. Plenty of new factors have already been brought to reality from imagination. In other words, as recently reported by the chief of a renowned intelligence organization, humanity will experience unstable upheavals in global scale for the next several decades. These upheavals are due to the inability of a society to cope with the increasing of complexity with time, just as what can be seen at a deltaic area of a river when the flowing speed drops.

Humankinds are about to understand the power in nexusity. Until now, systems and the system of systems have been recognized by most management and administrations. Recently, the theory of networks has drawn attentions of many scientists. This research proposes that the phenomenon of systems and networks are in fact undividable and are originated from the same source, the phenomenon of humanity or the phenomenon of personhood.

The omega path we are passing would be owing to individual nihilism infiltrated into most global tenants, the broken communality mainly due to globalization, and the lack of managerial skills for utilizing the potentials in local or regional resources. Self-destruction and the associated psychological symptoms are spread all over society. Such symptoms, I believe, are the source of terrorism, and are attributed to the implosion of humanity caused by the self-confinement stress and the severe and are hostility induced by the extreme cultural unbalance.

Globalization has broken various values of communality. Globalization of power regimes and multinational enterprises gave people unforgettable shocks. The local residents who have been under those excessive shocks have received no better solutions than instant allopathic treatments. The whole world is changing; the whole world under a transition. The Wealthy metropolitan areas are also going through abrupt changes. The people in these areas seem well prepared for the future except their poor management in balancing between family and career, between personal and organizational lives, and more.

How can we start all over again? Nobody can escape from his past and present environment. Even the most powerful Wall Street was free from their fault and mismanagement in 2008. An even more severe omega path is about to become present. However, it is worthwhile to note that the end of the omega path always generates of a novel alpha path for the next step.

The alteration of humanity caused by the internal morphosis has led us to a concluding part of a historical cycle from the omega to a successive alpha path. The apparent change of society in science and technology, law and culture, politics and economics and so on are the fruit of the transcent of humanity.

2-5. Major Sources of Global Anomalies

Three Major Sources of Anomalies in the Transcent of Humanity
Great Epsilon (Є) in Oceanic Environment and Grand Chi (X) in Continental Environment

Three Major Sources of Anomalies in the Transcent of Humanity

In Chapter 1, we have defined the realistic vision of the creative peace. It would be proper to comment here the three major sources of the global anomalies in the perspective of humanity in the author's point of view: the first from the inside of humanity which is the way of understanding the world, the second from the outside of humanity which are the global planetary conditions, and the third from the phase space of human sphere. This viewpoint is summarized in Table 2-5-1.

Table 2-5-1. Three Major Sources of Anomalies in the Transcent of Humanity

Sources	Causes	Practice	Our Concern/Interest
From inside of humanity	internal morphosis	fair balance of personhood	Integrism
From outside of humanity	planetary conditions	Grand Epsilon Grand Chi	Menesys
From the phase space of human sphere	apparent change of society	influence of political economic history	Cruxology

The first and essential source is the conflicts between centripetality and centrifugality of personhood. In most cases, these contradictory characters are not directly related to actual conflicts. That is affirmative. However, though these characters may not be the direct factors, but they could be the indirect and vastly influential factor in all the phases of humanity. That is the reason we cannot disregard the first source.

The centripetality and centrifugality of humanity are generated from the ontological problem of personhood. Centripetality is a trend of human mentality related to individual self and consciousness, communal systemicity, and ontological aspects of human phenomenon.

Centrifugality is related to individual subsciousness, communal nexusity and epistemological aspects. The bipolar tendency of humanity was portrayed in the various unnecessary conflicts in human history. Most failures in negotiations are due to the lack of the mutual consensus and mutual sharing. Mutual consensus is derived from the centripetal attitudes, and mutual sharing from the centrifugal one.

The balance, or combination, between the centripetality and centrifugality of personhood is thought to be a tentacle of the human race, even in the individual level. The life and history of communal and individual humanity is a game of extremity. The tentacle of human history to move forward in the field of unknown extreme is a healthy and well balanced personhood.

This fair balance of personhood is closely related to the concept of integrism, a major concept in this research. Integrism is the function of the balance in personhood, which will be used in this research when defining person and personal phenomenon at a higher understanding for a more advanced scientific expression.

Grand Epsilon (Є) in Oceanic Environment and Grand Chi (X) in Continental Environment

The second source, of the external conditions of humanity is mainly derived from geographical specificity. In general, the Grand (Continental) Slant and the Grand (Oceanic) Epsilon present in global scale as they are seen in the Figure 2-5-1.

The Grand Slant is the belt of mineral resources, including natural hydrocarbon (oil and gas) which exists in a slant shape from the tip of Northeast Asia to the end of Africa (i.e., from Kamchatka to Namibia passing through the Central Asia and the Middle East). The belt in general has relatively higher concentration of the mineral resources and considerably sparser density of population. Grand Chi is crossing the Grand Slant and forms a line of densely populated area in Eurasia.

Epsilon usually denotes a tiny object. We have a grand scale epsilon on our planet. The Grand Epsilon is made of the three branches of oceans on the planet, starting from the sea area which includes the Falkland Islands and the St. Georgia Islands near the Antarctic continent.

The west wing stretches to the North Pacific and the east wing reaches to the Malacca Strait, The East and west wings meet as they reach East Asia. The centerline of the Epsilon gets to Britain, Europe and North America. The most perilous parts of the planet are positioned at the west side of the Grand Slant. They are from north to south Central Asia, Kafka, Middle East and West Africa.

Robert D. Kaplan has traveled most areas where conflicts are happening in the globe and left impressive books with profound insights. This present research has a slightly different integrated idea in approaching the issues on peace, but with same realism that Kaplan intended.

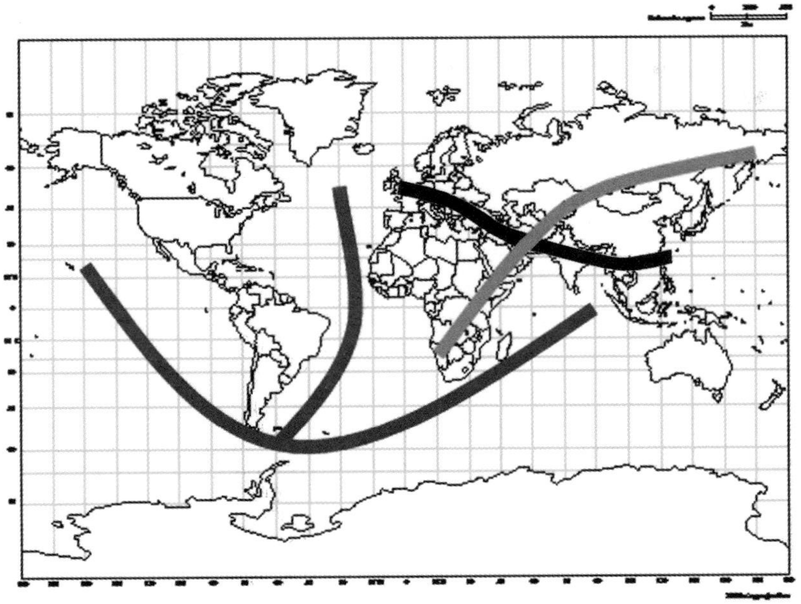

Figure 2-5-1. Grand Epsilon (Є) in Oceanic Environment and Grand Chi (X) in Continental Environment

The third and eventual source of conflicts is from the phase space of the human sphere. The human sphere can be mapped in terms of potential differences and ever-fluctuating flow of political economic power. Thus such phase space can be described with power variables in tensor forms and presented with several characteristic zones of initiative and positive anomalies, intermediate buffer zones, and negative anomalies.

The zone of current initiative anomalies are some countries, urban areas, and even smaller communities in the Western Europe and the East Asia; the negative anomalies zones at this moment are areas like Kosovo, Kurd, Sudan, Afghan, Kafka, and certain countries in West Africa. These anomalies will be continually changing due to the imposition and redistribution of power variables of the time.

The new development of anomalies especially contains innumerable minor fluctuations these days due to the low intensity wars all over the earth in which unconstrained mutual infiltrations take places.

It is believed that creativity and peace combined could be the best answer for the global political economy, regionally speaking, in the Korean peninsula, wherein the author resides, the Balkan Peninsula, the Middle East, the African regions, Afghanistan and other disputed regions in international high-density conflicts.

The inhabitants of those regions are sincerely desiring for some quick and solid solutions for sustainable peace, in the writer's point of view, the creative peace. The writer intends that his study to be not a mere peace study for verbalism. It is a hard task to arrive at agreement from contrary ideas and to discover coincidences out of the wide spectrum of human thoughts.

Nevertheless, this study is aimed to establish a methodology to define a number of basic components of the creative peace, to frame the multidimensional phase spaces in which the peace takes place, and to demonstrate how the dynamics work in the spaces as clearly as possible.

Chapter 3. Infinities: the Secret of Humanity

This chapter will cover the fundamental ideas in developing the theoretical framework to achieve the creative peace and to examine their feasibility in adopting for theorization. The main theme of the creative peace should be played around humanization and personhood that seem fast to a complete collapse of humankind if not properly treated in the present century.

Humanity is the essence of the creative peace, which should be kept at all cost. The theme of this Chapter 3 is extended to the Chapter 4 in Part II to establish the foundation of the science of humanity. We are going to deal with the theme of infinity, which is the secret of humanity.

3-1. The Change of Paradigm

The Main Line of Thought
The City of Alexandria on Pharos, over Night
The Agnostical Realism of Science
3 Worlds, 4 Minds and 4 Annulus of Cognition
The Third Level of Cognition; Transciousness

Complex Adaptive System is the main theme of the science of complexity. A considerable part of this writing relies on the way of thinking in the science of complexity. However, it is noted here that adaptation, self-organization, criticality, emergence, and other ideas of the science of complexity have been ruminated and changed the basic stance from the past establishment.

In this writing, a shift of paradigm implies the transcent of humanity that surpasses the adaptation to the present environment towards transcendence to new situations to come. "Transcent" is a novel nomenclature to denote a transcendental stair-wise leap using the third kind creativity of humanity. (See Chapter 1 and Chapter 7 for the definition of creativity.) Humanity is about to leap for transcent not for mere adaptation. The following are the comments on how the change of paradigm will be made.

The Main Line of Thought

There is a strong desire to establish sustainable peace in reality. Such peace does not stand for a perpetual and perfect tranquil peace, but a productive state of humanity, on which human race can live together, based essentially on tolerable conflicts. People in today's society are the tired pilgrims of the long human history, who can bare the tolerable conflicts with low levels of violence and of disturbances against their everyday life.

I believe our first target is to achieve a 200-years-long undisturbed peace. Long and healthy dynasties have been maintained for about two centuries, even though further studies on historical facts should be made to verify such statement. So far there have been comments on past global peace regarding Pax Britannica which lasted for 150 years, Pax Romana for 200 years, and so on.

In fact, those comments do not help this study, because all the past Pacis achieved by the governments of the various political subjects are not in range of the definition of the creative peace. The creative peace is defined in Chapter 1 as "the activated state of humanity in a stabilized condition."

The practical concerns and queries are whether good and plain citizens can live in freedom, whether every tenant of the world could be benefit from the reasonable level of humanistic education, and whether all inhabitants can reside in "a healthy economy with the worldwide continuum of the appropriate circulation and distribution." This book does not cover the long debates between so called 'isms' but proposes deliver the most reliable scientific way and opens a room for further advancements. These practical queries call for invitation to methodological developments with the respect for individual and communal personhoods.

Vocabularies such as personhood, person, personage, personism (not personalism), personality, personage and etc. have been used in this research to follow the above line of thought. The century we are facing has various potential perils to risk our personhood not because of collectivism as what happened before but because of the twisted understanding of the personhood of human beings. However, a new generalized definition of person was made in this research to establish a scientific description of humanity.

It is also important to mention that I tried to start from and conclude at the most problematic lines of thoughts appeared in humanity. It was intended so because the terminologies needed to be normalized as if our civilization is being founded again from scratch.

The starting line of thought would be at the three ideals: the first is **nihilism** as the null point to start from and advance for the realization of ethics and other means of life at a communal level, the second is an **agnosticism** in a positive attitude that can lead us to introduce scientific posture denying plain beliefs not verified with experimental evidences, and the third, an **anarchism** to overcome the scattered discontinuities in the current humanity towards a reasonable continuum.

These attitudes are believed to exist behind epistemological endeavors of the general scientific intellectualism. F. W. Nietzsche maintained a nihilistic view in his pursuit of Übermensch to give a birth to the new phase of humanity. Whether his philosophy is right or wrong, humankind is at the threshold of the era in need of new attempt for survival.

In this work, I endeavor to understand the real human, as well as the superman. The notion of this real-man is used to describe anyone who has lived on this planet leaving merely humble and shabby traces.

Nihilism is a sound base of thought but has to be overcome. When nihilism meets with substantialism, nihil or nullity in human intellectual reveals its character as a cavity as seen in the case of "*punctum caecum*," the blind spot in the structure of human eye. It is a metaphoric expression that I explain that nullity is not just a static cavity but behave as a creative and productive vacuity when a nihil becomes a part of a wholeness of the substantial universe.

Agnosticism is the blindness itself. The blindness disappeared when some medusoid, usually called jellyfish, has attained the earliest photic sensor in evolutionary process. The most primitive pallial eye, or stigmatic eye, has been first equipped to medusa. If we have a body with no sensing devise, we are supposed to go through the greatest stress to understand the world as Helen Keller did. The power of agnosticism stands where humanity is vacant: no knowledge, no emotion, no discernment, no planning, and no nothing. However, it is found that humanity has a vector: power and orientation.

Anarchism, when understood from its original fundamentals, is not of a destruction and terrorism as generally recognized nowadays. Its main theme is to start all over again from the bottom. In that sense, the human race could mend a number of discontinuities among different racial, cultural, political and economical entities.

The goal of our investigation can be achieved with the following three attitudes: the first is an **ultimatism** in humankind to mobilize all capability in an earnest and thorough manner to surpass the present omega path, the second is an **integritism** (different from "integrism" successively used in this research) to activate humanity based on personhood to attain the integrity of humanity, and the third is a **continuitism** in which discontinuities are discouraged to achieve effective *koinonia* (Greek biblical terminology adopted to represent mutual communication, interchange and association; here we understand the term to have a mode of a dynamic flow) globally in external conditions of Earth and internally in the structure of humanity.

These attitudes generally appear in the ontological pursuit of humanism. In the ontological understanding of the wholeness of the object of our concern, the overall "elasticity" of memory is necessary. The elasticity of memory is a terminology adopted as the contrary to "plasticity", which is frequently used in brain science and thought to be emerged after "dynamic modularity," which is a mechanism of higher hierarchy in brain activities, and is spread over all the phase space of cognition.

Dynamic modularity(動變 分格性) is an essential terminology of this writing to denote the intermediate dynamic process in hierarchically structured entity. The concept is adopted and developed from the idea of dynamic core raised by Christof Koch (The Quest for

Consciousness, 2004) and also from the idea of modularity in the development and evolution of the natural complex systems (Modularity, the Vienna Series in Theoretical Biology 2005, editors, Werner Callebaut and Diego Rasskin-Gutman, supported by Herbert A. Simon by his foreword).

The City of Alexandria on Pharos, over Night

Pharos is the Egyptian seashore where the famous Alexandria was built. The Alexandria was the center of Hellenism among more than two hundreds cities of Alexandria built in the conquered Greek territory in the 4th century B.C.

A thinker, if properly trained and educated, can create a well organized frame of thoughts in a single night. The systemicity of knowledge is responsible for the phenomenon. Actually, the systems of knowledge, developed or undeveloped, can be built in innumerable ways, because the knowledge system is an output of the matching efforts among presuppositions, experiences, hypotheses, desires and etc., that rest on the ground of the memory map from the actual environment. The Matching maps change the characteristics of thought by the location of its central idea and the binding coverage of the maps.

When Alexander the Great designed the city of Pharos in his mind in one night, he had probably done a lot of combinatorial thinking for his selection very fast before dawn broke. The selection process is comparable to the building a system of knowledge. The world of thoughts and philosophies are enormous. However, they are no more than describing the megalo experience of humanity of mankind in a respectable fashion. This is why we have to figure out what the whole picture looks like when we combine to integrate all ideas.

It is always essential to consider building the system of knowledge and philosophy from the ground, as the city of Alexandria was designed on the scenic island of Pharos. The ground should be the bottom of undisturbed bed rock. There are two comparable concepts in Buddhism, "null view (空見)" and "viewing nullity (見空)". Null view is not healable. Viewing nullity is a good foundation for further enhancement in learning.

The writer had an accident of being pricked by an injection needle into the apple of his right eye when he was thirteen years old. Such accidents are almost impossible to occur in reality. The author, in his days of middle and high school, was badly damaged by the serious agnosticism. Nothing was reasonable or predictable. There was no reason for the sidewalls built along the road to his school to collapse on him in his subjective world. He became a forgotten vagabond even to his family; he was mesmerized by a vast desert of null view, which means no bases for thinking.

When he had reached an understanding that cause and effect still works wherever, and is transformed into every phenomenon that is connected with all the others, viewing nullity and the fullness of the Providence. This conversion, very slowly accumulated within him after internal battling, supplied him with a highly firm and steadfast foundation of philosophical attitudes. People usually call it "(the hardest) adamantine mind," while I call it "(the most brilliant) diamond heart." After the conversation, he understood the blind spot in sole rationalism and sheer intellectualism.

The Agnostical Realism of Science

In this context, the standpoint of practical approach towards the creative peace should be an agnostical realism. I used Agnostical realism to introduce the scientific attitudes declining absurd beliefs not verified with experimental evidences. Agnostical realism seems to not maintain a firm stance of belief but has to reach to the basal foundation of a certain belief to derive action in any circumstances.

Recently schools of various scientific disciplines appear significantly damaged by the flow of research revenues and swarm assertions of affiliated discipleship. In order to mobilize all the necessary resources of humanity, we need to dig into the bottoms of our heritages again.

Recent development of evolutionism presumes no precondition for the orientation of evolutionary process. External environments and internal mechanisms interact with each other to produce evolutionary processes. However, even evolutionism cannot disregard the reality of the unrealized. Evolution takes place under the total influence of reality. Evolutionary biologists unconsciously recognize that C.E. (Caenorhabditis Elegans) under their microscope is, without exception, subject to the influence of the experimental condition.

Humanity has been affected by tremendous governance of the change in the greater galactic conditions. The reality of the unrealized, not the reality of the unconsciously-realized, has mostly been the subject of religion and culture with different set of explanations.

Then, why some disciplines of science, which are regarded well-defined and methodologically robust, imply that ultimate truth should be outside of the intellectual realm? Do humans recognize that the introduction of language, or equivalent linguistic tools, to humanity had established dipolar compatibility of dual mechanism, as what can be seen in the old brain (representing sensciousness in the author's terminology) and the new brain (consciousness, which seems overly simplified though) that compete against each other?

But such mental duality usually appears in our lives as a duel between instinct and rationality. This is related to the essential existential structure of humanity. Human existence suffers from the controversial confrontations of realism vs. idealism, of introspection vs. imputation and of belief versus reason.

Humanity is full of bifurcation that has to be effectively reconciling with each other if an urgent survival is required. We find the same in the scientific sphere. We call it agnostical realism for survival. Agnostical realism is that we have to do anything for survival in the real world when we do not have a logical frame or sufficient information to judge a situation.

A founder of mutual fund, if any, who has to manage the fund in an unpredictable market condition, would be basically agnostic, but he has to keep a firm belief on certain concrete contact point for his decisions. That contact point is his understanding the market that he has been saved bit by bit through his experiences.

I would like to comment how George Soros (The new paradigm for financial markets, 2008) as a student read Carl Popper. These two gentlemen do not appear to be agreeing with each other. Recurrence relation is the answer to a long search of Soros, the billionaire wealth builder. He extracts the recurrence principle working in the realm of financial market. Fallibility and recurrence undividedly exist in the human activities in a market.

Reading Soros idea, I felt he was telling a definite influence of "abnormal psychological expectation" in a market with enormously expanded bubbles. That is a human factor. Human factor is attributable to recurrence. Maybe the strongest human factor was revealed in Soros' intentions to purchase shaking amount of stocks in the Southeast Asian market in one occasion and to give away 300 million dollars to Eastern European countries in another occasion. It is said that "The Open Society and its Enemies" written by Popper had a strong influence on him.

The initial condition, the final condition, and the generalization of universal validity, which are scientific laws, are the three elements of scientific frame established by Carl Popper. Its three functions are forecast, explanation, and experiment.

Forecast and explanation are of symmetrical property from the same root of reasoning of a subjective entity. Scientific verification through an experimental procedure is derived from this symmetry of rationale. Proof and disproof produce asymmetry and lead to unbalance conceiving another initiation of a novel viewpoint.

These are good bases for establishing methodologies of science and technology. However, what could we do about human factors? Human factors are the prime crime of all problems we face in rational works. We allow some omissions, exaggerations, bypasses, and some superfluous but significant factors. Human factor is basically a matter of psychology. Psychology is similar to smelling scents. Psychology is probably too real to be dealt by only intellectualism. Thus, here, we face an agnostical realism again.

Agnostical realism is an instinct of a hunter. A prey is met when he forgets his own logic in many cases. Empty yourself. That is the advice to all generations who lost their grasp of reality by being caught in the past. We dare to see the realism of humanity.

3 Worlds, 4 Minds and 4 Annulus of Cognition

Agnostical realism is a willingness to build an epistemological frame of thought on an ontological thought. Karl Raimund Popper suggested a general structure of the world of humanity in a dynamic condition; there are three world expressions, which are World 1, 2 and 3. A diagram is presented in Figure 3-1-1 to show the inter-relation of the three worlds. Karl Popper called himself, as I recall, a metaphysical realist. Metaphysical realism is an understanding that the world is independent from people's minds. This philosophical attitude implies some strictness in including broad concepts and opens up for further advancement. However, a greater exposition to the ontological view would have led him to a more or less better explanation of the world.

John Eccles and K. Popper have been criticized sometimes for designing a neural unit of cerebral cortex (i.e., 40 MM dendrons linked with a mental unit, psycon, a unitary conscious experience). Hypothetically, psycons act on dendrons and increase the probability of the firing of selected neurons, while the reverse would take place. Such standardization process with quantitative units is a necessary step in science, but it seems more time will take to get to the process.

Sometimes propositions of a scientific party become a theoretical tool for the opposite scientific school in the next generation. In some cases, extensive experimental results change the whole structure of a technological understanding. Thus, staying in a demilitarized transition zone of theoretical antagonism may benefit when contradicting beliefs successively integrate.

Daniel C. Dennett, who proclaimed himself as a known decided materialist, also suggested an interesting model of human cognition and consciousness in his book, "Kinds of Minds" as shown in the Figure 3-1-2. He proposed the four evolutionary steps of mind: Darwinian, Skinner, Popper, and Gregorian. These models show the evolving mechanism of cognition and reaction.

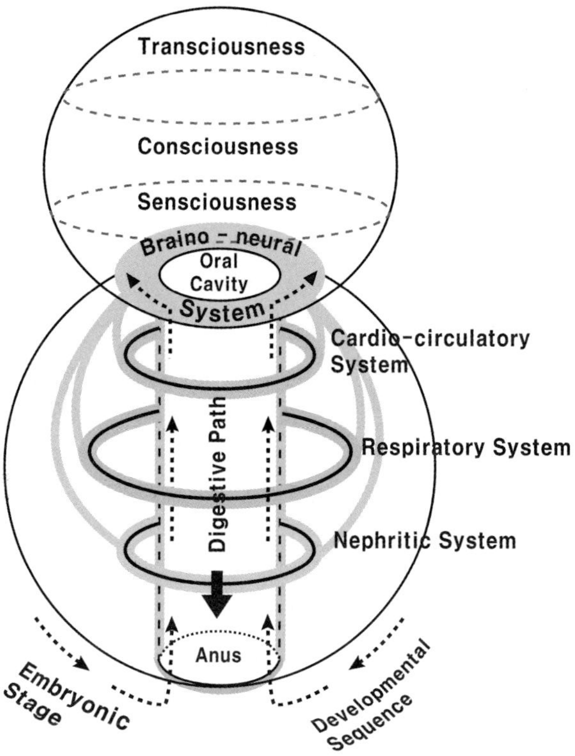

Figure 3-1-1. The Emergence of Three Layers of Sciousness in Cognition

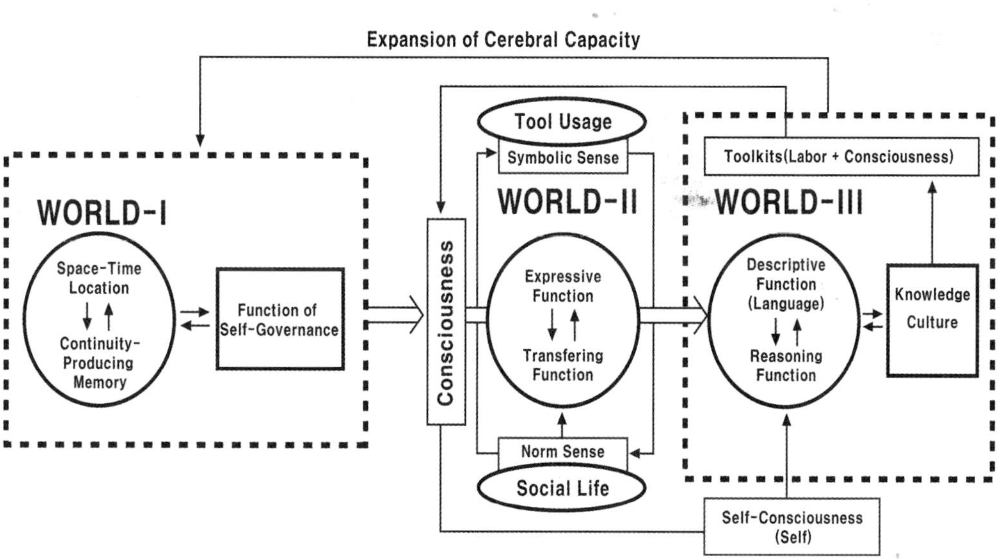

Figure 3-1-2. The Three World Expression of Karl R. Popper (illustrated by Yong Hyun, Cho)

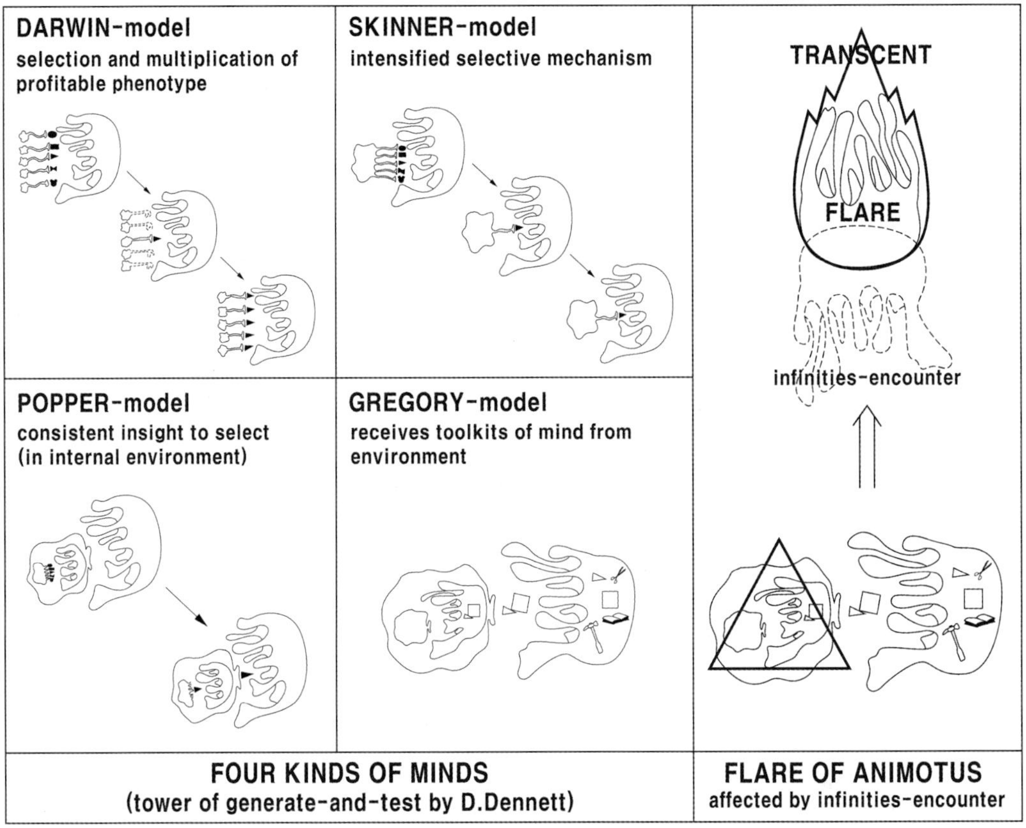

Figure 3-1-3. The Kinds of Mind of Daniel C. Dennett (Evolution of Minds) and the Mind of Animotus

The basic structure of human cognition used in this research is suggested in Figures 3-2 and 3-3. In this view, human symbolism is the highest mechanism, because it is based on the complex hierarchical structures of dynamic modularity in human brain activities and neurophysiology. Symbolism is not, however, caused only by dynamic modularity of physical neuronal cognitions.

Dynamic modularity generated from the domain of mentality looks much more complicated than its physical origin. Symbolism from mental origins happens to be magnitudes higher than symbolism from a physical origin because it has an unbound territory of infinite combination. Such combinatory outcomes of the mental origin causes ceaseless stress imposed on the human brain through amplified and metamorphosed forms of imagination.

This fact is attributable to the human exposure to the threefold of the ontological intrusions of nature, which consequently makes the three folds contacts with the complexity of humanity. The Metamorphism of human imagination is enforced to scale-down the scope by continuous episodic edition of innumerable pieces of memory and the imaginary derivatives of meaningful memory. Figures 3-1-4 and 3-1-5 shows a three dimensional illustration with its cross sections respectively.

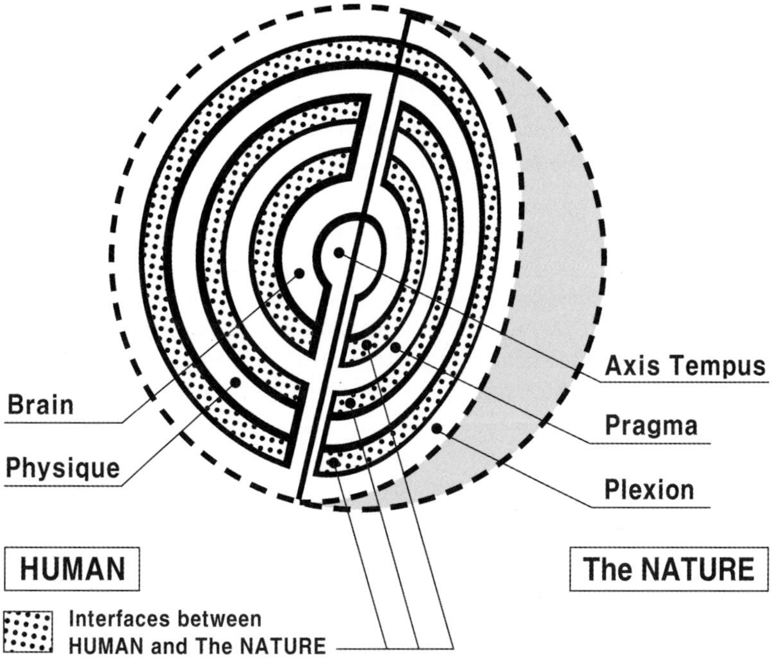

Figure 3-1-4. Three-Dimensional Illustration of the Threefold-Intrusion of the Nature into Human Cognition.

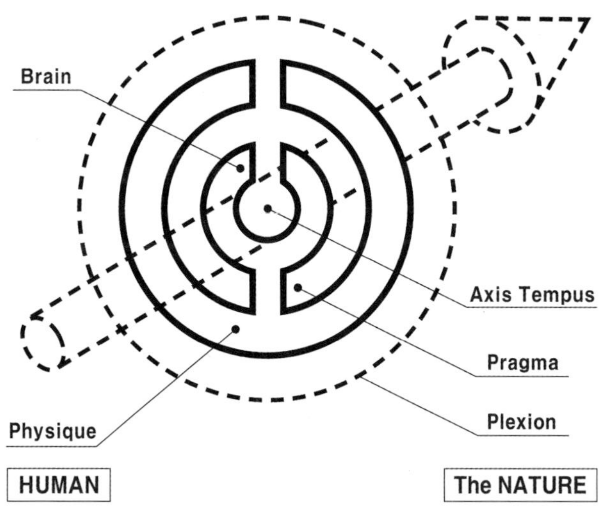

Figure 3-1-5. The Cross Sectional View of the Figure 3-1-4.

The hypothesis of the dynamic core (G.M. Edelman and G. Tononi, A Universe of Consciousness, 2000) has affected to the idea of dynamic modularity. Dynamic modularity does not stay anchored at a single spare as the dynamic core does but is freely out at the open spaces where the ontological intrusions of nature have penetrated into humanity. The intrusions of nature take place at three different levels of humanity at Plexion, Pragma and Axis Tempus. This will be explained in Chapter 4.

The Third Level of Cognition; Transciousness

Super-conscious-ness means elevated consciousness. "Scious" comes from Latin word "scire," which means "know." Transciousness is different from super-consciousness that is usually used in mystical para-physics. Transciousness is the highest cognitive domain that guides humanity to the transcent.

It is generally agreed that humanity has old and new brains. The old brain is the reptilian brain still working in our cognition. For instance, "blind sight" patients can pick up an object without using their ordinary eyesight. Getting one step further from the two-brain idea, it is required to imagine that a novel function of intangible third-cognition has been activated in an unusual way in some portions of the territory of new brain; a mechanism to deal with the problems of infinities.

Cortex is said to be the third development of a physical object and location. It looks reasonable that cortex does the central role of the third functional cognition, transciousness. However, the experimental research based on the neuroscience alone cannot pass over the obstacles of the complexity of human cognition. Certain overarching law should try to explain the origin of this complexity.

Usually it is understood that the trans-communication efficiency between neuronal synapses is around 30%. To avoid such inefficiency, brain has 100 billion neurons which are mutually connected 500 trillion times. Human brain is supposed to finish growth after birth, because the birth canal is too narrow for a full-grown brain to pass if the growth of the brain is completed in a mother's womb. It takes a considerably long time if 23 thousands human genes alone affect the design completion.

Human brain matures throughout a long enough period of a child's life. The design of neuronal activities is based on the reactions and experiences of sensory centers. This long and profound circumferential information seems to help human cognition to work in a more or less networking pattern and to provide ways to develop complex patterns of "dynamic modularity" among the billions of neurons.

It is suggested in this research that the dynamic modularity is the main character in the emergence of the three layers of cognition: sensciousness, consciousness, and transciousness. Most readers would find the first two layers of cognition, sensciousness and consciousness familiar. But sensciousness is not sub-consciousness, which we designed as the combination of sensciousness and major part of transciousness in our view.

The last concept, transciousness, has not been presented before. It is suggested by this work that transciousness plays an important role in the episodic memory combined with the dynamic modularity. The Dynamic modularity is a higher notion than the neural modularity. It would be convenient if regarded it as a kind of "cognitive modularity." (Modularity, Werner Callebaut et al. 2005, p17)

It is thought that the dynamic modularity can be extended to the theory of Infinities-Transfer Ring (ITRing). ITRing is an important concept to understand how the time axis

(Axis Tempus) intruded into the realm of humanity. The time axis is in the phase space of humanity. Metaphorically, it is a vacant tunnel, because time is an undefined axis through an ever successive integration of pieces of memory collected at about every 3 seconds interval as has been verified in neuroscience (E. Pöppel, 1989). This is illustrated in Figure 3-1-6.

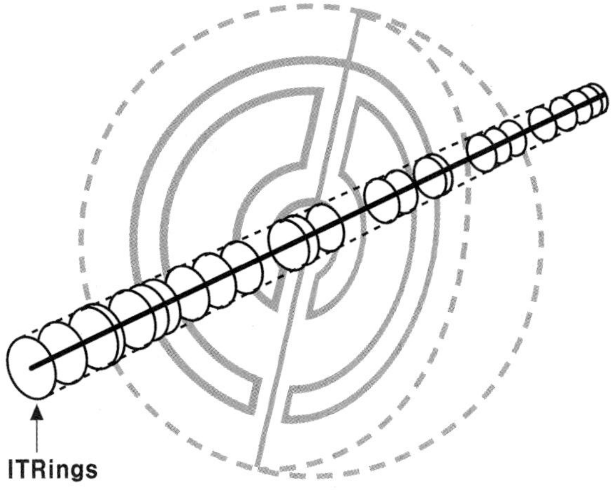

Figure 3-1-6. The Metaphoric Illustration of ITRing as a Vacant Tunnel Wrapped by Memory Pieces

Besides that, we have to consider the possibility that the time axis emerged into our cognition since the balance or the symmetry of chaos has been disturbed. Chaos in humanity is chaotic in a general expression. Chaos of our world is not symmetrical. The original chaos before the human existence was the undisturbed abyss (*tehom* in Hebrew meaning the deep) It was hidden by darkness but disturbed by the Breath for the very first time. The undisturbed chaos of the beginning is symmetrical in a sense that it was in the fundamental status of the non-accelerated and non-inertia condition in physics. I presume that such fundamental chaos is Rig-Veda; Wave generates time. Wave comes from the Law.

The Law of Heaven and Truth were born
Of conscious fervor set on fire
From this came stillness of the night
From this the ocean with its waves.

From the ocean and its waves
Then the year was generated
Appointer of the days and nights
Ruler of all mortal beings.
(from the Hymn of Origin, translated by Hae Choon, Choi, 1982)

It is not our intention to draw a scientific conclusion from the ancient religious thoughts. We intent to point out that a new ocean is open to us with full new waves. The author discovered through a deep spiritual meditation for decades that exaptation (脫應), usually meaning a non-adaptive evolution, has taken place in the transcent of humanity. It is especially done in a manner of transcendental intentionality to introduce a novel stream of creativity from the new dimension that gained in the course of overcoming situational stress. Transcendental intentionality will be dealt in Chapter 4.

Our research suggests that time is an agnostic factor intruded into the human cognition prompting an abstractive procedure in human mentality. It is not yet completely concluded into a theory in which whether time flows reversibly in a phenomenal world. We know that time flow is irreversible. It could sound a little strange to say that the irreversibility of time flow could simply be the result of irreversible edition of the human episodic memory.

When we see the world in the light of actual infinity and eternity, time does not exist. Humans can imagine or at least acknowledge the infinite possible extension of dimensional space, which we call in mathematics, the Hilbert space. It is strange how people can easily recognize and accept the concept of infinity, as well as eternity as the infinite time scale. It appears like human beings have an ability of shifting, projecting, and substituting their stress to the general public using an unspecified number of techniques.

The many unsolved ideas projected by individuals to the society tend to be arranged in the course of structuring pieces of imagination made for many encountered contingent situations. When we have the intention with enough energy to cope with our confronted struggles, our imagination tends to be elastic. The concept of elasticity adopted here assumes the role of mental propulsion and thus supply the energy to perform ceaseless powerful imagination until it forms a significant structure. (We call it "postefference" in Chapter 4.)

Macaques, a kind of Japanese monkey, learn how to clean sweet potato with sea water from the members in their group. The first Macaque that started a new technique is known to be a young one. It is well known that researchers suggest that human language has been started originally by kid(s) groups as well.

The biological needs for episodic edition are the reason why the abstractive ability was developed among humanity. A baby, whose metabolism drives a periodic hunger, has a subconscious need for episodic arrangement of its memory. It is an arrangement of affairs such as mother's approach, breast-feeding, following drowsiness, uneasy stomach, and bowel movement. When editing episodic memory, a baby enters into a novel world of non-substantial reality, which is the abstraction.

Human inference has a peculiar pattern of fitting engraved abstractive pictures in physical process. This abstractive picture is a sequential but massive conglomerate of feeling and can be specified into qualia and quantia. (Quantia is defined hereafter to be a kind of qualia that respond from our neuronal sensation for mathematical measure, logic, and complex notions.)

Inference stems from abstraction. Inference has three distinctive fields of logic, relation and space (the Math Gene, Ch. 7, 2000, Keith J. Devlin), but they all exist based on procedural chain affected by cause and effect.

Why are inference and abstraction important for the realization of the time axis? Irreversibility of time regulates the formulation of inferential and abstractive ability in humanity. Time, inference and abstraction are all non-substantial. If we insist that a human brain is wider than the sky, non-substantiality of human mentality enable humanity to embrace the sky. Non-substantiality creates a blank endless tunnel of time piercing through the entire human sky.

ITRing is placed over the surface of blank time tunnel to make a strand of episodic memory with an emphasis of the self-interpreted summit experiences in a row. This self-interpretation is related to the people's experiences of infinities. I suggest that the blank time tunnel in hypothetical imagination could be an expression of rolled-up hidden dimensions manifested in a quantum conditions.

Now, it is an assumption of this research that the boundaries between three layers are not as clear cut. Such is due to the fact that cognition is based on emotional reception. The emotional reception is not easily illustrated as in rational explanations. The Dynamic modularity generally supporting transciousness activities does cover sensciousness and consciousness altogether to produce an adequate integrated "gestalt" for transciousness.

Anyhow, the unclear emotional basis of humanity allows human beings to behave in accordance with the principles of the science of complexity. The mixed up apparent states of emotional context of humanity enable human beings to produce profound sources of nexusity, complexity, then of creativity.

Henry Poincaré commented that human intuition on natural numbers is the basic mental activity, from which matters and thoughts are created through consequential analogical inferences. Such feeling is generated when experiencing the flow of time in a fixed direction. (John D. Barrow, Pi in the Sky, 1992, Ch.5)

Among numbers, "three" comes from the meaning of many (over, beyond, afar). The psychology of three is based on a very primitive emotional frustration willing to avoid counting. In French, we find troi (three) and tres (very), and in Latin we have trains (beyond) and tres (three) (refer to Ch.2).

I believe the most fundamental relations found in humanity are bilateral. Two persons have the basic relationship: You and I. This I-you relations do not generate a division in consciousness. It provides a good environment for concentration, correspondence and direct reversibility. "I" need only one map at the first stage for "you." Let's suppose the relation between a man and wife.

When we get to the next step of a three person relation, man-wife-baby appears. "I" happen to need two maps. The First map in a full scale represents the relationship corresponding to "I". The second map is a little bit vague and is of less detail, for it needs room to admit new information about the baby.

The 'baby map' emerges out of the noise in consciousness to represent the most energy-absorbing entity of the couple. There is a similitude between a fractal in our consciousness and an ovum starting to grow in ovarium. It allows a triangular irreversibility for unanimity in decision if the communal relationship is firmly structured. Our transciousness has such internal structure in logic and supply dynamic change in our humanity.

The Trinitarianism of Christianity has some hidden logics for the fundamental development of human communality. The Trinity is the essential presentation of communal development. The embryonic development of "the fractal-third" in the buffer zone of human cognition generates room for asymmetry which in turn gives procedure, order, orientation and the emotional feeling for the time flow.

The fractal-third could be the source of triviality, the noise in our consciousness. It is noteworthy that triviality (three road junction in Latin) has a deep connection with the introduction of the novel dimensions. The Trinity, triviality and the Tri-great-absolute (San-Taiji, or tri-Taiji) have the picture of the fractal growth from a new dimension. That coincides with the human transciousness.

3-2. Humanizing

Humanization, Person and Integer
Anthropomorphosis
Humanity and Personism

Before delving into the topic of humanizing, it would be helpful to clarify the difference between humanification and humanization. Humanification is a new vocabulary to denote an excessive systemization toward human-centric humanity, while humanization denotes the proper way that leads to humane transformation of humanity.

Humanization, Person and Integer

Since neurophysiology and neuropsychology have emerged, "humanization" has not simply remained as a humanistic view but has become a scientific terminology. Philosophy has also been altered in its content and purpose since mind-body problem in scientific expression has been raised. Humanization will be treated hereafter as the methodology which will give firm ground in strategic approach towards the fulfillment of the creative peace. At the center of the concept of humanization the topics of person and integer are laid.

Why 'person' came to exist in reality, and how it is, deeply shows itself and behaves as a governance factor in the phase space of humanity; these are great questions that must be answered since it is our sincere conclusion that the person is the seed of the creative peace and provides its the frame and the foundation of the forthcoming creative peace. It is imperative for the human race to pay deep scientific attention to the phenomenon of person and to keep an eye on the threat of dehumanization process emerging in a wide spectrum of our civilization. When we deal with the problem of person or personhood, because of the complexity of the concerned properties and the ambiguity of preferential methodologies, the notion of integer will be utilized in explanation.

Integer is well defined in mathematics. In the realm of mathematics, integers are meant to indicate numbers that include natural numbers, zero and all natural numbers of negative value. An integer is a representing unit (a person) in a society; negative units and nullity are not considered in this study; however, as we may consider imaginary levels of units, which may be used in the discussion of later for the personal axis of virtual reality.

The following image is a picture of an integer person which has universality, inherency and self-centeredness. Inherency is a part of each and every integer person, provided that personal qualities belong to the internality of the person. Using the term immanency in the place of inherency may be a bit heavy to use in this context. Self-centeredness is to justify for an integer to behave as a determinant and to allow the existence of an alien objective world to an integer. A person, as an integer, is able to activate his intentionality and to transcend onto the next phase (not necessarily, in a teleological direction).

A person is a matrix of infinities. If humankind were to be distinguished differently from other animals, it could be done in relation with the infinity. Matrix of infinities is an essential nature of humanity. It will be explained in more detail in the following sections.

Anthropomorphosis

The key notion of anthropomorphosis(人態形成) is the recognition of infinity through the cognitive system, which includes sensciousness of the neural system and bottom-level unconsciousness of quantum status. (In ontological thought, we have to consider all actions and reactions of the universe and our entire physical condition.)

The process of anthropomorphosis is one of the keys used in unlocking the idea of the creative peace. The theme of recurrence is usually found in a closed system with iterative routines. However, humanity is not a closed system. Humanity has many ceaseless, active and quick interpretation cycles in its cognitive process because of its frequent interactions with other individuals.

It happens almost on the level of the swarm intelligence of bumblebees if we disregard the refined micro-processes of humanity. Anthropomorphosis is an emerging process of intrinsic attributes of humanity, either individual or communal. The intrinsic attributes of humanity lead us to a strange attractor of unknown (but predictable within certain probability and within certain constraints) that make anthropomorphosis possible.

Why can't the human race help but follow the path of anthropomorphosis? We know that the meaning of anthropomorphosis is transformation of becoming human or of taking the resemblance of a human. Is it merely the result of our feedback? I believe we may be able to find the reason from Popper's interpretation of his World 3. The objectivity of human knowledge is justified by the reflection of reality to World 3, which we believe to be is equivalent to "Anima Annulus," the fourth ring of human cognition. Anima Annulus is largely correspondent to the central processing unit of human cognition except when working on conclusions that involve innumerable qualia, images, sensations or symbols.

The qualia complex, emerged from the variety of dynamic modularity occurring in brain-neural environment, is, in our hypothesis, governed by so-called universal grammar. (I would confine it to only universal syntax, which has an origin from the most profound confrontation with the implicate order of nature.)

As Chomskey asserts, a language faculty is species-specific. Phonetic and other descriptive languages are not only patterns of communication. Varieties of language forms are an introductory conduit of human existence into the real world.

"The imagination (or inclination) of the thought of the heart" (or of the mind, since heart is a description of more emotional qualities) is a converged consequence of brain activities, which affects the directives (or inclinations) of humanity to be intensified for further convergence. This converging process is named anthropomorphosis. Left here for our further discussion would be the question of whether or not anthropomorphosis could cause cancerous symptoms to the phenomenon of humanity.

It must be noted that anthropomorphosis can be called a convergence process only if such convergence is made in a state of "relaxed selection." Relaxed selection is well presented in a case of babbling babies trying to learn language. They do not concentrate on learning various vocalizations but on relaxing while babbling (or babbling while relaxing). The highest level of creativeness appears when two seemingly contradictory concepts come together: convergence and relaxation. This state of mind is very much similar with the relaxed attention of driving on a highway. We can observe the road and surrounding environment in a relaxed condition with undisturbed concentration at the rear of the car and front of us.

Creativity in essence is a picking-up convergence in a relaxed selection. The brain alone cannot bring along healthy creativity. Creativity is in need of healthy physicality and healthy (not necessarily, balanced) emotion.

In Figure 3-1-6, Anima Annulus is shown to be supervenient (隨伴的) to Corpus-Pragma Annulus and Pragma-Cerebrum Annulus. Corpus-Pragma Annulus is the source of emotional aspects, while Pragma-Cerebrum, the source of rationale. Creativity necessitates sound communality to survive in social conditions.

Anthropomorphosis implies feedback effect of past and present humanity to the formation of the forthcoming nature of humanity, where creativity is the essence of both anthropomorphosis and humanity. Anthropomorphosis will be dealt again in Chapter 4 on the topic of the science of humanity.

Anthropomorphosis also has the same logical structure as that Kurt Gödel expressed in his Incompleteness Theorems. In our opinion, the Incompleteness Theorem suggests that a system is systematic and closed only in the intellectual structure of humanity and that, in nature, a system is open and thus not self-complete. The Incompleteness Theorem tells of exo-systemic self-reference made in a system which includes self-denial. Our intellectuality in some sense rejects the self-referential paradox. However, we find that the nature is open to the infinite Plexion in which universal nexusity is essentially allowed to sustain its original infinite nexus. This original nexus of universal Genesis intrudes into all the existence.

We have found indicative evidence even in scientific findings such as quantum physics which show that in the universal beginning the nullity is an opening that can be named vacuity, which is vacant to introduce substance into being.

The actual infinity exists in an infinite superpositional matrix form, which is non-interpretable, a mere single vacant opening like a black hole that sucks all the discernment out of the intellect. I would like to ask talented mathematicians on how this concept could be expressed in mathematics.

The entire academic system, as long as it is artificially orientated on non-exceptional logical system, is apt to confront natural phenomena. But the phenomenon of humanity follows natural phenomena; academic schools form islands of logic lest they avoid self-conglomeration by their finite logical process; the plexus of historical humanity is unavoidably anthropomorphosis as it aims at further comprehension of more perfect and wider reality (in a form of menesys dealt in Chapter 7, which is the compound of system and nexus).

Novel propositions that are not provable in the currently existing system of knowledge will be added successively. These will be related to anthropomorphic phenomena, which could be named as dimensional leaps, dimensional split open, or introduction of new dimension. The dimensions of humanity are tensorial and easily extendable as found in brain science.

Humanity and Personism

In order to clarify the phenomenon of humanity, it was deemed necessary that the articulation effect of humanity by embedment of mentality to the human physical conditions be clarified. It is without argument that the brain and neuronal system are the source of mentality and that they are vital to the governing of physicality. However, humanity has profound derivative aspects which we call mentality (which contain spiritual aspects that some people cannot give up). The dependence is rigidly fixed to the individual physical condition including the brain. Thus it is necessary to straighten out humanity and personism, starting from terminology.

The usual philosophical tendency of dealing human personality is called "personalism." The extension of personalism dealing with communal aspects was called "communism." I believe someone might be opposed to this classification and prefer "communalism" for the extension of personalism.

There is another issue with terminology here. Before this research, the term "personism" has not existed. It is intended to focus the aforementioned four philosophies into double file. It would be two lines of thoughts; one line would be that of personism-communalism, while the other line would be that of communism-personalism. (This definition supposes that the previous definitions of personalism and communism should be replaced in principle.)

We need to pay attention to personism. Personism is primarily concerned with the basic contents of a single person. It has an extension to communalism which deals with communality, the characteristic issues of communities.

On the contrary, personalism deals with human characteristic personality as the main concern of the philosophy. Personalism can be considered to be a derived idea, purely in the line of naming structure of thoughts, from communism. In this line of thought, a commune or a community is the primary concern of the philosophy. (Here, I emphasize again that this communism is not that which has existed for the previous hundred years in real history.)

The condition of a single person is the basis of thinking when communal or social considerations are analytically made due to the importance of human physicality in comparison with other factors of humanity. Humanity stands on and appears with the articulatory effects of its individual physicality. That is why we consider the personism-communalism line of thinking to be reasonable to handle the phenomenon of humanity.

3-3. Infinity and Infinities: Encounter, Transfer, Exile, and Matrix

Infinity and Infinities in the Cognitive Mechanism
The Actual Infinity and Pseudo-Infinities in Human Cognitive Conditions
Infinities-encounter and Humanity
Hyper-Topological (HT) Conversion
Infinities-transfer
Exile of Infinities
Infinities-matrix
Infinities-matrix and Human Abstraction
Flaring Tetrahedron Matrix and ITRings

The recognition of infinity is on the summit of humanity. Humans are nothing more than animals if the content of infinity is lost in the human mind (or, if acceptable, in the human heart). The human heart of this day is deprived of, or forfeits, its feeling of the infinity. Do we really possess the sensation of the infinity? Is it a mirage? If it is merely a mirage, where does it come from?

Symbolism and semiotics are tools to explain the relevance of human cognition to nature. Naming natural objects of the singular or plural objects, of groups or of individuals, objects tangible or intangible and so on are formulative fundamentals of humanity. Human language has contributed to the fixation of this trend in humanity by patterning and the consequent mapping of patterns.

There are traces of brain activity concerned with infinity largely found in the realm of religion and philosophy. The limit of symbolism and semiotics is found in the corner of mysticism. Mysticism, real but not verbal, has a complexity level comparable to the intuitive capability. It has yet to be clarified if the human race has been successful in maintaining of the comprehensive intuitive capability that we have the potential for.

The ideal of the growth of humanity toward infinity and the non-complementary indemnity for humanity up to infinity is indeed an incomparable blessing to humanity. (We are not talking about the Tower of Babel trying to reach to the utmost height but the orientation of humanity toward the infinity and eternity.) It is very interesting to observe how deeply this absolutely illogical loop of thinking is engraved in the human mind.

The concept of infinity is said to have been invented by Greeks about twenty five hundred years ago. Xenon was the first written scholar who cited the concept of infinity in his famous fables of the running match between Achilles and a turtle. As I know, Xenon was of a Semitic origin. Our concern here is how humanity has connection with infinity.

Studies on the temporal lobe report that special overwhelming mystic experiences were undergone when the lobe was stimulated. As we deal with the infinity problem, we will not to touch on such religious aspects but will try to focus our consideration on the cognitive mechanism.

Infinity and Infinities in the Cognitive Mechanism

What is infinity, and what are infinities? Some people propose that infinity exists in reality. Others do not agree. Infinity is the reflection of the nihil sense. Wholly, vacant feeling come when human needs are not sufficiently satisfied as expected. Need is a kind of blank phase space that has been formed for which desires to be sufficed by what is needed. When we lose a precious person from our lives, we feel a deep nihil. It might be true that the prototype of nihilism was originally the sense of hunger.

The sense of hunger can be extended to the emotion of eventual death, which is most often a fear. It may have originated from the unconscious fear of instinct at the level of unit cells in our body extinction. It is thought that recognition of infinity originates from the shocks and paralysis due to extraordinary peak experiences; something like the paralysis of an inchworm before probable death.

Psychological peak experiences can be undergone even at the human level when the human psychological container has been filled to the overflow. This phenomenon occurs when an excessive shock signal greater than his receptive capacity hit a person in a pulse-like pattern. Figures 3-1-4, 3-1-5, and 3-1-6 are illustrations that will help dig into the infinity problem.

In these figures, "Plexion," the combination of *plexus* and *onto*, is differentiated from the real infinity (or the actual infinity) which contains the self of the observer. In the dynamic viewpoint Plexion is the plexus of an infinite number of affairs. Here, the whole Plexion exists at the external realm of humanity. Such external Plexion is understood by the self as a potential infinity or pseudo-infinity, but not in wholeness but as an infinite number of superposition of infinite combinations of affairs. (Superposition used here is of a similar usage as that used in quantum systems describing multiple overlaps, not as in mathematics. P164, Einsteins Schleier, 2004, Anton Zeilinger)

There could be serious debates on the issue if humanity has *a priori* senscious recognition when it comes to the infinity. It is the stance of this research that *a priori* sensciousness on infinity comes from the environmental structure of which humanity is known to exist. In that environment, human cognition is a connector between Plexion and "pragma."

Human cognition behaves as a connector between Plexion and pragma, by the interpretation of the sea of received information introducing Axis Tempus. Plexion is entirety of nature apart from the cognitive self. Pragma indicates the range of human cognition, in which a human can perceive or feel. Human cognition consists of sensciousness, consciousness and transciousness. Transciousness is a peculiar part of human cognition, an aspect not found in other animals.

It was Georg Cantor who opened new gates of intellectual thoughts in his mathematical interpretation of infinity or infinities. It was the set theory of infinity that enabled humans, Cantor proposed, to rationally express all aspects of abstract human mentality. The hypothesis consisted of the thoughts on the set of infinities and the superposition of infinite sets of infinities.

These thoughts were originated from the fact that human psychological attributes, which are the basis of human intellect, is of an infinite continuum and of discrete infinity. We are confronting a knotty subject dealing with the continuously changing emotional phenomena which exists at the rear of human psychological attributes.

It was deemed that the issue on the superposition of the sets of infinities must be treated in whatever way available in order to handle the infinity appearing in the phenomenon of human mentality, or in the extremities of human emotion. In this context, it is thought that the infinite capacity of the human recognition relevant to all possibilities of infinite superposition is supported by the reservation or abandoning of recognition than by instant recognition, as it is usually considered to be. How could such reservation or shelving behavior come to reality? Suggested here is a hypothesis.

All Human encephalic neurons go through periods of development, settlement and death. In the cell death mechanism, it is assumed that infinitesimal influences and traces are left behind in individuals or in networks of encephalic neuron. They experience "fire and wire" together and settlement in the mutual vicinity of other neurons in accordance with concentration gradient. Glial cells help the settlement of encephalic cells.

However, the death mechanism such as apoptosis is not only effective in just the process of development, settlement and death. They are also proficient in the reservation and abandonment of information, the production of superposed information from the primary sense of nihility generated from a physical origin, and the formation of high-dimensional cognitive mechanism of nihility around the primary nihil sense.

This respective superposition can possibly be distinguished as a separate set of experiences in cognitive activity. The process of reservation or abandonment of a specific superposition due to the emergence of nihil sense can be seen as a sort of information enfolding. Extremely minute electrical signals, or the superposition of a number of those minute signals causes intense enfolding of information that is converted to various kinds of signals, which are later stored in adequate physical conditions.

Such stocks of signals must take on a recognizable form in order to be detected and connected if needed to the activities of the cerebral cortex. The process of detection and connection is treated as long term storage of emotional signals in combinational cognitive mechanisms of neural cells dependent on electrical signals and in that of neuro-glial cells dependent on relatively tedious chemical processes.

Infinities-transfer is of a different notion from ultimate experiences, but has the potential to become a generalization of ultimate experiences. (For ultimate experience, see Wesley J. Wildman and Leslie A. Brothers, 2002, Neuroscience and the Person)

The Actual Infinity and Potential Infinities in Human Cognitive Conditions

The reason why the actual infinity (the mathematical terminology for the real infinity) does not exist in the enclosure of human cognition is because the self of a human individual, the part of the actual infinity, is cognitive as an observer. The Absolute Other (Das Ganz Andereheit, DGA) is a potential infinity (math. term. for pseudo-infinity) which is a deduction of the self from the actual infinity.

At least, there are no other possibilities of cognition at the vicinity of the self. These possibilities of cognition are the content of cognition of the self in *a priori* and *ad hoc* bases, which are innately specific to individual sensation. The self and infinities are separated into mutually different sets. Thus, though they may not be mutually susceptible (or related), they are mutually correspondent. The self desires to become a small universe because it has correspondence with DGA.

Deeply engraved into the self is an emotional mechanism which is a complex cognitive mechanism formed through acquisition by learning in addition to the *a priori* frame. This emotional mechanism produces a responsive confirmation between mutually correspondent systems. This confirmation does not have to be of an emotional quality and in most cases can be of various differentiated computational measures, which is able to generate a reflective virtual reality (反響假象現實).

This reflection could be very messy but still be workable for maintaining mutual responses. Such a messy state can be rearranged in an efficient manner through the training of individual and communal cognitive process. The ceaseless arrangement brings about the emergence of Anima Annulus (the cross section of Anima sphere in Figure 3-1-5). This research proposes that Anima Annulus is related to "the default network." (Douglas Fox, Nov. 8, 2008, New Scientist)

The actual infinity is not an object of cognition like Ein Sof in Kabbalah. A cognitive status in itself has already damaged the actual infinity. Therefore, the world, in the sense of the total number of accidental affairs which are experienced and recognized by an observer, is nothing more than pseudo-infinity. All pseudo-infinities have a cryptological relation (or simply a functional relation) to the actual infinity. It could be similarly compared with the ultimate cryptographic understanding of DNA (deoxyribonucleic acid) in relation to the history of planetary evolution.

Let us take a few hypothetical leaps forward to achieve our academic purpose. Our study proposes that the ebb and flow of accidental affairs in the structure of dynamic ontology are pseudo-infinities in the pattern of specified topological information. We also propose that all unidentifiable impacts not caught or absorbed by human consciousness, are mere pseudo-infinities. Contrarily, almost all pseudo-infinities except for a finite number of selected experiences are merely the unidentifiable background of human consciousness. This finite number is considered to be close enough to infinite to be still uncountable.

These accidental affairs in the set of pseudo-infinities within the vicinity of human physicality and within the respondent range of consciousness are called "pragma", and are shown in Figure 3-1-4 to 3-1-6.

Hence, most of the aforementioned infinity and infinities are supposed to be realized within a cognitive mechanism but not by consciousness; those include the actual infinity that is all the accidental affairs including the self of an individual, the infinity of Plexion which is a deduction of the self from the actual infinity, parts of the infinity of Plexion, and all pseudo-infinities affecting the self within realizable range in the pragma region.

Infinities-Encounter and Humanity

Being human is initiated by the encounter of infinities (including eternity). Infinity is real at least in the realm of humanity. But infinity is not exactly realizable. Two of the most renowned genius mathematicians, Cantor and Gödel, became insane when they tried to figure out the real infinity, some people believe, with their own physical brain. Primitive people would have been paralyzed experiencing thunder bolts striking within a foot from them, which might be regarded as an "infinities-encounter." However, paralysis is not deemed equivalent to an infinities-encounter.

It is reported that macaques (Japanese monkeys) tend to recognize their "self" by learning tool handling, such as rakes. They start to make cooing noises when no rake has found food to eat. The cooing noise is a reference call, a vocalization for specific objects. A "self" could be regarded as the commencing point for the pursuit of infinity. Could it be a sort of a meme that will evolve the complexity of humanity? I believe it is of further importance than just a meme.

Why does humanity dare to seek after dialogue, or maybe monologue, with "Das Ganz Andere" (The Absolute Other)? Is it because we possess self-recognition? Or, is it because a human being has a container of infinity built in from the start? An archaeological finding at certain seashore in South Africa shows many indications that the symbolization process started around 70,000 years ago; accessories and personal ornaments that Cro-Magnons had made to develop abstraction were excavated. Then, is the abstractive capability so essential to the differentiation of humanity from other animals?

It is my basic understanding that the encounter of infinities are achieved by the human mapping ability of cognition in time-series, with which the matching among various cognitive maps should be interrelated, and eventually the piercing of those maps with a fully developed temporal axis extending into eternity. The time axis is presented in Figure 3-1-4 as "Axis Tempus."

Infinities-encounter leaves the highest part of human experiences to be reserved on separate shelves of cognition for further reasonable interpretations. Getting along with the Axis Tempus from the past to the present, rings of infinities-encounters are recorded on the time-series episodic memory, similar to the annual rings of perennial plants. The time-series episodic memory yields an infinities-matrix to be explained soon after.

Hyper-Topological (HT) Conversion

"Hyper-Topological (HT) conversion" is different from the Möbius transformation. HT conversion is suggested for a general understanding of certain phenomenon such as the Möbius band and Klein bottle. Maybe, some scientist has already thought of and materialized this kind of concept. HT conversion is believed by the author to be a higher step in abstractive metamorphism in human cognition. Thus it is proper to classify it hyper-topological abstraction.

It could be regarded as an unforgivable primitive misunderstanding of a non-mathematician, for which the author requests a pardon of contemporary intellectuals. In order to continue my thinking, I have no choice but to continue this.

The Möbius band and Klein bottle both show equivalent "hyper-topological intrusion." The Möbius band is a case where 2-D linear dimensional piercing has happen, and the Klein bottle shows piercing through a 3-D areal dimensional obstacle. I believe this is an idea that has not yet been dealt with mathematics. Anyway, to find some solution to our question, the most probable and logical answer is as above.

We know that HT conversion is possible only in human abstractive process. Abstraction can provide for thinking based on piercing mechanisms through dimensional obstacle. However, it is also observed that blocking occurs in human cognition in interpreting the actual infinity which is not piercible with only abstractive comprehension.

It is also thought intuitively that the knot and entanglements are identical in hyper-topology. The knot is abstracted in 3-D space and entanglements are abstracted in 4-D spatiotemporal continuum. But these are the examples that do not pierce the dimensional obstacles. The author believes things that do not pierce dimensional obstacles are eventually verifiable in logical processes supported by experiment. It means the knot and entanglements are real while the Möbius band and Klein bottle are merely abstractive notions.

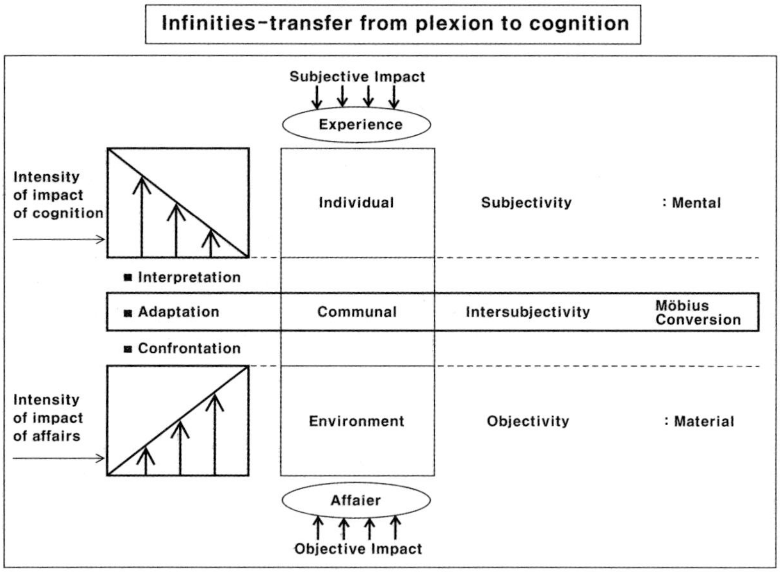

Figure 3-3-1. Infinities-Transfer from Plexion to Cognition

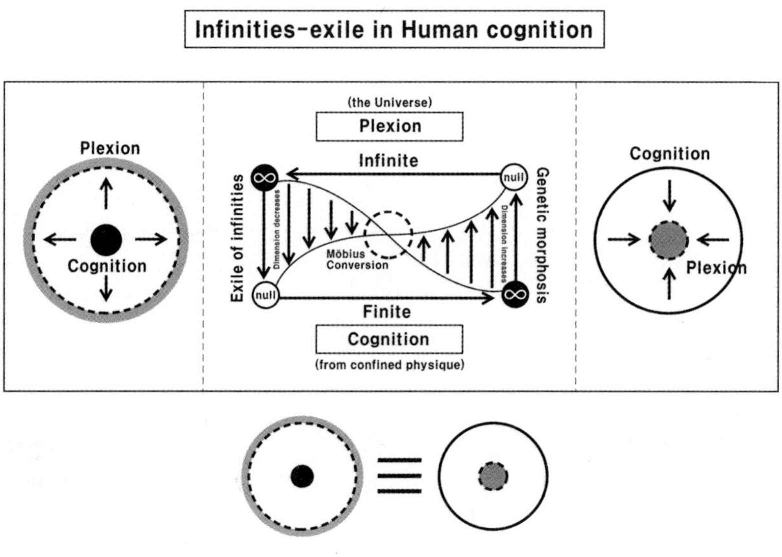

Figure 3-3-2. Infinities-Exile in Human Cognition

Hyper-Topological conversion helps the acquisition of contrariety between contraries in the later discussion for infinities-matrix. Contrariety is the relation between two contraries. Contrary and contrariety exist in the real world. We do not need to quote the parable of the spear and buckler. HT conversion has caused a peculiar abstractive function in human cognition.

Infinities-transfer (無限轉寫)

Infinities-transfer to humanity from the Plexion region is a more or less philosophical subject. However, when we happen to consider infinities-transfer to be a sort of cryptographical model in the Plexion-Corpus annulus, a new way out is discovered.

Infinities-transfer is not an irrational notion. It is rather a more reasonable explanation of humanity in the light of cause and effect. This is because it gives a foundation for ideas on how humanity has evolved. The speed of evolution of humanity is exceptional in nature. It depends on the relevance of humanity to the infinities factors.

Now, it has become our basic concern how infinities are transferred. The ontological infinity in the entire spatiotemporal continuum (which is equivalent to the infinity as existence, the actual infinity, eigentlich, and proper infinity) is thought to be the infinite plexus of dynamic affairs, which are respectively accidentally characterized.

The human brain instinctively (not a fully adequate word because in that situation brains do not exist) understands that the actual infinity cannot be realized by computation. It is thought that the actual infinity and nullity have been registered respectively as just singular and peculiar vacancies in the neural informative phase spaces.

But as soon as the actual infinity divided into two entities, the Absolute Other (DGA) and the self as a recognizer of DGA, the pseudo-infinity as DGA turned out to be the object of the cognitive mechanism of the brain. As far as the self becomes the recognizer, the self theoretically happen to possess the capacity of the reflective pseudo-infinity in its intrinsic structure as a reflection.

The pseudo-infinity of DGA is changed to become the infinite due to infinite superposition, infinity due to the infinite number of combination, or infinity due to the infinite number of pseudo-infinities. In other words, the infinite container which can receive Hilbert space of infinite dimensions is formed. (I would like to use above mathematical language without permission for propositional examination.)

This infinite container presented as special cognitive mechanisms, which could be enfolded and omitted an infinite number of times if appropriate information to support it does not present itself. Deviated from sound rationale could also occur, which could be simplified through hierarchical variation and categorization that could induce an immediate junction between corresponding elements through mapping correlation even though mutual gaps of relevance are too large. It was thought that infinity could not be transformed into information.

But, in the above cases, all the combination and superposition of all the pseudo-infinities, which appear as to be objects of humans, will be interpreted as the change of information (or, if for the taste and opinion of some readers, it would be the change of the Absolute Will or the change of spatio-temporal curvatures). The information is that which is allocated to a specific phase of DGA pseudo-infinities from the actual infinity and the phase itself also changes continually in dynamic condition.

If we reach the conclusion above, it is natural to think of a possible frame for transferring respective flow (or flux) of accidental affairs into the internal system of cognition from the outside. But such allotted information is regarded again as pseudo-infinity. This pseudo-infinity holds its own characteristic information. Transfer information from the real world into the cognitive mechanism shows peculiar patterns comparable to the transfer of DNA through a geologic time frame.

The characteristics of the transfer of information from Plexion to human cognition is similar to the instant freezing dynamic phenomena (in connection with time axis) into a snap shot, which is the static object of cognition.

Pseudo-infinities are transferred in cryptographic expressions. Any combinations of pseudo-infinities of any superposition are transformed into information, and the functional correspondences in the transformation are recorded into patterns of a cryptogram.

Cryptographs could be expressed fundamentally in digital patterns, in variable patterns of a wider range of digital information (including nullity, infinity and etc.), in well designed quantitative patterns of emotional expressions, or in any equivalent expressions of dynamic affairs to be included in a broader sense.

Sometimes when seeing quantum physicists who regard the universe as a computing entity discovered in the state of quantum entanglement, tries to explain a special pattern of ontological existence on reality.

Anyways, infinities-transfer is actually a pseudo-infinities-transfer, the cryptographic patterns of which are transferred into the cognitive mechanism through the principles of reflectivity and correspondence as found in genetic transfer.

Exile of Infinities

All the infinities encountered by individual humanity are exiled to form a container-interpreter of infinities, the "Infinities-matrix," because they are unable to be computed by brains due to the excessively intense signal of peak level experience. The difference of humanity from other animals would be the principle of "exile without the deletion of the indentations of infinities."

This exile process for humans is made possible because of the profoundly developed emotional mechanisms in the human cognition. The emotional mechanism has been developed through innumerable indentations of fresh summit level experiences. On this subject some may insist that complex emotional mechanisms were manifested in a remarkably short span of time in which only a single generation was concerned.

Some others might ask where this infinities-matrix is located. The infinities-matrix is the source of emotional outcomes as we experience fear, panic, frustration, etc., usually associated with rationales of nihilism, agnosticism and anarchism in any negative manner as conventionally defined. (This book also emphasized the positive functions of the above three isms.)

If explained by evolutionary perspective, we have to test the susceptibility and capability of single neurons. But if cortex has the exceeding susceptibility, god-modules have to be discussed. The so-called three God modules of different source of peak religious sensation have been experimentally examined. (Andrew Newberg, 2206, Why We Believe, What We Believe, Ch.7) It is interpreted a little bit different from the experimental conclusion that the three proposed God modules are interrelated with three sub matrixes to consist the infinities-matrix.

The module cognitive of the others is connected to the tongue experience of Pentecostal Christianity and has to do with the onto-matrix; the spatial brain module of Tibetan monks, which caused the trained sensation of ayatana (處定, which means location, sense spheres), is related to the topos-matrix; and finally, the linguistic module tested with

Carmel nuns has a deep connection with the logia-matrix.

However, it is most probable that the exile takes place not at a designated material location in the brain but in the form of a dynamic modularity of the entire nerve system as we feel chilly, shudder with horror, or shiver. I use the term "modularity" instead of "module," because it is not likely that the module would be static in its composition with varying time.

Amygdala is a small but important source organ in all kinds of emotional aspects. Hippocampus is another core for emotional control and the emergence of temporal sensation. It is not yet known how experience is stored selectively in an excited conditions and how a certain detailed part of the brain participates in certain designated networks.

A four year old boy went through a panic experience when a person in his car was burned to death in a car crash. It was observed by computer planography that metabolic activation in his temporal cortex strikingly decreased but with no anatomical damage. This shows that a shocking experience can cause damage to brain substantial in measures.

In a severe emergency, all other information flow seems to be stored and memorized incorrectly. Objectivity disappears and only the remnant of the panic impact make that specific impact become independent. They call it a "flashback," a weak stimulus that works similar to how a small wave at the seashore can revive memories of a Tsunami. (Verner Siefer and Christian Weber, 2006, Wie Wir uns Selbst Erfinden)

Flashback is a connection between two different memories. But when an initial conceptus occurs to compose a spermiovum, there is not yet a referential, reflective, nor projectional opponent. Therefore human cognition starts from an immeasurable shocking impact. That initial shock seems to affect all cells produced along the division process.

Division itself should come from the direction recorded in DNA, but the shock of penetration is the initiality of sensation without sensory organs. This part has to be solved theoretically; maybe at the quantum level, through ontological perspective, by the theory of molecular biology, or through any kind of non-neurological explanation.

(To the author, it is most likely that persontology has the right answer, because person-ontology to be discussed later in chapters to come would yield a room for non-neurophysiological elucidation.)

Figure 3-3-3. Visualization of the Emergence of Manifold (from the accumulation of transferred pseudo-infinities due to summit experiences.)

First manifolds are structured by infinities-transfer while creating potentially effective territories around it, and spreading branches of connection around the blank. The essential difference of humanity from the other animals is the ability of maintaining infinities-encounters, and/or the will to maintain it. The will to maintain the trace of an infinities-encounter originates from the spare capacity of the human cerebral brain, associated with the faculty of the right brain that has to reserve information with unidentified addresses and for integration.

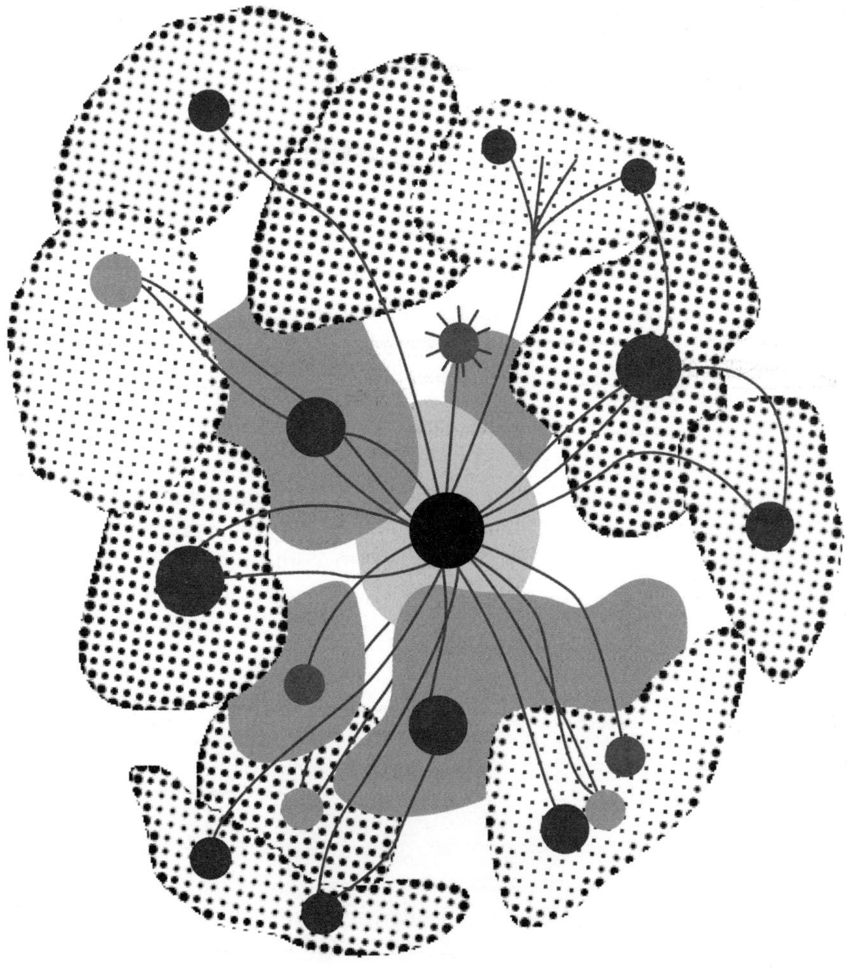

Figure 3-3-4. Visualized Illustration of the Patchwork Mechanism (among manifolds constituting infinities-matrix constructed in exiled status of infinities. Patchworks are done by connecting the adjacent manifolds at different localities.)

Infinities-matrix at the center and other manifolds around it shape a greater ball type integration due to the blank centers of manifolds that are connected together effectuating survival mechanism of the concerned entity. The blank centers enable the raising of proper freight-flight/fight responses for survival mechanisms through connections made by strong flashbacks.

It is all mixed up at the first stage but gradually restored by experiencing reasonable affairs understood afterwards. Initial infinities-encounters are clearly defined but after intellect gained through education, blanks are not experienced as easily.

Our present interest is of how infinities are exiled after their transfer. The exile of infinities being executed resembles reservation by shelving (most likely in an inert part of the neurophysiological condition like glial cells) during emotional computations. The process and condition of shelving depend on the mechanism of dynamic modularity.

Once again it is emphasized that this research is founded on hypothetical assumption that there exists a cryptograph of human physical existence in dynamic patterns of emotional expression and that the essence of brain computation relies on emotional signals. But it has yet to be clarified in various aspects how to understand the practical method of computation in a scientific way.

Anyway, it is thought that the human brain computes by topological correspondence between cognitive phase spaces, which is formed through the continuous mapping process since the days of early conceptus. We can simply understand it better if we look at it as figuring out a path to the washroom with the lights off, trying to get a shirt on properly in the middle of the night.

Infinities-exile is a blank for each summit experience or equivalently for single infinity encounters and single infinity transfer. These blanks successively connect to the episodic edition of the brain to form an understanding of time frame.

The time frame is well mended by the various intrinsic rhythms of an individual's body and brain. The arrangement of blanks generates an enfolding mechanism. This enfolding constitutes the attributes of a person.

(The author thinks that the exile of infinities appears as blanks in phase space of cognition and constructs the feeling of time. The blanks cause a kind of enfolding (of dimensions), which produces the attributes of a person.)

The interpretation of transferred and exiled infinities is made of the transcious level. Transciousness is strengthened by the scratch in human cognition beginning with the emergence of a human embryo.

However, the emotions of a person at the senscious level have functions to construct smooth combinational manifolds (in topological expression) for which irrational operations of a brain for rational arrangement can mend the discontinuities among them using various patchwork mechanisms. Consciousness and sensciousness collaborate with transciousness.

Infinities-transfer is made in manifolds of emotional environment. It is assumed that the human emotional phenomenon could be interpreted into the complexity of topological manifolds as can be imagined in the case of the human's five senses. The emotional manifolds are in essence presented in the matrix pattern.

Suppose that we disregard various cognitive mechanisms in sensciousness, which are beyond the governance of the autonomic nervous system among bowel senses, and simply concentrate on only the five senses; then they would be represented as 5 x 5 x N matrix (N; variable for stress status).

Orthogonal terms between two different senses are defined as to not having a mutual stress condition. But if a combination of two different senses gains the synergic multiplication effect, it might cause extreme stress well over the susceptible limit of the neuronal system in most severe cases for N. In such situations, extreme stress cases would be implicated to the infinities-matrix, which is the combination of pseudo-infinities.

The interesting thing is that the manifolds of emotional conditions might have significant influence on integrismic condition. Emotional conditions fluctuate to connect or disconnect the connectivity with the internal and external domains of humanity. Emotion is founded on the integrative ability to communicate with external situations. Integrism has to do largely with ever changing internal systematic conditions to sustain major functions which search for or are affected by the least possible stress inflow from external situations.

Let us consider 5+ mus; which stand for the emotional status of the five senses + μ (μ; the top emotional capacity over which infinities-transfer is activated). 5+ mus is thought to consist of the combination of emotional manifolds in a specific condition and usually ends up at the highest frequency with three representative patterns from the exile of infinities.

They are three patterns; subjective self reflection of onto dimension, objective spatial distribution of topos dimension, and relative logic networks of logia dimension. These constitute an infinite set of extreme experiences of intellectual and emotional conditions. This infinite set is the true source of infinite the human desire has. In the early stages of humankind, as logically inferred, they might have felt only chaos from the infinite set of extreme experiences.

One more point must be discussed. It looks like there could be a strange effect like the Möbius band after the exile of infinities. Möbius bands have a single surface mode by clipping an ordinary band and turning an end inside out and pasting it with the opposite side. Ordinary bands have two surfaces.

When the extreme of interpretative capability of the human cognition has been depleted, the junctions of infinities can occur. It is thought that blanks in manifolds made by the infinities-encounter tend to be mutually dragging other similar blanks of manifolds in the vicinity, when it is necessary due to the situation given by the reaction of amygdala. The theorization of infinities-junction would without doubt yield to the most complicated topological issues. With rudimentary knowledge, the writer cannot even select the right terminology.

This realm of extremities in human cognition seems to grow ceaselessly. It is most important to find how the human brain or human cognitive mechanism takes care of various pairs of opposite notions (nullity and infinity, reality and virtuality, reflectivity and resonance, yin and yang, convergence and divergence) in fusing and welding together in order to mutually identify each other in their extension to infinity.

Some times they use circulatory thinking, in another case just irrational welding, in some special cases something like the zygotic process is used. It looks to be most likely that the polarity of summit experiences or of exiled infinities vanishes in some dissolution region which exist within intangible interpretation mechanisms.

Although quite improbable, it might be caused by certain local effects of neuroglial cells which functionally cannot reserve the polarity information. However, it is distinctive that the root of an infinities-junction is essentially the mechanism of survival.

Maybe, this function of dissolving the polarities of infinities initializes the points of bifurcation in cultural divergence. Thus the interpretation mechanism of the infinities-junction comes out of the infinities-matrix, which is formed through integrated interpretation of all the summit experiences.

But, as commented above, the interpretation concerning the issues of infinities is achieved in accordance with physical requirement and thus with the essential mechanism of physical survival (by welding, twisting, generating Möb points and processing for zygosis of different polarities at the tangential surfaces of Möb contacts and so on, if our liberty of thinking allows us to speak up our conjectures).

Rationalization of survival, in connection with infinity's interpretation of humanity, either individual or communal, has direct relevance with the peace of humankind even in present day situations.

Especially, the advancement into the solar system within the next 50 years would require formation of a complex megalo networked system (Complex Menesys) and its related zygosis of incompatible polarities to provide stable territories at the peripheries of a megalo society in order to admit a number of abnormal members of diversity. The above ideas might be valuable for such complicated situations.

Infinities-Matrix

The infinities-matrix is a completely new concept in the scientific expression of humanity. The infinities-matrix is suggested to solve the secret of humanity. Humanity is seen as the fruit of successive responses to unceasing external impact.

The zenith of humanity is uncertain but exists dynamically. The flaring dynamic zenith and stabilized static part of humanity are all the offspring of the mother's womb called infinities-matrix. Every human self has its own infinities-matrix. The infinities-matrix of humanity is considered to be definable too. The infinities-matrix is at the center of the phenomenon of humanity.

The main theme of the subject concerning humanity would be consciousness and self. Language is the essential manifestation of consciousness. The structure of semantics in human consciousness and the characteristic of individual intentionality also come from the infinities-matrix.

"Phenomenological self" is the variety of contents of self-consciousness which changes continuously. The "self" is not static but dynamic. "Conscious self" can be regarded as a kind of virtual reality. Human consciousness is affected by emotional traps. It could be related to another important issue that human emotional traps are the source of the structure of human symbolism. Self-consciousness is a substantial core which governs the phenomenon of human society.

It might be meaningful to follow up with the population of groups in NCC (Neuronal Correlates of Consciousness). But it is not desirable to chase after the trace of, for instance, self-module or self-cell in cingulate gyrus. It is not significant in the ontology of relational dynamics for static self to incorporate the continuous variation of humanity.

In this context, we will come to favor the concept of the infinities-matrix which behaves as the skeletal structure of the individual self, when the infinities-matrix is the residual of integrated indentation of past experience. The mechanism of dynamic modularity secures the stability of individual self as a complex adaptive system. This is why we will take a look at infinities-matrix.

"Infinities-matrix" in human cognition is different from the mathematical terminology, "infinite matrix." Infinities-matrix is a kind of normalization process of the human brain to compose a reasonable phase space which consists of information obtained from the summit experiences of an individual, usually interpreted as the infinities-encounter. Normalization is a part of theorization process in physics to exclude the effect of diffusion from the infinity.

According to the Incompleteness Theorems of Kurt Gödel (Rebecca Goldstein, 2005, Incompleteness), the discussion point of 'self-referential self-denial' might have originated from the connectivity of exo-systemic source with the present system. It is my belief that an observation cannot be perfect in any circumstance towards our universe, because it can only be partial within the sets of all the observatory points.

Exo-systemic self-reference presented in a system contains self-denial, which is the opposite concept of self-reference. Thus, self-denial is not a nullity in the system. A nullity does not exist but as a hole, a vacancy or a vacuity in human consciousness. Nullity is the expression of the vacancy. Nullity is an opening in a manifold. Nullity is caused to exist as a hole in the emotional manifold created by overwhelmed impact to induce an infinities-encounter and its consequent infinities-exile.

Exiled infinities organize a matrix. This opening is an emotional exit to escape which enables the dimensional jump. Through the opening, humans can escape from their dead ends. The escape is achieved by transferring to other manifolds through the connections made among exiled infinities.

In my personal meditation, it was interesting to discover my emotional openings. My mind remembered and put together several pieces of my memory. The stuffiness of my life caused my mind to picture a scene, where I was driving on a narrow two lane road in a dark mountainous area of Kangwon province. From this scene, I was able to recall my infant memory when suffering from a severe cough. I could even feel my malnutritious body in my mother's womb because we had from the threat of communist revolt in North Korea.

When a manifold of memory is sufficiently formulated, I believe it becomes a dimension which can interpret affairs in its own logical way. When two manifolds are integrated into one manifold, they lose one dimension in the memorial classification as well as in the logical explanation. Then we think we are more systemized. This idea contains a much deeper implication than we can simply imagine. Humanity has a tendency to proceed to a lower dimensional structure of cognition. But in working life we have frequent chances to develop our new dimensional cognition. When we deal with the dimensional issue of the brain-neurophysiological experience, it is right to say that it is wider than the sky. As far as the conscious interpretation is concerned, these mental dimensions are treated the same with the space-time dimensions in our cognitive world.

We have to be cautious when dealing with the dimensions in mental sphere; we can take a look at Möbius band and Klein bottle for instance. Strictly speaking, these two objects are not of substantial reality appeared in the dimension of our cognitive world. They are just metaphors to satisfy the human intellect. The continuous surface in Möbius band has its thickness and the continuity with adjacent surfaces. It is merely an abstractive illusion of our brain activities.

Anyway, it was assumed that the human brain tends to smooth out combinational manifolds to ease the communication between the sets of unique experiences. These combined manifolds are dynamically changing modules of neural affiliations, respectively built up around their corresponding "infinities-matrix" (composed of the similar kind of unique peak experiences surpassing the sustainable ceiling of the susceptibility of an individual brain).

Dynamically changing modules are basically organized by the fire-and-wire principle. It is called dynamic modularity organized into several hierarchies of modules that are responsible for what the corresponding brain has experienced and directed to.

Dynamic modules are suggested to be almost identical with the dynamic core but discernable by their hierarchical and multiplicative processes to construct matching relations with a variety of symbolic expressions of transciousness. The Hierarchical process of the dynamic modularity has to do with the depth of abstraction and multiplicative process related to the planar complexity of experiences.

Here we are going to comment on the transcendental intentionality. Intentionality is a well known concept as defined by Brentano. It is the "aboutness" of where the human thought is directed. I believe the intentionality has an aspect of the dynamics. When the necessity of dynamics and the evolutionary tendency of intentionality in humanity are issued, it is believed that we need "transcendental intentionality" to illustrate the tensorial phase space in which all meanings of the human world reside in. If the intentionality lacks the transcendental character, humanity would be restrained by the constraint of driving energy.

Dynamic modularity is of a notion to which the concept of a strange attractor is applied. The Strange attractor has to do with the elasticity of neurophysiology of the brain to sustain memories or intentions as stable as possible. Such stability maintains the modules and their activities to get along with the transcendental intentionality. Intentionality is a tendency related to the elastic property of the human brain to be directed consistently toward a certain orientation predetermined by the past affections of overwhelming experiences.

The prerequisite of the transcendental intentionality is a considerable measure of stability supported by the sustainable elasticity. Each manifold in the phase of human neurophysiological activities is centered at a spot with the high concentration of peak experiences.

We can imagine that a unique manifold in phase space is specifically centered at a unique set of infinities-transfer (i.e. set of summit experiences categorized by interpretation in similar characteristics), which can be demonstrated by a plane with punctuated blank spots around the center due to transferred infinities.

The punctuation of blanks – the number, size and intensity on a manifold – is thought to cause the enfolding of cryptographic information generated from the unique manifold. Such enfolding would raise some psychological problems such as blockings in free association, emotional alienation, and, in my opinion, the maximized intensification of this enfolding could even result in autism.

In the human mental activities (cognition, memory, attention, learning, and thought), the basic information system is founded on the mapping processes. Mapping process is far more dynamic and mobile in comparison with existing mind maps which are in a static condition.

The basic rationales of the dynamicity are constructed on the relationship between the peak experiences and emotional manifolds. Here we can raise a question on how the emotional manifolds have been established. Human emotional manifolds are originated from the genesis and the development of human physique and mentality.

The manifolds are considered to have a frame which consists of three patterns; the pattern of the other, the pattern of space-time, and the pattern of relations. The pattern of the other is deemed to be developed to onto-matrix as infinities-matrix mature. The pattern of space-time happens to develop into topos-matrix, and the pattern of relations to logia-matrix. These three most fundamental sub matrixes of infinities-matrix will be discussed later.

The frame of the modern psychology is insufficient to deal with the concepts of sudden volatilization, diffusion and explosion in human cognition and consciousness. However, similar phenomenon as paralysis of inchworm happens in children's mental shock cases when an abrupt increase of information and instant excess of allowance in human cognition takes place.

This kind of accidents imply similar sorts of instant extreme experiences might exist in our everyday lives. It should be seriously considered that in intangible circumstances of the human intellect as well as, or perhaps with greater frequency than in tangible conditions exists such extreme experiences, either with or without realization. It is significantly probable because human intellect has been developed to be highly complicated and extraordinarily sensitive.

With such assumptions, the structural formation of the sub matrixes and matrix of infinities would need higher degree of freedom to reach a null stress or a nearly null stress condition, which promises the dimensionally higher status. Attaining the dimensionally higher status refers to (considering the dimension as the rank of tensor expression) the new variable dimensions being introduced to the existing status.

We can see the human infants starting to babble when phonation is enabled by the growth of the organs. At the initial condition, phonation would not be restrained by any precedent constraints so that the process of relaxed selection of baby babbling may be well balanced structure but open to select by liberation from stress condition, until they find obstacles through social contacts with parents and family members to give a specific shaping of ethnic vocalization.

An individual human being gradually establishes and widens his emotional territories by strengthening the foundation on the initial emotional boundary attained by the peak experiences from the day of conception. The emotional territory could be regarded as a combination of emotional manifolds usually not yet refined (should be sufficiently matured when the integration of personhood is well managed).

These emotional manifolds are formed around the earliest indentations of peak experiences, having been mutually connected locally due to successive stress situations. (We do not know if such earliest peak experiences have to be counted from penetration of a sperm into an ovum.)

These combinational manifolds start out from a dot, gradually turn out to be a line, and then develop into a face, gaining better smoothness. The degree of smoothness would be reasonable and natural to define in the light of the consumption of energy in communication with information.

This research suggests infinities-matrix to also be a psychological terminology. We have a viewpoint that there exists in human cognitive realm a certain brain-neural mechanism which constitutes emotional infinities of peak experiences into infinities-matrix. In a sense, person is an infinities-matrix of the individual. The personal attributes such as personality and its emotional aspects which we recognize in everyday life could be the attributes of the world which are interpreted neuro-psychologically based on the physicality of a human being.

Infinities-matrix and Human Abstraction

V. Ramachandran (2003, The Emerging Mind) commented about cross activation and synkinesia when he suggested the bootstrapping theory on the origin of human language. Cross activation presents between specific domains (e.g., hand and mouth) and visuosensory domain and Broca's area.

However, it is deemed that cross activation could be a kind of global phenomenon in a human neurophysiological condition. The intensity of such activation must be very much faint. Without profound cross activation, it is not possible to construct all variety of qualia and qualia associations that provide a complete symbolization of our thought process of physical and abstractive conditions.

Let's consider a basic abstraction of humanity in the following three viewpoints: onto, topos, and logia. Onto-viewpoint is existence-centric and gives a view on wholeness. Topos-viewpoint is three dimensional space-centric and provides a view on diversity and logia-viewpoint is centered at the logics of relations, processes and generates a view on time-centered episode.

The abstractive imaginations of humanity tend to be made along with the following three axes: onto-topos axis, onto-logia axis, and topos-logia axis. Onto-topos axis implicates the relations of unity-diversity from the match of wholeness and diversity. Onto-logia axis induces all universal affairs into a unique logic of an integrated humanity from the match of wholeness and time. Logia-topos axis deduces the illustration of the relationships among diversified entities and affairs in a network of processes.

These comments are simply intuited insights, thus it is eventually necessary to develop some propositions for the scientific considerations about abstraction concerning humanity. At the moment, onto-submatrix, topos-submatrix and logia-submatrix require respectively distinctive infinite extensions.

Human cognition starts from an immeasurable shocking impact. That initial shock seems to affect every cell made by the further division process. The division itself should be performed in accordance with the direction as recorded in DNA, but the shock of penetration is the prime sensation at the initial instance as there was no sensory organ.

If every cell possesses its record of information, it can be registered for the vertex value of onto-submatrix, from which onto-submatrix would grow. It is thought that there are two more matrixes as we discussed for the prime viewpoints of abstraction of humanity; topos-submatrix and logia-submatrix.

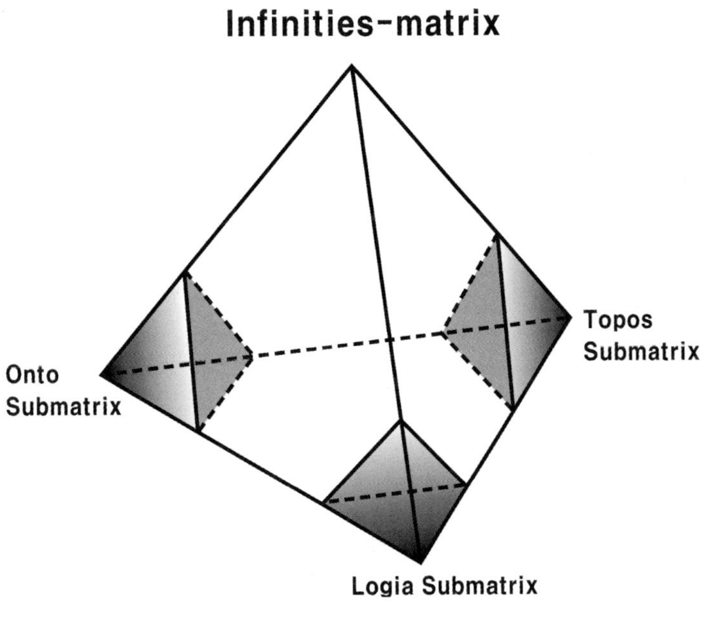

Figure 3-3-5. The Three Major Submatrixes Consisting of Infinities-Matrix (and the way they develop from their initial vertex points.)

It is proposed that the onto-submatrix is related to self-infinity. Self has a connection with the pattern of the Absolute Other (DGA). The thought has a Semitic origin and is well manifested in the Hebrew religion. In our view, the onto-submatrix is the expansive interpretation of the observed eternal Axis Tempus to the Pragma which enters into one's own realm.

Topos-matrix is a matter of phase-infinity or topological infinity. It is interpreted in accordance with the DGA-pattern. It is also similar with the Huayen Buddhist thought. It represents the tendency to assume self as the absolute centric position. It tends to expand oneself to the isotropic spatial infinity.

Logia-matrix is related to logic-infinity. Logic has to do with a relational pattern. The thoughts from the Greeks are deeply connected with a logical attitude. Logic is at the foundation of linguistic syntax and thus affects language, time and episodic history.

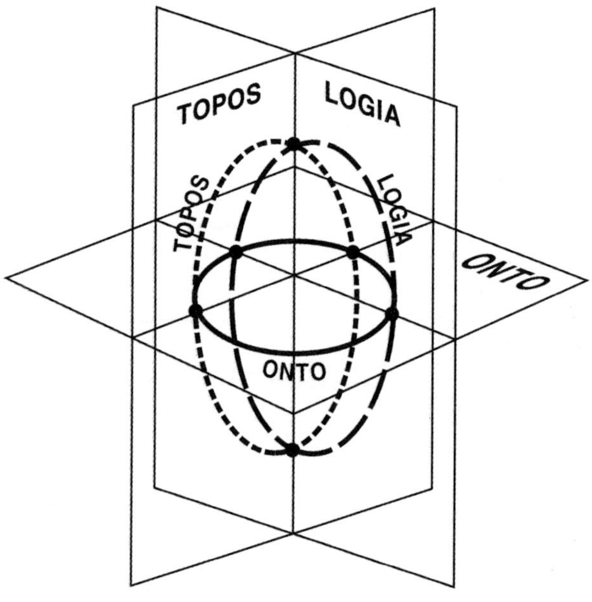

Figure 3-3-6. The Basic Orthogonality in the Triple Matrixes of Fundamental Humanity

It is suggested that humanity has developed basic matrixes of abstraction and universal language. Language, in major part, consists of abstraction especially in syntax. Linguistic matrix has three sub matrixes: onto-submatrix, topos-submatrix and logia-submatrix.

Consciousness and language are critical subjects when social problems are dealt. However, we focus on transcendental intentionality compared to consciousness and language in order to achieve introspection and the transcent of humanity. The transcent of humanity means the shift and entrance of humanity into a novel phase. The shift follows prior consecutive steps of transcendental intentionality and encounters with a new transition zone to make a shift to an essentially different pattern of life. If humanity had only intentionality, transcent would not occur.

Transcendental intentionality takes consciousness and language as intermediate stairs of systemicity and then utilizes the effect of nexusity. Systemicity and nexusity are retained in the process of preafference and postefference in human cognitive mechanism. Transcendental intentionality is activated in order to build human personage with higher degree of freedom. Preafference and postefference will be dealt in more detail in Chapter 4.

Preafference is defined by W. Freeman as the process of concentrated attention to assimilate oneself to an object. Postefference, defined by this research, is the process of concentrated attention to assimilate objects to cognitive self. Transcendental intentionality, equipped with preafference and postefference, activates deep and wide facilities of the dynamic modularity that encompasses from sensciousness, consciousness and transciousness.

The central axle of humanity is transferred from consciousness to transciousness by strengthening transciousness through the achievement of integrity in a person. Apparent languages are originated from the frame of the universal language produced by infinities-matrix. Infinities-matrix is at the center of integrity of a person.

When a human being concentrates on the subject of eternity and infinity in a harmonized way as a whole cognitive system, integrity becomes of a greater concern. The importance of integrism increases in human life and the transcent of humanity become highly focused. Integrism is the measure of human balance between systemicity and nexusity and will be dealt in Chapter 5.

Human transcent changes the human character. Human character starts from a social relation usually initiated with the mother-baby interaction. However, in the transcent of humanity, we find indications that a greater importance is imposed on the individual value of humanity because it is the individual leap that brings humanity to a creative breakthrough for the entire society.

Flaring Tetrahedron Matrix and ITRings

Three sub matrixes generally construct an individual infinities-matrix. It is made of three vertexes of the bottom plane and an Animotus matrix as the zenith, which is visually shown in Figure 3-3-7, which composes a tetrahedron. Tetrahedron has a stabilized geometry which is not easily changed by interrelationships.

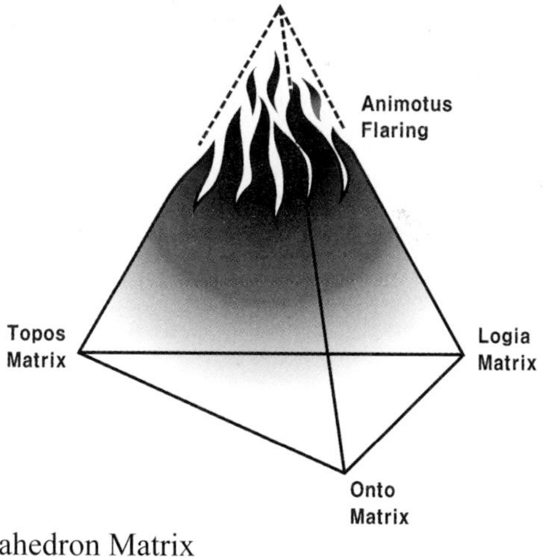

Figure 3-3-7. Flaring Tetrahedron Matrix

Animotus matrix behaves like an unstable flare in normal experience without the implied infinities effect on human cognition, but gets fixed (consolidated in neural firing) from time to time at summit experiences. When fixed it becomes a part of infinities-matrix.

Thus, it would be helpful to understand the Animotus matrix if we imagine a clear indentation imprinted on transciousness like an annual ring left on a cross section of a perennial tree. It is named ITRing as defined in the Section 3-1 and Figure 3-1-6 of this chapter. ITRing is shown in a different manner in Figure 3-3-8.

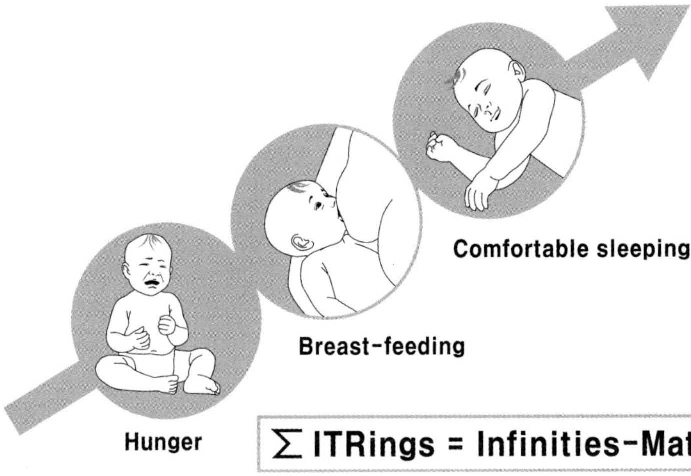

Figure 3-3-8. Infinities-Matrix and ITRings Gradually Coated on the Funnel of Axis Tempus

We need to introduce another line of thought to complete our discussion. This research would like to pay attention to the concept of the "loop space" generated from many ideas along with string and M Theory. Loop space appears to be the most promising and probable candidate of the scientific foundation for relational Dyon (dynamic ontology). (The Hole in the Universe, K.C. Cole, Ch. 6)

Loop space takes only "relationships" into consideration. The universe is only of relation, and merely of geometry. An elementary particle or even a spot in the universe needs and reflects all the other particles or spots, which are consequently correlated and inevitably connected as a whole. Beings are all interrelated.

In other word, being is interconnected, but non-being is the interconnection itself. Non-being or nullity is "the time" manifested to human consciousness. Asymmetry of time is attributable to the second law of entropy, which means disorder or chaos increases with the flow of time. Time, null and vacant, in common sense, or Axis Tempus, like a vacant funnel, is the basis for the interconnection and interpretation of human sensciousness, consciousness and transciousness.

The intrusion of Axis Tempus into human cognition had caused transciousness. Transciousness had initiated transcendental intentionality. If human beings had sole affection of consciousness, they would have been satisfied with plain intentionality. Intentionality alone is said to be an aboutness that human cognition perceive under "thrownness" into a situational environment. The intrusion of Axis Tempus into human cognition had initiated the suction of human cognitive energy to generate the transcent of humanity.

The flaring of Animotus matrix, which we can express in a tetrahedron matrix as in Figure 3-3-7, takes significance in reality when the flare is frozen by a summital experience. ITRing is the frozen substantiality in cognition. Freezing is caused by excessive information or an emotional impact which human cognition cannot sustain.

When an individual person get a sufficient acquaintance with a series of ITRing under significant interpretation in a specific social context, the intentionality of human cognition is going to be influenced by the infinite horizon of the confronted universe. This causes transcendental tendency of humanity.

The transcendental capability combined with intentionality makes everything different. Transcendence generates a novel direction of cognition in the phase space of human thinking. Humanity certainly has an orientation of cognition fundamentally sprung from the senscious instinct level.

Going back to the idea of loop space, it is interesting that many physicists think of building blocks. Such building block is the essential monad of universal existence and eventually supports the idea of discontinuity of the universal space. But the discontinuity has never been supported through on experiment. (Lee Smolin, 2001, Three Roads to Quantum Gravity, Ch.8)

Building blocks are vacuity that exist at the crossing position of the universal nexus of the point of matter (質點). Vacuity is the vacant emerging point of matter as we know in the concept of Planck threshold value in quantum physics.

The reality of dynamic ontology is in fact the dynamic relationships among the entities. Existentialism, centuries ago, started to depart from the ontological thoughts when they met epistemological influence, largely affected by the works of Emmanuel Kant, and changed its focus from the first principle of ontology to the individual existences.

Individual existences were unable to bear the load of universal totality. However, the thought of integrism takes its objects from variable attitudes of humanity, varying from the complete ontology, through a number of combinations between ontology and epistemology, to complete epistemology.

Because integrism is represented by the compound of sysy (the ontological attributes of an integer) and nexy (the epistemological attributes of an integer), sysy (systemicity) indicates the centripetal effects while nexy (nexusity) indicates the centrifugal effects to compose integrism of an integer, who or which has to achieve integrity in order to generate permunal (personal and communal) creativity through the activation of humanity.

The usage of sysy and nexy implies that there is somewhat different point of view in them from the ordinary notion of Dyon (dynamic ontology). Dynamic ontology in nature is interested in relationships between real entities. If we take steps further from Dyon to relational Dyon, we can find that the relational Dyon is interested in the plexus of accidental occurrences which consist of the reticulations of interconnected relations and the causality of successive events. The relational Dyon brings us closer to the understanding of Karl Jaspers' concepts of *"das umgreifende"* (the Encompassing, 包括者) and "*Chiffrelesen*" (Decoding, 暗號 解讀).

3-4. Dynamic Relational Ontology and Persontology

Imagine the Transcent of humanity
Dynamic Relational Ontology
Novel Way of Thought, Universal Persontology
Integrism and Complex Adaptive System
Humanity and Cruxology
Unitary Crux beyond Integrism, Cruxology and Menesys

Imagine the Transcent of Humanity

P. T. Chardin asked how "unanimization of humankind" could be realized and how "formation of noosphere" could be accomplished. His idea was focused on the idea of "Omega Point." However, it was thought that the idea of omega point may be an artificial connection of theology and science.

Considering human condition, I prefer a concept of "alpha path and omega path." Alpha path is a lowest stress condition with innate complexity in general region of initiation, while omega path is a highest stress condition generating apparent complexity in near-completion region of a "crevolution" process.

Crevolution(創命) is not a hybrid concept of creation and evolution. Crevolution denotes the way in which human creativity creates revolutionary changes. The last part of the "transcent" of humanity at its every remarkable step exposes a crevolutionary leap due to the subsidence of old dimensions and the imbibition of new dimensions of existential phase space.

This concept of transcent could be regarded similar with Chardin's omega point. It is suggested that transcent, crevolution, and alpha and omega paths are at the center of the conceptualization of forthcoming humanity issues.

In order to "scientize" the overall humanity issue, it is needed to rearrange our philosophical foundation on ontology.

Dynamic Relational Ontology

Ontology is supposed to be understood in a relational complexity. Being is relating. Relating in this real spatiotemporal space is inevitably dynamic. We can say dynamic relational ontology is dynamic ontology. Dynamic relational ontology will be named in short "Dyon."

Community in topological expression would be a piecewise linear complex of concerned individualities. If smoothness is attained piecewisely in complexes, the complexes are regarded as "integer communities."

Dynamic ontology is relational as long as it maintains smoothness between individual integers. Connectivity itself between integers is only a prerequisite of relation. The relation developed stepwise from linear, areal, and spatial; this relation is actually all spatiotemporal.

Michael Esfeld (Entangled World, Ch. 10, 2006) comments that there is a model for a philosophy of nature in terms of relations: the quantum correlations are another example of relations, which not only require no intrinsic properties as basis, but also exclude such a basis, as can be argued based on Bell's theorem. (Since Bell's theorem justifies the claim that no intrinsic properties exist beyond the correlations that are manifest in the experiments, Bell's theorem opens up the way for a particular realistic or ontological interpretation of quantum theory.)

The understanding of Sartre (Phenomenological ontology, 1943) is cited that, because the phenomena do not stand for anything beyond them, the elements of the ontology are of the phenomena type, namely relations.

It is a suggestion of an eventual change of our viewpoint to dynamic relational ontology. In this research, we would make a new definition for nexusity and systemicity. Nexusity is concerned with networking characteristic and systemicity with logically relating character. Nexusity and systemicity are just characteristic understanding for the ontological relation. Epistemology is a kind of distorted ontology that alters things in a way that humans can accept. The epistemological view is prone to omit, reshape, and add up, in addition to many other operations for humanity to understand the natural world.

Therefore, a person, either individual or communal, has two distinctive bipolar characters. One is systemicity and another, nexusity. Without systemicity, a person is unable to formulate his own software and hardware to govern his entire life. Nexusity is a parameter to connect the person with external factors and environments. In many circumstances, nexusity is a source of breakthrough when internality alone cannot solve the encountered problems.

A person is essentially a mixture of systemicity and nexusity. Systemicity and nexusity are found to be a different kind of ontological relation. By adopting these concepts, we can define a new phase of person, personhood, and personality. These two complementary and contrary tendencies, as counterparts, provide essential complexity in personhood.

The thought of integrism is introduced to denote the ratio of combination between systemicity and nexusity with which we can see the distortion of ontological status. Based on this thought, we will be able to understand the world in the manner of humanity.

However, it is recommended hereafter to discard our way of thinking to mix human factors with person factors. Person is a more universal way of understanding of the scientific and rational affairs and does not carry even a slightest bit of humanness.

Novel Way of Thought, Universal Persontology (格體論)

Persontology is a composition of person and ontology. The terminology is to reveal the ontological essence of person and contrarily to show the character of person manifested in ontological aspects. Persontology focuses on the balancing of dynamic harmony of the two complementary factors: systemicity and nexusity.

The main stream idea of dynamic relational ontology, or equivalently persontology, comes from the non-linear inflow and outflow of energy and other tangible and intangible factors. Nexusity, which determines the dynamics of flow in a concerned quasi-system (quasi is attached to clarify the universe is not a system), is most important in connecting external relationships.

Persontology is the total concept of the entire universe which maintains the eternal dynamicity of change and interflow. Actually, the dynamicity is the eternal property of humanity and a person.

Now we are saying that the universe is wholly dynamic, regardless of what the theories may be: inflation, Big Bang, Standard Model, etc. The universal dynamicity is due to integrism, the compound of systemicity and nexusity. The building block, which exists at the threshold value of Planck condition, is understood to be vacuity in human understanding.

We do not know exactly what the spins of particles are, what the gravity is, and so on. If we understand the relativity, that is because we are human. If we say human understands the wholeness and the parts, it is because we are both a human and personal entity.

Being person is originated from the character of the universe. What protects us from thinking like this? It is an anxiety that we might be misled by our self-understanding of being human and person. Humankind has to rethink on the definition of a person. It is not a feeling we have for ourselves that is a person. Person is a universal phenomenon which is not dependent on human feeling and social knowledge but rather a scientific concept which shall be clarified by strict academy.

Not only person, but also the concepts such as virtue, value, and liberty, which are familiar to humanity have some other implications in science and technology. It is a strong belief that human factors have to be redesigned through a special treatment to fit our forthcoming civilization based on various disciplines of strict science. Our value systems ought to be altered. Person invites us into the novel understanding of the universe in persontology.

Integrism and Complex Adaptive System

Integrism is the basic idea which makes the extension of the concept of Complex Adaptive System possible. Phase behavior has been an analogy that many researchers in complex systems pondered into, in order to illustrate the phenomenon in phase spaces. Complex adaptive system (CAS) has been chosen by Santa Fe Institute as a major research subject among those in the science of complexity.

It is known that Santa Fe focused on CAS mainly due to the practicability and applicability in factual world. It is suggested that CAS is most frequently met in transition zone which varies from systemicity to nexusity. New terminologies, systemicity and nexusity, were defined and added to express the behavior of "integrism" in phase spaces in complexity.

When the transition zone transits from nexus-prone to system-prone territory, there in whole scale exists an integrismic behavior. Integrism could be developed to present a measure of complexity emerged by the interaction of systemicity and nexusity.

It is realized that systemicity and nexusity could be reflections of the essential trends of natural behavior of universal existence, which is known as "yin and yang (陰陽)." Systemicity is a reflection of yang, while nexusity is that of yin.

X chromosome, that is responsible for creating female human, does nothing but wait as it is programmed by original system of human embryology. Y-chromosome for creating male induces additional development in initialization process of gonadalis by activating Sry gene. Sry stands for "Sex determining Region in the Y-chromosome."

It is certain that there are serial actions of Sox9 and MIS genes afterwards to complete a testis. However, the point of view is that Y-chromosome is deemed to behave as a nexusity factor in generation of the form of humanity.

Yang strengthens a system to survive and induces more folding mechanism to occur in order to simplify the chains of decision, and thus conserve more energy, while yin opens ways to probable introduction to outer world influences, inducing unfolding mechanism that widen the probability of variability.

It is a noteworthy connection of ideas to integrate many different ways of human thought, even though a number of discussions and debates would occur in response to such points of view.

(But in case of human, the correspondence of yang with systemicity and yin to nexusity was induced from the basic idea of genetic study presented in "How X chromosome connects our lives" by D. Bainbridge. In our discussion, it is necessary that yang and yin has to be defined with better refined rationality.)

In ancient thought, yin and yang were mutually opposite and in some ways expressed in a manner of mutual complementary alternation. Yin and yang respectively mean shade and light, literally. Shade and light come and go during the day, substituting each other rhythmically in intensity, thus generating wide transition zone associated with sinusoidal variation between them. The principle generating the change of yin and yang itself is "the great absolute"(太極, Taiji or Taegeuk). (The aforementioned definition of taiji follows "Lexicon of Human Nature and Natural Law" by Chen Chun.)

If such a mutual complementary alternation occurred in the physical world, there would be merely an appearance of dual entities, in a manner of continuing expansion and contraction of two opposite parties or parameters. However, when it is applied to phase space consist of various parameters, the transition zone in the phase space, where similar phenomenon like yin and yang exists, has to show continuous change from a substance to another substance which are presented on both sides of the transition zone.

Yin and yang, complementary to each other to construct an integer, say yin + yang = integrismic unity (provided that integrism is a factor of generating the effect of transition in phase shift from systemic solidification to nexal fluidization), are recognized to cause phase behavior. Concentration gradient in genetic embryology of neurosystem could be one of the cases which the mechanism is presented in nature.

Integrism is the capability of an integer that can change oneself to cope with any universal environment through introducing and thence digesting external conditions. When external conditions are dealt most adequately to generate best possible communality, it would be called the state of integrity. This integrity is achieved when best combination of systemicity and nexusity was attained in certain condition.

Humanity and Cruxology

It is suggested that we need to open a way to the novel methodology named cruxology to reveal phenomenal events of the universe in a general method of presentation based on integrismic rationale.

Crux (幾微) is the dynamic unit concept of the cruxological thinking. Crux comes out of the dynamic generation of minute initiality. It is an initial movement of an event. It is sometimes infinitesimally comparable to the magnitude appearing in quantum physics, and has the least level of energy. Sometimes, it affects general orientation to become as seen in microscopic fracture of lithological structure. Crux is a concept that can manage even infinite cosmological expression. It is independent to the size of events, but it is unrealizable.

Cruxology (機틀) usually deals with matters and events of considerable stability in sustainable structure. It has correspondence to the realm of classic physics, realizable energy. The size and orientation are recognized very significantly in our usual life. The scheme of cruxology is designed to treat humanological and sociological problem. Humanology is thought to study even individual and physical relevance of humanity to sociology.

Thus, it is reasonable to think that crux must be capable of bringing entanglement of emotional aspects with rational development. Emotion of an entity has its subjective (individually characteristic) structure, which is originated from minimum-stress rule expressed in orthogonality of tensor relationships in phase space, and founded on the physical constraint of the menesys of human sensory instruments including splanchnesthesia (or viscerosensory). Viscerosensory is supported by more than enough physiological evidences, to the extent that it constructs a backbone of the substantiality of persontological theory.

The network of neurons is completely different from the network made by human experiences which is called a person. It is proper to use terminology of menesys to describe the network which constitutes a person.

Simon and Ando (1961) used Nearly Decomposable System (NDS), for which the author would like to consider a more general terminology (Modularity, Werner Callebaut et al.). The connection between the two networks mentioned above remains as an extraordinary project. This book is another trial in that sense.

If we want to suffice the condition for epistemological inspection, the world of mental experiences must be clarified through communal interaction between more than two people or through interacting with external environment. We do not need to worry about reductionism in epistemological process, but must adopt topological approach in order to effectively utilize the traces of human mentality. Therefore, it is intended to establish a method of cruxology that enables to connect between the individual physicality and mental reality through interposing the phase space of symbolic tools such as qualia and quantia.

For instance, amygdala activates CRH, the most important stress gene, in hypothalamus and vitalizes all the hormonal stress axis of our body. Consequently cortisol, a stress hormone, is then secreted. At the same time, being alerted by notice of amygdala, the central nerve system of brain stem sends out stimulus transmitter like noradrenalin and acetycolin, and thus, activates its own contingent genes such as thyrosine and hydrosilase.

Noradrenalin activates heart beat and metabolism. Such preparatory "panic orchestra" allows a person in a specific situation to decide whether to "runaway or confront." In this context, a trauma (an external emergency condition) constructs an ultra-contingency picture through utilizing the cerebral cortex and the neural networks of the human body.

The fundamental of humanity is the physical situation explained above. However, the activation of hormone axes, secretion of stimulus transmitter, and activation of contingent gene can be thoroughly understood. They can be regarded as the overall phase of humanity when explained in a well integrated manner; when the manifolds of emotional reactions and logical procedures are induced by the neurophysiological network in the cerebral cortex are they presented.

This complicated procedure encompassing from brain-neurophysiology, the other physiological process, and the activities in intangible realm of transciousness could be unified in the cruxological domain, in which many systems and networks can be considered together in the menesysal form.

Unitary Crux beyond Integrism, Cruxology and Menesys

The concept of crux is explained in the sections above. However, there is the origin of manifested cruxes.

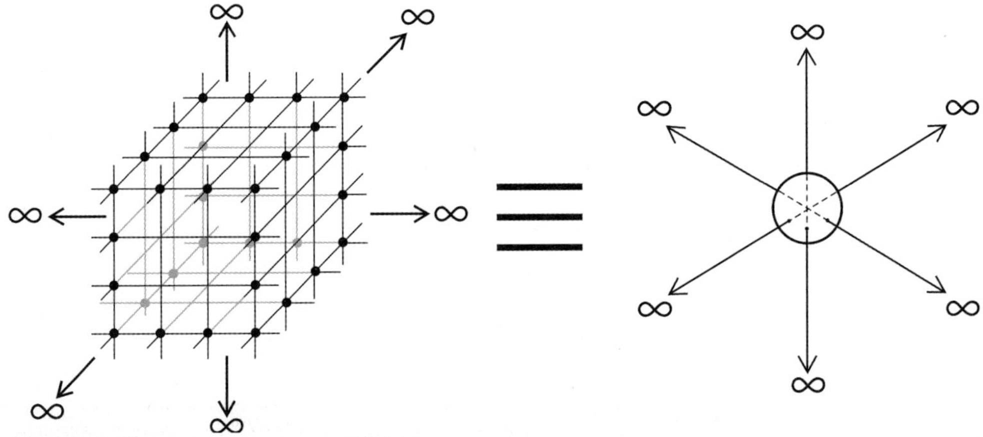

Figure 3-4-1. Visualization of the Unitary Crux of Stressless Status at the Origin of the Universe (Think of vacant set, vacant crux, full crux and unitary crux.)

The original crux is the unitary state at the Genesis of the Universe. The origin is indescribable. However, in human recognition, it is the cross between space and time. The time is expressed as a blank. The blank is the reality of Deity, and, if unacceptable, it is the reality of the Will. If this is still unacceptable, it is reality of the unmoved totality. It is unmoved beyond the existence of the Universe. It is in the perfect state of stress-free condition. (Maybe the condition of orthogonality would be more meaningful to some of us. The definition needs to be discussed further).

When the slightest initial move occurs, the blank of the unitary crux conceives the minimal stress. We can call this unitary crux of stressless status in the origin of the universe as null crux or full crux, which might be correspondent to null set in set theory. The stronger the stress, the more the Will is activated. As the unitary crux turns out to be the building blocks (or building cruxes) of the universe, the blank takes the form of the empty sets of cruxes. The empty sets that appear in all existing sets possibly have a philosophical basis as what mentioned above. The web of infinite number of building cruxes creates the manifested spatiotemporal continuum. This idea is more deeply dealt with in Chapter 5.

Such continuum of cruxes is subject to communication and interaction as we now understand in the quantum states of infinite particles – GHZ entanglement. Quantum mechanics supports the ontological integrity. Amir D. Aczel states that entangled particles transcend the space and that a substance made of two or three entangled particles are actually a part of a system that behaves as a single substance. The author perceives this as an ontological integrity.

The nexusity is the original pattern of the moves of the universal substance. The systemicity came later as the cruxes of spatiotemporal continuum developed in complexity. (We are not taking additional enfolded dimensions into our account on this discussion.)

As we have discussed before, integrism is the compound of sysy and nexy. Sysy intensifies the degree of selection and stability, while nexy strengthens dynamicity and the degree of freedom.

Cruxology is the morphology of cruxes in various situations. Menesys is the outcome of cruxology in accordance with the hysteresis, or path-dependence of cruxological development.

These notions are the main themes throughout the whole book.

3-5. Triradixism (三本主義): Value Theory

Ancient and Modern Triradixism
Interaction among Three Values
Interchange of Three Values between Respective Value Domains

In the neo-classic economics on which most of our economic interpretations rely, there is only one recognized value system – i.e. the value of labor. Economics did not admit nor acknowledge any other value systems, thus consequently narrowed the width of correlation among profound theoretical resources of economic phenomenon. Triradixism is to revive some of the ignored parts of values.

Through triradixism, value, person, and liberty are to be restored. Triradixism opens the sky not in the brain of a human but to the infinity connexion. Triradixism changes the natural environment to a very intimate object of humanity that should be preserved in its own value.

The idea of triradixism in connexion with the issue of infinity is introduced here in Chapter 3 for the first time in literature.

Ancient and Modern Triradixism

Triradixism, thoughts on the three roots(本) of values, has a long tradition in human history. Ancient and modern, eastern and western, south and north, wherever the region of human dwelling is settled, people never forget that they have the sky over their heads, the existence of other human around them, and the land they stand on. The three sources of values that humanity founded on are heavenly-root (天本), human-root (人本), and land-root (地本). From ancient days, people thought that the rain drops from the sky, collaborated for labor from neighbors, and grew crops and found mineral outcrops from land.

Modern Triradixism changed a little bit; the value of creativity replaced the heavenly-rooted, the value of labor instead of the human-rooted, and the value of rent (as an opportunity rent of environment, land, natural resources) in the place of the land-rooted. Creativity, labor and rent interact and compounded into economic utility. There are some difficulties to define the boundary clearly between creativity and labor. But, we may use the concept of mental labor and physical labor for their distinction. Triradixism is illustrated in Table 3-5-1.

Currently, humankind experiences the fast expansion of the domain of creativity, which will soon change the notions of vocabularies. This will be dealt in Chapter 7.

Table 3-5-1. Summary of the Conceptual Correlation of Triradixism

Ancient Roots	Three Domains	Humanity Factors	Values
Heavenly-rooted	Creativity Domain	Transcendental Intentionality	Creativity Value
Human-rooted	Communality Domain	Intentionality	Labor Value
Land-rooted	Terrenity Domain	Utility	Terrenity (Rent) Value

Interactions among the Three Values

The three values interact mutually. If the creativity value in designing potteries gets devaluated by the market as time flows, the decreased portion of the creativity value is transferred to or soaked into the communal value of the concerned society. It means that the society has increased the level of communal value of pottery and decreased the creative value in design (having been dispersed the concerned value into the whole society) so that every member of the society may enjoy a superior level of the communal value than other societies. Nobody in the society would pay a market price for past pottery designs.

The communal value of a society is valid until the composition of the current population is maintained. When a change has been made in the distribution and characteristics of the current population-mix, the communal value should follow the change in trend. Theoretically, the fundamental communal value of a society should be the accumulation of labor value of the society. However, the condensed portion of creative value added up to labor value thus increasing the total communal value.

The communal value is decreased due to depreciation of time and change in population. The rent value is also changed in accordance with the change of potential opportunity cost of the nature and habitats. For instance, if we plant a forest, that increases rent value. The labor value may alter rent value in many cases. The accumulation of the condensed creativity value influences the labor value by stimulating morale and enhancing productivity.

The relationship between the three values can be expressed in Gaussian plane in the following:

$iT = G(C, S, \theta, \varphi)$: in Gaussian Plane, *i* denotes imaginary value ....(Eq. 3-5-1)

,when G = Relationship between the three values in Gaussian plane
T = terrenity value
S = sociality or communality value
C = creativity value
θ = angle between S and C axes in Gaussian plane.
φ = angle between S-C plane and T axis

The letter *i* is attached to T, because terrenity value is the major source of imaginary expectation, but exactly speaking there are impacts of imaginary influence in communality value and creativity value. The above Eq. 3-5-1 is a conceptual expression, which cannot be visualized. Thus Figure 3-5-1 is only a trial to visualize the relationship between the three values in Gaussian plane.

If an imaginary value is input to any one of the three radixes according to the measured value (when exaggeration or devaluation is observed attributable to abnormal psychological expectation), it changes the function to coincide with the change of situation. But in our world of materiality, the above expression would be the best fit.

It was thought that the three-values relation is better expressed when we apply it to the dynamic development of tropical typoon.

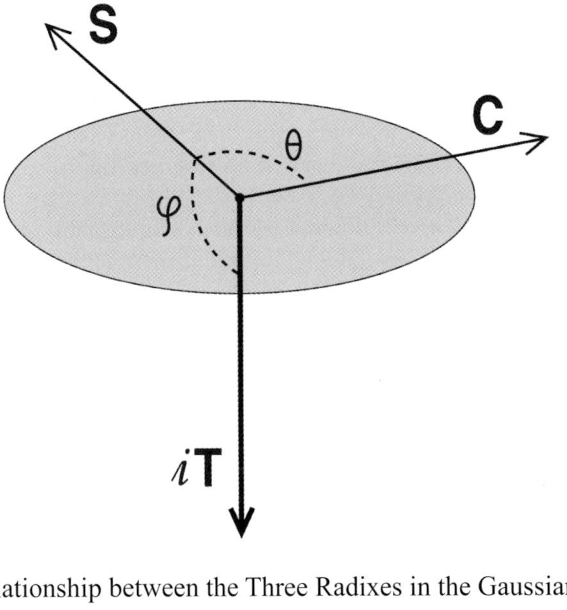

Figure 3-5-1. The Relationship between the Three Radixes in the Gaussian Plane.

Interchange of Three Values between Respective Value Domains

It is frequently said that the human brain is messy and unorganized. The advantage of the human brain is that it can place different lines of thoughts and different lines of stimuli in parallel format. Humans can see those numbers of multiple thoughts and stimuli in one eagle-eye picture and organize them with reasonable rationales. Hippocampus can acknowledge seven parallel lines of matters all at once.

There is an interesting story cited by a brain scientist when he referred to the seven buffers of hippocampus. The story was of an ancient prince in Japan who had the ability to receive seven different reports at the same time thus admired by his subordinates. But human body and its emotional networks can acknowledge much more dynamic modules of sensation connected to the decision making process. Using emotional modules, we can compute in a much greater capacity.

This mechanism is presented in the brain-neural physiology of the human body. It is this mechanism that enable humans to collect and deal with many flutters of transcious qualia into the most probable episode. Without "transciousness," humanity cannot evolve further nor create a novel future as it has been doing in the past.

The three different values are effective in three different domains respectively as shown in Table 3-5-1. These values can be integrated into the same basis in a utility viewpoint. Utility is originated from integrated emotional criteria for satisfaction which varies depending on different situations but still should always be practically useful for the benefit of the subjective self.

Value, utility, emotional intensity, knowledge, intelligence (in a more comprehensive perspective) and market price are all related to each other as we can see the correspondence in the mapping of correlated parameters. This mechanism helps to incur the interaction of the three values. Triradixism stands on this correlation among the differentiated levels of cognition covering from materiality and mentality, and from emotional computation to imaginary computation. Three value of triradixism are mutually transferable by the intermediation of the material measurement of money and the mental measurement of intentionality. Intentionality can mediate the value transfer because of its aboutness – feeling and its decision to select where to put personal preference.

It is not yet clearly proven through experiments why and how dreaming fills up the gap of missing memories. The author suggests, the process happens because cryptograph, embedded in human cognition, is in patterns of episodic edition of affairs. Human linguistic structure has a hierarchy. The author believes that cryptography in the episodic pattern of affairs is the most complex form of cryptogram. Such cryptogram must be equipped with an efficient rounding procedure in every calculation in our imagination process.

It was explained that the linguistic hierarchy comes from logia matrix, which is relevant to the tendency of being articular in a syntax formation. Articularity is an important factor in the value system. A greater value and less-than-expected energy for the concerned value bring a stressful condition, which in turn reduces, the articular movement in psychological domain. Such articularity in the value system appears from differences among creativity values, communality values and terrenity values. The creativity value has the least stress and the least articularity. The terrenity value has the greatest stress and the most articularity among the three. This line of thinking paves a way for Triradixism.

Personage is directed to energy saving, concentration and transcendence, in pursuit of the transcent of humanity. Homimmunage is directed to the conservation of proved stability. Creativity is driven by personage and examined by homimmunage.

Personage is related to delight, ecstasy and "flow" (of Mihaly Chikszentmihalyi's theory), in greater extent than homimmunage does. While homimmunage, being protective in character, is related to fear, panic and blocking. Flow comes from open and relaxed selections. We know that blocking originates from stressful conditions.

Flow, when less examined under communal conditions, would generate abnormal expectations about unnecessary false creative values that cause bubbles and evaporation. Flow, communally examined but deeply combined with corpus conditions, would bring creativity with pragma-prone tendency. Pragma-prone means solidly attached attitude of a person to only tangible values, disregarding infinity factors by which human beings can experience the valuable, the precious, and the intangible world. Pragma-prone tendency reduces the renormalization of the phase space of utility-knowledge into that of intelligence-pretium, a status where no unstable fluctuation between infinity to nullity exists. The renormalization of humanity comes from social stability.

3-6. Infinity Factor and Current Phase Space of Humanity

Key factors of the creative peace, which are humanity, integrity, and creativity, should be initiated and developed essentially by infinity factor.

The questions remaining unanswered during the next half century will be the questions on how the phenomenon of person generated by infinities-encounters came to exist, and how deeply personhood is going to be reflected in the phase space of future humanity. Individual person is a subjective figure who experiences infinities-encounter(s).

The personhood factor may be identical with integrism which consists of systemicity and nexusity. Personhood in itself is systematic and nexal. Personhood behavior as an individual in a society creates general and fundamental phase space of humanity. The phase space of humanity is fractal as seen in the science of complexity. Societies of various levels resemble individual people. Humanity is fractal because it is a personhood with integrism.

However, we are afraid whether humans will lose its personhood within a century. That is one of the probable futures we discussed in the peril of singularity. It is true that human will not be able to sustain its identity if we leave this planet. Liberty from Earth's gravity will cause us osteoporosis. However, it seems to be imperative to go out to the solar system in this era.

Person is the seed and provides the structure for the forthcoming creative peace. New dimensions, such as global space development project, will have to be introduced to current phase space of global humanity in order to give an integrated objective to human kind. We need to invite the factor of infinity to the current world of humanity.

It is recommended to consider certain policies for the global continuum of economy; curtailing the excessively amplifiable future value of cutting edge industries derived from the past space industry, and returning current of curtailed portion of value and wealth back to the unprivileged part of our planet. It is extremely important that the human race has to govern the continuity of the political economy of this planet in order to establish global continuum. This is what we would like to name "the administration of economic growing point" as seen in plant biology. The recent financial crisis has occurred due to the lack of administration on economic the growing point in the financial system.

We need to determine certain method and criteria on how to distribute future value hidden in the triradixes of value, to share the fruit of creativity value, and to allocate terrenity value and resources from the land.

This, however, is only one arm to save the human kind and protect the continuum of the planet. For the other arm, we have to exert human kind has to find and create a megalo project to integrate and concentrate human capability. The author would say we already have it, the adventure to the solar system, although it is the hardest problem we can tackle.

The real problem before us is not the technology part but in the software regarding on how to manage the unpredictable human activities mostly focuses on their own interests. We are not looking at a mirage, vanity, or avarice for the future, but they are the most realistic subjects at our imminence. We must proceed to acquire the management skill to utilize resources and environment of the Earth in the most effective way.

PART II.

The Science of Humanity

Chapter 4. The Science of Humanity

The science of humanity aims at the scientization of humanity. Scientization is defined in this book differently from scientification that describes the scientific change of expression in life styles. Scientizing is the restructuring of the internal attributes of humanity to achieve real scientific conversion of human life.

The valuable works of G. Edelman, P. Churchland, A. Damasio, V. Ramachandran, C. Koch, W. Freeman, K. Friston and other respectful researchers are sincerely appreciated for their theorization and contributions to the science of humanity in their research.

For the most part, Darwinian implications have tried their best to be well absorbed in the formation of contemporary mainstream thought. The methodology of the science of complexity has been adopted to some extent in order to interpret the non-linear mechanism of the real world.

However, when it came to the restructuring of the knowledge mechanism of humanity, it was inevitable that some important and reasonable methods of contemporary thought would be disregarded or enfolded. The purpose of this book is to provide an integrated overall reinterpretation for the survival of humanity living in and moving towards a scientific future.

4-1. Uniqueness of Humans and Humanity

This study intends to clarify that humans and humanity are unique entities in their relationship within the universe. The origin of this uniqueness is the infinities-encounter of human beings, as discussed in Chapter 3. Personhood originated and thereupon transcended a long way to fulfill its internal teleological aim in accordance with the mechanism of anthropomorphosis. The internal teleological aim has not been given from the outside but entirely from the inside, in accordance with the development of humanity commencing from and upon the impetus generated by the initial circumferential condition.

Personhood: Personage and Homimmunage

Humanity is unique in the sense that the behavior of human beings has immense variety. Such variety is attributable to human complexity; the origin of which is the variability of human identity. A Human changes oneself through intention or is changed by circumstances. These changes are governed by scientific cause and effect. Nothing appears fixed in the phenomenon of humanity. However, the major varieties come from the inter-intrusions between nature and humanity within individual bases embedded upon physicality. Physicality in the epistemological view is the source of mental and spiritual conscious sequences. Mentality and spirituality are not the source of the physical condition, because they are the product of humanity, or equivalently, of neurophysiologic activities. However, it must be noted here that, in persontological, or broadly from an ontological perspective, the source of physicality is the interactive relation between the whole dynamic universal continuum (= Plexion) and the partial dynamic universal continuum around a definitely limited individual entity (= pragma).

Before getting into the whole of the discussion, the issue of personage and homimmunage must be is dealt with first. Humanity is not a unified entity but is directed to integrity. That is how personhood behaves. Personhood can be thought of as the compound of personage and homimmunage.

It is observed that personage and homimmunage are developed dualistically in a personal life. Duality of personality in its definite notion is not attainable in human life, but it seems that there are people who have social lives that are developed in a bipolarized manner, distributed between personage, which is the amalgamation of positive pursuit and positive feedback, and homimmunage, which is the combination of protective pursuit and negative feedback. Thus, human personality in its abstract has a shape similar to that of the double humped Saharan camels.

For instance, the "*amae*" personality, which appears in discussions regarding Japanese deep psychology frequently found in Far East social circumstances, is a good indication of this line of thought. A person is socially defined based on what he or she thinks of him or herself. The duality of the Japanese is frequently discussed through the combination of "*honne* (real mind hidden)" and "*Tatemae* (shapely appearance)" However, this duality is not just a matter peculiar to the Japanese but, to the entire human race.

Such duality is found within societies containing a high density of population, greater imperative for external regulation, and the confused coexistence of distinctively different cultures. It is a psychological imperative to reduce repercussions and secure self-protection in crowded societies such as those found within the Far East and China. The current world will become increasingly more crowded leading to this duality becoming more pronounced.

Thus it would serve well to acknowledge the existence of this duality of personhood. This, of course, is hypothetical and has yet to be proven by a sufficient quantity of empirical evidence. However, we have decided to include it for the purpose of our present research toward the formation of a global design of the creative peace.

Anyhow sociability is more prone to homimmunage while individual pursuits are, in a greater extent and intensity, concentrated to personage. It is thought that personage in most cases tends to exhibit multiplicity effects as demonstrated within innate ovarian conditions in accordance with the variety of personal inclinations. So, if multiple personages manifest in some personhood, they may be thought to have developed from certain initial micro cores of each personage from the beginning of the human embryonic stage. It is necessary to mention here again that personages are not the whole personhood but partial trends of the individual personhood.

It is assumed that personage in its earlier stage is supposed to have frontal effects of multi-centered developments that later converge into an integrated and stabilized form. This integration process takes place through a balancing between homimmunage in a bipolar pattern. The shape of the bipolar distribution will be integrated into deeper intrinsic structural coherences following the path to stability. It is apparent that bipolarity decreases as inner coherence is improved.

The shape of the bipolar distribution of personage and homimmunage will gradually draw closer to one another and eventually merge into a single bell shape as a person matures and gradually becomes well accepted in a communal environment. Personage usually takes on plurality; the center of personhood being a major one, while homimmunage absords singularity.

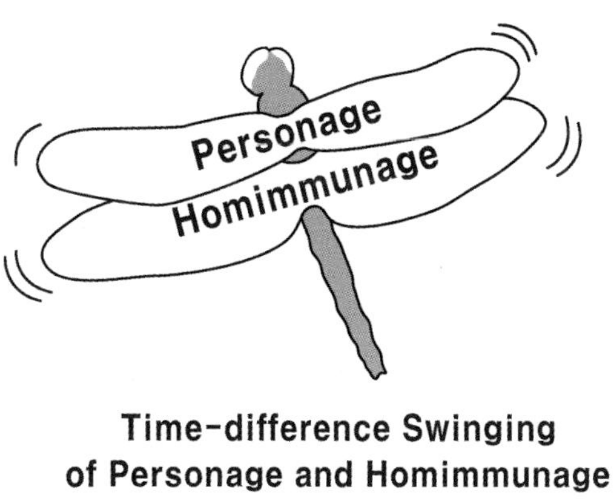

Figure 4-1-1. Personage and Homimmunage Expressed in a Metaphoric Illustration of a Dragon Fly. (Adequate time-different swinging motion generates best lifting force.)

It could be imagined that even a single hump case would have a distinguished personality according to the polarization of the bell shape. For example, if a left-prone bell shape were to appear, it would be indicative of a more individually prone person. In the case of a right-prone, the person could be considered to be more socially or communally characterized. For instance, without wanting to oversimplify too much, a left centered individual could be considered more western, while a right-centered individual could be considered more eastern.

Personage is the source of modularity within the frame of personhood. The modularity of a person, one of the rings of cause and effect in a person, leads to multi-personage if it ripens upon an individual following and independent path. Modularity appears when a person reveals his internal sincere wishes and conflicting emotions at the same time.

Language modules of a premature bilingual person may be the simplest case to exhibit the presence of such modularity. A completely matured bilingual individual would not disclose such a trace of modularity. The modularity of the person mentioned above differs from dynamic modularity to be discussed later.

Another good example of the coupling of two similar intrinsic structural characteristics can be found in the induction of stress conditions between the dual presentations of personhood in order to attain the collaborative mechanism in raising a stronger lifting force. In Figure 4-1-1, we see the case of designing a four-winged robot with a pair of wings located both in frontal and rear positions as found on a dragonfly. The frontal and rear wings in pair generate an excellent dynamic lift by combining the two swinging motions in the most effective rhythm allowing for a proper time difference.

(I believe the distinction between personage and homimmunage and grasping the slight difference in dynamic change of social hysteresis would help facilitate the next step of quantitative analysis following this research in.)

In addition I will add that there is interesting research regarding solar and lunar consciences. "Solar conscience, lunar conscience" are somewhat similar to the line of thought developed along with a metaphor to Greek myths (Murray Stein, 1993). But personage and homimmunage have a lie on a different set of notions from the above mentioned solar and lunar differentiation.

Uniqueness of Human Cognition: Transciousness and Dynamic Modularity

The human cognitive model in this research is in part defined differently from current popular understanding. It is believed that personage exhibits an inclination to become transciousness. Transciousness is a vital realm of symbolization and abstraction which enabled the formulation of "Language," labeled following the universal grammar of S. Pinker. Universal grammar is Language before languages.

But transciousness is of a higher and more complicated mechanism than Language, consciousness and rationalism. It is thought that the magnitude of information quantity has to be far greater than even that of the full capacity of the human brain if cognition is to deal with infinities and the superposition of infinities.

I believe humanity has emerged from the point in time where it encountered infinity in its emotional reception. It would have been an entirely overwhelming experience to encounter the Absolute Other: everything else other than oneself being engulfed in the entirety/ of infinity. It might have been similar to current religious experiences. Early infinities-encounters such as these would have required extreme energy that the human brain and neuro-system could not have sustained. This may have encompassed a period of time longer than a dozen millenniums ago, as evolutionists presume.

The territory of transciousness in human cognition was excavated to solve the energy problems associated with infinities-encounters by conserving the physical energy of brain activity that was increased through the many magnitudes of difference. Success in conserving energy was achieved by utilizing the effective means of symbolic omission from the information produced in conscious cognition. Symbolic omissions were made possible by developing skills of dynamic modularity within the brain.

Dynamic modularity appears to be successful in creating the simplest fluttering of symbols that are able to organize various scenic and episodic representations of human memory. The simplest and most effective fluttering of symbols is supplied from the most significant symbols stored within the series of ITRings recorded during the formation of an episodic memory in an individual.

The portion of transciousness in human cognition appears to have emerged most recently in the long history of human species, since personage happened to have attained dynamicity in the intentionality of consciousness. Human intentionality had to be equipped with a dynamic process in order to quickly enhance to greater survivability. The admissions of dynamicity into human intentionality made transcendental intentionality emerge, which induced the transcent of humanity into reality.

The uniqueness of the human cognitive model is found in intentionality dynamics, transcendental intentionality, dynamic modularity, and transciousness. The dynamic modularity and transciousness between a couple provides room for a deeper and wider interpretation of human cognition rather than simply focusing on just consciousness.

Transciousness is a tool used to attain the infinities-matrix as quickly as possible with the least consumption of energy. Another advantage of transciousness is the compressed presentation of symbolized sets of information within a voluminous memory. If we recall our whole life in a few seconds, we are utilizing our ability reserved in transciousness. This is why dynamic modularity has brought about a great leap in humanity.

Correlation between Intentionality and Homimmunage /Personage

Personage relies more heavily than homimmunage on the formation of relationships between the cognitive self (which implicates sensciousness, consciousness and, transciousness) and the other (which philosophically/ means the entire cognizable environment). The conscious self is a part of the cognitive self, and exists and behaves under indivisible connections with sensciousness and transciousness.

Personage is a term to describe the socio-psychological entity with a centripetal "self" that has independent interpretability of the environment and autobiographic memory. Personage is different from homimmunage in terms of their existential boundary already aforementioned in Chapter 2. The functional character of personage is frontal-advancing and positive while homimmunage is tail-protecting and conservative. Homimmunage and personage are considered mutually contrastive and in another sense mutually complementary.

In this context, homimmunage, in comparison with personage, depends more deeply upon, and is inclined to the physical self. Homimmunage has a function where the embedment of mind in a specific body becomes an important factor in deciding the singular and anomalous situations of human society. This dependence to physicality sometimes allows for an unexpected occurrence of affairs affected by environmental factors to cause many accidents like the sand slide model in a complex system.

If humanity had prevailed through mental causes only, the relatively closed circuit of actions orientated by human systematic and rational mentality would not have permitted enough variation we enjoy nowadays. Systematic ideas would have individuals follow a pre-designed route of burdensome lives that can be seen within collective societies. (It is also true that mentality, if well adjusted, would yield an amicable life as those found in free and wealthy societies.)

The current frame of brain science begins from the fact that the internal chemical balance of the brain sustains human life. The chemical balance is managed by the brain stem and hypothalamus. All the physical changes of human life could also be presented biologically in detail with specific, identifiable patterns of expression through these organs. Nevertheless, such detailed expressions do not give answers of how the biological self could be converted into the conscious self and how the self and the other could be distinguished.

Furthermore, the conscious self is not as dynamic as opposed to if it were in a transcious condition. Humanity is more than the self in transciousness. This means transciousness has a directive function of the conscious self which behaves in a logical way. It would describe ontological status better, not in a linguistic pattern but in a substantial pattern of communication from the whole cognitive network.

This is why people cannot easily describe their truthful opinion with only language and at the same time feel that they are deceiving their own honest emotion.

Intentionality and Preafference/Postefference

Intentionality has an intense connection with the afference and efference of the limbic system. The limbic system is considered to assume the function as the center of the integrating function between consciousness and transciousness. For consciousness, it is good to use a physical understanding of cognitive maps based on neurophysiological study, while for transciousness a fresh approach in topological understanding is needed.

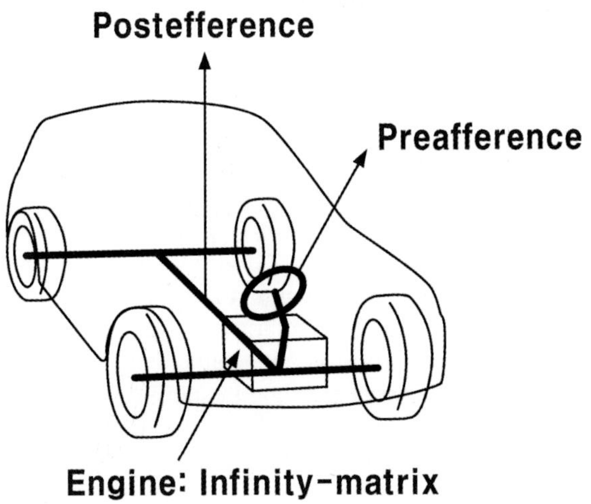

Figure 4-1-2. Preafference and Postefference

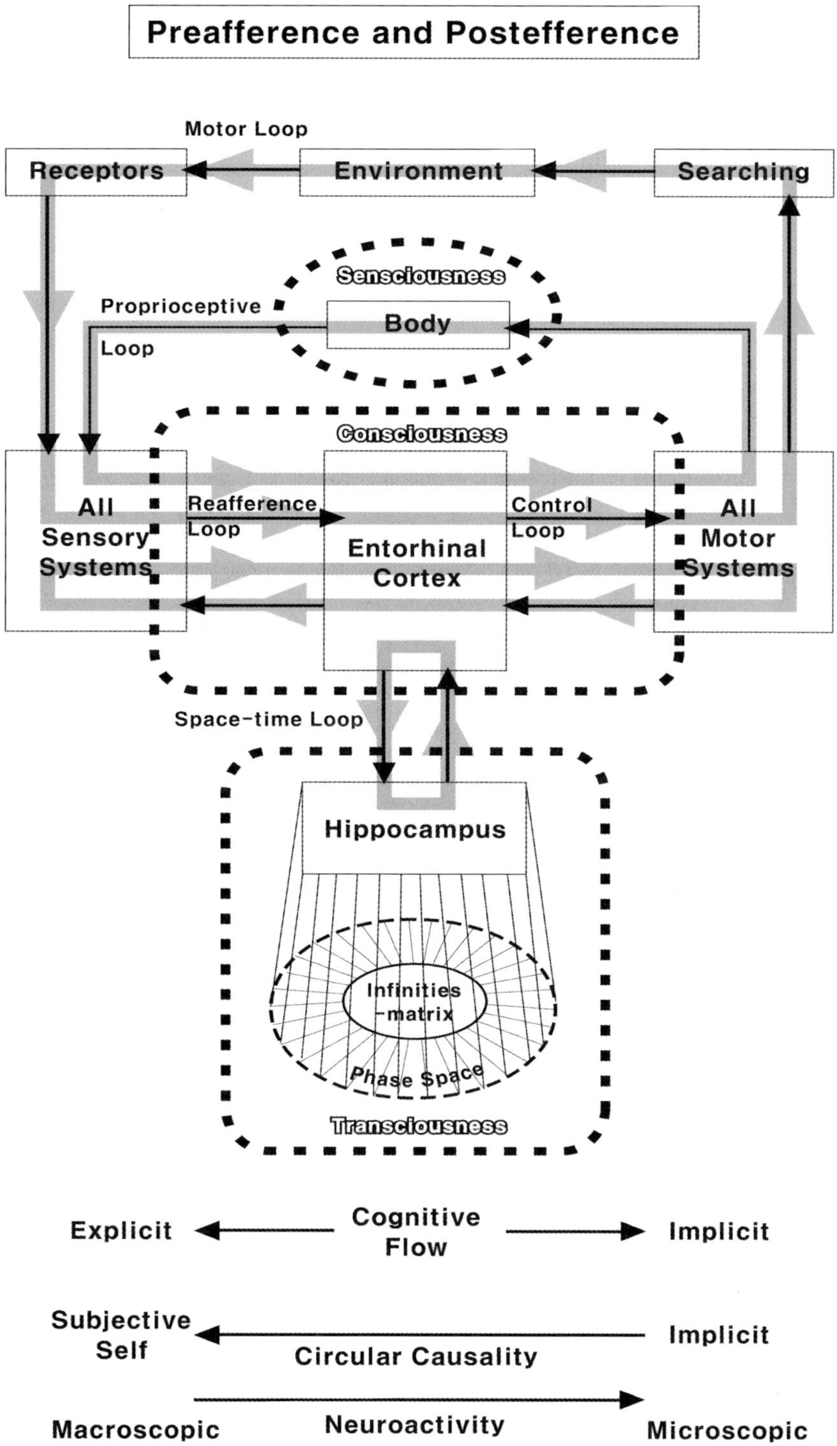

Figure 4-1-3. Preafference Diagram (modified from Reclaiming Cognition, Walter Freeman, p150)

We are not yet sure if mathematical methods can be directly applied to brain science. But it is known that non-mathematical problems can also utilize the concept of manifolds and tensor forms in rationalization. The thorough inspection of theoretical propositions is left to respectful mathematical specialists.

It was originally reported by W. Freeman that preafference is present in the human cognitive process. "Preafference provides the multimodal perceptual processes of expectation and attention. Without this configuration, there could be neither search nor perception. Without sensory recursion, there would be no intentional action. Without emotion, there would be no remembering."

The preafference process is executed in the limbic system, mainly the hypothalamus and the brain stem, by selecting out of various inputs of expectations, through chaotic dynamics of intentionality sustained by neural populations. These dynamics are supposed to provide a biological basis for the flexibility, creativity, and meaning of human behavior.

The limbic system is essential for all intentional action, perception and learning. The endorhinal cortex is at the center of the preafference process connecting sensory systems, motor systems and the hippocampus.

However, it was thought that transciousness would start from the loop between the endorhinal cortex and hippocampus (Figure 4-1-3), and occupy intangible phase spaces formulated inward from there. The intangible phase space is a tensor space with many dimensions (equivalently, with many ranks) of concerned parameters representing the indicative parts of the limbic system and their variable connections made by dynamic modularity over the global brain.

Since a person cannot escape from his past, the matching trend of which is recorded in his own episodic memory, he cannot change the facts of his experiences. The specific parts of the activities of dynamic modularity sustaining transciousness are believed to produce phase spaces of quasi-virtuality, while those for sensciousness remain at the level of plain feeling.

Quasi-virtuality indicates they are most likely to operate in a manner of simplified "gestalt" virtuality for presentation efficiency. The intangible phase spaces called transciousness in human cognition is very much similar to virtual reality in computer games, except they look like the least symbolic gestalt with maximum omission.

The outputs due to the efference generated from the infinities-matrixes of a person cannot be altered. The infinites-matrix is basically unalterable since it is formulated by indentation of the past and thus by the relatively stable mechanism of elasticity, which is the backbone of the matrix. (But, strictly speaking, the infinities-matrix looks alterable through certain processes caused by unusual influence from the Axis Tempus.)

Furthermore, it is thought that the infinities-matrix is a structural compound of three sub matrixes: the onto-submatrix, topos-submatrix and logia-submatrix, formulated from the human conceptus.

The onto-submatrix is thought to have its origin in the initial penetration of the sperm into the ovum. The topos-submatrix has its origin in the general trend of reverse intrusion of the anus into the oral cavity in early embryonic development. The logia-submatrix starts just after the topos-submatrix through the correlation process among the general diversification of initial organs in the earliest developmental stage occurring around the oral and stomatic cavities.

It is still as of yet speculation to be further examined whether this general trend of thought could be scientifically admitted. The rationale of the writer is to assemble the idea into the framework of the infinities-matrix so that the essential Language structure is intuited to consist of the three submatrixes.

It is said that human sensation and emotion are a kind of filter that omits and allows only a part of the external signals in for an extensive interpretation. Rene Descartes mentioned that physical affairs generate thinking and thinking affects physical process. His theory suffered criticism for a lack of explanation on how pure thinking influenced neurons. Although this study does not explicitly support dualism, the thinking process influences physical neurons through postefference.

The mechanism of postefference is a function of human transciousness and has yet to be examined through empirical observation. Transciousness is the conclusion of the hypothetical confidence that humanity is based on infinities-encounters, infinities-transfers and consequential formation of the infinities-matrix.

The infinities-matrix and its resultant ITRings are supposed to decide patterns of the fluttering (and also the complex combinations of fluttering patterns) in transciousness and send them out to the consciousness. The submission of fluttering patterns (singularly or in combination extendable up to even episodic symbols) is characteristic of the individual person in consistency with their past experiences. For this reason the content of efference would be directed preferentially in a positive manner or directed protectively in a negative manner, respectively.

Usually postefference, following the actions previously described, should consecutively mend the result of preafference. Postefference might be very influential and intensified through recurrence if certain mechanisms of volitional decision intervened. It is a very familiar occurrence in our practical life.

In Figure 4-1-2, we see that postefference is transmitting the engine power of the infinities-matrix to the steering section. There is another important metaphor for the relation of these two mechanisms: the collaboration between the two mechanisms allows humanity to maintain stability as that observed in a helicopters double set rotor blades assisting one another in the generation of dynamic lift.

Postefference is deemed to originate from the repetition of rhythmic propulsion in cell-level hierarchy. Human tendency to pursue ecstasy through repetition of the same physical movement has to do with the postefferential phenomenon in developing the abstractive ability of human mentality.

This may be the right place to deposit the hypothetical message that DNA in each cell creates physical consonance through mutual resonance with other DNA found in the whole body. DNAs of the same minute rhythms in every cell are the key to solving our query on how an individual person creates abstractive propulsion inducing transcendental intentionality. Abstraction comes from the repetitive ecstatic response of human physicality.

It is again hypothesized by intuition that DNA operates as a tool of communication between cells constituting the same body. The elevated minute rhythm of a specific body should be produced by DNAs during cell segmentation. However, in between cell segmentation, DNAs radiate in stable, minute rhythm. These two minute rhythms of stable and elevated conditions which are in contrast to one another are deemed to have originated in the oscillatory difference from a molecular level. The presence of such rhythms should be discovered by more basic studies on the primary reasons and elementary factors for generating human rhythmicity. This fundamental and specific rhythm of a specified body is believed to provide the basic condition of consonance that enables preafference and postefference to work together in the field of consciousness and even in the whole human cognitive process to a certain extent.

Let us get back on the postefferential issue. When searching for a new dimensional extension needed for survival or for new creative solutions, postefference could be extraordinarily strengthened. This is related to higher hierarchical creativity on a greater scale to be discussed within Chapter 8.

Postefference and dynamic modularity are dedicated to building phase space consisting of every move of transcious activities. This phase space is considered to match the Li (理) of the duality of Li-Ki (理氣) in Neo-Confucianism. (There have been long and heated debates between monistic schools of thought and dualistic schools of thought in regards to the Li and Ki. It is of some benefit to use the term duality instead of dualism.) The intangible phase space could also be interpreted identically (more than similarly) to the eternal object denoted within the works of A.N. Whitehead.

I believe that these philosophies did not possess the current information regarding brain science and as a result were not able to identify the part of pure logics found within human consciousness. Thus, logic in essence is no more than description of dynamic relations. Dynamic relations can take the pattern of the collective entity as imagined in the plexal form of the unity through intuitive projection. It would be interesting if anyone should delve into the theorization of the topological understanding on transcious phase space.

It is known that the hippocampus contains 10^7 neurons that can yield innumerable combinations of relational dynamics. Basic patterns found in the cognition of a cat are numbered to be more than 500 kinds. If we combine just 10 patterns among them, we will elicit more than 10^{20} outcomes.

In the case of a human, a far higher amount will result when we take into account all the intangible patterns implicated together. The phase space constructed by human transciousness would be so profound that the infinite sets of information may even be visualized in imagining the wholeness of nature. (One thing to be noted is that this fact alone does not indicate whether the Almighty really exists or not.)

With just the relationship between the representation mechanism and intentionality dynamics in a brain, one cannot easily describe the transcendental intentionality of humans. The extension of this idea is mentioned in the following explanation on the penumbra of consciousness in which phase spaces are generated.

A human being possesses the basic conditions to generate virtual reality alone, individually. In the annulus of the highest cognition, named Animotus (animi motus) annulus, we have come to construct a particular form of virtual reality by combining self-awareness and dynamic modularity, through the establishment of the infinities-matrix (= set of ITRings of an individual person) and utilizing the mechanism of mimesis or meme achieved in mother-infant relations and other social relations.

Penumbra of the Spotlight of Consciousness, or of NCC

C. Koch (concerned with Neuronal Correlates of Consciousness) and W. Freeman used the terminology "penumbra" in a mutually distinguishable manner. Penumbra is the fringe region of a half shadow. This terminology is also carefully considered in this research while studying transciousness.

Walter J. Freeman (*How Brains Make up Their Minds*?, 2000) made a comment-about the definition of consciousness used by William James in his article "Are We Automata?" in 1879. "James concluded that consciousness is 'an organ added for the sake of steering a nervous system grown too complex to regulate itself.' But nowadays we realize that it is not an organ that is located at some part of a brain. It is a higher level of self-organization."

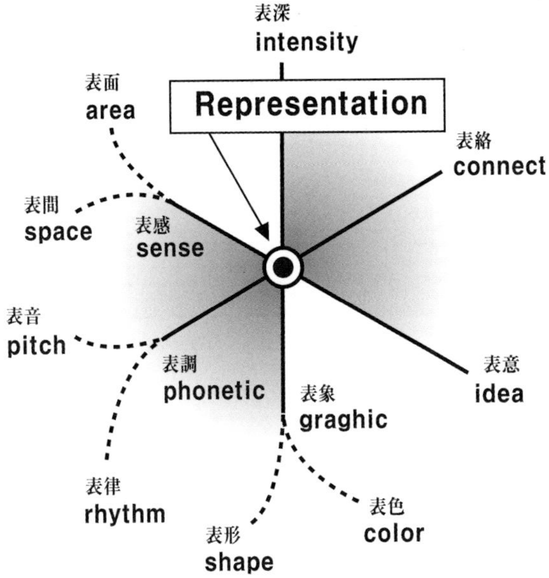

Figure 4-1-4. The Assumed Axial Parameters in Constructing the Phase Space of Transciousness

Freeman adopted the concept of preafference, originally suggested by Herman Helmholtz in the 19th century, with which human global neurobehavioral patterns allow sensory cortices to anticipate forthcoming stimuli from the external environment. He approached brain science with the idea of complex systems. He explained his experimental results from the brain imaging and neuro-dynamics with distinctive, ten building blocks to illustrate how neural populations sustain the chaotic dynamics of intentionality.

His theory starts from a pulse report at the level of individual neurons, and upon attaining attractor patterns through combining feedbacks from the sets of multiple excitatory-inhibitory cortical populations and at the level of whole hemisphere as well, finally ends up with a single global AM (amplitude modulation) pattern for one specific time. The neurons and local modules retain their autonomy. But they are boiling in unconscious processing, which is indicated by James as the penumbra of the spotlight of consciousness.

Freeman explained that the intentional arc of Merleau-Ponty is to achieve the assimilation of the self to an object through positioning the sensory receptors in efference and focusing the sensory cortices through preafference.

I think it is noteworthy that these frames of afference-efference, or centripetality-centrifugality, generally exist in human phenomenon. The frames are engaged with the problem of the self. I believe that the self is the centripetal essence which organizes all the internal and external resources of human beings into a systematic frame. However, the centrifugal connectivity with its environment and with the other beings has to be sustained for existence.

Afference usually maintains "pre"-afference in order to construct a quasi-virtual reality, which is one of the most important characteristics in human intelligence or human brain activities. Otherwise, the solid afference would have lost its flexibility. The flexibility in preafference, although known for focusing as its main function, allows human beings extra spaces for storage of relevant information and multiple intentions, within a confined limit which is associated with focusing.

In this context, the **penumbra** could be interpreted as a **transition zone between consciousness and transciousness**. If such penumbra, which is an anomaly on a human cognitive contour map, is made for potential distribution, active vitality would gush out. This inevitable quasi-virtual reality is, in scientific description, a phase space. The virtual reality in transciousness is maintained by the preafference. The virtual reality can maintain its flexibility by the changing and flexible afference. Such flexibility is the source of creativity.

The efference for flexibility and creativity is thought to be changing all the time following its original efference around the original location. The original location is determined by the afference and its preafference of external conditions, but the vicinity of the original efference is activated by the dynamic modularity producing effective and most probable combinations (for reality) of fluttering pieces in virtual reality.

Dyon and Trep

Dyon stands for "Dynamic Ontology." Trep is "Transcendental Epistemology" concerning neurophysiology.

Ontology in philosophy shrunk in its view to existentialism since E. Kant established a firm ground for epistemology. Existentialism has flourished in every direction of humanity. In some sense, we can say that existentialism has been focusing more in relational aspects of ontology in the modern age. Such ontology has been termed relational ontology.

However, it has been said that the contemporary ontological thought has entered onto a new trail; a new way has opened up to dynamic ontology. Information theory needs to focus on the dynamic ontology because information is regarded as the essence of the universe and the universe itself is dynamic.

The recent tendency of ontological philosophy is to find its place in information technology. Relational aspects in particular are the main concern in setting up ontological algorithms. The dynamic ontology studied in this research is focused with relevance and also dynamicity in ontological mode. Ontological relevance becomes important when dynamic states become present. In essence, ontological mode has to be dynamic, but recognition of it has taken time to come into the general realm of humanity.

The backbone of dynamic ontology will be dealt hereafter under the discernable short name of "Dyon." Thus, Dyon is specifically speaking the "dynamic relational ontology" which is able to treat ontological aspects in a relational plexus of affairs in the dynamic proposition between entities. Even the entities and the entirety of entities will ultimately be presented in dynamic relations only.

Humanity is close to moving from the static understanding of the universe beginning with the late part of the 19th century. It is affected by the increase of various speeds in everyday life due to the Industrial Revolution. In the 20th century, the dynamicity of common people has been magnified several times. Before the invention of the steam locomotive, it was an issue of debate whether or not humans could endure speed quicker than a horse.

Ontology and epistemology have been static and have not looked together until they met under dynamic conditions. The encounter of the two thoughts in dynamic conditions refers to the settlement of complementary relations in constantly varying connectivity.

The Transcendental character of epistemology was observed in scientific achievements and discoveries and verified the ontological modes of the universe. There has been an accumulation of vast knowledge and scientific data that led to the paradigm shift in science. It is similar to the overflow in a cup.

The communities of scientific study have major influences on paradigmatic changes. These phenomena are not solely exclusive to science but to almost every aspect of humanity. If transcendence is lost, human epistemology would not be better than the consciousness of animals.

As dynamic ontology must be interpreted in relation with transcendental epistemology, ontology and epistemology are thought to be intrinsically connected. Transcendental epistemology is named, in short, "Trep" (TRanscendental EPistemology relating particularly to neurophysiology). In Figure 4-1-5, we can see the double helix model implying the duality of Dyon and Trep. Most people position themselves where the depths of the conflicts between the two attitudes are not realized in their cognition as a stress condition. Dyon and Trep are in rivalry with and complementing one another. However, Dyon is the major character of this research because the integration of human thoughts will be pursued in this book.

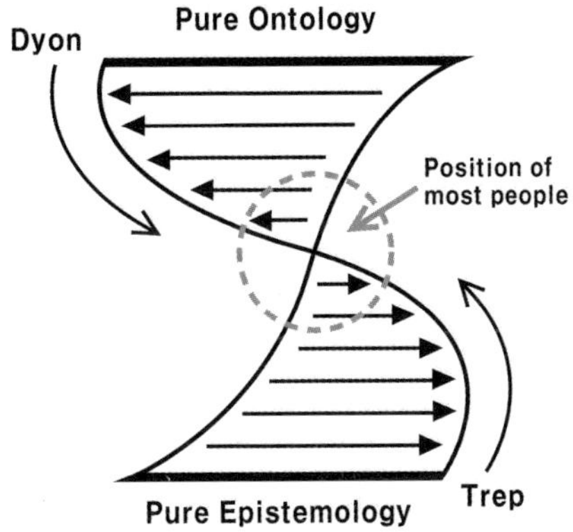

Figure 4-1-5. Dyon and Trep

Dyon is a way to produce the codes of context for accidental plexus of affairs. Gary Marcus commented on four factors, found in Chapter 8 of his book "The Birth of the Mind" (2004) that are responsible for creating the complexity of the human brain with a relatively small number of genes.

The first factor we can consider for our discussion is that the genome produces the cryptography of its structure coded by process, not by a bitmap of detailed design. The codification of the process is flexible especially when the process is very highly complex in the intensity of the communication and leaves many "holes," which allows the freedom of interpretation to some degree. Therefore it looks identical with the process of information storage of the potential zone of mutation, on which impacts from the environment give the directive for the next variation.

The second factor is that the genes are not isolated from each other but are mutually collaborative. Thus, he explains that one more gene added to the genome induces an exponential efficacy instead of a linear variation.

Following, the third factor is that the genome is extendable. Thus, the compression of the genome is made in such a manner that new functions may be freely added. It is analogous with the fact that the open territory for the potential transcent of humanity is hidden in the Plexion.

The fourth factor is that all genes can be used without any limitations in regards to the number of their usage.

Frankly, it is of my personal intuition that the genome is the codified results of communication between human cognition and the Plexion. The Plexion is concordant with the Absolute Other, with which humans can communicate under the quantum condition in the magnitude of time scaled to 10^{-20} second.

I believe that we have to seriously consider the opinions of D. Bohm and K. Pribram, with some corrections, which asserts the quantum level communication presented. My belief comes from the fact that time does exist in the nature of a continuum base and does not exist alone in human cognition. Time exists based on a more complicated mechanism than we can easily imagine. Time can be understood and perceived when the alignment and synchronization between Plexion and pragma have been completed in our transciousness. Otherwise, time is conceived fragmentally.

Thus, if we want to designate a time axis in the phase space of cognition, it has to be a blank. In my opinion, time is the appearance of enfolded dimensions, whatever the number of dimensions would be, eleven or twenty six, into each quantum building block, which I would like to call "*ontum* (*onto-um*)." *Ontum* is equivalent to "*bohu* (vacuity)" in Genesis 1:2 that are mentioned in the preface. Time might be carrying the residues of the dimensional disturbance caused by the surface perturbation of the Uni-crux, which will be discussed in Chapter 5.

Thus time flow is deemed to be the essential effectuator of the spatiotemporal continuum of our universe together with the gravitational wave which is orthogonal to the time flow. Time is of a fractal character as Isidore of Seville commented long time ago in the Sixth century. He said in his book "The fifth principle of time" meaning what partly implies the whole and that the whole implies the parts. (Sententiarum libri tres, Paris, Migne, 1845)

We cannot easily feel the fractuality of time in our lives but, time is fractal as we experience the living forms of the universe, whether they possess biological life or not. Time introduces causality into every being in order to make things dynamic. Dyon, the dynamic ontology, is owing to time.

Time causes irreversible collapses of symmetry. That is the source of entropic phenomenon. The collapse occurs at the hyper-topological surface, which I believe has a very thin transition zone, of the Uni-crux.

Dyon seems to have an indivisible connection with topological methodology. If we put the community in topological expression, it is thought to be comparable to a combinational linear complex of corresponding individualities. If smoothness is attained piece wisely in complexes, they are regarded as "integer communities."

I know that this part of thinking should be refined by mathematicians. In "A way Toward Phase Spaces," written by Honma Datsuo, it is stated that since the relevance between the locality and globality is essential to the topology; dynamic ontology would be relational as much as it maintains the smoothness between constituents.

In an "Entangled World" (2006, edited by Jürgen Audretsch), Michael Esfeld mentioned that the core of the philosophical challenge in quantum theory is to develop a philosophy of nature of the relations. "The chance for such a philosophy of nature is to avoid the traditional gap between the epistemology and ontology."

Frankly, it is the crude hunch of the author that the infinite number of entanglements might be the cause of the spin values of elementary particles. The general n-particles in GHZ entanglement could be the generalization of the Triple Boromeo Chain. (Shown by K. Aravind, quoted from *Entanglement* of Amir, D. Aczel, 2001)

Suppose there is no "I" but only the universe, the general GHZ entanglement of the infinity might be the pattern of the original universe. Our problem is that the human cannot "think" without his participation as an observer into the world. But since we are already participating in the reality, the ontology and epistemology are under a symbiotic antagonism not only in humanity but also in nature itself.

This section is added to give an idea, as crude as it may be, as to why this book suggests an eventual change of our viewpoints to the dynamic relational ontology, which is the persontology. Person could imply a universal rationale to be applied to every corner of the universe and every existence we can recognize, not in a form of received cognition in apparent expressions of human person but, in a form of essential mechanism of the universe.

4-2. Persontology

When we refer to the ontology and the dynamic relational ontology, it is necessary to consider one more point of view. That is the factor of the person in the ontological view. It is a firm belief that the factor of the person is not the scent, appearance or any other feelings we can derive from a person, but a rational and scientific measure to weave the factor into ontological thoughts. It should be a measurable and analytical mean in social relations and also in individual intrinsic conditions.

The ontology is an ideological view on the contemporary world as we choose to escape the intensifying humanification process. If we can establish a way of thought which combines the values of a person and the ontology, we would restore the human eye to humanize the world.

Persontology: Person-Ontology

What is the ontology? Ontology is a compound of "*onta*" and "*logos*." *Onta* means that something exists and *logos* means reason. Ontology has a long history in philosophy which traces back to Plato and Aristotle. However, the terminology was created by Scholastics and is known to have been first used by Rudolf Goclenius in the 17th century.

Ontological thought has provided a comprehensive coverage of the universal existence. The philosophical propositions of ontology were examined in depth by philosophers until the most representative anatomic philosopher Emmanuel Kant opened the gate to forthcoming existentialism.

The problems of ontology are identified in this research to be the premise of being a person(人格). A Person is the source of ego and self. A Person itself, not only ego and self, is the source of integration behaving as humanity. A Person is not confined to a mental concept. A Person as a whole is based first on physicality.

The Physicality would generate other consequent attributes, including mental and spiritual attributes. Phantom limb is not evidence of the preferential right of spirituality in human substantiation. If anyone should believe the superiority of human spirituality, he needs to understand that the integrity of a person is the real source of such superiority. The integrity is also the last step of human substantiation.

First of all, it is necessary to abandon existing concepts. Understandings of the person are usually established from our perception on the personal characters of the physical conditions, social appearances, sensuous appeal or other external expressions. The essence of the person is an infinite capability of humanity to extend to the whole realm of humankind even to the end of the universe. That is the extendibility of the person.

We are now about to start to cope with the extendibility to various living forms and environmental existence, which needs a substantial amount of time to examine ourselves. However, humankind has a definite idea of what nature is as a whole. Considering our discussion on the dynamicity of ontology, a person is the dynamics(動態) of the universe as a whole.

This concept of a person has room in interpretative rationale for some people to be regarded as the one and only Almighty. But it is not for the rest. The substantial body, Che or Ti (體), in the ontological space is presented as the relation or usage, Yong (用), in epistemological space. Actually humanity has its place in between the Che and the Yong and in between the ontology and epistemology.

If any one end is lost, human does not exist as a human. There is a centripetal factor of being human from an ontological perspective while the other end of the centrifugal factor of being in epistemology, even millennium before the establishment of two contrary ideas.

Ontological factors strengthen the systemization of internal attributes and the epistemological factor intensifies the networking nexusity for external connectivity. The brain activities need self-adjustment to make behavioral tuning between externality and internality which will be finished with adequate self-identification. The confirmation process of the self-identification is the essence of being a person.

In this context, the interchange of intellect, emotion, material, or any other form of values among people is the foundation of being a person. In order for such interchanges to be achieved among the disconnected systems, mutual adjustments between such systems (e.g., between the cognition of human brain and the environmental reality) should be invented and certain intermediate structures that enable mutual communication, contacts and exchanges should be constructed.

There is a proposition that human consciousness is an invention of the human brain to cope with the explosive expansion of external impacts, to which W. James commented that the copping is "for the sake of steering a nervous system grown too complex to regulate itself," meaning human consciousness is not a new organ but a new hierarchical self-organization of brain dynamics.

The human body is an open system to its environment as DNA is effected by external influences. But human mentality is a quasi-closed system: self-organizing itself and confining its associated cryptograph system similar to Language.

In this book, we would name it basic mentality, and also explore the additional enhanced mentality, a novel realm for creativity, flexibility and adaptability. The Basic mentality is a good foundation for humanity to take a firm root into a communal circumstance. The Enhanced mentality on the contrary makes its way to new discoveries of the realms of humanity, either tangible or intangible.

Thus, basic mentality should be institutional and organized, while the enhanced mentality builds its activities on the basic mentality. The basic mentality is mostly founded on logic and intellect; the enhanced mentality takes its advanced mechanism from the capability of humanity to adapt, change and expand quickly in accordance with its balancing sensation between symmetry and asymmetry.

However, humanity in the light of being the fruit of mentality, in metaphor, could be illustrated to be a precipitation following the flow of time. Therefore, it is not probable for an outsider, if not prepared with a logically acceptable reason or a compulsory influence such as threat, to intervene in the inner process of mentality.

Persontology has a firm view on what a person is. A person is a relational dynamic entity which can be and should be analyzed and presented quantitatively, qualitatively and scientifically even if it may take a long time to research.

For such approaches, we need to build a concrete and unshakable foundation. It is suggested here to design an enfolding and unfolding mechanism as seen in molecular biology and proteomics. Certainly, it has to be an intangible mechanism applied to logics, universal grammar, language patterns, and informatics.

For instance, we need to investigate in detail the linguistic expressions of babies babbling and how the expression changes its structure from plain phonation to vocalization. We need to study how cultural omissions take place when different cultures mix together. It is the same with politics and economy. I believe these enfoldings and unfoldings have a lot to do with the characteristics of humanity. I am not so sure if such an "accordion effect" could be related to the hypothesis of holographic memory of David Bohm, but there seems to be a strong connection to the process of inter-subjectivity of an individual and communal person.

The Accordion effect is the result of flexibility hidden within continuously "transitional" dynamicity of expansion and contraction. Being transitional for dynamicity represents the idea that the whole entity is maintained in a flexible range, thus not disturbing the wholeness. In this view we will not cover multi personage and stay within the integratable range.

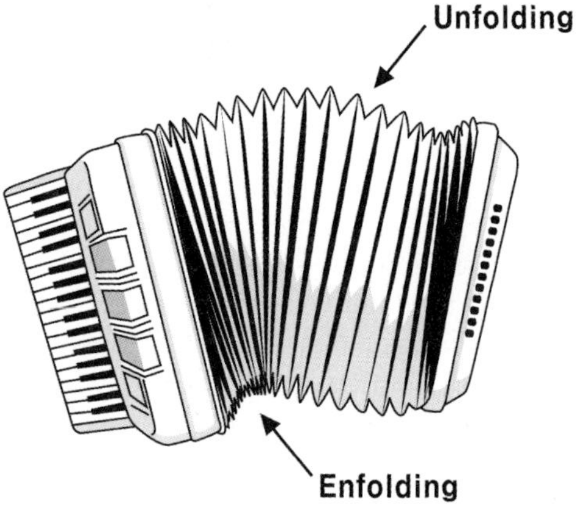

Figure 4-2-1. The Accordion Effect of Enfolding and Unfolding

The process of enfolding and unfolding occurs in accordance with the following description: enfolding happens along with Plexion-pragma-Axis Tempus of inter-intrusive direction of nature into human existence (Figure 4-2-1), and unfolding, along with the outward route from infinities-matrix to motor systems, then to action, transforming the environment of humanity. A genome produces a new life through an unfolding process. Unfolding is affected by environmental situation and regarded as the source of diversity in lives. To clarify the accordion effect of enfolding and unfolding is the essence of persontology.

In the last two decades, a number of articles on brain-neural science immersed in evolutionary perspective have been published. That research has achieved deep and groundbreaking work but has been relatively slow towards the urgent goal of providing a wider angle of methodology dealing with human history. The stance of persontology would be characterized by the advantage to open an integrated understanding of cognition and creativity.

The basic rationales of persontology are arranged as following:

First, persontology has a different basic frame from logical intellectualism. Emotional system of persontological frame executes high-magnitude computations due to its parallel connections. The parallel connectivity comes from the dynamic modularity, the content of which can be compressed by enfolding and can be decompressed by unfolding processes for inspection.

Second, the computation through the trial-and-error method is avoided, and the instant matching process of maps is executed for the confirmation of correspondence instead.

Third, the whole body is exposed to the illumination (equivalently, impulse, or set of wave equations) and thus synchronized input of information is achievable.

Fourth, human persontology implements intercommunication through the correspondence between the dynamic modularity of encountered individuals.

The correspondence of modularity originates from relaxed concentration of humanity (like the focusing of beams), which help to understand the plexal integrity of universal affairs in the ontological world.

Persontological Cosmology

A neuronist in 1898, Santiego Ramón y Cajal, intimated that the universe is a reflection of the human brain. It could be better described that the human brain is a reflection of the universe. Furthermore, a person is considered to be, as many lines of thoughts have presented, a small universe. However, this is not a scientific way of thinking. We have to conclude that metaphor cannot prove anything. Thus it is necessary for us to start from the viewpoint that persontology is not personal-ontology but person-ontology. A metaphor cannot be adopted in this discussion.

The universe is persontological. Chaotic character shows that it must be persontological. This may not be an intuitive line of thought, because it has only been from the latter half of the 20th century that humankind has developed a good understanding of ontology through the mathematical approach called the science of complexity. It is the author's opinion that humanity is a complex system (complex menesys) and that the science of complexity is the science of humanity. A human observes the cosmos to be chaotic because human observation is the source of chaotic interpretation.

In the preface of this book, an interpretation for Genesis 1:2 was introduced. In the interpretation was a pair of dizygotic twin concepts, *tohu wa bohu*: *tohu* meaning chaos, other side of the twin dizygote, *bohu,* meaning vacuity or nihil. This *bohu* can be interpreted to be, in modern physics, a building block of the universe, which has the quantum threshold value over which a photon is emitted. Vacuity is the resource, not resource material, of the universe. Some physicist like G. Schroeder regards *bohu* as building block of the universe.

It is considered that commentary on this verse in deep psychological interpretation coincides with that of modern physics. We are indifferent to whether or not people will acknowledge the presence of the One and Only Almighty from this verse. Our concern is this: personhood being introduced into science can be a new source of provocative understanding towards reviving and sustaining humanity in modern science and technology.

Chaos is a matter of order, and order is a matter of time. Chaos, or Chaotic Compressurement (*tohu*) is thought to be the infinite folding of "*totus nuntius*" (the total ontological information of the universe) due to time compression onto temporal nullity. Axis Tempus (the name of the time axis in individual humanity in this study) is a fantasy.

According to E Pöppel, "Sensing a continuous time is factually an illusion of the connection of successive states of consciousness, which generated in integrations of experiences of three seconds." (Zeit von der Urzeit zur Computerzeit, 2002, Klaus Mainzer) H. Bergson said that continuousness is our selves. Consciousness, either self-consciousness or temporal consciousness, is not an ontological substance and not to be regarded as a flow of experience which is impossible to analyze. Temporal consciousness is a macroscopic state that neural networks construct during the formulation of self-organized structure.

It is strongly suggested that the time axis could be mistreated as a fantasy, but it turns out to be real when we meet the time axis in our continuous, indented records of infinities-transfer. Infinities-transfer can be visually imagined as being recorded like an immense number of tight, dense rings surrounding the vacant time axis. It could be visualized just like a life-long funnel made of disconnected rings of experiences on the surface of a vacant circular cylinder. We call this object the "ITRing" (Infinities-transfer Ring).

G. Schroeder cited an important concept of Tsim-tsum from the Kabbalah. Tsim-tsum is in its essence nearly identical to the concept of Maister Eckhard regarding time to be the perseverance of God. Tsim-tsum is the Divine Self-contraction; for an individual integer, Divine loss of the state of actual infinite complexity is the emergence of a set of pseudo-infinite complexities of an integer and all the other integers.

Theoretically, in this context, the emergence of the epistemological view means the complete loss of the ontological view. However humanity must take both ideas into account simultaneously. It must be so because humanity has "Must-mist." (Alphabetical backward tracing of tsim-tsum) Although here is it expressed humorously, humanity must be mist or mystic.

Mixing in a mist is what humanity has chosen to do in its consciousness. Mixing up is the identity of humanity. Why do we mix? It is to select. All human conflicts come out of various selections. Our theories are the rationalizing process of a mixing mechanism. Meme of R. Dawkins is also the output of mixing. Mimesis of A. Toynbee is also a process of mixing; unlike the single wave propagation.

The universe is persontological since it was reserved by the contraction of the Wholeness for finite integers. Person is generated from the time of observation in the light of the infinity. The observer and actor as an integer is a person, who is equipped with a receivable container of matching correspondence with the infinite universe.

Personism

Personism is a perspective which deals with a person embedded in physicality. Thus through the thought may start entirely from a materialistic view, eventually a mental and spiritual prospect is established. The materialistic view of a person attaches too strongly onto the Che (體)-body factor, while the mental prospect of a person preponderates towards the Yong (用)-usage factor.

In a strict sense, personalism is concerned with real commune, a social body of living together, and with individual personality mainly related to the attitude of living together.

Within personism the concerns are focused on regarding the substantial individual person and relationship-based communality. Personism gives accent to human physicality as a commencing line of human rationales, while securing paths to the solution of mentality and spirituality in accordance with monistic dualism or monistic triadism.

Personism-Communalism Line of Thought

Personalism refers to the philosophical movement that stresses the value of a person and thus can be expressed as an idiosyncratic mode of behavior or expression. Communalism refers to a system of theory or government in which the state has a loose federation of self-governing communities, a system of loyalty to the interest of one's own ethnic group rather than to society as a whole, or the practice or advocacy of communal living or ownership. (Collins English Dictionary 21st century edition)

In the line of thought connecting personism and communalism, personism is the point of vacuity that introduces all the manifestation of humanity into reality and communalism is the pot in which the introduced contents are boiled. Therefore we can see that the personism-communalism line is able to effectively express the networking of ontological relations.

The extended continuum of Whitehead is the relational complex compound. The relation generated by an atomic individual in the continuum is of omni-direction and omni-penetration, as the universal continuum intrudes all the particular individuals. (Whitehead und Huayen-Metaphysik", 2004 UUP, Jin Kim) It is an extremely idealistic concept which disregards the articulative effect of real existence. We need to make an adjustment to the infinite extension of our thought in accordance with topos factors.

There is a good instance of the thinking method following the Huayen-metaphysik. In the author's view, the Huayen way of thinking comes from the infinite extension of topos submatrix. Shimiz Hiroshi has elaborated on the philosophy of locality, or the philosophy of topos, by thinking from the basic idea of bioholonics that describes nullity(空) as locality(場所) (Shimiz Hiroshi, Life and Locality, 1992, NTT Publishing)

The function of locality has a major significance in his philosophy. It is similar to the function of a gradient factor in the developmental neurophysiology used in brain science. Both of them define locality in particular relation to globality.

When information propagates in a hypercolumn within the cerebral cortex, energy is supplied to the major path. It is such a distribution of energy that decides the boundary of dynamic core. This boundary is realized by the difference in stress condition of brain cells, which is, in the author's point of view, governed by the inbalance between energy supply and information propagation. We propose that energy efficiency in supply and conservation depends on the neuroglial cell, as we can observe in the cause of the brain-efficiency of Einstein. We can imagine how a single person could recognize ones own local peculiarity in a communal society, as we can see in a single cell within a hyper column or in a dynamic module.

Table 4-2-1. Nomenclature of Personism-Communalism Line of Thought

Existing View	Remarks	New View	Remarks
Personalism	Personality is at the center of concern	Personism	Individual person is the subjective entity
Communism	Commune is the subjective entity in language structure	Communalism	Communality is the outcome of the interaction between individuals

It would be beneficial for readers to have discussions with one another regarding the terminology of personism as shown in Table 4-2-1. Personism maintains parity to personalism, as communalism is to communism. Person and personality are different to one another, as commune and communality are mutually different from one another. Person and commune describe reality, while personality and communality describe the quality and characteristics of reality. In this discussion it is not our aim to alter the existing names of existing philosophies but to try to express why this research intends to use unfamiliar terms such as personism and to express a combination how of philosophies can make a difference in human thinking.

Communalism

Communality is embedded in human physicality, the basic symbols and signals of which are largely derived from physical information. Human communality is under the governance of directives generated from a non-material phase space of abstraction, which has independently emerged and been managed by its own intrinsic attributes. Communal abstractive directives can be called "directives of expedient indication."

Since a Human being has intentionality, it can be suggested that transcendental intentionality is adequate enough to describe the mystery of humanity. Transcendence has abstractly been dealt with in width and depth by Aristotle, Kant, Hegel, and Sartre, philosophers of phenomenology, mystics and so on.

Even Buddhist thoughts have proposed a coincidental structure to support transcendence. The indication of transciousness found in Zen is expedient and has opened a route of alternatives by its emotional computation of digesting fluttering of artificially simplified signals of qualia. The directives of consciousness in decision making are affected by such expediency (方便).

Transcendental intentionality is meant to lead humanity to a well balanced, detailed correspondence between the map of human cognition and the map of the Plexion realm. Plexion maintains an expression of the universe in the form of an infinite dynamic plexus of infinite affairs. Thus we are meant to have an entity for a certain period of time during which the entity exists as an integrated integer. A Human can also exist as such an integeral entity. The communal entity, also known as community, is one of such an entity, even though the degree of integrity is relatively lower than that of individual humans. (The author has generally used community to express *civitas*, a body of people, not *urbs*, a geographically defined community.)

In the world of the plexus of accidental occurrences, tensorial forms covering the parameters of orientation, intensity, and frequency can show where the human "expedient indication" (which is the most convenient track) should be directed.

Information Barrier

An information barrier is formed when negative communality is built up within a community. Negative communality can be illustrated in a triangular interactive diagram. This interaction is shown in the lower diagram of Figure 4-2-2, with "deception" at the top edge and the other duality of negative communal reactions, "violence" and "decadence" presented at the lower two edges, which have been psychologically studied in depth to be known as sadism and masochism. It is one of the most curious parts of psychology in which their internal correlation is yet to be dug out.

Communality is defined to be the balancing between these interactive components. The imbalance of socio-psychological aspects is thought to have originated from the ignorance of personal boundary in the communal domain. Such ignorance of personal boundary stems from deceptive knowledge on one or both sides of the community.

Deceptive knowledge strains the truth and distorts facts and consequently changes the stable structure of cognition and knowledge into twisted forms psychologically. The psychological twisting of human intellect based on the cognitive process and knowledge production is largely attributable to the emotional process and in turn to the volitional process which comes thereafter.

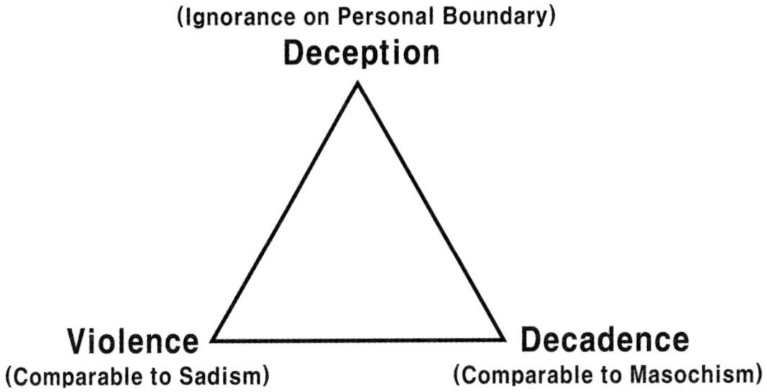

Figure 4-2-2. Negative Communality illustrated in a Triangular Interactive Diagram

The above diagram, Figure 4-2-2, presents deception to be the cause of violence and decadence. Violence is the most leading attitude of looking down upon and threatening the position of the opponent. In fact, violence is actually a form of sadism which violates another's mete and invades a neighbor's territory in communal living. Violence is realized in social relations in various intensities.

Contrarily, decadence is an apparent form of masochism which allows the violating landmark of another onto one's own territory without proper resistance. Decadence is also realized in social relations in various intensities, varying from excessive luxury to lewdness.

In modern society it is legally allowed for people to commit acts of adultery in most western countries. It is not the purpose of this research to clarify the irrationality of the present legislative system, but to show the structure of communal arrangement of human mentality as it is originally embedded in corporeal physique. However it is generally recognized that human living in modern society must object to violence.

Communalism and Multiple Personas

Communalism has a concrete association with the phenomenon of multiple personas. Since humanity established distinctive cultural deviations in differentiated environments of geographical and societal variation, the individual person has been changed in its internal mental conditions to possess tendencies of multiple personages in reaction to their social life.

This trend has been intensified due to the increase of societal changes which forced them to develop disintegrating tendencies of their personality in order to adapt to various environments of everyday life, home, workplace, social life, religion, and others. We are going to touch on the topic of multiple personas in the next section.

4-3. Multiple Personas

Multiple Personas
Multi-Personality, Multi-Personhood, and Multi-Personage
Multi-personage
Parallel Arrangement of Multi-personage
Multi-Attractor and Apparent Person

Person is not defined clear in science and even within arts yet. The etymological study of person traces its origins from "*persona*" in Latin, "*persu*" in Etrurian, and "*prosopon*" in Greek. *Persu* and *prosopon* both means mask, and sometimes face, which is similar to "*panim*" in Hebrew. Boetius was the first theologian who made the first ontological definition; "*persona est rationalis naturae individua substantia,*" which means that a person is an individual substance of a rational nature.

It is noted in this study that person is a composition of networks and systems. A human reacts to internal and external stresses by bypassing and making sense of internal resistance using whatever method available. A human sometimes successfully avoids and absorbs pain and shock.

What is strange, however, is that a person usually reveals a bipolar phenomenon within its structure. This may be due to the balancing mechanism within humanity. In brain science, a human is sometimes defined as a self-catalysis in the self-organization process.

Multiple Personas

In modern society, the problem of multiple personas has become significantly important. The social changes of the last two centuries have had a deep impact upon the nature of a person. As explained earlier, persona was oriented from the meaning of mask in theatrical performances.

The thoughts on persona have long and profound traditions within western Christian culture. The Divine Trinity has been the most outstanding subject of theology and philosophy for the past two millennia.

However topics on persona presently open a new field of treatment on humanity in accordance with new scientific findings on human beings. Thus the adoption of the new term "personage" would be appropriate for practical purposes. It is implicit that personage deals with defining the role and function of an individual.

Despite social diversification and segmentation being deeply advanced, the applicability these phenomena to individuals happens to be restricted to only a few that society or the community will permit. The human behavior of living in distinctively segmented societies shows that there are several distinctive personal roles for respective sub-societies. Multiple personas are the result of the by-passing ability of humanity against a stressful circumference within a complex society.

Multi-Personality, Multi-Personhood, Multi-Personage

Multiple equivocations will be the theoretical tool to express multiple personages in concerned parts of this research, promoting a good understanding as to the reason why a multiple number of organs and neurons mutually interact in relation with one another.

It is the author's opinion that the phenomenon of person can be treated in topology as an expression of manifolds. The connectivity between the regions of various manifolds provides the major causes and conditions of multiple personality which has a variety of hierarchical levels along with different intensities of manifold existence. Multiple personality is developed while being related to how human experiences are able to formulate their psychological phase space in accordance with the developmental process and the intensities of psychological affections of peak experiences, from the view point of depth developmental psychology.

It is thought that the phenomenon of multi-personhood appears because of the topomorphological complexity caused by the partially dismantled structure and fluctuating local density of the phase space of human mentality, which is usually due to the increase of social complexity and excessive stress overloaded within a person. It is necessary to practically deal with the issue of multi-personage in an individual utilizing scientific solutions. Since Multi-personage could also be interpreted as a religious problem, the author has no objection in treating this issue from either a religious or scientific viewpoint. However, the multi-person is recommended to be interpreted using psychology, and sociology based on the latest research on brain science.

Current economy is not easy to make an integrated diagnosis, so our only remedy is the contingent method, which we will call the modified method. In the last sub-prime crisis, we found that the lack of philosophy to diagnose the economy was the major reason that invited worldwide panic and global stir.

We believe that the viewpoint of a continuum economy can be adopted with which we can face up to the economic reality in light of the phenomenon of multi-personage. It is our intention to create a political economy generated through the combination of logic of multiple manifestations and the theory of complex dynamic modularity.

Multi-Personage

Multiple personages will be one of the key words for the social science of forthcoming generations. The integration of personage has been threatened and weakened due to the propagation of industrial developments, which have induced schizophrenic divisions of personhood.

Present cultural treatments and medical sciences do not present effective solutions to reintegrate personage problems. These multi-personage problems are in fact not solely of individuality but also of communality relation.

It may be an appropriate solution to blaze a new trail in the realm of living altogether with the multi-personage problem. Virtual reality or vegetable reality is where most modern people are caught up with.

These two VRs are major sources of post modern multi-personage problems.

However it is noteworthy that various setups of institutional arrangement of past modern society have provoked schizophrenic impacts on persons as well.

Since social sciences will be reinterpreted here based on facts found through psychological and neurological technologies, the focus here will be "technologies". However, the resulting theoretical development of economics, politics, sociology, humanities and etc. will be unavoidable within the overall discussion on the multi-personage problem

Parallel Arrangement of Multi-personage

Multi-personage raises the problem of monopoly or partial monopolies to the consciousness of individuals. (Sub-consciousness of individual will also be dealt with to a certain extent later on in our discussion.)

Monopoly causes troubles to the individual person largely due to the effects of series arrangement. Human consciousness is mostly apt to monopolize the imposed situation at one time.

When consciousness in strong concentration does not take advantage of the profound connectivity with overall sub-consciousness of the concerned person, personage plays the role of a mask in order to adjust to the external environment.

Therefore it is desired that natural ways of parallel arrangement for multi-personage be found to mutually collaborate the personages of a person. Such parallel arrangement of multi-personage is the most important subject which will be discussed next.

Multi-Attractor and Apparent Person

The apparent person we meet in everyday life is a well balanced compound of several branches of tendency within their personality. The branches are called personages because they do not represent integrated personhood but a part of the holographic presentation of the apparent person. Each personage is thought to possess a specific region of existence in the phase space of mentality, a particular attractor for the personage and representative orbit of living traces. Living traces are matter of meaning in a personal existence.

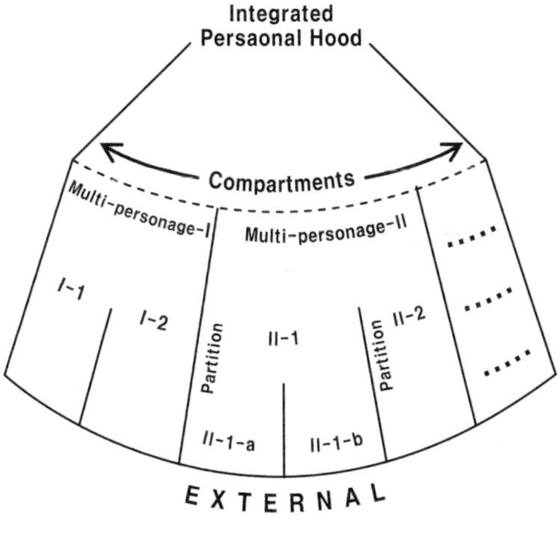

Figure 4-3-1. Multi-personage and Integration of Personhood

W. Freeman cited Merleau-Ponty that we are moved to action by disequilibrium between the self and the world. This disequilibrium is an endogenous instability that puts the brain onto an itinerant trajectory, i.e. a pathway through a chain of preferred states, which are learned basins of attraction. (How brains make up their minds, p121) The trajectory of meaning is a learned result and is thus isolated and stable to a certain extent. These concepts are shown in Figures 4-3-1 and 4-3-2. We can see a protruded attractor in one of the trajectories. The extended concept to economic continuum is shown in Figure 4-3-3.

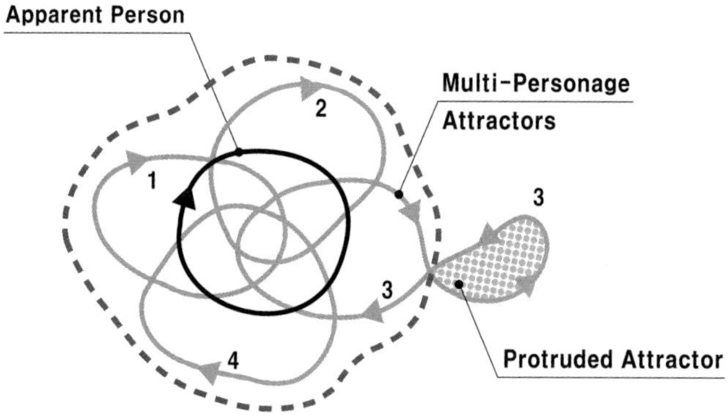

Figure 4-3-2. Multi-trajectory of Multi-per`sonage and Apparent Person

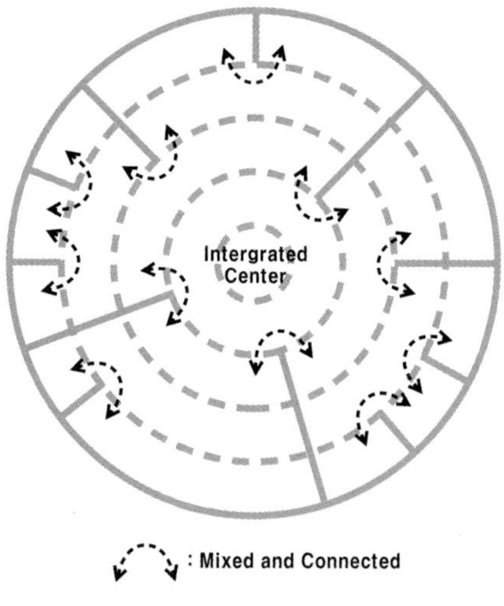

Figure 4-3-3. Continuum Economy and Connectivity of Multi-Personage

In the past, sociality surrounded us to protect our individualities. In the new, current era, a civilization is influenced by protruded attractors or imaginary parts of a person, such as virtual reality. There is an impending era in which we will have to regard omission and integration as a necessity. To sustain humanity, we need to find proper strategies.

Table 4-3-1. The Characteristic Change of Economy in the Last Two Centuries

Multiple Economy	Mixed Economy	Continuum Economy
Industrialization	Connection started in Internet and Virtual Reality	Continuum grows in Psychological aspects, Legal and institutional aspects Economical aspects, And finally in Political aspects

The continuum grows gradually from psychological aspects, legal and institutional aspects, economic aspects and finally to political aspects of power structure. Multiple economies are characterized by multi-centric economies in the global environment due to the bifurcation of economic thoughts and attitudes of global humanity. Industrialization has provoked this.

We are now passing through a mixed economy, which is strengthened by the Internet and Virtual Reality, and entering into the next step of X economy yet to be defined.

4-4. Dynamic Modularity (動變分格性)

The concepts of personhood and persontology have connected humanity and the universe. The material universe called out for the phenomenon of life to appear and life invited mentality to the frontal surface of the world. Life has transcended from a material existence.

The vitalized activity of dynamic modules within the human brain is thought to be the cause of the superposition and transcent: these are the essential mechanisms that separate humanity from other animals.

Dynamic Modularity: Focusing on Brain Elasticity

Modularity is the quest of humanity in post modern society. In fact, modularity has been the object of an endless search by the human race. But it is worthy to note that the concept of module does not provide a good enough solution to our questions from the molecular level, organic level and up to the super-organism or communal level. Modules exist in dynamic pattern related with nonlinearity. The dynamicity of modules originates from the nonlinear change of their internal and external conditions in accordance with the changes of their environment.

If there is a separate and independent entity in continuously varying environmental conditions that affect the entity by the non-linear interflow of energy and/or material, it means the entity is a dynamic module. For an entity, being independent means it has a systematic dynamic core that provides specific patterns of behavior. Thus, we can say there is a dynamic modularity in universal phenomena except in isolated or locally linear conditions.

It is essentially the same with the rationale of the science of complexity.

In essence, modules in nature actually do not exist. Module is a means for humans to understand nature in a spatiotemporal manner. Besides that, humanity is the cause of the modular presentation of nature, or if acceptable, of eternity. As far as humanity is concerned, modularization has special modes originating from the interrelation between the subjective entity and its pragma (the part of reality that is in contact with the entity, defined in next Section 4-5). The main theme of the symphony of dynamic modularity is the sequential development in the decoding of the cryptograph of nature by humanity. But there are variations on the theme due to environmental changes. And although very rare, some variations are found in humanity caused by internal influence on itself.

Human consciousness is the source for rationale, and thus the product of symmetry in the universe. Rationale is the output of the systematic frame of Language. Language written in the capital letter "L" is universal grammar and originates from the infinities-matrix of individual humanity. Universal grammar is thought to be engraved in the sequential presentation on the surface of the human Axis Tempus. As explained before, Axis Tempus is a vacant path to the innermost humanity. Axis Tempus is the cause of the existence of human consciousness. Consciousness is in most situations almost symmetrical due to rational character.

The offset part of asymmetry comes from transciousness. The effect of transciousness is connected to reality through its influence on consciousness, which is responsible for interpreting the message for the survival of the concerned individual. But the asymmetry of transciousness has the echoing response from its boundary of experiences. The boundary is the result of the pragma, or "percept" for all directly experienced objects as described in neurophysiology.

The modularity that really exists in space-time continuum is "dynamic" modularity. The composition of constituents of dynamic modularity can always vary. The composition is globally distributed within whole brain and body. So it contains all the domains of sensciousness, consciousness and transciousness. Physicists comment on the beauty of symmetry. But the real beauty realized by humans and humanity is the composition of symmetry and asymmetry, in most cases a little bit of deviation from symmetrical condition. It is an offset-beauty.

Anyhow, the human has symmetry in the body with a few offset organs like the heart for arrangement purposes, and serious asymmetry in the forward direction for action and movement. Dynamic modularity of human cognition has the resemblance of such patterns. I believe it is not simply a metaphor, but has a deeper association yet to be unveiled.

In my opinion, "gestalt grouping" should be illustrated by dynamic modularity. Hermann von Helmholtz mentioned that the grouping is an unconscious process based on abstraction. The difference in my thinking is that the grouping is a ceaselessly changing process of transcious processes based on the brain elasticity that is gained by the infinities-matrix of a specific person. The brain elasticity implies the presence of an elastic core, which is always subject to change but in fact remains stable in the dynamic module.

Dynamic modularity most likely has three hierarchical levels of modules at least: basic single fluttering level, combinatorial level of multi-fluttering, and episodic combinatorial expressions.

It is suggested to consider the source of social development as human communality. Human communality is governed by the quality, intensity, and character of mutual contact among individuals. Communality is thought to be mostly influenced by dynamic modularity which dominantly appears in human brain activities inducing brain elasticity.

The elasticity of dynamic neuronal modules can be understood by the durable activation of sensory neurons in charge of steadfast images when sparks of retinal neurons quit due to the discontinuity of information flow. Based on my personal experiences, elasticity can be well experienced by common people when they vividly imagine listening continuously to entire melodies they are fond of.

This is also true with the recollection of paintings in detail. Dynamic modularity is not yet well-defined nor clarified thoroughly enough, as I know of, in the present stage of brain science. However the following comments are made by the author in the hopes of a consistent logical arrangement, for the practical application to serve as the building block of logic within this book.

The Basic Theory of Dynamic Modularity

The 'Free Energy Principle' of Karl Friston has been reviewed by researchers in neurophysiology. Free energy means the prediction error of the brain, which is the difference between the Bayesian prediction of the brain and actual sensation. Free energy is thus said to be roughly equivalent to the difference between the total energy in the system and its "useless energy", or entropy.

I believe entropy is decreased by accumulating empirical data and is directed to enhance the synergy of prediction by formulating a dynamic mechanism in combining the fluttering of images and in generating the optimum construct of the gestalt combination.

We know that gestalt is not generated in identical patterns in each individual in regards to a designated object under exact conditions. So, the optimization process of presenting so-called gestalt varies greatly and has deep relations with energy management of the brain. The optimization process is governed by the mechanism of dynamic modularity.

It is worth observing here that the human society behaves like a neural society in principle.

1) Varying human modularity has to do with neural modularity, which is basically dynamic.
2) Brain modularity is the output of the relevance between the brain and the environment, both varying in dynamic conditions.
3) Human neural modularity consists of individual modularity and communal modularity. They have different principles from each other. Individual modularity is divided into physical and abstractive modularity. Individual abstraction is formulated by the function of post-efference which governs imaginary and symbolic procedures.
4) Communal modularity is apt to change more easily compared to individual modularity. But individual modularity in transciousness varies in greater range than do individual sensciousness and consciousness.
5) Individual modularity based on brain modularity shows an embedded articulation due to physicality. The articulation of the human individual modularity is governed by the patterns of human physical conditions.
6) Brain modularity is developed based on the Synchronized Fire-Wire principle (Fire together, then wire together).
7) Synchronized Fire-Wire principle generates contemporary modules in response to Effective Environmental Impacts.
8) Effective environmental impacts consist of a number of information streams to which

an individual as an integer can currently be susceptive. These information streams have to do with Plexus of Ambient Affairs. Effective environmental impacts are called "Pragma" in this book and are usually equivalent to Plexus of Ambient Affairs.

9) Brain modularity induces the modules of transciousness, consciousness and sensciousness. The induction process has to do with the interactions between brain modularity and emotional modularity. Modules of transciousness, consciousness and sensciousness are not separable from each other and thought to be in a concentric existence. A physically embedded module exists at the central part of the modules under consciousness. Consciousness is in absolute governance of the cognitive mechanism because it is in charge of the balancing mechanism of humanity. Modules under consciousness are positioned at the central part of the modules of unconsciousness. The only difference between the consciousness modules and sensciousness modules is whether they are at the summit (currently revealed) parts of human understanding in the environment usually for rational decision making. Rationality does not necessarily mean intellectuality. The rationality is the measure of consistency of a human life or equivalently human autobiographic memory. The maintenance of consistent rationality among the near-infinite flux of received information is known to be usually tested and well arranged during sleep.

10) Emotional modularity is the basis of the formulation of human modularity. Emod (the compound word for emotional modularity) needs to be mapped in its essential functions and interactions with other modularity, for instance, intellectually intensive modularity and comparatively stable physical modularity, etc. Emod is always variable within a certain stable range to be able to cope with changing impacts from Plexus of Ambient Affairs.

11) Consciousness is like a tightened bowstring that makes a whining sound when struck – a state of tension (or emotionally amplified stress). Deeper concentration than usual to a specific symbol anchored at certain physical experience is necessary to maintain the state of consciousness for a specific item or episode that is connected to the symbol. This concentration provides a source of energy and is generated from emotional modularity.

12) Elasticity is defined by Synchronized Fire-Wire relations. The Synchronized Fire-Wire relation is the source of hardwiring. Hardwired parts of the brain and neural nexus whimpers or vibrate as a high wind whistles through the air. Synchronized Fire-Wire relation based on material physicality provides ever-whining steadfast foundation to hold modular concreteness. On the contrary, plasticity gives greater room for the changes of modules with weaker joint strength. Thus, elasticity, which I believe is a newly defined terminology in this book among brain related research, is more strongly embedded in physical conditions with greater self-recognition. (It is understood that elasticity has strong tie with composite feelings of the entire bowel.) Plasticity allows greater variability in module formulation and dissolution in the nerve system, while elasticity permits lesser variability in module dismantling.

The Network and System (Sustainability and Tolerability) of Humanity

Nexusity and systemicity are two tendencies of human being. Nexusity has been developed for the sustainability of humanity. Systemicity is necessary for the tolerability of humanity. In dealing with nexusity, the internal ontological condition of the network is considered to gain greater significance (Albert-László Barabási, 24 July 2009, Science). In referring to systemicity, we have an extended notion of system to the system of systems to take care of the increase in greater complexity as society develops.

We are at the doorsteps of entering a new territory that allows an integrated understanding of the system and network issues all together in a single region. It comes from the balanced understanding of humanity possessing a dual scheme of systemicity and nexusity.

To take care of the above mentioned extending ideas, we need to introduce a concept of dynamic modularity that contains both mechanisms of sustainability and tolerability together. It is thought that sustainability and tolerability are observed in both cases of systems and networks. If we adopt the idea of folding and unfolding in proteins and so on, we can imagine a system as a folded network and a network as an unfolded system.

Connectivity is the result of an unfolding process and discontinuity due to an enfolding process between entities caused by the systemizing of an entity.

Factors that Affect Human Modularity

Modularity is generated by nexusity and systemicity. Extrinsic aspects of nexusity and systemicity affect intrinsic aspects of nexusity and systemicity, and *vice versa.* The mental status, for instance, the strictness and self-indulgence in life style influence the physical status of a person to become stubborn and rigid usually.

Systemicity is the major cause to formulate the core part of dynamic modules. Nexusity is the major cause that maintains the peripheral part of dynamic modules. The core and periphery of dynamic modules generally formulate the transition zone in between. The transition zone could be widely spread or shortly contracted depending on the combined effect of systemicity and nexusity.

In the dynamic modules of human being, transition zone is the most dominant, significant and anthropomorphic part. Living as a human is the alternating process of decision to choose systemicity and nexusity. The transition zone between core and periphery is stretched due to the wide variety between system and network, between contraction and dispersion, and between concentration and relaxation.

Human modularity has a deep relation with the emergence of multiple personages in our times. Modularity provides the means to multiple personages. As discussed previously in the above sections, all people, without exception in this era, have multiple "personages." The tendency towards the formulation of multiple personages comes from completely different sets of living circumstances encountered by an individual in one's life.

Modular dynamicity is affected by modular structure. The division of core – transition zone – periphery makes significant difference in modular dynamicity. It is deemed attributable to the different level of energy consumption within the three parts. When a human entity, either individual or communal, concentrates on the maintenance of dynamic core, it has to keep the energy level very high compared with the network condition of the periphery because the consistency of keeping high energy levels is the essence of a dynamic core in the sense that networks do not require repair and maintenance. The intrinsic property of the core part demands a lot of energy to sustain the logical circuits seen similarly within electronic devices.

It is suggested by the author that the energy supply in a network is maintained by a flickering mechanism. When there exists a sufficient level of nexusity within a network, lumps of energy travel through the nexal circuit of the network, merging and splitting continuously, exhibiting flickering mechanism here and there.

Information transference (and/or energy transference) in human dynamic modularity is deterred by the irregularity of nexal connectivity within the local segment of network. When a network has no deterrence in its territory, information or energy flow receives an even distribution of intensity throughout all segments. When the flow capacity of information or energy within the network reaches maximum capability, it should be called the state of ideal systemized network (=netsys), which is the best ideal pattern of a network possessing the best nexusity and the best intrinsic systemicity as well.

Practical Application of Dynamic Modularity to Social Development

We have two aspects of dynamic modularity in practical application to social development. One is the plane division of dynamic modularity and another is the organizing principle of hierarchical dynamic modularity.

The first part is related to the cruxological development of human society. The plane division of dynamic modularity in human cognition is characterized by the emergence of comparatively wide scale and profound content development of the transition zone between the dynamic core and dynamic periphery. It is illustrated in Figure 4-4-1.

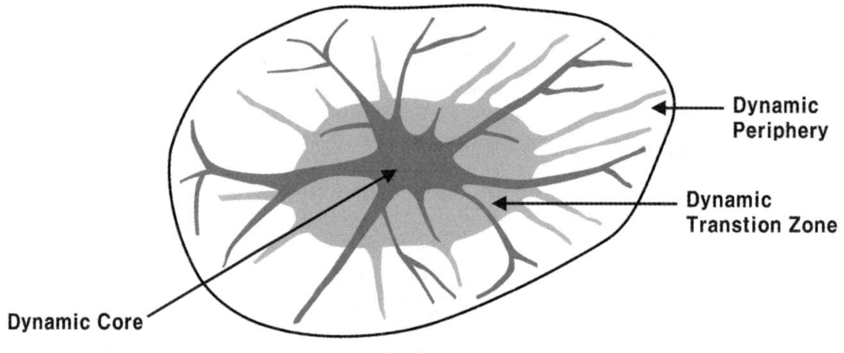

Plane Division of Dynamic Modularity

Figure 4-4-1. Plane Division of Dynamic Modules into Three Parts

The second part is related to intrinsic characteristic of a society. Dynamic modularity in humanity has established hierarchical organization through the experiences of innumerable social interactions of human history as shown in Figure 4-4-2. Sequential dynamic influence of lower level modules to higher level modules is proposed in Figure 4-4-3.

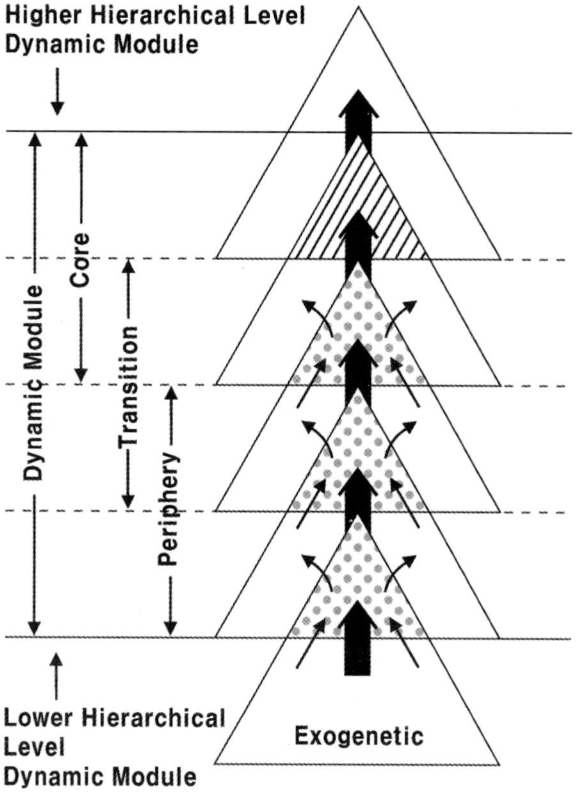

Figure 4-4-2. Hierarchical Organizing of Dynamic Modularity

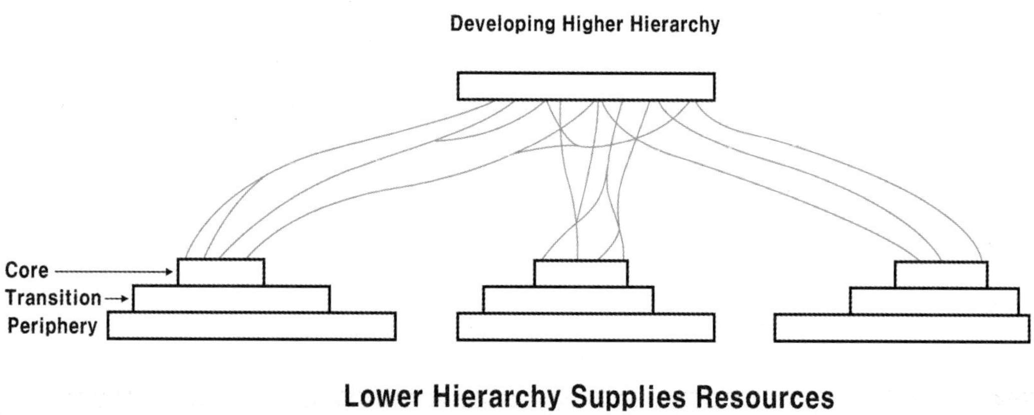

Figure 4-4-3. Sequential Influence of Lower Level Modules to Higher Hierarchical Level Dynamic Module

We can see in Figure 4-4-3 that present status of globalization should be enhanced by developing higher hierarchy of global dynamic modules.

4-5. Ruminating Mechanism of Anthropomorphosis

Anthropomorphism
Quadruple Annulus of Cognition
Inter-Intrusion
The Effect of Rumination in the State of Inter-Intrusion

Anthropomorphosis is a backbone mechanism that sustains humanity in its orientation of morphogenesis. The imagination process of humanity appearing in every aspect of life is somewhat similar to the ruminating mechanism of cows. The ruminating mechanism of anthropomorphosis is not as simple as ruminants, because human beings possess the complex effect of inter-intrusion within his or her environment of cognitive activity.

Anthropomorphism

Anthropomorphosis is in some sense the process of formation of humanity. It can be observed all around us within surrounding human creations. This includes the manufacture of furniture, buildings, and the establishment of cultures. All of these are the balanced symmetrical emergence of a cryptograph hidden in humanity.

Outputs of humanity interact with internal structures of humanity and interactions are the coding and decoding process of communication. Even in strict materialist view, some earlier writings of K. Marx intimated that the products of communities are the manifestations of the characteristics of those communities.

The definition of anthropomorphosis was originally implied within the Old Testament (Torah) found in Genesis with its ancient meaning of human resemblance to the Divinity. *Imago Dei* and *Similitudo* were used in Latin version and Christian theologians thought that they are the description of God's secret in His craftsmanship in regard to how He created human beings.

Most philosophers would not agree with this view, because they think anthropomorphism should be defined in a reverse way, i.e. gods' resemblance to human. Human generated images of God in their resemblance. The reason of the Christian belief in "human resemblance to the Divinity" is found in the key point of this concept and is founded upon how an *anthropo* (human) could enjoy the association with the infinite Divinity.

In a general sense, the infinite Divinity or simply infinity would provide a way to transcend the future of humanity in a theoretical structure. It is recommended to the readers that one not try to consider any specific religion with respect to "Divinity." Humanity is the matter of infinity, which is the essence of an ancient Divinity.

(To explain the aforementioned again, the image and similitude of God was originally the directive of human design. So, human beings are supposed to be a reflection of Divine character in their internality and appearance.)

Anthropo and *morphosis* (transformation) construct a concrete manner of intercourse and mutual communication between God and human. Recently it is generally being used to denote "the transformation into human" form discarding the other possibility of interpretation concerned with the bothersome idea of transformation from Divinity to human centric lines of thoughts.)

However, there is a point of view we have to revive in the usage of this vocabulary. The Divinity represents the wholeness. At least we can say the concept of the Divinity has been dedicated to reinforcing the concentration of human thought into the wholeness. Without the idea of wholeness, humanity would have been dismantled and consequently dissolved into bottomless nihilism. The idea of wholeness is well represented by ontology, which is the centripetal force of human thoughts.

It is my opinion the ideas of ontology have stemmed from several different roots of anthropological background. They are Indo-Aryan, Semitic and various sources of Shamanism. Indo-Aryan ontology is deemed to have dehumanized the essence of ideas in the course of theorization. We can find such traces in the philosophies of Buddhism and Greeks. Semites do not seem to have exerted much effort to construct an ideology until they were affected by Hellenism. It is my personal thinking that Semites' beliefs appear to have maintained a personal relationship with the Almighty through a human personality and they must have kept communication under relatively harsh conditions within life in Middle East.

Maybe, it is too simple an opinion on the subject of such great complexity, but in the days of mythology before they became mytholized, Greek myths and Vedanta myths also maintained human factors until the days of intellect had started. However, there is another lively query concerning anthropomorphism in a modern sense.

Joachim Bauer wrote a book entitled the *Memory of Body* to clarify the fact that mostly in the domain of human relations are those factors that influence the regulation of genes and affect our health. Genes do not live alone in an autistic manner. Genes and environment should be regarded as a unitary condition for our survival. The function of genes are considered in two ways; one is the usually well recognized genetic text of DNA sequence and another is concerned with the regulation of genetic activities. It means that the experiences of an individual organism will influence the formulating of patterns of responces and signals that affect genetic regulations. In other words, it is experimentally proven that specific patterns of genetic response are produced through specific human experiences. (Joachim Bauer, 2002, Das Gedächtnis des Körpers, preface)

Bauer's idea coincide with the assertion of Jantsch. Erich Jantsch commented in 1979 that we as individuals live in coevolution with ourselves, that is coevolution with the mental product of ourselves. It is an alternative expression of anthropomorphosis. Modern anthropomorphism is wholistic humanism.

Quadruple Annulus of Cognition

It is an extension of discussion from Section 3-1 about human cognitive model. From an ontological viewpoint, it is noteworthy that the wholeness of existence, if any be comparable to the entirety, has been directing the transformation of humanity. Here in Figure 4-5-1 showing the ontological mechanism between nature and human. Embryological process of humanity into nature, which can be understood through interpenetrations (or, inter-intrusion) into one another, establishes contact spheres between nature and body and again between body and brain. This idea proposes the supervenience of human cognition to brain-physiological physicalism at the base, but that the strong survival intentionality of humans in inter-intrusive structure sprouted out mental and spiritual aspects of humanity. It is a completely new phase of development, which we can label the emergence of humanity. Humanity is mental and spiritual although it is not yet so clear how to define the mentality and spirituality in comparison with what we have vaguely meant until now. In the author's viewpoint, mentality and spirituality were caused by the encounter of humans with infinities, which was most likely provoked by the continuance of the time axis into what is generally called 'eternity.'

Interaction due to inter-intrusion between the nature and cognition

Figure 4-5-1. Diagram of the Basis of Humanity: Constructive Illustration of Inter-intrusion

In Figure 4-5-1, the nomenclatures of four rings of human basics (appeared after the inter-intrusion between human integer and the nature) are illustrated. Table 4-5-1 presents the domains generated through inter-intrusion.

Table 4-5-1. Domains of Humanity Generated by Inter-Intrusion between Nature and Human

Domain	*Affiliation*	*Remarks*
Plexion	Nature	Equivalent to the universe excluding human cognition
Corpus	Human	Human body receiving all universal signals
Pragma	Nature	Extent to all objects of human cognition
Cerebrum	Human	Human brain
Axis Tempus	Nature	Time axis omnipresent in the universe and in human body

If we put our understandings together, a line of orderly thinking can probably be given as follows; <Plexion> - Axis Tempus- brain- neuron- postefference- elasticity- limbic systems of veins and networks (neural, hormone, acupunctural meridian and etc.) - plasticity-preafference - sensor- keratinosite(skin)- pragma - <Plexion>.

It is suggested that the region of skin and brain-neural system began division just around the oral region of phonation. It is believed that the subconscious feeling of ontological wholeness has to do with the opening of the placenta and umbilical cord, our bellybutton, while the human epistemological reality has connection to the path from anus to oral cavity.

Table 4-5-2. Quadruple Interfacial Annulus between Inter-intruded Domains (Annuluses are cross sections of concerned spherical views.)

Name of Annulus	*Remarks*	*Time Scale*
Plexion-Corpus Annulus		10^{-20}sec quantum process
Corpus-Pragma Annulus		10^{-3}sec neural process
Pragma-Cerebrum Annulus		10^{-1}sec preafferential process
Animotus Annulus	(within Cerebrum) Infinities-Transfer Ring; ITRing surrounds Axis Tempus	Real life time scale in formulating Infinities-Matrix by freezing Animotus Flare, ITRing $\geq$ Infinite-Matrix In number
Axis Tempus(vacant)	Void funnel axis	No time scale (Eternity), blank axis into which surplus dimensions are implied, (Surplus dimensions are the intrinsic attributes of Axis Tempus)

It is worth thinking that the above surplus dimensions are the cause of the unbalance in quantity between muons and negative muons in their emergence. Surplus dimensions are the cause of the generations of gravitational waves. Surplus dimensions and gravitational waves are variable by locality within the universe. Gravitational differences bring complete difference in the space-time continuum of respectively different localities. The locality in the universe does not mean the spatio-temporal locality but hyper-topological locality. Hyper-topology is concerned with the decay or collapse of dimensional condition within the universe.

Time dilation means that time must be flowing more slowly in a moving frame of reference. It is thought that Axis Tempus of a person might originate from the time dilation when hyper-topological locality makes infinitesimal difference in generating a niche that a cognitive entity can intervene in.

Relation between Plexion and pragma appears to be related to the identity of noise. Noise is a cognitive pattern in brain of complex un-interpretable signal. Discordances due to resistant humanity, both individual and communal, generate noise patterns without order that come from a local rationale in the pragma which resists the global rationale of the Plexion. Resistance is a kind of stress, the cause of node formulation in the menesys of humanity. The resistance and stress stems from the unexpectedness. But natural noises make trends of 1/f, as the pragma constitutes a plexus of 1/f pink noise. Individuality is embedded within the human physique. Such embedment is the cause of anthropomorphosis.

The issue of noise becomes significant if we want to understand the extreme condition of human cognition. In Chapter 5 noise phenomenon will be discussed in connection with the science of integrism. In this section, the indiscernible complete noise, which is a manifestation of the Plexion, cannot be interpreted by the pattern of our past life episodes. This means that humanity needs to invite futuristic interpretation.

Such futuristic interpretation includes the integrated interpretations on a number of different episodic entities, the reverse interpretations from the future to past, and/or the combined strategy for interpretation of the integrated and the reversal. Transcent of humanity is actually a kind of escape from all known interpretations paving the way to the third kind of creativity. (Refer to Chapter 8)

It appears correct that the spatiotemporal continuum is not a static matter but is rather a dynamic affair, and this can be expressed as the set of relations of affairs. Furthermore, we can say that we are not tangible beings but better defined as not more than dynamic affairs. In this view, the Plexion is the entire set of the universal affairs. We can see why humans as a dynamic plexus of affairs have to ruminate in a vein of the un-interpretable signal from the Plexion under the living condition of pragma.

The phenomenon of humanity has both feedback and feed-forward effects. The feedback effect is found within the intrinsic attributes of humanity changing as time flows. The feed-forward effect is reversely performed as if the intrinsic attributes of humanity decides the contents of anthropomorphosis, which is formulated in the past.

We proposed that humans have three cognition levels – sensciousness, consciousness and transciousness. Transciousness is most heavily related to dynamic modularity of humanity.

The basic cognition and intention of sensciousness is generally comprised with bowel feeling; in sensciousness, it is likely that the highest consumption of physical energy is used in receiving and mobilizing one unit of information, which is yet to be defined by experiment.

In consciousness, the human uses language and linguistic tools; the level of physical energy consumption looks intermediate to process equivalent amount of information

In transciousness, it is likely that the rhythmical fluttering movement of flaring fire in the "Anima annulus" region is full of imagination; qualia, sensational symbols used to appear in dreams, illusions and visions; projection from the modules of complex synaptic connections.

The pattern of the least consumption in terms of unitary physical energy occurs in a standard leaflet fluttering with an emotional constituent (not yet well defined quantitatively) in transciousness in comparison with the other sensciousness. Thus, transciousness enables to maintain high efficiency of energy consumption in performing the extraordinarily vitalized dynamics and complex structure of transciousness,

"Animotus" is named after the combination of animi motus (anima + motus), which means "ruah in humanity" (ghi, 氣) plus move. Ghi is related to topos submatrix. Topos submatrix has to do with Animotus annulus because topos submatrix makes composition of symbolic fluttering in patterns of spatial occupation.

In Neo-Confucianism, Four Inceptions and Seven Sensations (四端七情) is one of major issues. The debate was shortly called Four-Seven Debate. Simply speaking, the Four is the pure essence of human emotion equivalent to Animotus and the Seven is the practical emotion distorted by the need of an individual.

Animotus is not bifurcated emotional energy but is basically equivalent to the Four Inceptions, which is the self-organized emotional reflection mostly concentrated within the Animotus annulus. The Seven Sensations are the integrated projection of strange attractors induced by a number of need-occurrences occurring within the communal region of humanity and named consequently as "commotio" (excitement or commotion, similar to Seven Sensations originally connected to the self-centered individual bowel feeling if we want to march this idea into the Freudian theory.)

Most activities associated with creativeness are performed in transciousness that governs the creative activities at the center. The activation of creativity was so highly energy consuming that the human brain had to introduce a new way of thinking which initially utilized temporarily simple symbols, images and combinations of the least number of neural signals that represented the entangled association of related episodic memories. Later it seems that symbols and images have been upgraded to complicated and complex forms of dreams and illusions, etc.

Inter-Intrusion(相互浸入)

Inter-intrusion used within this book is expressed in a differentiated manner from the inter-penetration concept of Huayuen Metaphysik (華嚴形而上學) found in Buddhism. Inter-penetration is quite conceptual, abstract, and idealistic. However, inter-intrusion in the realm of humanity is restrained by the physical condition of humanity.

For instance, nature is spatially limited by constraints as seen in the intrusion of neuronal plexus into the eyeball. Because we live in a spatiotemporal continuum, we cannot avoid having a sequential order of spatial intrusion. It protects the nature from taking one locality of space on by multiple material intrusions. It is, I believe, the evidence of a time-intrusion to space. That is why we believe time is of a step higher dimension than space. It is more adequate to comment that time is the highest essence of being.

The most interesting point is that time intrudes into any kind of dimension that constitutes any aspect of our thinking. That is why we can make recursive thinking for all the phases of our past, future and present. It produces solutions of superposed manner which allow us to detect the density of thinking which influences our decisions and intentionality. Transcendence has an orientation determined by the intentionality and how well the tensorial direction of intentionality matches together with the density and rate of dynamic mental flow.

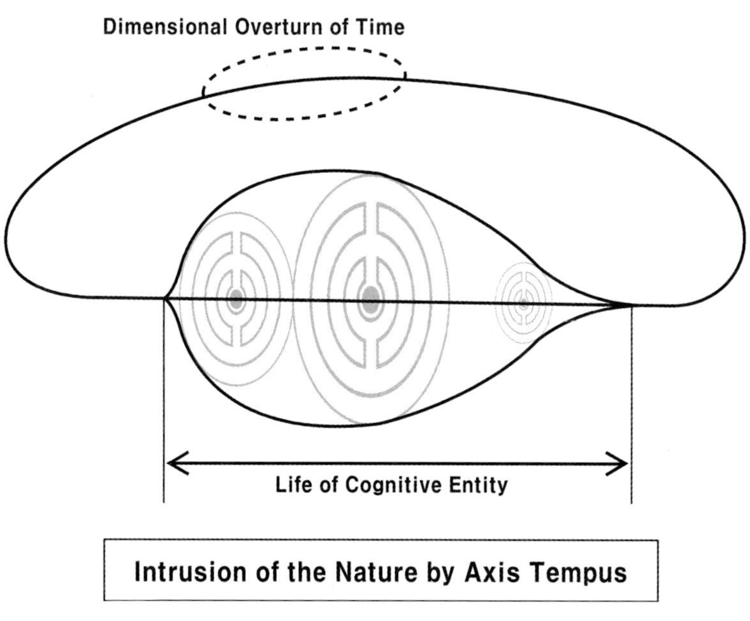

Figure 4-5-2. Inter-Intrusion between Humanity and Nature (Time penetrates the inter-intrusion between human and nature)

Transcendence is a situational phenomenon generated by inter-intrusion. Inter-intrusion is the relation formulated between nature and a human. The nature, as a whole dynamic existence, and a human, as a part of the dynamic existence, form a mutual relation physically penetrating and intruding to each other, from those on stemmed relation are cognitive processes and inter-subjectification (or, equivalently, inter-objectification).

The notion of intrusion is not to be smoothly generalized in the embryology of animals. However, if a specific kind of frog is picked up for illustration, an oral cavity is developed at the location of the first contact and intrusion of a sperm into an egg and at the opposite location of symmetry on the egg surface an anus develops. The easiest model case to explain this is the earliest embryological development of an Echinoidea. Initial embryogeny produces a round sphere, then filling its internal cavity followed by subsequent reverse perforation made from the bottom location of the anus-to-be toward the top location of the oral-cavity-to-be, producing body cavity. It is observed that the initial development of the neural system and other visceral organs take place along with the annulus of the oral cavity.

In this case of intrusion,

(1) it is thought that the first contact with external conditions at the annulus of the oral cavity causes the greatest stress to the concerned ontogenesis. Embryonic cells should have experienced the greatest impact at the moment since the time of fertilization. The mapping of internal responses due to the stress of contact is similar to the fundamentals of the emergence of neural system.

(In intrusion, stress conditions develop around the peripheral boundary of intrusion. If the stress conditions are plotted on the Cartesian grids, greater concentration would be found at the vicinity of intruded spot than comparatively remote locations. It might be generally anticipated in connection with anomalies.)

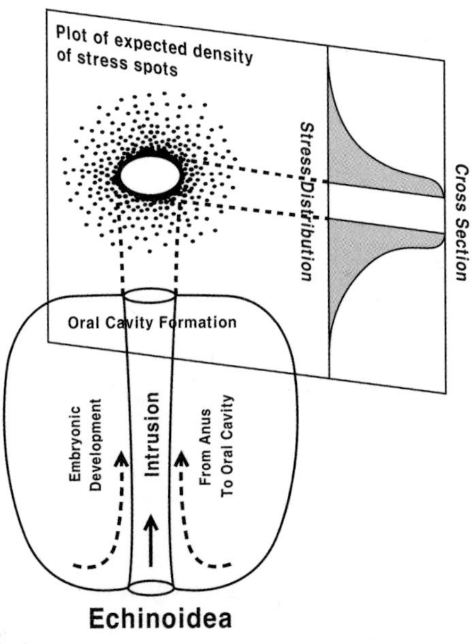

Figure 4-5-3. Plotting of the Density of Spots around the Oral Cavity at the Moment of Intrusion (Provoking the emergence of neural system)

(2) it is also thought that a very primitive temporal feeling could have been attained in accordance with synthesized endogenous rhythms caused by digestion process made by the vertical transference of admitted materials through changing conditions of internal cavity and other associated processes for treatment,

(3) that the fundamentals of spatial feeling could have been formed by the accumulation of exogenous objective perceptions in accordance with the continuously altered feelings of contacts at the external boundary of skins.

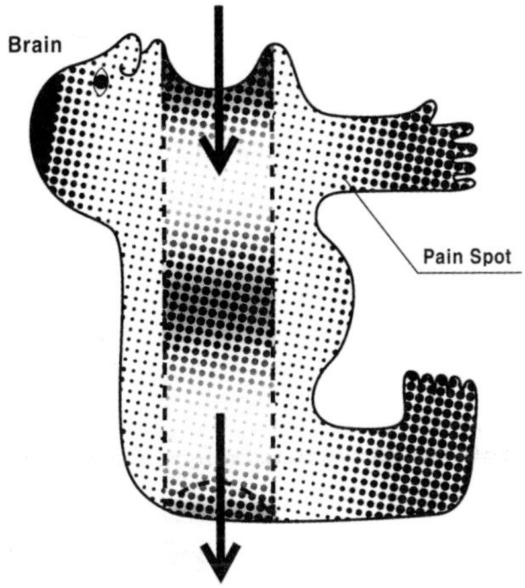

Figure 4-5-4. Basic Understanding of the Generation of Spatial Feelings

The concept of inter-intrusion is thought,

(1) that, at first, a human is an existence, whose cognitive system has been arranged pivoting around it's experiences, and thus the plexus of infinite number of affairs in the Plexion induces wholesome ontological recognition by the reflection of individual self-recognition. This situation is considered to be an intrusion of human understanding (similar concept to *noesis* or *nous* of Plato, the intuitive understanding of the idea, not following the concept of *Verstand* of Kant) into nature.

(2) Various neural systems in and around bowel and limbs, developed in phylogenesis originally for the purpose of metabolism (and physical movement), are considered to be an intrusion of the phase space of pragma into an individual physique.

Pragma is a part of the whole plexus of affairs in nature and composes the external environment of an individual person. These neural systems are proprio-receptors (receptor to sense the head location), neuro-receptors for the strains of muscular systems and bowel intestines, receptors to sense the concentration of oxygen, etc.; meridian and vein systems in acupuncture, which could be considered as one major but differentiated senscious receptive system.

(Physical movements are considered to be a secondary physical function to metabolism which supplies nutrition.)

(Acupuncture points are selected from meridian and vein points which appear on the skin surface and in the hypoderm. The healing mechanism is not yet well clarified. However, it is reported that Encephalin, an endomorphin, is produced by an electric flow of 2 Hz, and

Endorphin is composed by the flow of 15 Hz. “FATA(華陀) complete guide of acupuncture,” 1991, Sung-han Publishing Co. Ltd. p20-21)

(3) Brain-neurosystem and other dependent and relevant systems to it are generated for the treatment of stress conditions in the internal and external physical environments.

These systems are considered to be an intrusion into the phase space of pragma, which is a bundle of affairs experienced by an individual person. The intrusion of the cerebrum into pragma is formed mainly by the five major sensory receptors.

(For instance, qualia, quantia and various symbols are encoded in episodic combination with other fluttering patterns of cognitive expressions to be transferred to superior cognitive mechanisms, which I named transciousness in this book and where they become a part of meta-representation.

The fallacy of homunculus seems to be revived with a novel meaningful function when the secondary parasitic brain or process is commented in “The Emerging Mind” by V. Ramachandran, 2003)

(4) The Secondary parasitic process is the virtual reality in a tensor space that a human has constructed. It is a phenomenon of humanity existing in the deepest part of human, disregarding good or evil and right or wrong.

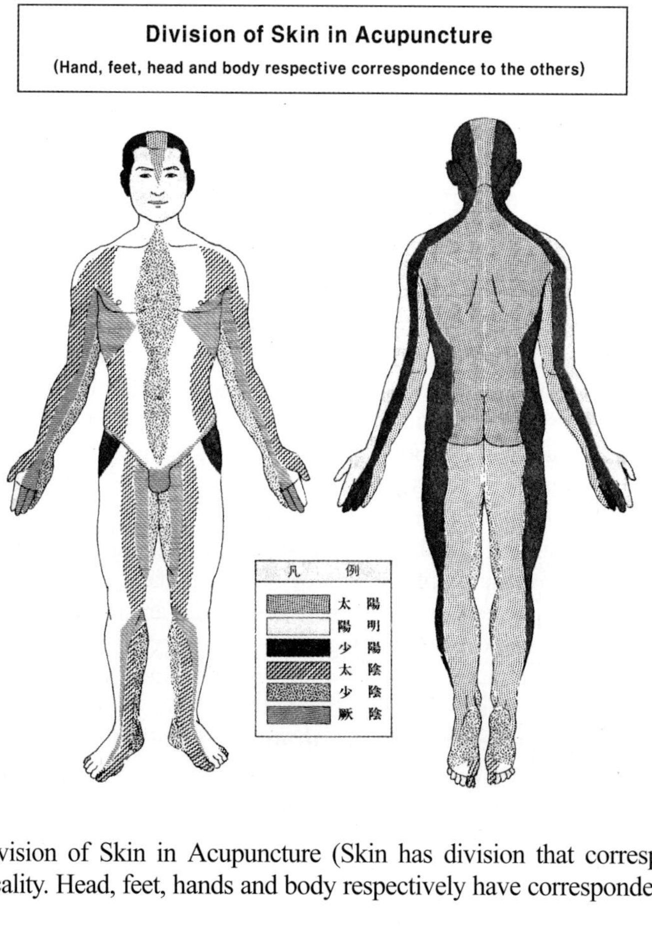

Figure 4-5-5. Division of Skin in Acupuncture (Skin has division that corresponds with brain locality. Head, feet, hands and body respectively have correspondence to the others.)

The Effect of Rumination in the State of Inter-intrusion

Rumination has a good effect that has saved humanity over as long as the past ten thousand years. In shorter terms hundreds of years, humanity has had to clear up remnants and junk that amassed. Thus, some have said a victor has survived to reserve his name and reasons losing greater benefit of diversity. But it looks like without competition and struggle, refined humanity and concentration of energy and resources would not have resulted. It is without a doubt an important decision for the human race at this point in history to find out the range of balanced and energetic drive regarding anthropomorphosis.

Rumination is a process to confirm the stability for a system and extendibility for a nexus. Inter-intrusion makes disturbs a system by supplying new impacts externally. In the long run a system is strengthened and refined due to the ceaseless interference of external factors. Thus we can say that rumination is a process of reconfirming the potentiality of a system. The comparison of potentiality among different systems appears in the competition. Potentiality is an outcome of the symmetrical stability of a system.

Intentional movement of an entity is very accidental and comes from asymmetry of the inner system. The imbalance caused by internal asymmetry appearing as the intention of an entity.

Neuronal embryology of humans shares a similar mechanism with the embryology of human society. Societal embryology shares resemblance to synaptic connections. Market pricing is done through the rumination of value assessment (or pretium assessment to be introduced later) of the market continuum based on the systematic knowledge and communal intelligence.

If we want some definite quantification of value concerned with socio-psychological aspects, land price is the most probable scientific candidate that we can obtain from our present economics. Individually, we can measure the response of encephalonic activity. Socially, we can measure the response from market, which is the container of the communal response from people.

Key factors of the creative peace are humanity, integrity, and creativity. The orientation and transcendence of humanity in our days are the beliefs in the universe, science and persons.

Transcent of humanity is achieved by and after the innumerable rumination of external impacts, which are recorded in the phase space of human cognition. This rumination leads humans to the omega path which is an exceedingly hysteric procedure. In the contemporary era we have encountered the logical labyrinth of the omega path.

Transcent, as I reckon, is a consequence of intrusions as found on the blind spot wiithin visual sensation, or inter-intrusions. Furthermore, it is said that the first eye evolved in nature must have opened a vast new niche for all animals. (Andrew Parker, 2003, *In the Blink of an Eye*)

An eye is also an intrusion of external light into the animal life. An intrusion of external environment into a system looks like inducing an excessive concentration such as hyper-connectivity around the intrusive spot. V. Ramachandran (2003, *The Emerging Mind*) utilized an idea of cross-modal synthetic abstraction and bootstrapping theory to explain the abstraction and syntax of language.

4-6. Transcent of Humanity

Postefference and Transciousness
Transciousness and Ontological Loop
Transcendental Intentionality
Transcent of Humanity

We are going to deal in this section with the transcent of humanity, not of a human. Descent or transcent of a human is not the concern of our study. Transcent is a created word from transcendence. It is strange why people have not used such prospective vocabulary in current English.

Postefference and Transciousness

There is no need even to cite Oswald Spengler, the historian and philosopher, to state that there is a cycle in any civilization. We can observe that, besides civilization, any type of community has certain patterns that resemble the aspects of human life.

Such resemblance can be sensed intuitively considering the fact that both civilization and community are developed by humans and thus characterized by humanly features. In other words, as humans have limited lifespan, civilizations, nations, and other kinds of communities also share such end; in a way there is always a cut in a noodle.

However, the dissolve of community is not due to the limitation of human lifespan, but because of the discontinuation of mutual relationships. This mutual relationship is the base of Homo plexus—the dynamic human relations.

To put it differently, the reason that Homo sapiens were wise and great is not only because they started using tools and fire, but also because they developed a cooperative community. When considering only the skillfulness in hunting, humans would be no better than wolves or hyenas. Nonetheless, humans are the greatest species on the earth because we cooperate to establish a better place to live and a method to achieve such goal.

If humans dealt their wisdom with the vague subconscious emotional feeling, which we call intuition, the conversion to the societal symbolism that gives power to overcome the insubstantiality of wisdom by transmitting it to the semiotic domain, such as languages or letters, thus to the improvement in humanity, the history of humanity would have seen continuous halts.

Humans have recognized the difference between life and death and, with the positive psychological mechanism, chose to fight for life. In the course of the fight, they achieved an inter-subjective cultural process that was realized through the "memory and inference" in the individual's consciousness level and the "record (language) and communication (negotiation)" at the communal level.

As memory and consciousness of a human are based on an endo-biophysical condition, those of the community are based on physical life itself that becomes molded into civilization and history, but not into mystical collectivism or any other concept. However, the most critical aspect of the communal consciousness is its supremacy through language, symbols, and mental expression mechanisms.

(Feelings are like symbols that are written in our body. This is what as I believe the adherents to ontology in the information sciences have been trying to clarify. The complexity of physicality has been transferred to the vitalized phase space through the formulating mechanism of the exo-biophysical development of human brain; it is the secret of the communality in humanity.)

Going beyond establishing languages, humans have developed a wide variety of symbols and the ways to correspondingly express different symbols. These symbols rely on memory or language, which then again relying on impetus and indentation. This impetus and indentation remain in the sentiments similar to the endo-biophysicality, which can be represented by the neural systems.

In fact, the reason why wisdom or intelligence has maintained throughout time is the multi-strata characteristics of the intelligence system. The entanglement between the residues of endo-biophysical memory, the correlation of affairs in the emotional system, the intensified selection within associated thought process, and other types of brain or nerve systems have achieved a strong structure that can bring out and maintain personal or cultural foundations within individuals or communities.

Learning among humans establishes very gradually, which can be understood as the way of stabilizing the fundamentals to connect consciousness to endo-biophysics. This is so because the process of endo-biophysics that relates to the nexal system of the nerve systems and the development of consciousness in the nexal system of glial cells - though it has not been proved yet - realizes through its communication with endo-biophysical environment.

The foundations of human memory and consciousness has developed with pain because of the very instinct of self - protection in humans. When a new born comes out from its mother's womb, it faces such pain and shock that this experience defines the initial environment for cognition towards the environment outside of mother's womb (胎盤 外界).

This experience also serves as the moment when humans face the highest ceiling value of sensation in their entire lives. This maximum sensation becomes the standard of the potential distribution within the emotional part of one's life.

Philosophically speaking, the emergence of humans occurred from the communication between subjectivity and objectivity. Neither subjectivity nor objectivity can be separated from biophysics, for humans cannot escape mortality. The consciousness related to biophysics is even more prompt, implicit, and problem-solving. Such aspect can be metaphorically described with a cat running up to a tree in the face of a danger. (It is ridiculous why a monkey does not escape to the tree on which it can survive best with its refined grip within Aesop's fables.)

And a biological man acknowledges himself that he is an individual with limitations, who has to interact with others in order to constitute a community. From the bio-evolutional point of view, humans come into existence in the state of larva, in which the growth of the brain is especially strengthened.

From the viewpoint of cognitive process of human brain expressed in Chapter 3, the exile of infinites-encounter comes from the peak experiences. Then the vacuity of cognition is originated by infinites-encounter and manifolds are formulated and mutually connected surrounding the point of vacuity.

The infinites-matrix (that the exiled infinites-encounters has formulated around Axis Tempus) and various patterns that infinities-matrix has generated are presented to the human consciousness, and humans tend to select the patterns in an abstract state.

This kind of pattern is influenced by and based on the many different types of past experiences (spiritual, psychological and physical) and the decision-making process. An abstract state means the process of constructing a gestalt feeling through the patterns mentioned above, and it also means the complex process of choosing and joining the fluttering. That complexity differs from person to person according to each personal characteristic.

If we follow the long-term geological chronology, we can deduce that human consciousness has developed very rapidly and that it continued to advance promptly even after development. But the duration and continuity of this kind of situation cannot be easily predicted.

Considering the tendency until today, we must regard the phenomenon of human-consciousness to be a very peculiar one that appeared in the Earth ecology. Through the formulation of human consciousness, numerable drives and impetus have been generated, such as the vague hope that we cannot suppress down, the mixed forms of every kind of beliefs, various creative attitudes, the awkward excess of positiveness and the excessive self-conscious opinion. It is also observed that there are untouchable dark sides of human consciousness that make the nihilistic abyss absorb the psychological energy of humans.

The negative factors that exist in the consciousness and the susconsciousness of a biological human, such as the various kinds of mental diseases, schizophrenia and autism, are other types of peculiar phenomenon that, once combined with the positive factors mentioned above, they manifest the high potential and the vibratory influence to human society.

This dynamic human consciousness is the secret of humanity. But the true mystery of humanity is the human nature that makes one maintain their personality in a consistent state even in an inconsistently changing world. The word, person or persu (from the Etruscan language), is thus the personality derived from consistency.

The consciousness and the culture phenomenon of humans is the byproduct of the process of developing humanity for survival. This means that the secret of humanity is a life phenomenon.

As we have seen, the development of a human's humanity is generally indivisible with the degree of his maturity. An individual's human degree of maturity is either objectively or subjectively related to the richness of his life. If one's life is mature and thus has a rich soul, mind and body that do not lose its balance, we can deduce that he has a determined self-consciousness, a harmonious relationship with others and a prolonged life span.

From this, we can infer that the development of humanity is also related to a community's maturity and lifespan. The researches about the intentional communes show us that the mentality of their members is the key factor for determining a community's life span. The intentional communes that existed in the USA after the 1960's were more than 10 thousand members. They were studied and concluded that their life span was determined by their members' psychological factors, such as self-sacrifice and communication.

And maybe it would be reasonable to extend and apply this result to higher levels, such as a nation or a society.

Transciousness and Ontological Loop

The major reason that human beings have developed transciousness is that human cognition is intruded upon by the infinite scale of the time axis. Such intrusions have been processed and will progressively be continued into the future. The intrusions are seen in their major portion by the activity of SCN (Supra-Chiasmatic Nucleus), which is behind the temple and consists of 10,000 neurons. But we know that there are dispersed minor peripheral clocks in our whole body. These dispersed oscillators vibrate independently with an independent cycle.

We have communal processors in the neuro-physiological network of our bodies that is known as the mirror system. Mirror systems nowadays are known to be a super neuronal network that can integrate complex activities of the brain. The author understands that the human mirror system appears to ignore the deviation in cognitive process when concerned nerves work together as a system but sometimes confirms the offset deviation of each neuron from the representative value of the dynamic system when individual neurons are concerned.

Ken-ichiro Mogi commented that the mirror system in general is a kind of (social) processor, which manifests the universal contingency（偶有性）that is the (compositional) relationship between (incoming) accidentality and (internally originated) certainty. (Nou to Souzousei, Cha.4. Parenthesis contains the interpretation of the author on K. Mogi's logic.)

It is generally acknowledged that the brain structure of humans is designed to close the loop after output. The interchange of information is not allowed between output module and sensing modules. Therefore it is considered that transciousness and post-efference were developed originally due to the presence of Plexion and infinities-transfer. Transciousness is the result of senseless imaginary symbols that has the least substantiality and post-efference is directed to a predetermined orientation due to the successive teleological enforcement.

Transcendental Intentionality

Transcendental intentionality is also thought to be developed under the major influence of the infinite scale of the time axis. There would be no transcendental intentionality without the intrusion of time into humanity.

Intentionality is a notion introduced by Franz Brentano in the 19th century that refers to the relation between mental activities and external pragma. Intentionality in philosophy is said to be simply "aboutness," a sort of aiming of human cognition. The concept was first found in Scholastic philosophy. Brentano introduced the idea in his book titled "*Psychology from an Empirical Standpoint* (1874)." His definition of the idea was a mental phenomenon, which is different from a physical phenomenon. His definition of intentionality is the direction towards an object, or the immanent objectivity.

Every mental phenomenon and every psychological act has content. Such an act is directed at an intentional object. Every belief, desire or intent has also an object. Brentano commented the "intentional inexistence" to indicate the status of objects of thought in mind. (It would be interesting to link this idea with the orientation of this research by thinking of another possible expression, such as "the (efferent) **imagination** of the (rational) **thought** of the (emotional) **heart**.")

Being intentional, or equivalently having an intentional object, is the key to distinguish a mental phenomenon from a physical phenomenon. Physicality lacks intentionality. (But let us start from considering physicality at the senscious level and as the data base of unconscious memory stored with tensor coordinates which is combined with emotional aspects.)

We purpose hereafter to enhance the idea of intentionality further towards "transcendental" intentionality. Intentionality of animals can be exemplified by the wooing howls of wolves or vein of plants such as pumpkin shoot feeler. It means the notion of intentionality needs a definite connection with the reality. Wooing howls may easily find their pairs but would not necessarily end up with intercourses. In the case of humanity, a person does not usually remain at the intentional level. The secret of a person is the ceaseless overcoming of self in his everyday life. It is comparable to the accidental evolutionary processes of other animals. Humans make evolutionary processes even in their conscious life. The infinity, whether true or false, seems to cause the process.

"Transcendental intentionality" is a slightly altered concept to explain the human cognitive process. Transcendentalism is well propagated by Ralph Waldo Emerson. But a variety of transcendentalism has been expressed in various forms in the past works of global humanity such as self-overcome, self-denial, self-accomplishment and etc. Transcendence needs to be interpreted "achievement of liberty from the self-system or, if acceptable, from the prior human-restraint system" for human beings residing in finite conditions. Transcendence is the key word for "freedom and liberty" for which the human race has so long struggled.

Transcendence is one of the essential mechanisms of humanity in creating a new foundation to interpret the whole scheme of their present environment on a higher level of hierarchy. One has to escape from present confinements to a wider and higher place to make the bird's eye view possible. It is sometimes illustrated in literary works "cracking out of the shell of an egg."

We have an interesting excerpt from an ancient document well over 4,000 years old. Transcendental cracking of the embodied mentality is very well expressed when Elihu, the youngest friend of Job, a renowned patriarch in the Old Testament, expressed his irresistible emotional outburst in his upright and earnest shouting, "Does it not make your heart leap to your mouth?" or "My heart jumps out of my mouth!" "A heart that out of the mouth" would be a slightly frustrating expression to modern people.

However, it is said that the ancient human had a stronger physique-based way of thinking (in a broader sense it could be called a physique-based way of mental processing) than the present human being has. The physique-based mental process is actually an emotional way of thinking. Emotion provides the highest degree of concentration, for instances from the level of the "fight or flight" reaction up to the level of voluntary martyrdoms, which is when a person is honest and has the least mental blocking. The emotional process usually presents better concentration than the mental process because of its so called "state-dependence" quality that originates from its physique base. (M. Solms and O. Turnbull, 2002)

State-dependence is directly related to the notion of the plexus of accidental occurrences. Human mentality is state-dependent; which means that the human being operates in a situation-respondent adaptive existence. The situation is temporally sliced into vertical sections of the plexus of accidental occurrences around a person. A person is equipped with life-long developed, susceptible and interpreting mechanisms to interact with circumstances.

The equipment is what we call mentality based on the bodily neurophysiologic mechanism. The neurophysiologic mechanism is connected to the cognitive process mainly managed by emotional processes and further to eventual electrochemical processes. There could be some variations of interpretative theory, but the general look is most similar to physicalism. (J.K.Kim 2004) (It is my opinion that perfect physicalism would yield perfect mentalism, the best and vice versa.)

Furthermore, emotion is the key of the human transcendence of the ordinary person out of his boundary. It is thought that transcendence emerges when a person is emotionally concentrated. The Emotional condition of a person is the overall integrated expression of an individual existence. Emotion is an integrated receptacle of external signals to a person. Certain parts of the emotional aspects could be expressed from time to time, but our main concern here lies in the whole emotional status of a person in his physique.

Motherhood is the special carriage of personhood in the manner of white noise. White noise is a random signal of a flat power spectral density. White noise is a status of an adsorbing medium that we usually experience when we admit a specific signal of life, holding our breath. We do not concentrate but we keep our emotional status in balance to receive external signals.

A balanced feeling full of noise is what I believe happens in the reflective mirroring between mother and baby. White noise or "flat noise" is the neuronal expression of the entirety of personhood. Communication between people looks most efficient when flat noise formulated on the surface of the emotional sphere. Flat noise is the secret of motherhood by which a mother transfers her personhood to her baby in a neuronal expression.

The most delicate maternal discernability in the choice out of the innumerous emotional signals in her noise set is the modulating between two personhoods of the infant-maternal relationship. It wouldn't be farfetched to say that this close relationship also has to be studied in fetomaternal situation as well. The Baby's experience of transmitting its intention to the surface of maternal flat feeling is learning the basic scheme of transcendental intentionality of humanity. Transcendental intentionality is in turn the base for the transcent of humanity.

As we can understand in thermal noise, electronic noise and the other noises, emotional noise contains all impacts of Plexion intervening into pragma.

When we come to the ultimate point of the thought of transcendence, Divine Transcendentality is the essential attribute of the one and only God together with Divine Ubiquity and Divine Universality. Actually, transcendence is a manner of human approach to solve the problems of encountering infinity, such as Divinity and the other impressions of infinitely extended ultimata. It is thought that brain science leaves a reserved realm for a number of experiences with unsolved "infinities-encounters." "Infinities-transfer" and "infinities-matrix" are introduced in this research.

As discussed, we need to take a differentiated path of explanation from that of sheer intentionality. The diagram of the four annuli of humanity in Fig. 4-1-1 shows that our humanity consists of two distinctive parts: the part of the corpus represented by the bowel and the other part of cerebrum represented by the brain.

Simply using an analogy of an automobile, it can be thought that a brain generates intentions to motivate humans focus on objects, environs or affairs, just as an engine produces self-generated power to transport the body of a car. But a car has different movements in accordance with the proposed paths and the patterns of roads.

Human physique (corpus) is a decoding device of the plexus of incidental affairs of the universe (the absolute other, from the universe of which an individual observer is excluded,) as a brain is another decoding device of adjacent environmental pragma. Modern brain science has yet to achieve a deeper investigation into the unconsciousness and emotional mechanisms.

The reason we have to comment further on the transcendentality of humanity is that human is in a state which is connected to the universal environment named "Plexion-Corpus Annulus."

Transcendence is well represented by the existence of the human DNA; it is an expression of the billions-year long historical encounter of humanity. DNA possesses the chronicles of history (or in my vocabulary, unalterable human hysteresis). As revealed through recently discovered DNA facts, a human is supposed to have certain directives constrained to his past annals when a specific situation and environment meets at present.

It is well known that baseball batter with 40% success had disappeared (Stephen Jay Gould, Full House, 1996). Today's average batting is 26%. Best batters have now the average of 35% and had decreased from 40% of 1930s since the last best rate of Ted Williams of 40.6%

exceeding 40%, which is human physical limit confronted since 1940s.

Transcendental intentionality is one sort of losing concentration under tremendously stressful condition as batters experience harsh competitions. "Transcendental" means variation that started by introducing the easiest or the smoothest influential flow of motives, mechanisms, dynamics, force, affairs, matters, -etc. from adjacent or communicative manifolds. S-curve is attained, differing from Aristotle curve.

Transcent appears after that. Transcent is a phenomenon arisen by tunneling or imbibitions effect when the summit touches the boundary of the next higher hierarchy.

Transcent of Humanity

Transcent is a new vocabulary born along the lines of "ascent" and "descent." It is designed to mean that plain "transcendence" needs in addition to emphasizing incessant mobility to overcome innumerably confronting boundaries and obstacles in the future of humanity.

The ascent and descent of humanity could be the description of the orientation of human beings historical aspects. But a transcent is suitable to describe the renormalization process of humanity in accordance with the successive variations of the scope and the content of humanity.

The descent of man is a part of the transcent of humanity. *The Descent of Man*, the title of Darwin's book, is an evolutionary theme represented most recently by the concept of selfish genes of R. Dawkins. *The Descent of Man* is more meaningful when combined with the concept of the transcent of humanity.

Someone might point out that it is a little bit too teleological view. The transcent of humanity originates from the ontological structure of humanity. A centripetality of the ontological structure of humanity provides an integration of human activities.

Why do we have to consider the transcent? Humans actually experience the affair-wise leaps in his/her epistemological existence. An old physics professor wrote that an affair in physics is not an accidental event but defined to be a physical condition (or physical quantity) of a local point in a spatiotemporal continuum.

The transcent creates a leaping from the inside. When humanity has achieved a certain summital accumulation up to a point very close to novel alteration, there is a mechanism of "uplifting in-draft." How is it possible? The exponential increase of the changes in culture or civilization comes up when diverse factors are activated attributable to the sufficient attainment of complexity of the system. A Self-organization mechanism works in such a situation. It abruptly reveals a hierarchical leap without a prior external indication.

In the author's opinion, if a human hologramic plate zone exists, it would be an emotional plate zone. In Figure 4-6-1, we might have to think of some probabilities that hologramic memory is a candidate model to explain the mechanism of human memory as asserted by Daniel Bohm and K.R. Pribram. Hologramic plate zone (輪帶面) could exist as a continually varying global emotional status in the phase space of human neurophysiological condition expressed by ever dynamically changing modularity with the relatively stable cores.

It is an interesting thing that human brains do not recognize the internal and external changes and concentrates only to compute the change itself in accordance with the change of situation.

The computation of the neurosystem could be concerned either with incremental differences from the previous signals, or with the deviations from the reference signal values furnished from the infinities-matrix.

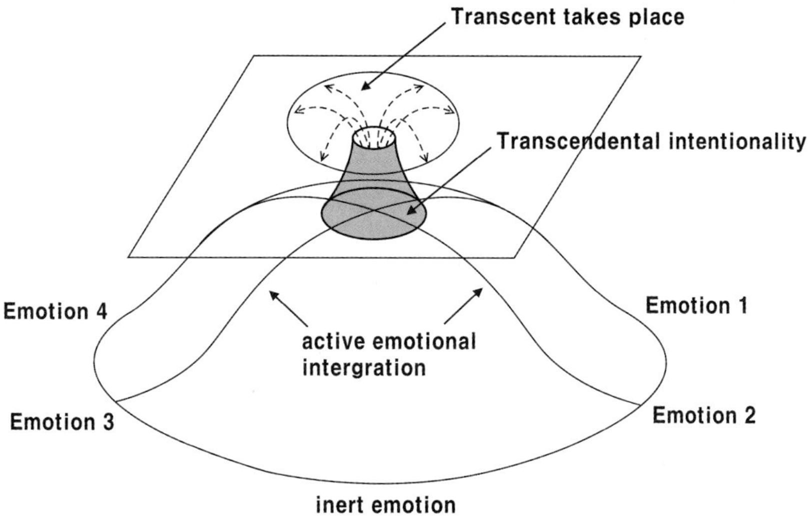

Figure 4-6-1. Upsurge of the Sensation of Humanity Introduces a New Dimension. (Summital accumulation of the feelings of extremities comes sufficiently close to a new dimensional extension and an uplifting indraft takes place.)

It was thought that transcent is a mechanism frequently observed in the phenomenon of creativity. Creativity process has to have an "Aha!"-experience at the final moment when all figurarity of the new idea finally shapes up. Communality is the result of individuality's transcent. Referring to the findings of computer engineering, computers cannot follow humanity, because genomes are extendable for the compression mechanism with which new functions can be added to the existing genes and every gene can behave as a cornerstone for new extension (*The Birth of the Mind*, Gary Marcus, Ch. 8). It means that humanity has opened a realm for the transcent to the infinite Plexion.

Chapter 5. The Science of Integrism

The Science of Integrism could be called, here in the Asian atmosphere, Tong-Ghi science (統幾科學, or 統合能科學). Tong-Ghi or the capability of integritization explains how human society can be united and how it will be sustained in a unified state of integrity.

Although the Integration potency is an opposite notion to the differentiation potency in the biological process, integration and differentiation coexist with one another in life forms. They are partners for life. The Integration potency is a firm background and foundation for the differentiation potency that allows it to pervade into the active biological processes.

Life is an achievement of nature's ability to differentiate within its integration potency. The Differentiation potency and the integration potency exist together in almost all life processes, but were originally under contrariety in the human abstraction.

Life is an isolated systematic entity in the reverse process of entropic degradation (or entropic increase) of the universe. The origin of life is the integration that was achieved through potency, which is within the process of design of life such as in the DNA, RNA synthesis. Integritization, as we are going to deal with here, is the restoration process of the integration potency of human society.

Integrism is a measure of integrity of humanity. At the individual level, the integrity of a person is the governing factor of a person as an integer in a community. When a person is considered to be at his most effective condition as an integer, it indicates that they behave better in their designated communal level; their degree of integrism converges to unity within community. Contrarily, when the degree of integrism goes to nullity, the whole integer is considered to be autistic and closed to external surroundings.

The openness and irresponsiveness of an integer, or equivalently of a person, depends on how it manages its communication process between its internality and its externality. Thus a theoretical methodology was designed for an integer-person to consider that integrism consists of systemicity and nexusity. The ratio of integrism between systemicity and nexusity gives insight into the internal attributes of a person. Once internal attributes of an integer are clarified, the structures of personal and communal behaviors can be understood in an easy manner.

5-1. Search for Integri-ti-sm

An idea on Integers

Social sciences required a firm basis of theorization in dealing with individual identity. Identity must include a scalar mode to describe its capacity, a vector character to direct its tendency, and a tensor expression to indicate its dimensional positioning. An integer is very clearly defined in mathematics. If possible, social sciences would also require a clear mode of expression of the integer which can be agreeable to the public.

At this moment, allow me, the author, the opportunity to talk about my experiences regarding integrity. About 23 years ago, I decided to search for self-integrity. Most people have their own pictures of integrity, which would be the ceiling of what they could attain of integrity. I believe the Bible sings about integrity as below;

> Blessed is the man
> who does not walk in the counsel of the wicked,
> or stand in the way of sinners,
> or sit in the seat of mockers. (Psalm 1:1)

The terminology "integri-ti-sm" came to me during one of my personal meditation sessions on the patriarch Job and his life story as he was considered to be "a man of integrity." I also reflected on one of the most respected people in the Far East, Confucius, who was also a man of integrity encompassing wisdom, scholarship, virtues, emotion, music, state-administrative insight, etc. Soon after my time of mediation, it came to me that a community or a society could also take the path to integrity, which has an unavoidable relation with the creative peace I have been looking into for so long.

The society I live in has experienced the merciless oppression of Japanese imperialists over the course of three decades. Very soon after liberation, there was a period of indescribable civil war, where people killed their own brethren shedding their blood on local soil because of ideological antagonism. My people desperately needed an unshakable ideal and methodology for peace. I have cried within myself for decades because there was no way out, with a longing that no one in the world would pay attention to. I have tried to write about how to reunite southern and northern Korea.

In my long search for integrity, it occurred to me that the main theme of study about peace should be a global interflow of everything without sacrificing the entity of the respective person's character. This entity should be able to be equipped with the right and responsibility as an integer (正人). As was mentioned above, an "integer" has been originally and until now a mathematical notion, yet is believed to yield more profound implications on sociological importance.

The science of integrism aims to be a practical science that achieves integrity of both internality and externality within an integer person. Usually the integrity of a person is related to his internality. For instance, perfection is a goal of life for a certain perfectionist, and then he himself will have to be his own subject to the accomplishments of his perfection in the light of his own perfect criteria. The man of integrity, as we believe, must also meet the requirements of his external surroundings, which extend to the entire reality of one's life.

Community or society as an integer can be better understood when its internal mechanism is well described by certain means. An integer has its integer way called integrism, which is an integrismic measure that is expressed in this Chapter 3.

In this context, to avoid confusion, we must differentiate "integritization" from "integration." Integration is used for "the act of combining or adding parts to make a unified whole." In psychology, it means "organization into a unified pattern, especially of different aspects of personality into a hierarchical system of functions."

On the other hand, however, integritization focuses neither on the unified pattern nor simply on the unified whole. Integritization is a series of activities of an integer to fit into his surroundings, i.e. adaptation, provided that the integer maintains its systemic character which should be already developed. The integritism we are going to discuss next has to do with integritization.

Integri-ti-sm and Integri-sm

Integrism is often incorrectly referred to be a terminology of the Roman Catholic Church. Apart from that context, I have used this word for more than a dozen years alone in order to establish a scheme of thought which I have kept in my mind. It was only one year ago, presumably in 2007, when the word "integrism" was found in Wikipedia. I have known for decades that there was the word "integralism" which was an active political movement of Catholic Christians in the southern part of Germany.

Integrism, as I use it, is a special mode of the word "integer" to represent the personhood and its associated human aspects. Being human is a process of establishing the self first and maintaining connectivity, which includes communication and human relationship with other member of a community.

However there is one more important factor which is usually disregarded when it comes to being and living as a human. The mechanism of the change of humanity is hidden in integrism. This change deals with flexibility and creativity, and finding new dimensions of the structure of meaning. Integrism has the ability to influence human efforts in achieving integritism, which is the pursuit of the integrity of a person, can be a very important factor of integritism if managed properly. The practical applicability of integrism is centered on how we define and utilize it.

Integritism is the attitude that represents the human effort to overcome present constraints and obstacles in order to accomplish human integrity. A human is a being who can think of oneself thinking. Some philosophers like D. Rosenthal call this phenomena HOT (higher-order-thought), which implies that human consciousness has a sphere of transcendence. Self-consciousness implies that one can think about his own psychological state and his own experiences, so if we combine these two definitions with integrism, we can state that integrism has a close relationship with the HOT characteristic of self-consciousness. Integrism is concerned with both internality and externality, and with both subjectivity and objectivity through the formulation of a peculiar phase space of inter-subjectivity. That is why integrism is expressed as the combination of subjective and internal systemicity, and objective and external nexusity.

Search for Integritism; Cruxology and Integrism

Integritism of an integer is deemed to be achieved by adequate cruxology fitting into concerned circumstances. Cruxology is directly related to the developmental process of an integer, similar to the developmental embryology in the case of human conception. There are innumerable developmental processes for the corresponding innumerable kinds of integers. Cruxology covers the broad span of the life of an integer, from its initial embryo state to its death.

Integrity of the human race can be attained if cruxology maintains a balanced combination of self-building and connecting-with-others. An integer builds itself by systemization, omitting or enfolding its unnecessary parts for adaptive teleology. Integers, the essence of humanity encompassing the universe to an individual, are governed by cruxological principles, which are essentially influenced by integrism of the concerned integer.

Integrism must govern all processes of humanity, on the level of physicality or of mentality. When integrism is applied on an individual basis, an intrusion of nature onto it could best be visualized by the example of colon bacillus as it connects with the outer environment. If integrism were to be applied to the entire human race as an integer, a good example would be world governance, since world governance is a complex mechanism in which dynamic modules in the world are all actively and implicitly implicated. In this book, a vision for integritium is introduced in contrast to the notion of imperium.

In order to introduce integritium in a balanced and efficient manner, what are absolutely needed are an integrismic preparation and a cruxological provision. Integrism would provide, if integritium is to be intentionally planned, the complex prescription of external or apparent nexusity and of internal systemicity. In the provision of the cruxological situation the generalization of scientific thinking on a global scale is required, since it is the backbone of internal systemicity ($S_{i,\ i=\text{internal},\ S=\text{systemicity}}$), and the interflow of specified humanistic activity, which is the spread of apparent nexusity ($N_{a,\ a=\text{apparent},\ N=\text{nexusity}}$).

S_i and N_a are related to the value interflow among different value systems, value covalence among value subjects, and value transference among value spheres in Triradixism. Karl Marx believed that he had succeeded in finding the value expression that enabled the interpretation of utility value into exchange value. The problem contained the issue of how labor can be interpreted into the value of labor product and how labor hours can be expressed into the value amount of labor products. Such labor products and commodities mutually compose a phase space of value correlation in a specific society, which in turn expresses the relationship in market prices in monetary value. Thus the morphosis into monetary expression is thought to be the matured value expression.

When we consider socio-psychological aspects, it is deemed that pain, as an expression of senscious and emotional measure, should be considered the major factor in the conversion and transference of value. Ecstasy and pain are mutually reflective contents of humanity. Utility value is the expression of such neurophysiological states. Such insights will be applied in the legal consideration of economic valuation.

Scientific ways of thinking and humanistic approaches in activities have been intentionally improved in the last half century on a global scale in contrast to the increase of systemized violence. Integrism represented by S_i and N_a, and cruxology, are factors that affect the topics on value.

Terminological Context of Integrism

We know that human language is on the road to perfection. English is also on the way. I was surprised when I heard around two hundred new words, including "international", were invented by an individual writer a few centuries ago.

Table 5-1-1. Terminological Context of Integrism and Personism.

noun	organ	human	person	integ(e)r	commune
adjective	organic	(humanic)	(personic)	(integric)	(communic)
-ization	organization	**humanization**	(personization)	(integrization)	(communization)
-ation	(organation)	(humanation)	(personation)	integration	(communation)
-ification	(organification)	**(humanification)**	personification	(integrification)	(communification)
-age	(organage)	(humanage)	**personage**	(integrage)	(communage)
-ity	(organity)	humanity	(personity)	integrity	community
-ality	organality	(humanality)	personality	integrality	communality
-ology	(organology)	humanology	**(persontology)**	(integrology)	(communology)
-alism	(organalism)	(humanalism)	personalism	integralism	communalism
-ism	organism	humanism	**(personism)**	**(integrism)**	communism

In Table 5-1-1 , the terminological context of integrism and personism in the English language is presented. Words in parentheses are not in use. I was unable to find the right terminology in English to express my way of thinking. I wanted my own way of expression. The emphasized terms in the table were chosen to bring a novel set of ideas into our journey of thought. Cruxology is from an entirely different source of terminology for which definitions will be made in Chapter 6.

(I know that this comment does not excuse my incompetence in English writing.) Personism and integrism are not commonly used terminology even on the Internet which is full of new innovative spirit. The following vocabularies are checked as to their usage by Collins Dictionary, 21st Century Edition.

Activation of Humanity Generates Creativity

Generally people say that the human being is destined to create. Humanity is an interactive process and its resultant fruit. Activation of humanity gives vitality to a society and generates creativity as usually known.

However, abrupt fluctuations of excessive energy to a system generally destroy the adaptability to the environment. We count stabilized condition as beneficial to our lives. Nations seek for stabilized status, too. Our wish would be to acquire both activation and stabilization.

Activated and stabilized humanity is the answer to the dream of the human race. Stabilized humanity does not mean there is no allowance for fluctuation. Sometimes threat to our stable life provokes daring spirit to break through.

Activation of humanity is made by a number of effective contacts for proper stimulation. The stability of humanity is maintained by the simplified chain of logics intrinsically embedded to cope with environmental situations most likely to be met. That is why nowadays we focus on the DNA and physiological evidences to find chains of logics that humanity has conserved.

It is the chain of general emotional reactions that support in a programmed manner the logical consequences in a programmed manner of our everyday life. Our fashion of logic is the result of our style of emotion. Emotion is the body, and logical intellects are the clothes. Humanity is mostly a dynamic dance of this couple. In that dynamicity, we see unceasing never ending creativity.

Creativity is the seed and the offspring of humanity. Humanity plays on the ground of the integrismic theatre, exposing the shuffled combination of self-centeredness and hand-in-hand relation.

Creativity is the game between the individual person and the others, the society. As Apostle Paul once said, "Our conscience belongs to the others, not of us". Inter-subjectivity belongs to the others. Conscience and inter-subjectivity mature through contacts with others. Therefore they are "of" the others. Creativity stems from an individual but belongs to the others. Activation of humanity in a society generates creativity in individuals.

Integrity Permits Fluctuation within Tolerable Variance

Integrity possesses a well managed fluctuation within a controllable range of humanity. It is controllable both in subjective and objective ways.

It is generally accepted by people that stability stays within such range where irresistible impact is not imposed upon an entity. Stability is significant in either the physical or psychological domain.

Integrity comes when a human being entertains and achieves wisdom to manage the surrounding environment. The degree of integrity increases together with the tolerable variance of fluctuation. This is true at the level of the personal realm and also in expanding ranges as described in the consecutive process of "修身齊家治國平天下 (personal moral - family management – statecraft – world peace)."

In pursuing integrity, we often find some people fall into closed relationships with "the other" and end up isolated and stubborn in the course of moral achievement. It is a dead end. Integrism is a natural equipment to guide us to real abundance of profound humanity.

5-2. Integrismic Phenomenon

Humanity is personal. Being personal means communicating and interacting with others. Personal humanity comes from the social environment. Personal humanity possesses an integrating mechanism, in the absence of which humanity cannot develop, evolve nor maintain the social frame. Society is the concentration of anthropomorphic energy and rationale. The orientation toward anthropomorphosis is unavoidable but we are not going to be dragged along solely by the trend. The reason is that humanity has taken a course of self-adjustment by integrismic measure. We are going to talk about integrismic phenomena in this section.

Integrism (統合能, 統幾) and Qualities of Human Integrity

Integrism presents the idea of quantification of the qualities of the integrity of humanity. Integrism would be the objective phenomenon of social and human science, and eventually of natural science as well if properly designed. Integrism is a new term made to show the measurable quantities of the consistency and emergence of creativity, and even the capacity and connectivity of an integer entity expanding from an individual to the whole universe. However, integrism is just a relative measure to see the effects of systemicity and nexusity.

In the development process of political economics, we also have to discuss the philosophical aspects of natural science. W. Park cites on various theories for this issue. (*Essential Principles of Economics*, Ch. 3, 2005, SNU Press) For instance, I. Lakatos presented a methodology of SRP (Scientific Research Program). SRP is a method of blending 80 percent of K. Popper's theory and 20 percent of T. Kuhn's theory together.

In my view, Popper concentrates to keep systemicity for natural science while Kuhn is concerned with nexusity in order to have novel paradigms emerge. The two factors of systemicity and nexusity, at a ratio of 80 to 20, are the determining factors of integrism. This idea needs some more explanation for generalizing, though. W. Park assessed Lakato's approach to be most adequately balanced.

Integrism could be expressed in terms of contingency to connect accidental occurrences, and ingenious coherence to elevate the independent subject out of bounds which enables "system occurrence" done through pattern finding from the outside. The pattern finding is performed at the qualia level, which can show the corresponding interrelationship between external occurrences and the brain physiology.

Integrism and Partitioned Multi-personage at the Outskirts of Humanity

In natural circumstances, i.e. in the initial condition of humanity, the alpha path of humanity is open, as can be seen from the speedy and smooth learning process of a baby from its mother.

However, when stress is accumulated during its emotional growth, through experiencing unwilling situations and events, the infant absorbs the seed factors of division into its personhood, and indentation or precipitation, whatever the right name is, and it then generates the multi-personage process far later in its adult stage. It is not a desirable development for a well harmonized and integrated personhood.

Usually we value integrity very much. Integrity is certainly at the center of humanity and personality. However, it is noteworthy that humanity could not have survived without peripheral multi-personage in an integrated personality. We experience our multi-personage in our everyday life, altering every hour or two. Multi-personage is not distinctively divided in a person but it has some form of compartment in our selves. Multi-personage is not a spiritual phenomenon but, a psychological one. This is performed through brain activity.

Our habit, our thinking trend, emotional tendency, and trained volitional decisions are all a part of multi-personage. Multi-personage is a realm where we mix with others and the outer environmental flow. We admit the strong influence of the outer world and reserve it in our fashioned multi-personage. Sometimes, I believe, we design it.

It seems true that the multi-personage gives a person-like response to its environment and behaves somewhat independently in the thinking and in the emotional processes. It is a skill of the brain to avoid stress by making quasi-independent compartments, which is a distinctive dynamic modularity in neurophysiology. But it is supposed to be present at the outskirts of the phase space of our personal humanity.

Hypothesis of General Continuum: Cosmological Integrismic Phenomenon

Before reading this part, I recommend readers to see Figures 5-2-1 to 5-2-5. Integrismic phenomenon is cosmological. It is certainly hypothetical presently. However, we know that human abstraction was born at the impending need of survival, connecting our complicated situation to a symbolic emotional semiotics of the least complexity. Human mentality is filled and arranged with such symbols and signs of various kinds. Integritization is difficult because we have to pass through this world of abstraction in our intellectuality.

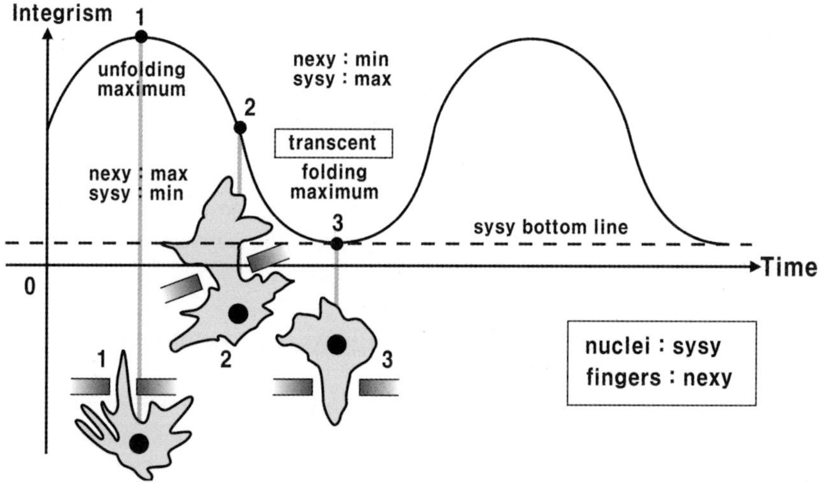

Figure 5-2-1. The Amoeba Integrism; Nuclei (Sysy) and Fingers (Nexy)

We will see hereafter how integrism works in universal, human and micro conditions. Integrism has the general mechanism of creating a continuum because it is capable of connecting between entities. An amoeba has a nucleus and fingers. Integrism is like an amoeba that systematically unifies its constituents with the nucleus and tries to reach outer entities with its fingers.

A human is a unified entity of the highest complexity. Human integrism expresses itself based on event-centric(事件中心), or accidents-centric(偶有性中心) methods. Humanity gained its highest level of complexity by introducing time factors into its realm. Emotional balances, whether they are painful or pleasant, are sought by the reservation of various emotional energies.

Leveling of emotional stress is usually done through proper allocation. Emotional energy is liberated in accordance with the releasing plan or catharsis process. Such a process is respondent following environmental conditions.

Thus emotion is extremely responsive to accidents. Rationality is an individual and/or communal interpretation on emotional reception or, equivalently, human susceptibility. Emotion is an interpretation of self (system) within the environment in the light of relational ontology.

History can be reinterpreted by event-centric views. Event-centric explanations lead us to the opening of a vast emotional world, because events are filled with emotional responses. Our memory is full of such history telling.

The boundless and extensive vastness of humanity, although it is indefinable logically and emotionally, originates from an unclear shelving (reservation) behavior of the human cognitive process. Cognitive processes need to be provided with certain bypass process against complete paralysis or destruction of entire system.

Cognitive process has been evolved and designed in such energy-effective manner that it may concentrate on the central concerns only and diversify the excessive flux of signals at the periphery of current cognitive processing into the very vague patterns that are mostly not hard wired. I personally believe this (shelving or bypassing) part of neural operation is deeply related to neuralgia activities which are known to be the least active chemical process and mainly good to conserve the intensity of neuro-signals.

The scientific research on the human brain during the past decade revealed that eventual rationalization works of neural process are inevitably adhesive to emotional process. Emotional process indicates modular behavior due to the plurality of more than five sensing mechanisms that are accepted generally. We also have internal and unconscious sensibility in our gut feeling and in the other parts of cognition.

Usually, modular theory of neurophysiology is based on hardware systems. However, the theorization of integrism, which is a part of the subjects this book deals with, takes a route of conceptualization of software modularity. Software modules should be metaphorically described as mirage-like phenomenon, meaning virtual reality.

The mirage can also be described through functional expression in a tensor form. The special point of this book about brain science is that dynamic modularity is theorized relying on super-consciousness, namely transciousness (newly defined in Chapter 3), which is different in a greater extent from other neuro-physiological theories.

We are aware of the old notion of apperception, and its relevant ideas are somewhat comparable with transciousness. However, the concept of transciousness and its relevant family ideas are based on a solid foundation of experimental and empirical findings in neuro-physiology. In this context, it is encouraging that we have had a great leap in building the neuro-physiological foundation over the past several decades.

"*Neuro Science and the Person*" (edited by R.J. Russell et al. 2002, p297) of the Vatican Observatory quoted; "Arbib's schemas and Damasio's somatic markers perform this mediating role between sensation, evaluation, judgment and action. Like Kant's schemata of the understanding, they operate at a level differentiated neural input, --- and ordering for the sake of determining the composition of subjects and predicates, evaluating evidence, and prompting action. What is doing the ordering? --- These schemata is the product of imagination, they provide the unity of "inner sense." --- They are the process through which the mind emerges as an agent. They also determine the inner sense of time; they do not exhaust the "transcendental unity of apperception." (from Critique of Pure Reason translated by N. C. Smith, St. Martin's Press, 1965)"

Stephen Happel, the writer of the article, expresses further "Arbib's schemas are processes, if I understand them correctly. --- Arbib emphasizes that both visual and bodily maps are coded in our neural architecture, where intelligence is a "multiple, layered plexus of properties." They exist not in a temporally linear string, receding into the past, but in a three dimensional lattice that models the subject's of being in the world. The network is an active information-seeking process."

I do not think of the network as only a process regarding the intangible information. We do not know why the general behavior of within networks is random and scale-free. However, we do acknowledge that the universe is actually information itself (Hans Christian von Baeyer, Information: the New Language of Science, 2003, Ch.25).

Nexusity of networks originally provides the base of universality that enables this world to exist in continuum and dynamically interact. Nexusity is the essential character of information, and it will have to expose its contribution to the universal structure of the cosmos in future theory.

Information should mean the intrinsic infinite number of dimensions hidden in the infinite number of cruxes of the Uninexus. The original universe, if understood as Uninexus, does not have space, time, and material, but is the unitary body of nexus consisting of infinite number of cruxes. Crux is a vacant space for action, which we call nullity.

Why do we need the hypothesis of general continuum? We need a profound understanding on continuum in order to stabilize peace status. Continuum is a body of connected cruxes. Crux in our world is a crossing of dimensional interventions. In the original universe, the Unicrux formed up of infinite number of dimensional intervention. What if we think that the unitary original crux, the Unicrux, is formed of the infinite complexity, which is the (absolute and original) situation equivalent to the null complexity of perfectly fixed condition sustained by infinite energy?

Anyway, after the *arche* of our cosmos, the dimensions of cruxes in our world had been diminished drastically due to the entanglement of less-than-infinite number of dimensional crossings. I would not regard the dimensions as those described in present cosmos physics and quantum physics. It is enough at this stage to understand it as a tensorial dimension. The tensorial dimensions can be applied in social science and even in brain science. I believe the dimension can emerge if sufficient accumulation is made by weaving or netting a large enough number of scratches into a manifold form.

The universe is the continuum. It is a gravitational continuum. It surely is a spatiotemporal continuum, but space and time is generally affected by the effects of gravity. In Planck scale, it is said that discontinuity of space-time appears like space-time foam. I would like to visualize it as space-time eddy in a complete turbulent flow. Eddy contains the effect of dynamicity due to energy variation. I reckon these eddies could be equivalent to the loops dealt with in loop gravity theory.

Maybe, the partner space of our universe, which is similar to the concept called "bulk" by Lisa Randall, could be full of turbulence that produces the compact eddies of turbulence that introduce building blocks to our universe. However, it is not easy to think of a multi-universe which drains my mental energy to the infinite depth as the actual infinity that drained all the mental energy of Cantor and Gödel.

(I think that there are healthy rules in a correct arrangement within human thinking. The rationale of infinity should converge to unity, the rationale of endless circulation of logic should be related to the duality or triplicity of monism and the rationale of nullity or imaginary entity must appear with real and integer entity.)

Planck's condition about our universe is the result of losing the fixed arrangement of infinite dimensions caused by the allowance of infinitesimal break of balance. (The break of balance or symmetry caused the dynamicity of the universe and the integrated vectors to be induced.) Thereafter, the flow of time, the dimension of which is not yet clear to physicists until now, was introduced. That broken balance is one billionth of the universe in matter-antimatter occurrence from the beginning.

It is known that gravitational potential energy is negative and the energy in the universe come from the gravitational potential energy. It is thought that the sum of matter and energy in the universe is equal with the amount of negative gravitational energy. If we define gravity as the restoration power of the universe back to the state of nullity, we have to consider the possible collapse of dimensional structure of the origin of the universe in order to generate our universe. The origin of the universe should be dimensionless (by this, I mean, infinite dimension) which is contradictory in our condition but non-contradictory at the origin.

Crux is not the point of matter(質點). However, a point of matter is equivalent to a building block in cosmology. I would name this point of matter or building block as "ontum (compound of onto + um)." The point of matter formed is of such a property that a multiple occupation of space is not possible, as we find in fermion particles with 1/2 spins. On the contrary, crux has a concurrent occupation of many dimensional crossings at a spot of nullity as we see in boson particle with integer spins. I think the orthogonality in cruxological architecture which is to be dealt in the next chapter is analogous to the integer spin of boson particle.

Crux in our world can reach the level of the least unit of time and also the least unit of information. How could time and information be essentially the same? Time and information are not separable when they have initially appeared at the break of the original Unicrux.

This explanation is just to aid the visualized imagination in metaphoric comparison with our universe. Someone might be able to solve this problem in mathematical expressions. Frankly, I believe the spin effect is due to the stress condition of space-time continuum. It is thinkable that the initiation of time flow might have caused the stress condition of spatiotemporal continuum.

What is time, not in an epistemological sense of human cognition, but in the eternal scheme of the universe? It is the cause of relation, the original source of integrismic condition. Nexusity originally generated from the ontology of time, while systemicity stemmed from the topology of space-time continuum. Logic is the product of these two effects.

Temperature and entropy of *material* led to the discovery of the atom; temperature and entropy of *radiation* found the notion of quantum; and the temperature and entropy of *black hole* induced to the understanding of the structure of space-time foam, the space-time eddy, which in this book will be called ontum (間子).

In the author's point of view, ontum could be identified as the carrier of unit information, if the hidden dimensions are neglected. It can be interrelated with the notion of "qubit," or "cont" as contemporary scientists suggest.

The hypothesis of general continuum is based on the scheme of thought mentioned above. I would like to complete the hypothesis in the near future when we approach sufficient intellectuality to do such work. I summarized the hypothesis of general continuum by listing the major concepts that the hypothesis stands on.

First, there is an endless supplier of continuum to our universe, which is named Unicrux. Unicrux has an infinite number of dimensional crossings. Cosmological continuum is spatiotemporal and has additional hidden dimensions due to the difference of locality.

Second, the infinite number of dimensional crossing is reduced at the boundary of Unicrux due to surface perturbation. Surface perturbation is caused by dimensional disturbance. Dimensional disturbance is a frame of dimensional architecture at the boundary that drastically reduces dimensional density in the very thin transition from internal surface to external surface of Unicrux. The architectural frame is the scheme of supporting structure that continuously changes due to the alteration of our universal condition.

It is imagined that the gravitational wave is generated out of the surface boundary of the Unicrux. Gravitation is the intersection of time flow. Time flow is orthogonal to the surface of the Unicrux. Time is actually the manifestation of the Unicrux that forms the major axis of the continuum of the material universe. We reckon that the gravitational wave is the cause of the spatiotemporal continuum.

The Minkowski continuum of spatiotemporal manifold has an imaginary time and a real space of three dimensions. The author understands the quaternion of William R. Hamilton as an expression of a spatiotemporal continuum with the axis of the real time and the three imaginary spatial dimensions. We believe that there is a general correspondence between algebraic notions and reality, the theoretical stance of which is not a popular topic among most scientists, except Gödel and a small number of people. It is thought that quaternion and Minkowski continuum are intrinsically connected concepts.

To Maxwell, space was an elastic continuum. The elasticity is essentially a reserved quality of restoration potential. If all materials are generated by the quantum particles, equivalently by waves, and again by equivalent perturbation from the surface of Unicrux, our universe is subject to alterations based on the conditions of the dimensional degradation at the surface of Unicrux.

Third, the general continuum is the mix of the simplest spatiotemporal continuum and a higher dimensional continuum with the hidden warped dimensions released from the surface boundary of locality. We believe the complexity of various hidden dimensions produces the black materials and black holes.

Integrismic Phenomena Appearing in Every Universal Hierarchy

In the last decade, many writers published popular books in their own disciplines and specialties in plain-writing for the general public to understand the latest development in the academia. Such good will and deeds of scientists have helped the author who has far less expertise dare to about this topic.

We believe the universe had an original status before the initiality, which is presumably the commencing of time. If we put the aforementioned contents into a similar explanation again, this original status must be close to the Absolute Monad.

The Absolute Monad (equivalent to aforementioned Unicrux) can be expressed as a single point of whole existence or equivalently as the Infinite Nexus (equivalent to aforementioned Uninexus) with an infinite number of vacant points as shown in the Figure 5-2-2. The Absolute Monad is the source of vacuity of our universe. Unicrux is the expression of the entirety of connectivity (relationship) in Uninexus. Uninexus is the expression of the unity of substantiality in Unicrux.

It is present consensus in physics, as I know, that the original status reduces into the size of 1×10^{-9} when our universe happened to appear (Physicists claim that the original size of the universe was 1×10^{-9} of its current size.) It is due to the disequilibrium of the Universe. One billionth is the remnants of the matter-antimatter collision. Our universe is shown in the Figure 5-2-3 as an infinite number of the points of matter, which does not have nexusity. Nexusity is gaining gradual governance again as observed in entropic procedure with the overall increase of universal disorder.

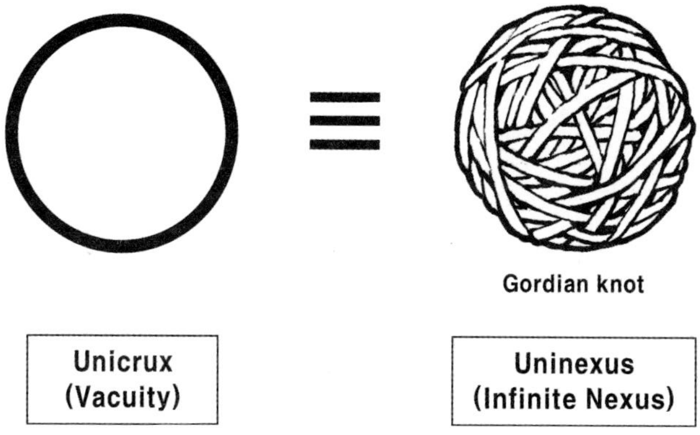

Figure 5-2-2. The Absolute Monad and the Infinite Nexus

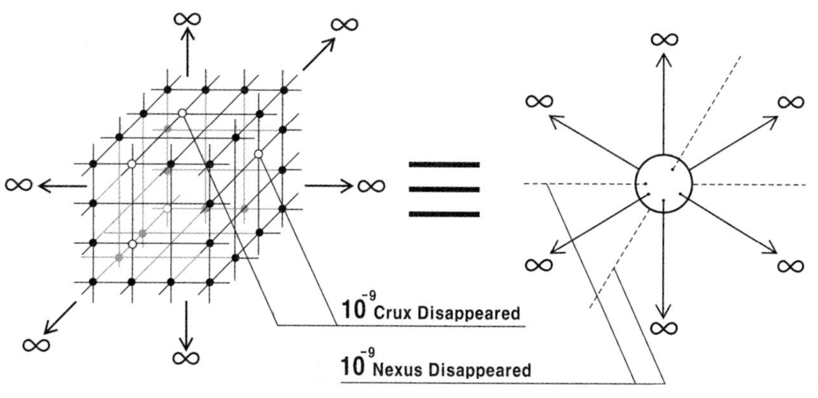

Figure 5-2-3. The Present Universe with an infinite number of the Point of Matter

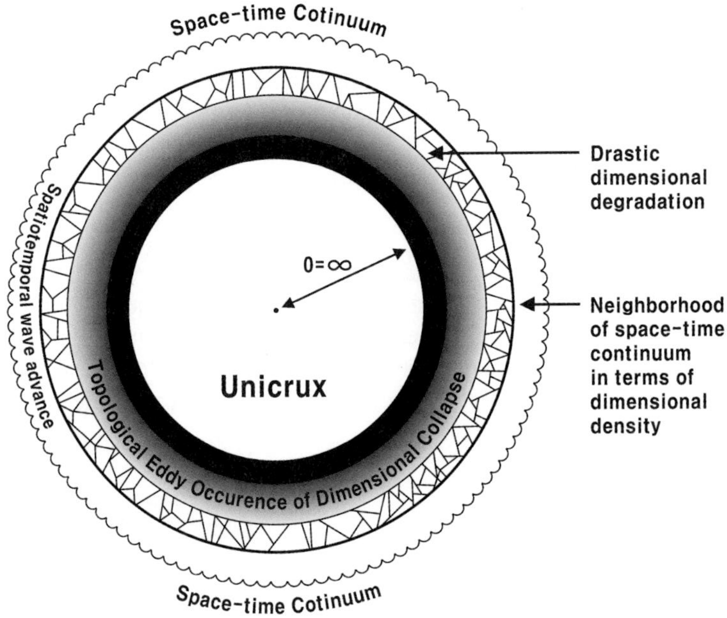

Figure 5-2-4. Occurrence of Eddy Building Block at the Boundary of the Unicrux

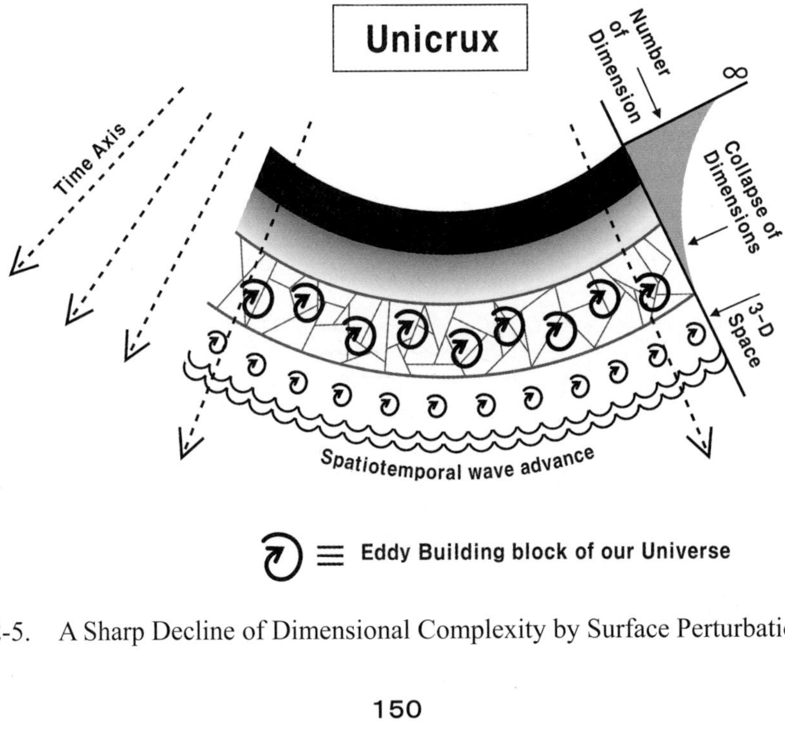

Figure 5-2-5. A Sharp Decline of Dimensional Complexity by Surface Perturbation

Integrismic phenomenon is the ambivalence of systemicity and nexusity in universal existence. When systemicity presents, materialization governs. When nexusity is revealed, continuity is attained by connectivity. In fact, the quantum entanglement of N particles can exist (when N>3), which implies that an infinite number of particles in entanglement could exist in the universe.

In a similar point of view, we do not need to deny that the whole universe at the point of origin was orderly connected as an entirety. In our degraded universe with various dynamic activities, our observation which is the intrusion of self-organizing process into the universe should have broken the entirety and brought systems with degraded cruxes and enfolded dimensions. The enfolding of dimensional intervention in a crux enables it to make the logical sequence simple and clear to sustain a system with less energy.

There is a notion of degeneracy in quantum physics and in molecular biology. This notion can be generalized for our discussion on the entangled state of existence in topological phase space in which the existence is not evenly spread. We would recommend a terminology of "generacy" to denote the evenly spread ideal status like Uninexus and Unicrux. The start of our universe indicates that the degeneracy was initiated in the original universe due to the asymmetry placed in the original perturbation.

In my imagination and understanding, Unicrux (and Uninexus) is not an ideal status but a real one (but not as our material reality) and absolutely stable with infinitely conserved energy that propagates gravitational influence over all hidden extra dimensions, disregarding how many it is, on which our universe orthogonally stands on.

In a fresh way of thought, Unicrux could be considered as the externality of our universe. It is more than the infinity and the absoluteness, which our brain has invented. Our universe could be the result of epiphenomenal morphosis from the external influence of Unicrux. It can be compared to the epigenesis of mineralization process, in which the internal surface of mineralized cavity behaves in igneous environment. It is the same metaphor as the growth of ova in a matured female ovarium. The Epigenetic mineralization produces beautiful amethyst crystals around the internal surface of a cavity.

The original perturbation of the external surface of the Uninexus (equivalent to the internal surface of our universe) is caused by the vacuum which is full of relations and null of existence. It is full of nexusity and null of systemicity. Systemicity is attained in our universe as the material mass in the course of increase of disorder in a degeneracy process.

The mass is thought to be the inertial effect of dimensional intervention. Gas flow in permeable rock experiences an inertia effect in the plexus of micro conduits when a branch junction is met. Gas and solid rock are different in density, but nonetheless both are matters.

Likewise, the universal play of dimensional intervention is the manifestation of the dimensional density problem. I believe that the dimensional density problem appears at the boundary of Unicrux. At the boundary of Unicrux, there is a surface perturbation due to the degradation of the dimensional density as seen in the Figure 5-2-4.

The existence in Unicrux (equivalent as in Uninexus) is the single endless flux of the entire nexus crossing the single crux in the infinite number of dimensions. The endless flux passes through the single nullity of the single crux, which produces null stress.

However, it is in a changed situation of the vacuity (transformed from nullity), where the single endless flux of the entire nexus causes drastic changes to all the segmented fluxes, and consequently changes into complete turbulence of fluxes of infinite number.

The unitary flux is from the infinite number of dimensions and becomes infinite number of fluxes of unitary dimensional continuum in our space, which generally shapes into spatiotemporal continuum, regardless of how many hidden dimensions are with them in different locations.

Such surface perturbation causes infinitesimal eddies of complete turbulence to produce building block that possesses some resistance against the inflow of quantum as shown in Figure 5-2-3. I think each building block consists of each topological eddy of dimensional turbulence. Each quantum building block has a resistant threshold obstacle beyond which existence as a matter is allowed.

The interpretation above is thought to be the Higgs mechanism, which is known to be the only theoretical way of generating mass in Standard Model of particle physics. I wish my intuition has worked right.

L. Randall and Andreas Karch discovered that "not only can there be infinite extra dimensions, but physics in different locations can reflect different dimensionality. Gravity is localized near our universe in such a way that it's the only the region near us that looks three-dimensional; regions far away reflect a higher dimensional space.

It may be that we see three spatial dimensions not because there really are three spatial dimensions but because we are stuck to this bane and gravity is concentrated near it, while the surrounding space is oblivious to our lower dimensional island.

There are also possibilities that matter can move in and out of this isolated four-dimensional region, seeming to appear and disappear as it enters and leaves our domain." (p326, the New Humanists, 2003)

The presence of initiality does not indicate whether or not the universe has endless rhythmic expansion and contraction and whether or not it had a single Big Bang and is advancing to eternal expansion. It is recommended to not think in a way of absolutism nor teleology, even though the initiality is given in a deteriorated form of infinite dimensions. It is thought that dynamic eddy blocks of the dimensional disturbance at the surface of boundary are the actual form of building blocks that we have in quantum physics.

Building blocks enable quantum photons to appear in this universe. It is conceivable that the opening of building blocks was made by a certain initiation of the universe. The threshold of opening could be the Planck's conditions of quantum level building block.

In this frame, the entanglements of particles are the interaction of residual dimensional connectivity. Entanglement is the immanent property of our universe which originated from the dimensional infinity. At the quantum level of the universe, integrismic phenomena are presented as we have seen in the expression above.

Let us consider at the molecular level of our universe. We have many evidences that molecules interacts with each other. Molecular interactions are presented in the world of chemical elements. Due to such molecular interactions, recent interests are focused on molecular biology.

"Human brain is not that complicated" and is "characterized by a genome of only 23 million bytes. The genome is 6 billion bits-that's 800 million bytes-and there are massive redundancies. One pretty long sequence called ALU is repeated 300,000 times. If you use conventional data compression on the genome, we get about 23 million bytes, which is a level of complexity we can handle." Thus, R. Kurzweil concludes that we do not yet understand the principle operations of the human brain. (p230, *The New Humanists*, 2003)

We need to accept that humanity is capable of solving all kinds of problems not through our systematic intellect, but rather through mostly depending on our operational maneuver, harmonizing with circumferential conditions. It is a subject that tests our intellectual capacity to think in the line of relational ontology.

The universe does not operate in an isolated condition depending on the location. The manageable range of genomic information of human being could imply the vast extent of operational connection with infinite variation of circumferential condition. I would like to comment that the genome should be regarded as a communication system between internal integrism of human and external plexus of universal affairs. The presence of junk DNA may contain certain communication manual in a method of cryptogram.

At the level of neural systems, human beings consist of hundreds of different modules. Recurrent neural networks and relevant modules provide surroundings for spontaneous consistency, which is the foundation of stabilized conditions. Based on this consistency, communications with external environments are made continuously.

At the level of organs, we find that neural networks of our body strengthen the systematic condition as well as integrating the collaboration among organs. Integrism has been embedded in human physicality for millions of years.

At the individual level, we can easily imagine that individual entities show the most distinctive phenomenon of integrism in comparison to other levels. At the communal level, basic connections among individuals are shown best at the nuclear family level. Communal entities are ready to collaborate and interact with other communal entities founded on their own internal regulations.

At the level of states and nations, self-recognition of nations and states have played a major role in organizing citizen's power, and networking among nations has come in sight to prove the nexusity that forms part of integrism. It was the same trend at the level of confederacy or consociation of states in their early appearances following the nations. At the regional level, we have seen integrismic effects observed in international affairs since the Thirty Years' War.

At the global level, it is recognized by the public that globalization has been progressing since the end of colonization era. At the planetary level, the human race is just entering a new era to consider planetary situations in reality.

At the universal level, the integrismic phenomenon revealed in the universe is the consequence of pers (=French abbrev. of person) + ontology, (or, person + ontology), along with how the dynamics of the universe behave.

Integrismic Factor and the Coefficient of Strange Attraction

In the management of integrismic phenomenon, an indicative factor, and an integrismic factor is considered to be utilized in discerning the variation, the scope, and the mechanism of the complicated phenomenon.

Integrismic factors must be relevant to the coefficient of strange attraction. Strange attractors can be described as a predictive pattern of affairs' present development. There must be some jointly owned terrain in their functions because integrismic factor and the coefficient of strange attraction will both stabilize a certain pattern in the transition zone of the phase space. Such stabilization processes in both cases are to some extent relevant to the systemization in accordance with intrinsic attributes of central mechanism.

We have many patterns of networks; lattice networks, small-world networks, random networks, modularized networks, scale-free networks and so on. These networks or parts of the networks are subject to transform into systems through enfolding process of dimensional structure that is presented to us in the content of information. This concept will be expressed in later chapters too, but such transformation would allow the loop of logics to become a system more effective and simple enough to maintain the entity economically with less energy.

The coefficient of strange attraction is responsible to find the territory of habitat range of Complex Adaptive Systems (CAS). Integrismic factors find the range of transition zone between system territory and network territory. Integrismic factors will also be useful in defining the range of dynamic core formation and the range of peripheral development. In peripheral development, nexusity helps an entity to develop connectivity with the outer world. Connectivity, in most cases, stabilizes the network and then the foundation of system which possess the accompanying networks.

For both human beings and humanity, the strange attractor is projected to the anticipated future under the pre-afferential/post-efferential forecast, by which present patterns that would influence future figures through forecasting (or, pre-casting) process.

We have explained that the pre-afferential function of the neuro-physiological process is a way to find the best mix of actions for the immediate future. However, it is the opinion of the author that the pre-afferential mechanism can be activated in the situation of virtual reality where the post-efferential mechanism is prevailing in an imaginary scheme (very similar to the dreaming process) in a concentrated simulative process.

We know that we can activate full simulation of our emotional mechanism of our body as if we are experiencing a future situation; our heart pounding, fast flow of blood, sighing, and so on, only by concentrating on imagined situation.

The fundamental difference between humans and the other species is said to be the substantiality of a knowledge system that develops and evolves generation after generation. It means, in another word, a concentrated post-efferential reality. The knowledge system supported and sustained by post-efferential mechanism of our brain is actually the product of a process of concentration and it is the combined residue of our concentrated imagination.

It is necessary to understand that the process of concentration and post-efferential mechanism are not the only the actuators of the human knowledge system. We examined the concept of white noise which has a broader coverage of the cognitive knowledge than the intellectual knowledge.

There is the λ (Lambda) factor which is related to the degree of Complex Adaptability (CA) of a Complex Adaptive Systems (CAS). It was reported in an early stage of the conceptualization of CAS. It is known that the habitat range of a CAS is when λ is 0.273. The number is the critical value to determine the transferring status from stability to chaos.

Integrismic phenomenon is wider than the complex adaptive system because integrism encompasses from all the range from networking to system. The system is an integer, a metamorphosis of network, emerged by the phase shift. A CAS is a semi-stabilized strange attractor between the ideal network and the solid system. Therefore a CAS has residence in upper side of transition zone, between network zone and system zone. In the lower side of transition zone, various networks reside in theory.

It would be beneficial for humanity to develop a new Lambda-plus factor to judge the critical conditions for the territory of integrismic entities, whose concept will cover the variety of universal existence.

We would like to suggest also Ω (Omega) factor which is related to the degree of uncertainty of a CAS in interacting with its environment. An Ω factor intensifies the exertion of humanity to overcome an uncertain situation through its systemicity, by–using profound rationalization of the logical process and generations of detailed internal systems of knowledge. Ω factor is related to the presence of omega path to be discussed later.

The interesting thing is the hypothetical idea of homeochaos (Complex Systems: Chaos and Beyond, Kunihiko Kaneko, Ichiro Tsuda, 2000). Homeostasis is well presented in the general metabolism of living things. Homeochaos is more concerned with dynamic stability whereas homeostasis observed in the metabolic process is very stable because of the embedment to the physicality.

Kaneko suggests "The notion of homeochaos can be used for a network like the brain, immune systems, ecosystems and chemical networks in metabolic systems, where the network should be adaptive, flexible and intelligent. The hypothesis states that the weakly chaotic states are naturally selected to acquire the flexibility in growing and evolving systems; thereby homeochaotic states become invariant, and among their systems the dynamically controlled system like cardiovascular systems is stabilized to maintain the chaotic states, namely an achievement of homeodynamic control." (p236)

Ω factor, embedded in fixed topography, measures if we can come back from the omega path which loses the dynamicity, and thus take liberty to the alpha path to regain the dynamicity.

5-3. Generation of the Science of Integrism

The science of Integrism has some aspects of the Theory of Everything (usually called among scientists as TOE). It is based on the feedback circuit of humanity which intruded into the nature. However, it is rather more significant to say that the feedback chain effect of the nature intruded into humanity. To support such connections, we have a time factor which has intruded into human cognition.

Anthropomorphism and Integrism

Human beings are different from the other species of animals and primates. Unlike other species, we have an endless transcendental intentionality toward eternity and infinity. The delicacy of human mentality is attributable to the desire for infinity and eternity. And that stress-tamed character of humanity, which seeks for spatiotemporal infinity and eternity, is the one that has provoked in humans the endless pursuit for the delicacy of human mind.

Intellectuality is the fruit of the endless transcendental intentionality toward eternity, the process which we would call anthropomorphism. Intellectuality is not, as usually thought, the source of the anthropomorphism, but rather the result.

Anthropomorphism means the attribution of human forms or behaviors to deity, animals, or anything else, which here we prefer, to future humanity. We have said in Section 3-2 that anthropomorphosis is an emerging process of intrinsic attributes of humanity, either individual or communal.

The intrinsic attributes of humanity lead us to a strange, an unknown but predictable attractor within certain constraints. Within such constraints, as seen in a steamship that follows a predestined route to fit the draft depth and the marine topography, anthropomorphosis of the path of humanity is anticipated.

"The imagination of the thought of the heart (not of mind)" is a converged consequence of the brain activities and it affects the future of humanity, to be more intense for further convergence. Since the recent findings of brain science get closer to the association of thinking process with undividable effect of emotion, it looks inevitable that the mind will be replaced by the heart in a long run.

This accelerating and intensifying process of humanity, tedious and steadfast though, is anthropomorphosis. Anthropomorphosis has feedback effects of the past and of the present humanity that lead to the formation of forthcoming nature of humanity.

Creativity is the essence of anthropomorphosis in a sense that creativity comes out of humanity and changes the contents of humanity. Creativity usually works for the acceleration of feedback effect; nonetheless, when the third kind of creativity (that is the highest and most complex form in creativity process) takes the role of inner regulation to modulate overall abnormal outputs, various healing processes for humanity can also be activated. This third kind of creativity can manage the situation whatever the remedy is. (e.g., feed-forward, feedback inhibition, or anything else.)

The plexus of historical humanity is unavoidably anthropomorphosis aiming at further formulation of more perfect and wider menesys, which is the compound of system and nexus. Catastrophes can hit the human race if humanity takes extremely drastic changes, however social regulation should take part in designing a stabilized future.

(As we discussed before, novel propositions that are not provable in the existing system of knowledge will be added successively. It is related to anthropomorphous phenomena and can be named dimensional leap, dimensional split open, or an introduction of new dimension.)

Anthropomorphism is still hypothetical but practical enough in our life and for the humane civilization. Anthropomorphosis is the outcome of human integrism. The strong point of human integrism is that humanity may seem to be wandering yet successful in finding the stress-less alpha path as far as the circumference permits.

Human beings behave as if always strengthening their system only. However, such system-prone tendency is not likely to prevail even when human intellect seems to be strengthening the only systemicity. It is due to the fact that the so called human system is not wholly system-dependent. Human intellectuality has an integrismic effect in its energy allocation. Our brain energy is well distributed to give the best balance to our activities for both systemicity and nexusity.

The human brain sees what it wants to see. We have to reach to an understanding that the stress for survival is the unalterable ultimatum presented to us from millions of years ago, which is thus engraved in our physicality. Even the undefeatable brave martyrs had such physical stress most of the time.

We do not yet have a thorough understanding about the identity of humanity. The basic philosophy of anthropomorphosis is not about how humanity will profoundly change the future of the human race in every aspect. Anthropomorphosis tells only why humanity reacts passively in the whole arrangement of the universe. Humanity is passive because it is the offspring of the universe.

We have anthropic theory (hypothetical) in cosmology. Anthropomorphosis does not have a teleological orientation like anthropic theory. But humanity is under the feedback influence and feedback inhibition in the whole universal condition. Humanity accelerates feedback effect due to the addition of the ever-deepening internal systemization.

In the process of perceiving and receiving something, humans are directed to do so.

There is only one entity in the world that composes the triple loop (environment, person (brain-body) and noosphere.) The triple loop is the source of peculiarity of humanity. Noosphere is the virtual reality of human creativity. The human race is caught by the fatality of making its own cocoon.

The secret of humanity has to do with Karl Jaspers' concepts of *"das umgreifende"* (the Embracer) and *"Chiffrelesen"* (decipherment). Humanity, having been embraced by the Embracer, in other words by the Absolute Other (please disregard a bit of difference between the two notions), pulls out of the mouth the single endless line of silk extracted from the live existence to wrap itself up before maturing in the cocoon and soaring to the highest.

The noosphere, if we agree, consists of systemicity and nexusity. The external appearance of noosphere is governed by the local contents of spatiotemporal continuum. The texture of noosphere, however, is governed by the globality expressed in systemicity and nexusity.

In the author's point of view, anthropomorphosis is the result of the combined effects of sysy and nexy. Anthropomorphosis has a logical frame supported by the combination of personism and communalism, not by the line of the coupled thought of personalism and communism as explained in Chapter 4.

Denis Noble states in his "*The Music of Life*" (2006) that the reason we can naturally think of the self as an object is that the self is always associated with a particular body. David Bohm converses that "self-image has a sense of godlikeness," "time also bring in the notion of godlikeness," and "it's so exhilarating that the body itself wants to hold it." But Bohm said that insight is beyond time and dialogue is a gateway to insight. He explained that dia-logue means "through-meaning". Thus we can get into the world of the "meaning flowing", which allows us to co-exist in a coherent way, creatively. (*Thought as a System*, 1990, Chapter of Sunday Afternoon)

In this context, personism ideally establishes the individuality with the body (體) of thought mainly based on systemicity, and the communalism with the usage (用) of thought that is mainly formulated by the connectivity of nexus among person. In this connection, the key factor of personism and communalism is also the combined effect of systemicity and nexusity.

Now, let us think of autism to clarify the practicality of integrism. The symptom of autism is known to be attributable to the fact that the imbalance of activity between the right and left brain causes the reduction of sociality and that an excessively systematic brain activity takes its place. Such opinion supports that integrism is a judging parameter of personal manifestation. A person is manifested in a community by not only a systemized personal character but also a networking ability.

However, at this point of discussion, I would waive to say that the brain activity based on human physicality is the sole foundation of the integrismic phenomenon. I would like to keep such comments to myself until we get further support by more profound scientific findings.

Identity of Noise: Nexusity Connection

The world is filled with noise. Why do we easily ignore and neglect noise? Is noise of no necessity? Is not noise a real substance of the world? Why does noise-like nexusity not attract our concern when we are concentrated on the systemization of our life?

Background radiation of the universe has a similar image to the background noise of our everyday life. It originated from the instance of initiation of the universe. Background noise defines the character of our living environment.

Noise is similar to background radiation of the universe. Background radiation is the evidence of the residual indication of the earliest temporal moment right after the Big Bang. Noise at the bottom should contain the characteristic effect of the initiality. That is an igneous rock layer in the geological metaphor. On that basis, sedimentary rock layer precipitates. The sediments are the combined influence of the environment. It is similar to the fact that we have a background radiation and some varying radiations due to the local changes in the galactic journey.

Noise around us is the indication of our nexusity; it is the effect of nexusity that surrounds us, but identified as noise because the effect is observed by our systematic viewpoint. Noise and nexusity both are of complex property. We usually think that complexity is the property of systems.

If we go a little deeper with our thought, we can find that the whole world is in a full complexity that we cannot even describe. This means nexusity is a greater source of complexity than systemicity. Nexusity itself does not have an organized viewpoint unless the wholistic universal view is presented.

Herbalism indicates that plant chemicals sustain the complex condition of homeostasis of earth by responding to and adjusting to the changes in ecological conditions. Such biological feedback chains are generated from a complex connection of innumerable elements of the nature. Utilizing the herbalistic view to understand the concept of noise is quite helpful. Plant chemicals work as "a message from chemical communication" while prepared medicine is simply a noise to the ecology of the earth. The network of human system is a noise to the nature and vice versa.

I believe there is a primary nexusity coming from the infinite Plexion, the whole universe. This primary nexusity is not interpretable by our cognitive mechanism which is developed based on the inputs from our Pragma. We can imagine some effect from the secondary nexusity of Pragma that just surrounds us.

Noise is mainly the indication of nexusity around a specific subject. A subject communicates with its externality in its own internally originated systematic manner. Thus subject-to-subject communication is supposed to be made in a mutually refined manner, having been cultivated together for a long time and successfully creating an inter-subjective phase space.

Noise, in most cases, is the un-interpreted residues of our cognition, not much of significant trace of communication. Cognition is sometimes restrained by cognitive capability and in some other cases is limited by miscommunication coming from inability in maintaining a mutually communicative order and system.

The main issue with regard to noise is the identification of the source and the extent of the influence of the concerned noise. The next major issue is the relevant quantity of the noise, most probably the secondary nexusity. The relevant quantity means the intensity of intent, and whether or not the intensity of mutual communication would be communicable, which is directly related to the density of nexusity, more exactly the density of the flux of nexusity.

We can find another interesting remark on the noise interpretation related to network science: intrinsic and extrinsic noises. Csermely comments that various parts of network, like modules, motifs, or elements, may cause noise to another network segment, which we would call the intrinsic noise. The Extrinsic noise comes from the environment of the network (Swain et al., 2002).

In the Internet and in the microchip, the intrinsic noise can be 100 times greater than the extrinsic noise. There are many other systems in which two noises provide a similar level (Argollo de Menezes and Barabasi, 2004). It is noted that the total noise of a gene transcription set was found to be dominated by extrinsic fluctuations and was therefore a function of the network interaction (Pedraza and van Oudenaarden, 2005; Peter Csermely, 2006, Weak Link, Ch.3, p49)

We talked about the primary and secondary noise. The Intrinsic and the extrinsic noise belong to the secondary noise. Csermely states that what is good or purposeful is called a signal and what is disturbing, undesirable, residual is called noise. It is so when systemicity is not considered as an independent parameter for network. But our focus is to distinguish systemicity and nexusity in the integrismic phenomenon.

The Most representative three noises are known to be white noise, 1/f (pink noise or flicker noise), and $1/f^2$ (Brown noise, following Brownian motion); the three are different in the quality and character in the noise spectrum.

White noise is related to fully random fluctuations with no correlation in time. Brown noise is present in diffusion processes with no correlation between increments. Pink noise has correlation with all time scales.

It is interesting that pink noise occurs also in a power spectrum of human drawings and reveals the touch of humanity. 1/f pattern that appears in cartoons and Indian ink paintings has a balanced mix of unexpectedness and expectation and thus shows the world of human abstraction. The white spectrum has the maximum unexpectedness which brings no significance. Realistic drawings meet the highest expectation for human cognition. (Musha Toshimitsu, *The World of Noise*, 1993, Ch.3)

Musha suggests that noise is not oriented to a certain direction. When we feel noise, it is like–gripping a tentacle in darkness. I believe noise is the output spectrum of the boundary searching effort of human cognition. Search for the boundary of our cognition is related to those unknown perilous affairs that might happen at the boundary such as, expansion and contraction, drastic variation in density profile and reflection and permeation.

Noise is usually characterized as the sum of sinusoidal waves. The distribution of the constituent sinusoidal waves follows the below equation:

$$P = cD^{-\alpha} \qquad \text{.......................} \qquad \text{(Equation 5-1)}$$

Where P is the contribution of the sinusoidal wave, c is a constant, D is the frequency, and α is a scaling exponent. Pink noise has α between zero and two, which implies a fractal dimension. Pink noise contains disturbances equally on all time scale, which means that it is scale-free. (Csermely, ibid. p324)

A human being has a special ability of concentration and memory. If an event is strongly engraved into our memory, it will not generally be forgotten for the rest of life. A person cannot escape from it. We are the result of our morphosis process on all time scale. Each one of us has a personal metaphor that consumes us from the past. That is why we are familiar with the pink noise.

Pink noise is the status we create in our cognition, a status with a properly schemed condition characterized by the least level of concentration and a widely applicable memory. If we link noise with the concept of integrism, we would find it very useful in application. Noise is connected with the concept of nexusity.

Besides that, we have discussed about the noise from our communal relations and the origin of human transciousness in Section 3-1. The embryonic development of "the fractal-third" in the buffer zone of human cognition generates room for asymmetry which in turn gives the procedure, order, orientation, and emotional feeling for time flow.

The fractal-third could be the source of triviality, the noise in our consciousness. It is noteworthy that triviality (three road junction in Latin) is deeply connected to the introduction of novel dimensions. Triviality is a noise at the initial stage. Integrism is the way novel dimensions are introduced by connecting with other external entities.

The Generation of the Science of Integrism

The science of integrism is a systematic description about integrismic phenomenon. Integrism will be defined later in this chapter; at this point of discussion, however, integrism will be understood as a universal mechanism that manifests integrative existences in a spatio-temporal continuum.

The Quantum and molecular views have not found an answer to explain about how the infinite numbers of existences in the universe are connected and mutually relevant. Quantum entanglement has been discovered in scientific tracking. The relativity theory and quantum physics together is expected to bring a solution.

However, concerning humanity, which is basically originated from the relationships among carriers of individuality, science could not achieve a thorough grasp in the chiasmic ontology of the universal continuum. Such character of humanity would be the main reason many philosophical attitudes have been coexisting in the realm of humanity with unsolvable confusions. The science of integrism, despite its peculiar name, was thus designed to find a scientific foundation to treat the complex and confused problem of humanity through a logical approach.

As the first step, we have listed the basic rationales of which we approve of their credibility. In this writing, reasoning of the system of these postulates was made. We illustrated a view on the cosmological integrismic phenomenon. However, in order to launch the science of integrism, we assume the following three postulates.

The three basic postulates of the universe in the science of integrism are: (provided that these are deemed to be the minimum set of meaningful postulates for the science of humanity)

1. The universe is interconnected as the continuum.
2. The universe has ceaseless flows of contents within itself.
3. The universe has the flow of vacuity moving in and out.

The vacuity of the third postulate is not familiar, but it is designed to reveal the meaning of building blocks, which, I believe, is the equivalent notion of the Planck condition. Only vacuity can theoretically be transferred in and out of the boundary surface of the Unicrux.

Nullity in this writing was presumed to be non-identical to vacuity. In human mentality, nullity is regarded as the non-substantial virtuality in a phase space. Vacuity was adopted here to present the idea that nullity is nothing more than a referential point for human perception. However, in explaining the origin of the universe in this study, nullity has the definite meaning of none-ness allowing dimensional piercings without stress, which is the dimensional disturbance.

Vacuity permits an essential and deeper meaning to the thought of "*creatio ex nihilo*" when we combine it with the notion of nullity and *nihil*. Perhaps "*creatio supra vacuitas et ex nihilo*" (Creation beyond vacuity and out of nothing) would be a better description of creation.

Physics has ascended to the threshold of universal understanding in the last few centuries. During the last 100 years, the relativity theory and quantum physics have evolved together to become the M-theory and Standard Model, which have produced ideas of the hierarchy problem, super-symmetry, and Branesville.

Lisa Randall, in her *Warped Passages* (Ch.24), notes that the most important finding of theoretical physics in the last decade was the obscurity of the definition of dimensions and the appearance of extra dimensions. It is worth waiting for the experimental results of LHC of CERN as Randall and her friends are doing in order to take another step toward completing the theorization of localized gravity.

However, it is known that the general n-partite GHZ (Greenberger-Horne-Zeilinger) entanglement state can be understood as a generalization of the tri-Borromeo loop. (Ch.17, *Entanglement: The greatest Mystery in Physics*, Amir D. Aczel; Ch.9, Entangled World, edited by J. Audretsch)

I would be pleased if no fallacy is found in my assumption that N-partite situation is on the whole no different from the infinite-partite condition. If the infinite-partite entanglement is deemed to exist physically, ontological reality needs to be understood differently from past interpretations.

Existentialism has been developed since the birth of the epistemological standpoint. Existentialism is actually the epistemological revision of ancient ontology. Ancient ontology has given various implications depending on the differences of cultural, temporal and racial aspects.

(It is merely the intuitive thinking of a plain intellectual that the Planck condition needs to become the subject of deeper research to clarify the variable intrinsic boundary conditions of the universe in terms of dimensions as I discussed before in this chapter.)

But, it would be enough in this piece of work to realize that ontological reality of the universe could be reinterpreted in a totally different way. This means we need to find harmony between ontology and epistemology.

The conflict between ontological and epistemological attitudes has been present since intelligence appeared initially in *Homo sapiens*. They originate from the nature of the human brain.

Carl Sagan (*The Dragons of Eden*, 1977, Ch.7) expressed that the dextrocerebral (or the right brain) even perceives unrealistic patterns and has a potential discomfort to the real world when compared with the sinistrocerebral (or the left brain). It was interpreted that the right brain provides institutional foundation for intuitive understandings by setting up a boundary of phase space in the thinking process and constructing a mechanism of correspondence between two different mappings and between localities of mapped information.

The right brain tends to be in anxious states since it feels greater stress due to the fact that it has to use more energy caused by the accommodation of more data and information to designate a new change of locality and also by the addition of new dimensions and sometimes even coordinates.

The quantum wave does not need a medium to propagate as needed with classical waves like sonic waves. It is thought that the Planck threshold value might behave like a medium which generates the pseudo-inertia effect for quantum waves.

The Planck threshold value might present omni-oriented ontological tension that creates equivalent effects to the hypothetical and long forgotten ether. It is known that the light has only two elements of transverse wave without longitudinal wave.

In this way of thinking, the etheric phenomenon was considered to have noncompressible properties that do not derive from longitudinal waves and possess solid-like properties equipped with configurative elasticity which permits only transverse waves. (Takutsi Tsuzuki, 2002) Anyhow, this is an adequate place for commenting on the representative theories about the universe.

In the last few decades, some of the most leading scientific thinkers apparently have gotten closer to certain integrated pictures in their understanding the universe through information theory (H. C. von Baeyer, 2003, *Information; the New Language of Science*). Information is said to be the essence of the universe. It means that realities originate from non-material source and cause.

They come from yes-no (yes or no) queries and their subsequent answers. This query and answer is information content that eventually came to connect with meaning, thus being named "semantic information content." The unit of information content, a qubit, is suggested by A. Zeilinger and C. Bruckner. Qubit is defined to be an ultimate constituent of inanimate material.

Once semantics has been admitted into consideration, the total structure of the system is supposed to be realized. The structure of the universe as a system has a lot to do with the relativity principle and quantum entanglement. The information theory is regarded or expected to be one of the most representative developments of modern science; quantum physics, the relativity principle, and the information theory.

The information theory, as we see it, does not give sufficient room for humanity and even tends to show the lapse of humanity. Humanity is something more than information. The science of humanity, if properly designed, must be greater than the information theory, encompassing and even surpassing, the ultimateness of it.

The information theory looks like it is getting closer to Huayen-metaphysik in Buddhism, which is prone to the pure logic of mentalism. Oriental thoughts have experienced millennia-long systematization processing based on the intensification of logic, which had alone led the civilization into a mental shrinkage as the result of disregarding materialism.

Good balance and exact symmetry of logic are found in the Book of Changes (I-Ching). This book led ancient oriental intellectuals into traveling on closed circuits of thinking, which alone did not permit people to venture for the dynamic frontier utilizing human free will and decision.

I-Ching presently has several versions like Chou-I (周易, being the most ancient), Song-I (宋易, Song dynasty) and Cheng-I (正易, Korean) and probably could become as variant as the human race changes in the future.

The above variability of I-Ching is caused by simply altering the method of the procedural combination of major variables, the eight signs of divination, in accordance to philosophical stance. (Lee Shen, *The Illustration of the River of Chou-I*, 1991, Ch.5) I would not ponder into the depth of I-Ching, because I always prefer to stay at the initial status of thought-development of any schools of thought in order not to take the burdens of contamination from precipitated humanity.

A similar comment could be made for Huayen-metaphysik, from which many contemporary physicists are inspired through its smooth logicality. Through these genuinely logical ideas the ultimate development and extension of human thoughts up to the total inclusion of human mentality were made possible.

The pitfall of the information theory and some other "theories of everything" is that they avoid dealing with the issues of the person. The importance of this research lies on the fact that it tries to theorize person-related problems. If we frankly consider the scholastic products of current neurophysiology, becoming a human is factually a transformation process through humanizing the unprocessed raw information received by sensory organs.

The humanization process engraved in DNA is not only a hysteresis or hysteretic record of a human entity, but also includes a tensor in a manner of intentionality coping with encountered environments as a person. This essential subject of humanization is far from the debate between evolutionism and creationism.

Is there any way we can cope with the human alienation problem? Human beings are bound to individuality, except for those who do not hesitate to abandon humane physical nature in order to attain super-biomechatronic abilities. Humanity is essentially communality. Hence, the communal process must be discussed before the scientific consideration.

In this context, we have to think about the human "integer" problem. Integer(正人) is relatively well defined in mathematics. In Latin, it is defined well in usage. But, in English and in general sciences, the integer needs to be redefined especially in humanity terms which we are going to restructure. Integer is a whole entity who can be restored to integrity.

The integrity of an integer is subject to the command of systemicity and nexusity. We agreed to simply call systemicity, 'sysy' and nexusity, 'nexy'. Why do we have this dipolar tendency in humanity?

Bipolarity in humanity is shown most simply in the fact that our bodies are symmetrical left and right. Symmetry is more than deeply engraved in our mentality due to epistemological systemization. Our body was given its symmetry in order to equip the orderly organs in symmetry with a specific orientation. We walk forward keeping balance. We observe our surroundings with our forward eyesight and thus we interpret it symmetrically.

Echinoid (sea chestnut) gains embryonic development through the upward growth of its back passage to the oral cavity. Nerve systems are developed around the oral ring. I think complicated human development progresses in a similar manner essentially to bring the growth of sensory and brain organs around the oral cavity. Human systemicity is the result of adaptation to earthly conditions including downward gravitation.

The life form is a local expression of the person in a local adaptation. Earth life forms are not built for Mars, although the two planets have remarkably similar environments. The person is now seen to be the characteristic integrism that makes the concerned entity resides in the local condition.

Usually we consider systemicity to be the main character of humanity. Sysy is the expression of condensation without volatility, which denotes here the ability to hold each integer's share in the society, such as thought, commodity and design.

Sysy has significantly less variability than nexy due to the enfolding of dimensions of nexy, being degraded down to the dimensional simplicity of sysy in response to the environment. Although the dimensional simplicity of sysy may be presented in a clear chain of logic and well governed rationale, sysy is complex in a sense of its manifested pattern which is the object of human intellect. That is why we say that systems are more complex than networks.

Anyhow, CAS will be free to transit to the phase shift to the extent of variability attained by the gain of nexusity. The degree of complexity is matched with the participant degree of nexy observed from the viewpoint of sysy because the complexity is at a degree that we cannot understand, like in the case of noise. We have now come to an opinion that sysy (systemicity) is not necessarily essential, whereas nexy is innately essential.

Therefore, sysy must be a derivative of nexy that the human intellect can understand within its dimension of rationale, the dimension which is simplified from the full dimensional complexity of affairs that exists within nature. The dimension of rationale is digestible by humanity because the human brain diminishes the dimensional complexity from the noise level to the pink level. Humanity halts its advancement further into too much systemization when its tentacle recognizes or finds a reflection of its own self-pattern in its cognitive process.

We believe that such a reflection of self-pattern is the cause of anthropomorphosis. The reflective self-pattern is not necessarily so clear as to halt systematic processing. The strong power of imagination in transciousness radiates the reference wave, which is the compound of directive indications of our self-pattern. This reference wave can be superficially realized in our feelings and emotions in the compound pattern of breath, heartbeat, appetite and so on. But more deeply, we find the reference wave in the abstractive structure of our intellectual and volitional pattern.

This reference wave becomes important when objective waves are attained in the course of analyzing data and information from the outside environment. Reference waves and objective waves are used in the hologram technology, which is, in our opinion, a strong candidate explaining human cognition concerned with transciousness. We can call this pseudo-holographic.

A gene is a seed, but it is the activated network around the gene that decides what kind of variation will emerge in the life to a newborn organism. Ideas on loose connection, decisions of mutual agreement, and so on are thought-provoking in the sense that the gene itself is which has been deemed to be a loose cryptogram of developmental history which has been recorded in a sequence of information.

The looseness reflects situational adaptability, while junk sequences support the activities of nexy as seen in the noise phenomenon. The Original universe was a world of nexy and it will become once again a world of nexy, which is stronger than sysy. Nexy should be considered to have a higher degree of dimension than sysy.

The random entropy of our universe will eventually increase to its maximum. We have islands of negentropy in the vast sea of entropy. Nexy is the source of negentropy while sysy is the source of entropy, since sysy admits only necessary information and neglects everything else in its vicinity.

Unicrux (Absolute Monad) has an infinite and complete chain of nexus. The alteration at the boundary of Unicrux has induced the incomplete chain of nexus in our universe as described in Section 5-2. The Dynamicity of our world has emerged as the time vector has been produced.

The time vector can be considered to be either one or two dimensional. I believe the existence of spin could be an indication of two dimensional time, allowing symmetry in temporal balance. However, the main axis of time can be considered to be imaginary. The mutual offset effect of temporal balance brings a sinusoidal panoramic view of the sequential presentation of affairs.

Alpha Paths and Omega Paths

Omega path is an immersion into the endless endeavor of the fractal development of a never-ending logic structuring storytelling. Fractal effects come from the fact that a person is a fractal existence. A Person is fractal because time is fractal and episodes of life forms are fractal. Ω path is different from the Ω of the halting problem in mathematics.

The Halting problem is a query as to whether there exists an algorithm by which we can predict a specific program which will halt after finite steps of computation. Ω in the halting problem is defined to be the probability to halt the computation of an arbitrarily generated program for a universal Turing machine.

In our research, on the other hand, the never-ending Ω path is concerned with humanity and thus with the ontological state of emotions in humans. We have to consider that logic and rationale comes from the structure of emotions that a person wholly intends or is orientated around.

The Ω path stops when our transcendental intentionality finds a way out (exit) through stress relief. Otherwise, we would be stress-tamed and caught by a systematic circuit of relatively simple logic by discarding numbers of extra dimensions from our circumstances. Such neglected dimensions are what we call enfolded dimensions. The folding of a dimension, the loss of our freedom, and the gaining of more systemicity are what we have been doing since the Industrial Revolution and the development of national spirits in the last several centuries.

The Alpha path is found when new stressless dimensions are detected by the tentacles of human's transcendental intentionality. Alpha paths are considered as new allowances for space (new continents), opportunities (for new social classes), resources (including mineral resources), technology, religions, cultures, etc…

Humanity has alpha paths from both nature and nurture. The recent finding of Emma Young (p29, 08-07-12, *Newscientist*) implies that both genes and experiences of precedent generations determine our genetic make-up. Evodevo or intellectual design is not a matter of concern of our viewpoint. Evolution has been conceptualized by the sequence of the temporal causality, while creation should be reinterpreted by the persontological view into the whole plexal dynamic affairs according to our view.

The phase space of causal rationales has to find relevance with logical explanations of the infinite entanglement in the eternity (*infinitus implicatio in aeternitas*). DNA is the decoding device of a biophysical entity from its encoded past, despite not being regarded as a commander in position. If we accept that DNA is partly a tool used to communicate with the universal history manifested around us, we can be more confident of our existential destiny, since it would open up another dimension of the alpha path.

The evidence of the alteration of alpha path and omega paths by integrismic change is shown in a discussion about the change of DNA on an astonishing level of order when a cell is exposed to strong stress. For instance when a bacterium experiences a hostile and fatal threat, bacterial DNA organizes itself into a crystalline pattern, which provides greater chances of survival. When the threat is removed, the coil reverts back to its original state. (Joachim Bublath, Chaos im Universum, 2001, Ch.12, Der Zwang zur Ordnung)

Survival Interpretation of Preafferantial/Postefferential Leaps

Humanity has developed an unsurpassed toolkit for survival, which is the mentality of the human brain. Thinking certainly is governed by neuronal activities, which is materialistic and is considered also to be a computational process, with a contingency plan prepared for human survival.

We know that the hippocampus and the amygdala have instant warning functions for contingency based on emotional processes. From a different viewpoint, the neuronal computation of these warning functions is accelerated by introducing topological means for quick decision-making, using dynamic modules to match simple–warning symbols which represent perilous situations. Such warnings are emotional but must also include precedent signals.

Human brain constructs its own epistemological world by mending together the vacant blanks in the phase space of cognition. Such virtual reality of the human brain is the evidence of soldering or patchwork, but could also be an indication of the infinite extension of the temporal axis to eternity. Such an extension is another necessary condition of becoming a human.

The survival instinct of the brain has been widened to the entire scope of cognition. It strengthens sensciousness and consciousness, and has brought great advancements to humanity by introducing new dimensions of abstraction, post-efferential activity and transciousness. We believe that the human brain has a computational standard at its maximum alert, which is closely related to infinities-matrix. Harsh human competition is the result of the maximum alert, infinities-matrix is the survival interpretation of humanity in between Plexion and Pragma.

Infinities-matrix is responsible for formulating the standard wave, reflective waves given by situational adaptation, and the deviation between the two leading to hologram interference. Humanity as a mediator between Plexion and pragma takes the roles of interpreter and orientator respectively.

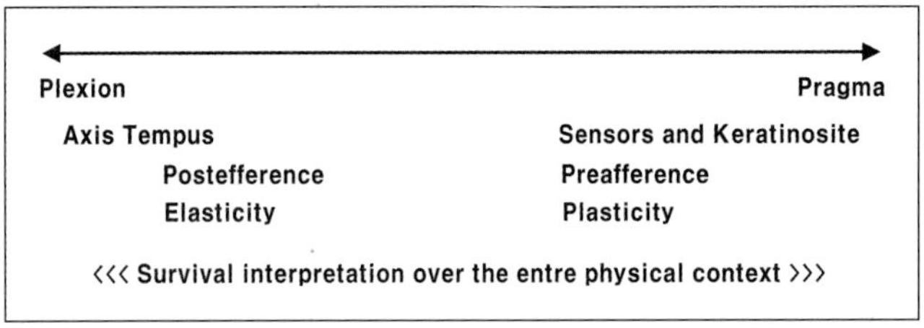

Figure 5-3-1. Survival Interpretation of Preafferential/Postefferential Leaps in between Plexion and Pragma.

Crevolution Process; Creationism and Evolutionism

Essentially equivalent are the debates on whether creation or evolution has the right answers regarding complementarities of some structural, ideological conflicts; the hen (chicken) or the egg, DNA or protein, logic or reality, normalism or realism, Alminianism (free will) or Calvinism (deterministic).

Let's suppose that the time axis does not exist, in other words, that the time axis is at the state of vacuity. If we hypothesize on a deeper level, supposing the time dimension to be at the exact threshold level of Planck's condition thus causing the universe to be presented in a two dimensional infinite plane, then all the causal hysteresis of the universal history would exist in a pattern of a compressed single plane. ("Hysteresis" is meant to denote the notion of the path-selection effect of decoherence phenomenon in quantum theory.)

It should be emphasized that the infinite loop of the cause and effect phenomenon is contained in the singular vacuity plane of a temporal axis.

This is a scientific expression of the Divine Volition and an un-manifested holographic step of the Creation. It looks very much similar to the loop of human DNA which includes all countermeasure solutions, which have been discovered based on the decoded interpretation in human physique of all the external influences which exceed the critical tolerance of manifested humanity in the history of the universe. (We believe that DNA is a tool with a communication mechanism of human Pragma with regards to the entire Plexion.)

In this sort of loop, the probability of inclusion of such paradoxical sentence such as "I always tell lies." is diminished. (We intuit that paradoxical solutions are left in the forms of intron-junk. If so, introns become the desperate last measure to treat untouchable and unacceptable irrationalities such as the "infinity matrix" and "paradox frame." Introns could perhaps be the linking parts of a function, in relation to the time axis, for decoding cryptograms, when communication is made with ontological noumenon (the whole eternal reality, considered identical with Jaspers' terminology to indicate "Das Umgreifende"- 包括者))

This cryptogram is not necessarily of digital but is communicative with Plexion in the manner of logic records of the individual entity and its logbook of experience of all accidental affairs.

The stance of this book is "crevolutionism." Crevolutionism is a complex of creationism and evolutionism. Creationism and evolutionism are of mutually different territory.

The realm of creationism, based on its exact meaning, is in the category of theophany. Theophany is revealed only if God intends to manifest Himself. No creature can advise Him in principle. As I understand, creationism is left safe, or critically unchallenged in the Creation. There is no theory in existence with proven evidence that could explain what happened at the start of the universe.

But in other innumerable occasions, creation and evolution could go together like both sides of a coin. That is what happened in the quarreling history between fatalism and free will. God is transcendental by deism. But God is able to intrude in this world whenever He is willing to.

In the realm of evolutionism, if evolutionists do not challenge the absolute position of the Almighty, would explain all but the Creation. Actually the Creation is not the concern of evolutionists but of atheists. Why do we not leave the Big Bang and Multi-Universe behind and focus on the creation of life?

The theory behind the creation of lives is debatable. Species are the source of discordances. Creationists would say that God gave unalterable differences among species. If the evolution is achieved in a stepwise or stair-wise pattern, which means that discontinuity is present in the process, should we really insist that there is no room for the intervention of a complete outsider? Let us stop here to think in this line of thought. We had better make a fundamental common ground.

I would dare to recommend another way to change the meaning of crevolution (創命). As I mentioned above, crevolution (創進) equals creation plus evolution. Evolutionary creation and creative evolution are not enchanting words to me. Hereafter, I would put **crevolution (創命) = creativity + revolution**.

That is not only crevolution but also the Crevolution in which the Person, who is not necessarily human but strictly the universal integer in the persontological view, has the lead in the universal history. It is a biblical stance we find in the Book of Proverbs of the Old Testament, Chapter 8 commenting on the Wisdom as the Creator of the universe. If anyone does not want to use capital letters, so be it as they want. No problem (The usage of capital letters for this terminology can be left as a choice for an individual to make.)

On the side of humanity, evolution does not work flawlessly because frankly speaking, it is not teleological. Humanity is not teleological even considering just the history of Galaxy, which is far shorter than the whole universal history.

However the non-extinguishable desire of humanity aims for survival and anthropomorphic achievement. That desire has been revealed in our creativity. The number of revolutions and leaping exertions to overcome our harsh reality tells the truth about us. These leaps in revolutionary adaptation without losing our essence of humanity will endure in the futures to come.

5-4. The Science of Integrism

Definition of the Science of Integrism

Integrism is the capability of integritization, which means how human society can be united and will sustain the unified condition under integrity.

Integrism is a measure of the integrity of humanity. At the individual level, the integrity of a person is the governing factor of a person as an integer in a community. A person is most effective as an integer when his degree of integrism converges to unity. When the degree goes to nullity, the integer is autistic and closed to external circumference. The openness and closure of an integer, or equivalently of a person, depends on the communication process between its internality and its externality. Integrism consists of systemicity and nexusity. Integrismic ratio between systemicity and nexusity gives insight into the internal attributes of a person. Once internal attributes of an integer are clarified, the structures of personal and communal behaviors can be understood better.

Integrism governs the entire process of humanity in physicality and mentality. If the integritium is to be planned, integrism would be the complex prescription of external or apparent nexusity and internal or intrinsic systemicity. In provision of the future of cruxological development, we need the generalization of scientific thinking in a global scale, which is the backbone of internal systemicity ($S_{i,\ i=\text{internal},\ S=\text{systemicity}}$), and the interflow of specified humanistic activity, which is the spread of apparent nexusity ($N_{a,\ a=\text{apparent},\ N=\text{nexusity}}$). Integrism represented by S_i and N_a, and cruxology are factors that affect the topics on value system.

Integrism is a conditional compound measure of the integrity of an integer, which is an individual constituent of a community, in compatible terms of systemicity and nexusity. Community is a generalized notion of an intermediate set between the constituents and the whole.

We can describe integrism as follows; integrism has a combined mechanism of systemicity and nexusity, is a tendency to acquire dynamic balance and orientation in the transition zone, and is a qualitative and quantitative parameter.

We have discussed many times in this book that Integrism is a combined mechanism of systemicity (sysy) and nexusity (nexy). Sysy has the mechanism of nesting as we generally dealt in the network science. Nexy has the mechanism of extension and diffusion. The combined effect of sysy and nexy is presented as α, a scaling exponent, in Eq. 5-1. We need to find how to denote the combined effect.

Integrism is a tendency to acquire dynamic balance and orientation towards a strange attraction in the transition zone. Humanity and the human brain cannot sustain their identity unless they maintain dynamicity. If we are static, humans cannot be deemed human anymore. By remaining dynamic, human becomes the self. The self of a person changes but has a continuous successful orientation towards, the strange attractor in ever-changing environmental conditions of the transition zone.

Integrism is a qualitative and quantitative parameter. We can administer the acceleration and deceleration of human history, as a whole or sometimes in separate parts. Because of this reason, it is crucial to find out the mechanism and relevance of integrismic factors with situational environments in various conditions. The dimensional settlement in an integrismic environment is achieved by sysy whereas the admission of new dimensions is done by nexy.

Definitions of System and Network

We have used the terminologies of system and network for a long time, and intensely in the last half decade. However, it is important that we find the right usage of these two words with distinctive definition.

The word "System" derived from the Latin word, systema. System as a vocabulary has experienced a definite change with the enhancing life style of the human being after the Industrial Revolution, its meaning started including, also, the mechanical view on the universe. It is not only the responsibility of F. Bacon and J. Locke to have provoked the mechanical way of thinking.

Some states that the root of the historical development goes back to the fire and faggot for G. Bruno. The obstinate decision to execute Bruno, by the religious court, had influenced G. Galileo and some scientists to ponder solely into finding the cause and effect of the mechanical aspects of nature.

The notion of mechanism has survived and developed behind our subconsciousness for the past several centuries, even constraining the flow of our liberty within a conduit under the control of our subconsciousness. Mechanism is a character of a system. The current civilization is too firmly grasped by systemicity.

We are about to enter into an era of many alpha paths detecting many candidates of novel dimensions which will be introduced depending on the impending decision of the human race. The frontal parts of human society are about to pass the bottom of dehumanization process attributable to excessive logics and materialistic philosophy.

However, it does not mean that we are going back to the beginning of our civilization when illogical and superstitious trends were dominant. We have to revive only the essential vividness of life from the past.

A system is an isolated entity partitioned by boundaries. Precisely, we need to adopt another notion to proceed with our discussion. "Systema and nexus" are considered as parts of the whole.

Systema is adopted from Latin in order to disregard the effect of isolation and boundary in our ontological (or persontological) frame. It is used differently from the word 'system' in the sense that systema is an undividable part of menesys. System is a result of enfolding unnecessary dimensions of networking node to satisfy systematic function and efficacy.

Strictly speaking, systema is a peculiar kind of nexus that has a drastic decrease of dimensional complexity compared to nexus. The dimensional complexity is what we call nexusity. Nexus is identifiable by human cognition because it has sufficient dimensional degradation to be recognized.

If not realized by humanity, it is no more than a form of noise. However, we know that noise has many levels of differentiation. Brown noise is the most realistic and pink noise is the most familiar noise for humanity from which creativity is activated.

Nexus, the general and universal network realized by humanity, should be scale-free to include every cognitive existence in the human phase space. Such human phase space is what we call pragma. Nexus contains all systemas and partial nexuses which are fragments of nexus.

As discussed in Section 4-1, Dyon (dynamic ontology) and Trep (transcendental epistemology) has a structural correspondence to the contention between centripetality and centrifugality in human mentality. Humanity is the outcome of balance and antagonism. Dyon understands the complexity of existence. It is a way of human cognition finding parts in the whole and the whole in every part. Humanity is always concerned with the wholeness and sensory parts in physicality.

Human cognition grasps "static" things and affairs by its rationale that has been established in the past experience, and figures out "dynamic" affairs by its instinct tied up with the development of real situations. Thus, the human cognition eventually understands the integrated emotional cognition of the instinctive entirety of a physical surrounding.

Dyon is generally stronger in the centripetal reaction whereas Trep is stronger in the centrifugal manner. In short, the tendency of Dyon contributes to systemicity and Trep contributes to nexusity.

Systemicity(Sysy, 體系性) and Nexusity(Nexy, 網狀性)

What is the hidden secret behind the human cognition? As we have discussed above, Dyon and Trep is the answer. They have affected us to stay in an antagonism between systemicity and nexusity. Such contention makes our brain to generate various dynamics. The complex modes of dynamic modules of a brain are the source of nervousness of our brains. Dynamic modularity has led brains to consume more energy.

Sysy is under the influence of Dyon in the development process of the human being, while Nexy is under the influence of Trep. The idea of anthropomorphosis illustrates that human transcent converges in order to narrow the scope of development down to a certain destination that is predetermined by the intrinsic orientation. It is due to the fact that humanity has small arc of loci to evolve.

The arc of loci of human evolution has become narrow because the constituents of humanity have been collected to satisfy the character of human physicality and mentality. Human physicality and mentality gave rise to the emergence of endless marches in order to achieve a communal world. The essential factor of the physical constraint is the dimensional character such as size and weight of our body which is deeply affected by the earth's gravity and the material basis of our planet. The key factor of the mental constraint is the endless extension of human desire towards infinity and eternity.

The characterized physical and mental conditions forced the human being to develop intensive and extensive communality. Intensive communality is the pattern of humanity that gets to the bottomless depth of loneliness. We have to acknowledge that intensive communality is the root of "sensus communis" developed in Aristotle-Cicero line, "sens commun" of J.-J. Rousseau. Extensive communality comes from the endless extension of desire to contain everything within our territory. The communality in the human race has highlighted convergence trends to be fixed.

The human race takes the strongest impact from extrinsic factors concerned with sociality among other species. Human evolution is supposed to be influenced by both genetic and social conditions. Habit, emotional status, ceaseless concentration for specific purposes, excessively hard training and other activities may induce different outcomes from the same genetic sequence. Humans, in this sense, are especially distinguished from the other species.

It is generally agreed that all the species on the earth, about twenty million, use the same genetic code. It implies that all species on earth have the same origin of life. Only three percent of the human gene is active which means that the remaining 97 percent are inactive. The difference between human and fish is that human gene is significantly more prone to be activated by environment and experience. (Kazuo Murakami, Inochi no Baka-jikara, Wake up Sleeping Gene, 2003)

We know that when we fix our system and let network do its job to solve our problem, it could result in leading to one solution. But if `we cannot solve the problem within the given environmental conditions, we have to go back and find intrinsic solution from our inside.

In the Dolly project, omnipotence of the mammary gland cell was activated by applying hunger stress. It is known for a long time that if we want to see the bud of an oriental orchid bloom, the temperature should be lowered to impose stress on the plant. Concentration is not exclusive to human mentality. All living cells are subject to concentrate under excessive stress. Systemicity utilizes strong concentration to cut off trivial dimensions. We called it the enfolding effect of unnecessary dimensions.

Our phase space of integrism originally consists of nexusity. If some portions of nexy are changed to sysy losing the unnecessary dimensions by the treatment of stress, the volume of phase space is deemed to be decreased.

(By fixing the phase space, we might consider that some rooms are left in the phase space. I would call it "rumy" because it is a topological space and not real. Rumy has very similar characteristics to those of pores in a porous medium. It is purely a theoretical development of thinking.

Rumy is a kind of cavitation that is sustained by the systemized structure of dimensional phase space. Rumy is a room for nexusity. Metaphorically speaking, Rumy is a permeable pore space in phase space comparable with room for Nexy. Its systemized texture structure made by enfolding is comparable to phase space for sysy. This metaphor is analogous and similar to porous lithological medium in which fluid flows through its permeable pores, although it does not look matching exactly to each other in the whole frame.)

In many cases, sysy structure could be similar to a shape of a particle in an open space. It has a free space wide open to nexusity. Nexy in this case is not affected by the packing of the sysy structure. Nexy can be interpreted like a flux formulation. Thus rumy provides inter-connectable common cavitation among individuals in its vicinity.

There is a saying that we are all imprisoned in the past. It calls our attention to think that most influential sysy are produced along with the Axis Tempus. Sysy mainly appears in the episodic rationales which are generated by the logia-submatrix lit by the eternity in accordance with Axis Tempus. Sysy takes the central role of systemizing dimensional disturbances present in the cognitive phase space through the orthogonal intensification or relational expansion.

Orthogonal intensification is the initial cruxological systemization in stress-free cases. Relational expansion is generally done by nexy in stress-tamed situations.

Logia-submatrix in infinities-matrix shapes up proper episodes in the course of interpreting eternity. Nexy forms crux by collaborating with sysy where topos- matrix makes orthogonal crossing with Axis Tempus.

Then what is onto-submatrix? It is related to the self of human and it is most likely to be originated from the first infinities-encounter which is the experience of the initial conceptus, the sperm penetrating into the surface of an ovum.

Sysy and nexy in most cases collaborate together because they are born from the same root. Sysy is the condensed or solidified nexy by proper dimensional foldings in specific circumstances.

Functional Definition of Integrism

The functional definition of integrism ought to be presented in integrismic phase space. We have met the correlation of the concept of noise and nexusity in Section 5-3. Scale-freeness of noise matches with the manifestation of the humanity's characteristic. We have discussed several times about the most probable match among the spectra of humanity, integrism, nexusity and noise.

Figure 5-4-1 presents the scale-freeness of network in ln-ln plotting. The straight line displays power law. The same type of distribution would be presented at any scale, which means that it is scale-free.

Figure 5-4-2 shows both noise spectrum and integrism. Integrism is significant in the range of α greater than zero (α>0), but mostly significant in humanity where 0< α<2.

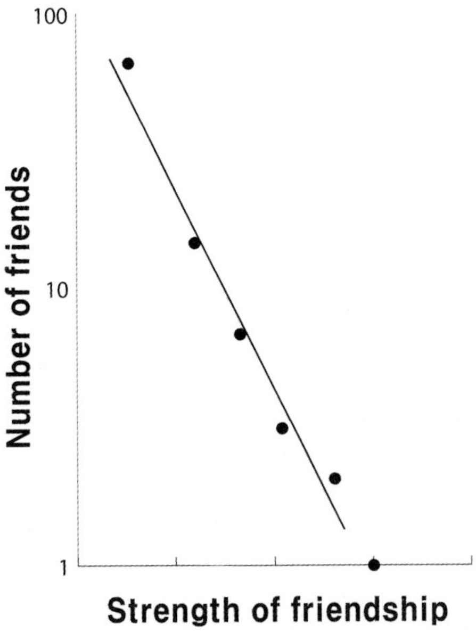

Figure 5-4-1. Generalized Presentation of Scale-freeness of Network (on double-logarithmical plotting, from p14, Csermely, Fig. 2-5-B)

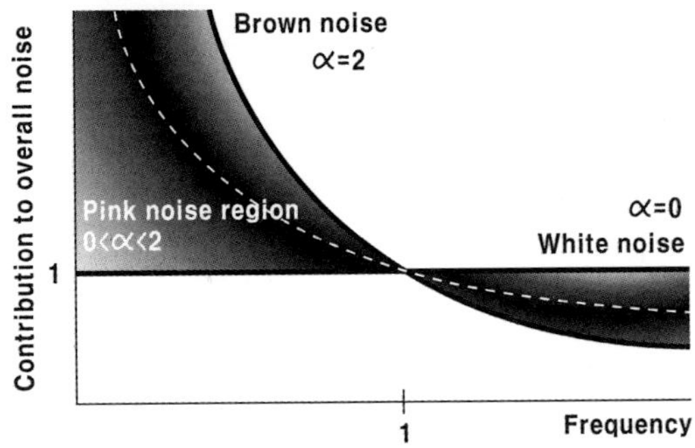

Figure 5-4-2. Noise Spectrum and Integrism (reorganized from p50 Fig. 2-5, Csermely above)

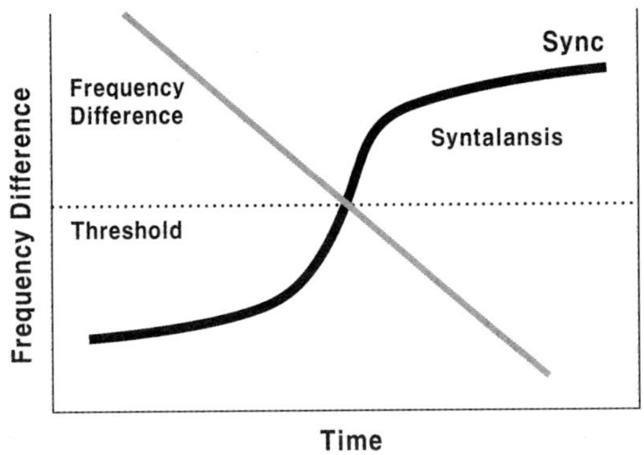

Figure 5-4-3. The Sync Phase Transition for Achieving Syntalansis (p83, Fig. 3-10 Csermely above)

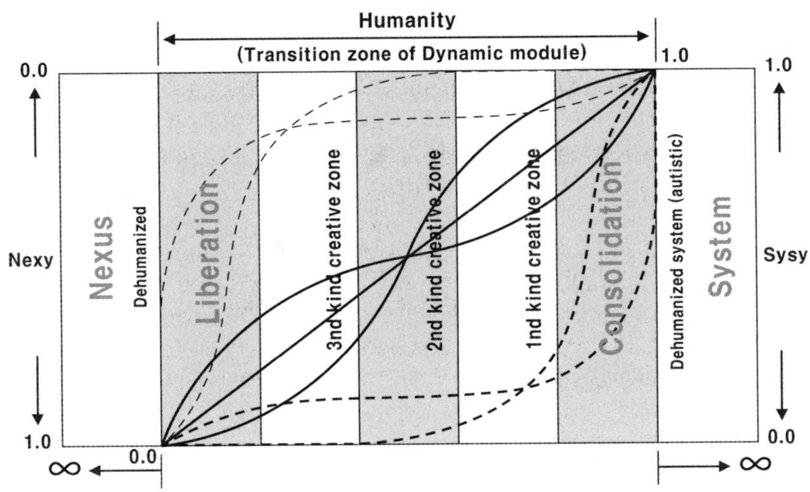

Figure 5-4-4. Definition and Territory of Integrism Related to Humanization

Figure 5-4-3 illustrates the phase transition of synchronization to achieve syntalansis. Syntalansis is defined as the extensive synchronization of the oscillators of network elements. The development of syntalansis influences the phase transition. As the frequency variance between different oscillators decreases below a certain threshold, sudden synchronizing happens, achieving syntalansis.

Figure 5-4-4 shows how network and system vary in changing the territory of integrism. Integrismic phenomenon is wider than complex adaptive system because integrism encompasses the entire range from the conception of networking to complete system. It is thought that a system is a metamorphosis of original nexus and was emerged by phase shift in the origin of the universe. Humanity is manifested in the transition zone of the dynamic module in human cognition. Various paths of dynamic modularity of humanity can be considered as seen within the transition zone of humanity which consists of the consolidation zone toward system, the first kind being a creative zone, the second kind creative zone, the third kind creative zone, and the liberation zone toward nexus state, consecutively.

CAS is a semi-stabilized strange attractor in between diffused networks and condensed systems. Therefore, CAS resides in the upper side of the transition zone, which is between the pure network zone and the pure system zone. Theoretically, various networks reside in the lower side of the transition zone and they are partially affected by systemicity of various magnitudes and styles.

It is observed that integrismic factor ought to be relevant to the coefficient of strange attraction. Generally, strange attractors can be interpreted as a predictive pattern of the present development of affairs. There must be some commonality in their function, because integrismic factor and the coefficient of strange attraction will both stabilize certain patterns in the transition zone of the phase space.

The integrismic factor is a relative measure to see the effect of systemicity and nexusity. It is our choice to use any significant formula that is designed to best fit the situation. Therefore, integrism would be the compound description of extrinsic nexusity and intrinsic systemicity. Thus, we have integrism as a function of S_i and N_e.

$$\text{Integrism} = F(S_i, N_e) \quad \text{.........................} \quad \text{(Equation 5-4-1)}$$

, where S_i is intrinsic systemicity (S=systemicity, i=intrinsic), and N_e is extrinsic nexusity (N=nexusity, e=extrinsic).

Integrismic factor which is merely a designed expression of integrism, could be simply stated as

$$I_l = N_e / S_i \quad \text{..} \quad \text{(Equation 5-4-2)}$$

, where I_l is an integrismic factor. It was thought that integrismic factors could be designed in various ways.

We had an equation for noise,

$$P = c\,D^{-\alpha} \quad \text{..} \quad \text{(Equation 5-3-1)}$$

, where P is the contribution of the sinusoidal wave, c is a constant, D is the frequency, and α is a scaling exponent. From Eq. 5-1, we have the equation $\ln P = \ln c - \alpha \ln D$, where α is the gradient in ln-ln plot. For the pink noise, $0 < \alpha < 2$. Pink noise is suggested to be a characteristic feature of a system complexity of various systems, by many researchers, such as Gilden et al, Gisiger and so on (p52, Csermely above). It means that α is an indicative parameter for the characterization of systems.

It was discussed before that noise has scale-free behavior like networks. The α parameter of scale-free network can exceed 2 and go further to 3.2~3.4 for the human sexual contact networks and even up to 4 in the power grids network (p16, Table 2.1, Csermely above). The higher number of α presents the trend of fewer hubs and more connections. It seems that such change occurs depending on the transformation process called topological phase transition.

Csermely explained that as α becomes a larger number, the degree of distribution shifts towards an exponential decrease. This implies the rapid number decrease in highly connected elements and at extreme case, the reach of the star phase. Csermely has a complete figure presentation between network variation and the factors of complexity; stress and resources.

I would like to discuss about this cruxological change to the best menesysal achievement in the next chapters to integrate the idea with integrismic condition. One of my beliefs is that complexity looks like the possession of an integrismic bidirectional character.

However, it is necessary to define the characteristics of complexity to describe the bidirectional effect of integrism. It could be seen as redundant with Eq. 5-3, but we can define

Characteristic complexity = characteristic nexy / characteristic sysy

, which is written as

$$C_{ch} = N_{ch} / S_{ch} \quad \text{.......................} \quad \text{(Equation 5-4-3)}$$

, where $(\ )_{ch}$ denotes characteristic. If characteristic complexity gets closer to unity, diffusivity prevails. If it is nearer to zero, the condensation of system occurs. Then, we would raise a question in mind regarding relationship between N_{ch} and N_e, and also between S_{ch} and S_i.

Csermely quoted the work of Derenyi et al. and Palla et al. that the global topology is best monitored by the measure G / N, where G is the size of the largest connected component of the network and N is the total number of its links.

In our study, G has to be mended by the additional effect of core-periphery (system-network) relationship for the simplest case, and N should be changed considering the effects of the diversified cruxological entities of node, module, system and so on. Cruxological entity is differentiated from genuine network by the effect of systemization.

Another way of thinking is to consider the capacity and connectability of our cruxological entity, where φ ($1 - \varphi$ is the effective size of systematic bodies) is the capacity and K (K is the effective surface area of cruxological entity for connection) is the connectability. It means we can use $(1 - \varphi)$ and K instead of G and N.

We have another implication from the so-called second part of Shannon's equation, which is the indication of the quality of transmission medium and it is written as

$$C = W \log_2 (1+S/N) \quad \text{..........................} \quad \text{(Eq. 5-4-4)}$$

, where C is the number of bit per second transmitted by certain medium, W, the frequencies range of signal, and S/N, the ratio between signal and noise.

If the Equation 5-5 is rearranged, we have $S/N = 2^{C/W} - 1$. There, S/N is considered to be related to the reciprocal of I_1 and C_{ch} as mentioned above.

For the further functional definition of integrism, it is necessary to develop a more delicate method to keep up with the new findings of network studies. The aim of this book is to partially modify and apply the technical method of fluid flow to the porous medium, as will be discussed partially in the last two chapters. A fluid flow in the porous medium has almost all the factors to be considered in the socio-psychological and political economy. How to simplify the equations of the model in order to be managed in practice is the next problem to be considered.

Inference of the Quantification of Integrism

The quantification of integrism is not so hard to organize if we succeed in collecting necessary data, analyzing them in the proper way. The usage of integrismic criteria has to be tuned up with reality in practice. We have to understand the hidden treasures of old wisdom.

It is interesting that Proportio Divina is recommended for the comfort of human sensation which is furnished by the balanced integrity of humanity. Usually people use Ø for Proportio Divina, where its approximate numerical expression is 1.618.

When we equally divide the line, we get Proportio Divina from the ratio between the whole line and the long line segment, thus the ratio between the long line segment and the short line segment is the Proportio Divina. The shorter line is the fractal-third, as previously mentioned, which will please us when it fits in to the whole line by the summation with longer line. It creates a fractal figure when it is presented in golden rectangles in combination with logarithmic spiral. (Eli Maor, e: *The Story of a Number*, 1994, Ch. 11)

The integrismic insight on the mechanism of concentration and liberation in neurophysiology provides a significant analogy to the mechanism of systemization and nexusification in the field of sociology. The social systemization necessitates the concentration process of the tenants within the national territory and social nexusification is subject to the relaxation and liberation of people.

The author states that we can measure the factor of integrism. In other words, we can find the ratio between sysy and nexy and thus can correlate the effect of integrism to social phenomena.

We know that the world is dynamic and nothing stays as we perceive and conceive through our cognition. It is only our interpretation that sees as if things are static and affairs are definite. Systemicity is a tendency of our interpretations to reduce the dimensions of our cognition down to a controllable range of action planning. Sysy prefers a static flow in its system boundary. Sysy behaves as a dynamic core of a dynamic module.

Nexusity prefers a non-static flow outside of the boundary. But nexy within the system boundary tends to behave as a dynamic periphery of a dynamic module. It implies that nexy has certain distinctive divisions of intensity as we can see in the division of the inertia flow (slightly non-static) and the turbulent flow (seriously non-static) and the non-Darcy fluid flow (non-static).

It might be possible to derive some functional relationship between these three groups of components: nexusity (flow), the characteristics of nexal condition (fluid), and the variable environmental condition of nexusity (medium of fluid flow). The environmental condition of nexusity would be actually an integrismic condition.

Systemicity is a drastically enfolded dimensional state of nexusity, which is analogous to the solid state. Nexusity can be analogous to the fluid state.

Please see the emotional aspects of integrism in the last section of Chapter 5, to acquire information about basic works to be done in quantification.

In the science of integrism, complexity makes a whole difference in the variation of the transition zone. Integrism is the cause for the complexity of humanity and of the nature. Complexity is not real but is only the result of our brain computation in reality. It means complexity is our understanding of the nexy computed by our sysy. Someone commented once that the complexity is the direct result of the computation of the cosmos. My opinion on this is yes and no.

It is thought that the integrism and the transition zone of the complex adaptive system have influence on one another as below.

1. Integrism has impacts upon the stability of concerned transition zone.
2. Integrism affects the spread width of the transition zone in terms of expandability.
3. Integrism influences the sensitivity of parameters in the transition zone.

Complexity variation in a transition zone, if adequately modulated, generates creativity. Transitional status is where self-organized motivations, thoughts, activities, and systems can emerge.

Integrism makes differences within the variation of the transitional phenomenon; the complexity of humanity, the infinities-matrix of humanity, and the complex adaptive system of humanity. The CAS of humanity is the various phenotype of humanity in human society and nature.

Integrism has a connection with cruxology. The cause of complexity occurs when crossing happens between sysy and nexy. Sysy is responsible for establishing systematic consciousness and the thinking process in rationality and logic. Nexy induces a system to extend to a new environment and dilutes the concentration for its logical imposition. Therefore, when a system of logic, sysy, meets a non-logical connection with externality, nexy, sysy produces many possible interpretations as our brain does if it encounters with new situations.

Sysy and nexy are not to be mutually allowed because they may end up with multiplication effects on complexity. Mutual crossing between sysy and sysy, or between nexy and nexy, makes complexity converge to nullity. Complexity is the matter of stress condition causing inertia and turbulence in a flow pattern. The ratio between sysy and nexy, the essence of integrismic factor, determines the cruxological development and embryology

Integrism has a connection with creativity. Cruxological condition determines the environment of creativity along with the character and the contents of creativity in a society.

5-5. Emotional Human Integrism.

Emotional Aspects of Human Integrism
Vacuity and Chasm
Void Set in the Persontological Universe
Socio-psychological Crevice in Propagating Ideas

Emotional Aspects of Human Integrism

Human integrism is directly related to the emotional aspects. Its keystone module is at the center of social nexus as we can observe similarities in the keystone species of the food chain. In humanity, family module (around mother module) and history module (around god module) are located at the center. The forfeiture of motherhood discussed in Chapter 9 is relevant to such a tendency.

Are these modules movable and subject to change as time flows? That is another issue. But I do not agree with opinions that say that the family module is a fascistic view and therefore should be altered. Such idea has been asserted by some socialist scholars without due discretion. We have to realize that we should not allow unnecessary chasm with nature to happen for it will hinder our advancing into the unknown future.

The family module and history module are tightly connected to the emotional phase space. Emotional aspects provide the integrismic mechanism to decide "nexy or sysy?"

Sysy is relatively easier to understand because the emotional mechanism of sysy stems from the centralization to formulate the "self" of the human being. It is thought that energy consumption for the sysy process is largely due to the consistent loss of energy. But, once systemization has been achieved, the unit rate of energy consumption in the dynamic core stays stable and changes with the variation of size.

The Nexy process has a greater variation than the sysy process; sometimes dismantling, and in other times just connecting. Nexy is a little bit harder to understand and to be theorized. We talked about the circle of accordion effect in Section 4-2. The accordion effect consists of successive unfolding and enfolding processes.

The accordion effect is the result of flexibility hidden in the continuously "transitional" dynamicity of expansion and contraction. Being transitional for dynamicity represents the idea that the whole entity is maintained in a flexible range which will not tear out the wholeness. The accordion effect is similar with the genotype-phenotype relationship but different when considering the energy loss and reservation during the chain reaction of circular emotional transition between folding and unfolding.

Unfolding is related to the process of manifestation and enfolding is done by the ability to disregard such things, like the exile of an unacceptable and numerical rounding. Emotional dynamics hidden in the unfolding and enfolding processes produce a circular path as we can see in thermodynamics. It is connected to the loss and gain of the emotional energy which has a functional relation with the physical energy.

We need to consider another aspect of emotional dynamics related with dynamic modularity. Dynamic modularity is maintained by the innumerable connections among manifolds in the phase space of the neurophysiological environment. Manifolds are connected, consuming energy due to the accumulated stress between transciousness and sensciousness.

As discussed before in Chapter 3, the manifolds of human abstraction are generated from summit experiences. Dynamic modules which consist of many such concerned manifolds are supposed to appear in consciousness only if they win in the competition against dynamic modules.

It is reasonable to think that a brain with high energy-consuming modules is going to endeavor to find stress-free dimensions using any method to reach out. Excessive stress is caused by uncompensated energy consumption. Uncompensated energy consumption arises when dynamic modules are not consistently synchronized. This kind of interpretation at the individual level could most likely be applied at the communal level.

Nexy starts with the initial complexity such as knotting and nodulization at the dispersion-relaxation status which has mutually orthogonal (no stress) dimensions. On the contrary, sysy achieves higher levels of rationalization by enfolding in the concentration-tension process. When fatigue has accumulated to the maximum limit as the sysy process progresses, the whole scheme changes its orientation to relaxation. Relaxation can experience the suspension-tension-suspension which is the nexy process between systems, or hierarchical incorporation between systems. This could be identified as similar to the nested process in the networking theory.

Integrism varies in the range of opposite orientations; dispersion and concentration, relaxation and tension, divergence and convergence, and decentralization and centralization. The cognitive system of a personal entity receives the signal of impact depending on the degree of communicational permeation of the dynamic core into the extensional edge of the dynamic module and thereby on the quality of the continuum (of the dynamic module) conditioned by the degree and complexity of connectivity.

It is thought that the interchange and interaction of Triradixism has to do with integrism. In Triradixism, we have value axiology for the three values of creativity, labor and rent for the respective domains of creativity, communality and terrenity.

The process of creativity generally occurs best when nexy is activated and the stabilized system is ready to support. We also acknowledge that, in some rare occasions, urgent and unusual circumstances permit sudden creativity to pop out – sometimes they prove themselves to be most valuable.

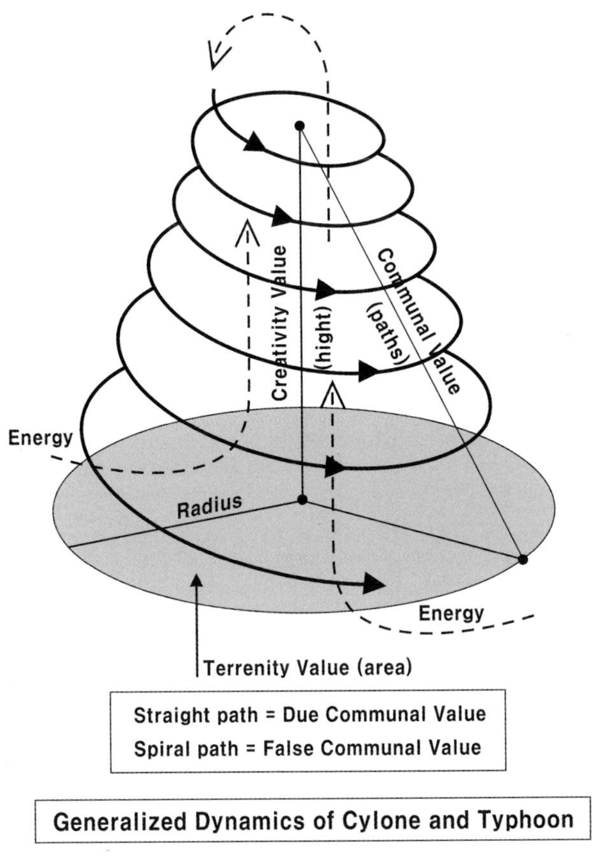

Figure 5-5-1 Interrelation among Respective Three Value Processes of Triradixism in the Light of Generalized Dynamics of Cyclones and Typhoon

The social process appears in the course of communal collaborating by (sweating through physical) labor. It is identified as a strengthening of sysy in the existing network. When a system is developed by gradually accumulated values of creativity in it, it will take the central core of the society using the accumulated creativity values. The special point of view maintained in this research is that labor is regarded as the psychological foundation for the social (communal) process. If labor in the community does not exist, the psychological tie of a community will be broken.

The terrenial process, represented by rent, is not likely to be directly dependent on integrism. Land value and land prices are determined in the complex relation between the existing system and network. The changing rate of the land price is attributable to how much sysy and nexy are activated.

These three processes are dynamic. A generalized dynamics of typhoon is useful to understand the mechanism. In Tropical Ocean, at the temperature of 27 degrees Celsius (80 degrees Farenheit), a vapor from the vast areal coverage starts to get to the center and circulate around to ascend in a spiral form. When it reaches the top of the convection, the air flow is supposed to scatter towards the periphery of the typhoon.

In this dynamic model, ascending height is a creativity value produced in the society. The areal coverage is a terrenial value. And the sloped descending flow is a communal value or social value which adsorbs the creativity value as the creative process loses the energy. The three values and their domains have an interrelation as illustrated in Figure 5-5-1.

Vacuity and Chasm

In Chapter 3 we discussed an interesting opinion about the cosmos suggested by a science columnist, K. C. Cole. Quoting Lee Smolin's *Loop Space*, she commented that the difference of the being and the non-being is in that the being is connected and the non-being is the connection itself.

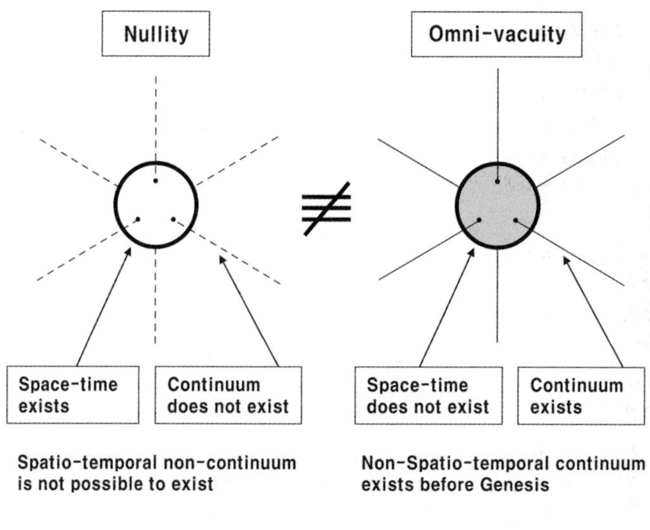

Figure 5-5-2. Non-spatiotemporal Continuum and Spatiotemporal Non-continuum.

The vacuity of the universe is omnipresent. We are about to find the Being, that matches the dynamic relation "*das Umgreifende* (The being of whole comprehension)." I believe this is an idea surpassing an invisible mending of logics. In my view, vacuity is the omnipresent introductory inlet of the symmetrical particles, at the destruction of which our cognition activates.

As we discussed in Section 5-2 as to the hypothesis of general continuum, Cole cited Alan Guth to express that gravitational potential energy is a negative value which would make a par with the occurrence of material and energy. I believe the gravity is possibly a candidate for the restitution. Gravity is a sort of potential wrapping or grasping of the manifested universe of spatiotemporal continuum. Partial discontinuities cannot cancel the grand continuum of the universe.

As presented in Figure 5-5-2, universal vacuity is an omni-vacuity. If the universe is in an undisturbed symmetry, there exists just the unitary vacuity. The unitary vacuity of undisturbed symmetry is equivalent to the omnipresent vacuity of non-spatiotemporal continuum. Figure 5-5-3 is drawn as I understand to illustrate K.C. Cole's expression as March-Russell explained to her.

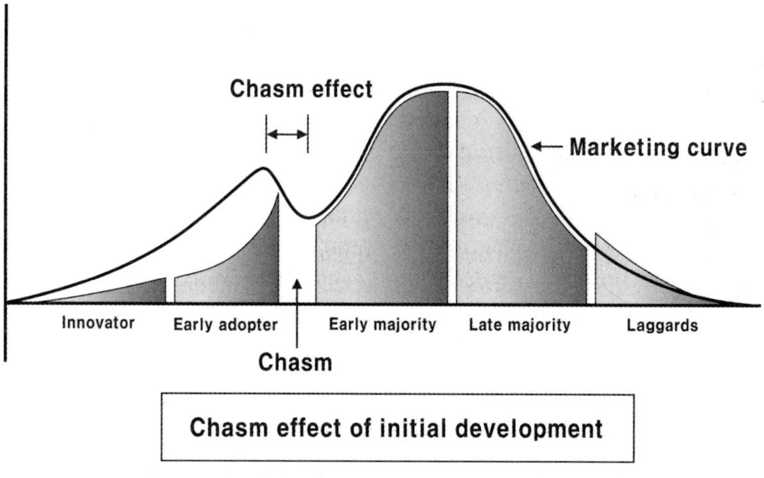

Figure 5-5-3. General Chasm Appeared in a General Continuum

Chasm is another concept to be considered for clarifying the characters of continuum. As shown in Figure 5-5-3, five levels of E. Roger's Diffusion Model have a good implication of how the psychological underground strata formed alike. The dispersion of innovation is described quantum than incremental. Quantum means an innovation diffused in a very short span of the time. And only a small part of new innovations is successfully admitted to the relevant market.

Rogers thought that five levels of diffusion are innovators, early adopters, early majorities, late majorities, and laggards; the corresponding percentages of occupants are respectively 2.5%, 13.5%, 34%, 34%, and 16%. It is said that there is a chasm between early adopters and early majorities. It is called a chasm theory (G.A.Moore). Propagation will not be successful if the flow of expansion does not move over the critical point in its initial stage of diffusion.

Volatility problem in creativity process had been recognized by the author in his value theory in 1987. Total 84% of creative ideas were rejected and buried by the market. It is largely relevant to the subconscious emotional cognition.

Socio-psychological Crevice in Propagating Ideas

It is interesting to realize that Oswald Spengler dealt with the number and mathematics in the first part of his "The Decline of the West." We are going to talk about void sets in the persontological universe.

In a universal cosmology, we had to deal with quantum physics in order to clarify the initiation of universal existence of space and substance. Well-known is the Planck condition as a threshold value of the appearance of micro particle into being.

We have the void set of omnipresence, $\{0\}$, introduced into set theory. Without void sets, both of concept and calculation in Set theory do not exist. Void sets could be the evidence of the incompleteness of logical system, so there always exists undeterminable issues in a system.

On the contrary, we can say a void set is the proof of the mutual intrusion between human cognition and natural plexion. It is said that void set is simultaneously self-referential and self-denying. This phenomenon has a resemblance with the human response to infinities-encounter.

Vacuity is similar to the void set which is omnipresent in every possible set of numbers. Nullity, void set, is the start of cognition. Human cognition is subject to and destined to be initiated by vacuity and void. Emotional aspects of numerology starts from void zero, the nullity. Most important seven numbers of humanity, 0, 1, 2, 3,$\sqrt{2}$, e, π are met within the magnitude range of π. The unity, 1, is the essential emotional figure of an integer. However the emotional base for social relation is nullity as everybody agrees to collaborate or fight for mutual destruction, which comes from noneness.

Numbers and infinities are psychologically founded on the bottomless fear of the void set. Moreover, the rationale behind mathematics, physics, and the other academic works stands on the subconscious feeling of vacuity. Without the vacuity being filled up by brain activity, humanity becomes unable to solve impending problems when encountered with pressing fears of extinction.

I believe the void set psychologically attracts our mental energy in various ways as infinity does. That is the first emotional threshold of the world of abstraction. We are all mortals destined to die, and power is one of the measures to distribute the social fear of death before the situation of panic.

The human is an IGUS (information gathering and utilizing system), as T. Siegfried explained (The Bit and the Pendulum, 2000). Not only are we IGUS but other species are essentially IGUS as well.

Then, what is the cause of the difference between man and the others? We need to explain what happened in the evolutionary leap between animal IGUS and human IGUS. This can be done by look at the void set settled in human understanding which provoked the starting journey toward infinity. The void set affects us while being completely hidden not even appearing in our dreams. The void set appears as the bottomless fear which pushes us to move forward ceaselessly in our lives.

The homogeneity of a society is neurophysiologically founded on the emotional homogeneity of its population. We know there is a crevice chasm in the model of Roger's Chasm, which can be interpreted again in the context of brain science.

It is interesting to think of linguistic phase spaces as a pattern of mapping correlation. Integrism has to have a correlation with phonetic vocalization. Physical conditions of vocalization are deemed to have pinpointed spotting mechanism for the correspondent correlation with the desired expression of the emotional state. Each specific emotional state of specific integrismic condition is expressed in a specified vocalization.

The flight is nexal reaction while fight is systematic action. The reactions will have to be done by the whole existence (body). The correlation between integrism and qualia (and quantia) is comparable with that between integrism and vocalization. Qualia and vocalization have ontological expressions connected to the onto-submatrix.

The relation between integrism and semantic positioning is comparable with that between integrism and symbolic sign. Semantic positioning in a map of meanings is not different from the positioning of symbolic sign in the sense that topos-submatrix is related. It is similar in the cases of hierarchical formation of syntax and also in the case of dynamic modularity. In these cases of further complexity, logia-submatrix has contributed in syntax formulation and in the activation of dynamic modularity.

The conclusive comment of Chapter 5 is that integrism reveals the hidden mechanism of the universe that all the phenomena in our planet, in the Galaxy and in the universe are subject to communication of many forms and manners with other existences. We are all mutually related and connected in a manner of integrism.

In political economics, integrismic view tends to represent the attitude via the media. Integrism is the comprehensive stance that admits controversy to the system and nexus together.

Chapter 6. The Science of Cruxology (I): Social Continuum

Cruxology is the mechanism for humanization. It is the science of cruxological framing in human society. It is named as the "Ghi-Teul" of science (幾틀 科學) by the author to reflect the conceptual mechanisms of the Far East. Ghi-Teul is a hidden "dynamic" framing mechanism of social continuum.

A crux is the earliest infinitesimal move of an activity eventually becoming that of a macro trend. Cruxology is a theoretical set of mechanisms that govern and determine the internal relativity of human cruxes, activities, and trends becoming the greatest aspect of human history.

Thus, the science of cruxology is supposed to thoroughly encompass all of the phase space of humanity. The major aim of the science is to understand how to decide when and how humans wait, see, plan, and take action.

Cruxology will be able to guide activated humanity into the consecutive chains of stabilized conditions, in avoidance of unnecessary fluctuations and upheavals. Cruxology leads and allows us to see and understand human behaviors and outcomes as menesys.

Menesys could be a flexible idea, regarded as a technological approach after the science of cruxology. But it is expected that menesysal management of the human phenomenon would solve the many problematic situations of humanity.

In global humanity of the East and West, various philosophies have prevailed. As the most critical trend, experimental evidence(s) have attained greater explanatory power from society, with such significance that it cannot be gained with a simple "generational" idea. Such proofs are based on material methodology ever since the days of Enlightenment.

However, the material view alone has difficulty in viewing the wholeness of integrated diagnosis regarding contemporary problems. Another important point is that humanity by themselves is not competent in analyzing and solving their own issues.

Thus the human race has had extremely complicated problems that need to combine with material science, the problem-solving humanity, and the holistic (wholistic, as I prefer) understanding of the world to aid in solving humanitarian issues. This combination seems to be all achieved except the wholistic view when a novel approaches the human brain and systematic molecular biologies emerge. Cruxology is designed to be a refined and delicate approach to link micro moves to the macro tendencies of humanity.

When developing an idea under the frame of science, a set of propositions has to be presented to show the disciplinary structure and the interrelations among the concerned propositions. In this chapter we are going to deal with the conceptual presentation of the science of cruxology until getting into quantitative methodology.

The quantification of the cruxological approach would require us to go through a greater time-consuming process to straighten out a complete quantitative theoretical foundation, measurement parts of science, and the practical calibration method in social relationships at both the macro and micro levels.

6-1. Minima (smallest, 幺) and Crux (crossing, 幾)

Minima
Minima (幺) and Emotion (情緒)
Knotting
From Minima to Menesys
Formation of Crux and concerned Orthogonality
Crux and an Individual Person
Crux as a Base for Elasticity of Brain and Society

Minima

A novel vocabulary "minima" is adopted from a Latin word. Minima stand for an "initial infinitesimal" move of dynamic incidences, specifically designed to describe the least emotional move. (Note that *emovere*=disturb, *movere*=move, *motus*=movement and emotion, *commotio*=excitement, 氣發)

Minima include logia-boarding (理乘), in other words, logia-accompaniment. These notions are adopted from the Hsing-Li philosophy (性理學) of the Far East and refined to fit in the novel approaches of contemporary emerging sciences. In fact, the Hsing-Li philosophies are known to not play with the infinitesimal scale of human emotion. But the general approach of Hsing-Li is applicable to the wholistic view of humanity and to the fundamental idea of the problems of a person.

(Hsing-Li Philosophy (性理學) is Neo-Confucianism developed and prevailed in East Asia in areas of China during the Song Dynasty (since 11th century), Korea of the Koryo Dynasty (since 13th century), and Japan (since 16th century).)

Plato, along with several well known Hsing-Li scholars, including Chu Shi or Zhu Xi (1130-1200) in China and Yi Whang (1501-1570) in Korea, and A. Whitehead in his early writings have emphasized Li (理), or an equivalent notion of idea, as the eternal object at the summit of the cause.

(In his book, Science and Modern World, 1925, respective eternal object have intrinsic relations with other eternal objects, thus making all the determinations subject to this relation. It is earlier than Processes and Reality. Discussions regarding this issue were made by Sang-il Kim in "Whitehead" and Eastern Philosophies, 1993.)

Minima are the essence of crux. It is suggested that scratches of minima, even when unrecognizable in consciousness, contribute to the emergence of the cruxological phenomenon in neurophysiology. Crux (幾) is a crossing of *nervus*, a Latin terminology which usually stands for a nerve but also could mean moving strings (幺, small threads) as used here. Thus, crux is the "intertwining of the moving string of senses."

Senses in human susceptibility are thought to be mutually additive, subtractive, commutative, etc.; they are partly similar to the vector and tensor systems if these senses are well presentable in coordination with proper dimensions and ranks. For instance, audio-visual presentations have interaction in an additive manner, in such a case when we are in a movie theatre. When delivering a soft massage, Rock and Roll would not be welcomed. Producing concerned maps and correlations to it would be a very complicated field of study.

Minima (幺) and Emotion (情緒)

Crux means dynamic ghi (幾) or static ji (機, 木+幾= tree + ghi). In modern Chinese, these two similar letters are pronounced in a similar way, "chi" or "ji." But in ancient Chinese and current Korean, it is pronounced "ghi" for both letters. In this book, they are given semantics as differentiated phonetic values and are distinguished by their "well."

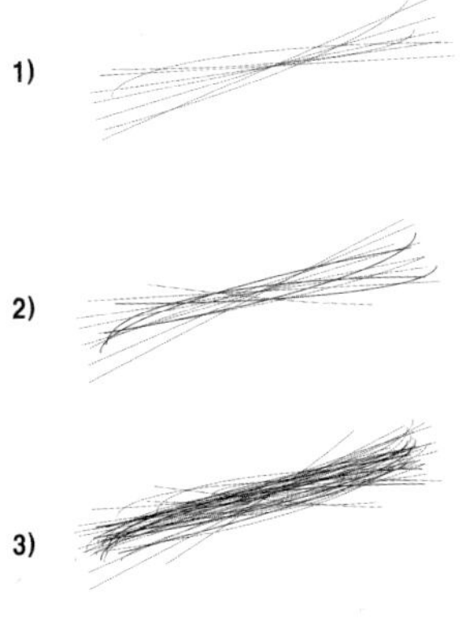

Figure 6-1-1. Visualizing the Formulation of Minima by Successive Indentations of Dynamic Ghi

Crux must be activated solely from the natures (性) of the nature (自然). But this is so when the entanglement of dynamic ghi causes a movement or stationariness. Ghi is thought to be substantialized by successive occurrences of minima. Minima means the least thing, "yo" or "yao" (么), are the least string. The modern semantic usage of yao, meaning kid, is not applicable here.

I feel uncomfortable to discuss the string theory in cosmology with my superficial readings of only a dozen books written by physicists for the general public. Nonetheless, it is important to mention that the emotional aspects of human cognition have a remarkable resemblance with the general string mechanism including crossing, knotting, interweaving, untying, etc. The string is a symbol for nexusity and when enfolded is representing systemicity.

According to A. Whitehead, reality is seen not as an independent and simple "static" location but as dynamic and interrelated events. Dynamic and interrelated understanding is what we are aiming to build up in this writing. It is thought that the redundant interrelation of plural minima is an important factor in order to make a significant path on the surface of human cognition.

Go Game and Knotting Effect of the Board

For an easier imagination, let us think of the board game Go, which has a nineteen by nineteen grid system. Crux, the intersecting point of two crossing straight lines, is actually an assembly of four half-lines. A Go-board has two sets of nineteen straight lines of array, perpendicular to each other.

Thus a Go-board has four corners, the twisted spatial character that enables a number of variations for games. The four corners can be considered as the inlet of topological energy. (Or, distortion energy of the phase space due to knotting.) Their edge lines and edge points are supplying inlet to the boundary of capacity that introduces inner energy.

The central point on the board is called Heavenly Prime (天元), which introduces the least undistorted energy. The Heavenly Prime would yield none of the distorted energy from the boundaries when the distance to the boundaries in all directions is infinite. On the Go-board, this distortion exists at all points of crux whether they may be dense or sparse.

Intuitively, integer individuals in a separate community located in a wider society is like a corner or an edge point of an edge line, and a kind of knot. The integer has to endure the external stresses imposed on him at the location it occupies. Geopolitics runs under the similar logic to this Go-board character. The knotting effect at a specific locality of the global territory of the system or network is critical.

The network has sparser and diffusive characters in its organizing mechanism; however, depending on the topological location, various connections create distinct modes, such as hub point or simple connecting point, and so on.

System is very densely organized and imposes heavy stresses on constituents to suffice the internal systemic aims. System presents in such situations the heavy entanglements that occur due to the necessary surveillances and on-time communication.

Knotting

The knotting mechanism is effective in the process from nexio, nodus, articulus, modus, figura, and even to systema presented in Table 6-1-1. They all partly include the knotting mechanism in activities. Knotting is an achievement of connectivity. The knotting mechanism has many variable functions and steps as we can find in simplex connection, Möbius connection, and the others that we have found and will find in the knot theory.

During the last several decades, profound progress has been made in cosmology. String theories are presented in various manners. One of the proposed models is a warped universe. According to the model, we live in an isolated pocket of space in which the universe is observed as the four dimensional world. The strings theory needs to be verified through experiments. (Lisa Randall, Warped Passages, 2005)

Another strings theory has been theorized for the science of humanity. It is still a matter of debate whether we can adopt such an idea of physics to humanity. However, it is considered applicable to the science of humanity because the idea provides the general explanation and possible application to the real world matters. It will be proven pragmatic in the course of understanding the scheme of idea.

It starts from minima, which is a minimal sense. Minima can formulate nexio, a uni-directed string that has the meaning of connection. Nexio could be one or two steps. The single step case delivers information from one entity to another whereas the double step case can be seen as the ecological building block of resources-consumers-predators. When we design the basic building blocks of ecological networks (J. Bascompte, p416, 24 July, 2009, Science), we tend to have nexio or nodus in our scheme of thought. In the basic "triangular" relation of the omnivory chain, we can see a peculiar case of nodus.

When nexio is sufficiently developed and crosses with other nexio, it will formulate nodus, an isolated conceptual node. The formation procedure is called nodularization, which by itself is a very simple relation. Nonetheless, if significant numbers of nodus relations are formulated, modus with modularity will be shaped. Between nodus and modus, articulus is thought to be formed due to the influence of the environment or other internal condition.

Emotional stresses, such as unhappiness, agony, hostility, etc. cause topological warp (crookedness) in the phase space of consciousness, usually recognized as emotional mood, physical condition, self-confidence, and more. Such stresses come from the concerned knot (node) of logical systemization, which is the settlement of systemic relation by the folding mechanism of the dimensions of the respective parties. This system emerges when modules sacrifice their attributes as confined system through the enfolding process to shape up the systema in an interconnected whole scheme.

Minima give slight but successive impacts, leaving traces of a trend in emotional influence. This basic mechanism is possible because the emotional indentations come with the memorization process that takes place in human neurophysiology. The embryology of menesys starts from such infinitesimal movements of emotional life.

In a complex market developed from intricate socio-psychology, the emergence of a bigger entity is suggested to have the following process: nexio – nodus – articulus – modus – figura – systema. However, market is distinguished from other mega-social entities, because it has prevailing impacts of nexusity, so to say networking.

Nexio is produced by a folding process (plico process) following apparent connectivity. Furthermore, nodus is produced by a multi-folding process (complico process). Modus produced by a level higher communication following intrinsic connectivity occurs. Figura is determined by a multilateral relationship and through consortis communication. 'Systema' is different from isolated 'system' because of the presence of external connectivity. From nodus to systema, we can compare them with nested networks of the theory of the "weak link" by P. Csermely.

Our question is given for a different purpose from the pure scholastic achievements; it aims to give practical values to the real impending political economies of this century. A menesys has its own process of conceptus, growth, and intrinsic phase changes. The embryology of menesys has to be systematically organized in theory.

Table 6-1-1. The Emergence of Socio-psychological Entities in Complex Market

Name of Move and Entity	Remarks
Minima (the least string)	"Yo" or "yao" (幺), socio-psychological effect, which is the least string. Scratch of yao.
Nexio (connection)	Emerged by the indentations of minima, formulated by the consistent activities
Nodus (node)	Knotting plural nexio with mutual commitments with at least three steps of circular logic in a network, (but mostly represented by individual in social life), the least nucleus family in Trinitarian interrelation
Articulus (articulation)	Sometimes effectively emerges with an articulated function like a member of our body, a nucleus family, and a basic commune due to special internal or external conditions
Modus (modularity)	Mainly functional and embedded to material and physical conditions as seen in biological DNA and socially organized under strict principles; Modularity is a lower level than system
Figura (pattern of diversification)	Characteristic pattern of changing alignment due to meeting and parting of herds and modules
Systema (system)	Called in Latin, because of its isolation and boundary effects, it is mostly disregarded in an ontological frame (different from system in a sense that systema is a undividable part of menesys); System is a result of enfolding the unnecessary dimensions of the network node to satisfy the systematic function and efficacy
Netsys (networked system)	When systems are networked
Nexus (network)	General network
Sysnet (systemized network)	When intrinsic attributes of a network is systemized, which means the network is systemized in nature
Menesys(megalo-network-system)	Contains all the variable entities discussed above

Since our proposition of theory is based on the field of persontology, there exists an embryology but not a process of senility. Senility, in the persontological scheme with an emphasis on the wholeness and the overall connectivity of the universe, is simply a process of path change from omega to alpha depending on the increase of stress situation.

The reason we partially disregard the senility at this stage of human history (race) is that the human race seems to have higher probability at introducing new prospective dimensions. The author approaches this book with the scope in mind for the next one thousand years.

Menesysal embryology is relevant to its integrism. The embryology is also governed by the cruxological mechanism. Integrism influences menesys by affecting the ratio between sysy and nexy. The integrismic ratio affects the status of menesys in combination with the effect of initial location in its environmental phase space. For instance, it is very much indicative from where a menesys started its embryonic growth, such as locality around a systemic attributes or that of networking attributes.

The cruxological measure would contribute from the earliest morphogenesis of menesys. The overall process of main body formations of minima, nexio, nodus, modus, figura, and systema is governed by cruxology. It was discussed that a systema is the persontological version of a system in the light of existence with overall-connectivity to wholeness. However, there is another line of cruxological formation regarding networking. Networking mechanism manifests the measure of the universality of the cosmos.

Thus, it exists in all the cosmos in a very similar manner to the cosmic microwave radiation. I believe it might be more fundamental than simply being compared with the above metaphor of background radiation mentioned above. This is due to the belief that the nexusity of the universe is the original essence of persontological universe. The original cosmos, without its implicated intentional beings that ceaselessly observe, think, and act, can be expressed as the infinity of a complete systematic network, where the whole network itself is a unit systema.

Such an ideal form of the universal ontogenic network, named the Uninet, is located at an end of the menesysal spectrum of our universe. Contrary to the Uninet, we have the Unisys on the extreme opposite wing of our universe. The Unisys is thought to present the conceptual status of the system with perfectly developed nexal efficiency in the system. However, it is not yet absolutely clear to me whether or not the Uninet and Unisys are actually identical in their internal pattern. My present stance is that system is developed from original nexus originated from the Unicrux presented in Section 5-2. Unicrux is the origin of our universe.

However, both of them have equally the most effective pattern in the light of integrismic criteria. The most effective pattern means that the menesys does not need any alteration of its sysy-nexy pattern to improve.

Formation of Crux and Concerned Orthogonality

It is desirable to define the proposition of orthogonality which states that the stress of human society is null in orthogonal relationship. Crux is formed through compound dynamics of minima from different dimensions.

The orthogonal relation of dimensions in humanity is manifested in infinities-matrix. Infinities-matrix generally consists of three major sub matrixes as explained in Chapter 3; Onto-submatrix, Topos-submatrix, and Logia-submatrix. Those submatrixes are orthogonal to each other. It is thought that other submatrixes that can be defined by mutual orthogonality and existing submatrixes can be added up to compose new infinities-matrix. Though, I have not yet figured out what they would be exactly.

Crux of multi-dimension is formed by expansion and multiplication of orthogonal relation. On the other hand, it looks also possible to degrade the cruxological relationship by the dissolution of orthogonality. The multiplication and dissolution of crux includes the prevalent affection of stress dynamics that is influenced by emotional aspects.

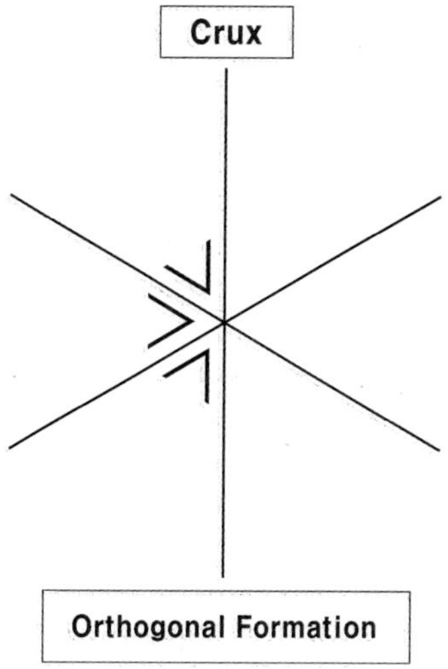

Figure 6-1-2. Visualizing Crux Formation Affected by Stress-free or Stress-tamed Orthogonality

Crux and an Individual Person

Crux appears to be a result of string coupling or the orthogonal coupling of minima. Sudden emergences of cruxes are attributable to the string coupling's condition of orthogonality, which is the null stress or the least stress relational situation. The part with the least stress is of the least concern in a cognitive system.

Human response to the environmental impact especially is dominantly due to the emotional aspects of human cognition. Cruxological change is made on individual basis because individuality is the only concrete foundation of connecting emotional-intellectual areal field. It is actually why we had to select personism-communalism line of thought in Chapter 4.

Crux as a Base for Elasticity of Brain and Society

I would like to share a short tale that I heard from my father during my elementary school years. In any country, there are some tales of hungry farmers. This is a story of a farmer name Dol-yi ("dol" means a stone, which is Einstein, and "yi" is a suffix denoting a person). He was very young and always hungry—hungrier than any one else in his village. Whenever he stretched his back to take a rest from farming, next to his working place was a large abundant river was flowing generously before his eyes. He would sit down on the bank, heaving deep sighs, while digging up the earth with his index finger, saying alone "Oh, my mother river, please give me a bunch of tducks (delicious Korean rice cake)." He repeated this every day, shouting "Tduck!" with his whole heart, while his strong inex finger would stroke the earth continuously.

The story continues to take place in a king's palace at the capital city. The king at the time was a deep lover of art and calligraphy. He issued the whole country a decree that he will grant a grand prize to the best calligrapher in the country. Dol-yi heard the rumor and traveled along the river bank and arrived at the palace to meet the king. He boldly introduced himself as a hermit calligrapher. He requested the king to allow him to have as many delicious tducks as he can eat before he demonstrates his calligraphy.

Oh, what an easy prize that the king can reward! And, what a lifelong desire for a hungry peasant that was to be fulfilled! However, when poor Dol-yi finished his pile of the most delicious tducks in the country, the only thing left in front of him was unfamiliar tools of calligraphy. With his eyes closed, he silently sat there sweating for so long that even the serving eunuchs started to think there was something wrong. Dol-yi realized that he had no way to escape this situation without leaving an art piece for the king. He had the eunuch open white drawing paper wide before him and got a lot of black ink on his index finger. Shouting with his whole heart "tduck!" he drew the strongest single line in his life from the left to the right. At the very moment the stroke has been made, he has died of the depletion of his *qi*, breathing his last.

When the king was told of this situation, he was disappointed but soon forgot the whole story about this hermit calligrapher since he had many other contestants to test. However, when the prime minister glanced at the stroke drawn by Dol-yi, he cordially asked his king to allow him to borrow the work for several months. It did not take long for the king to hear a strange rumor of his prime minister fishing in his study.

Prime minister caught fishes out of Calligraphic River

Figure 6-1-3. Fishing on a Calligraphic Picture

One day, the king saw his prime minister greeting with the calligraphy and fishing rod together in his hands and suggesting to demonstrate. With the calligraphy flat on the floor, the prime minister put his fishing string into the black finger stroke drawn by Dol-yi. It did not take long for a jiggle, followed by a fast snatching, a big fish jumped up out from the calligraphic stroke.

Several other fish were caught from the calligraphic river. The prime minister expressed a deep obeisance to the king and explained how he realized the picture was not a simple usual brush stroke of plain calligraphers. When he came home with the picture and spread it out on the floor, he heard sounds of the great river flowing and gushing and the sound of gentle winds to adrupt whirlwind changes. The king finally understood the summit of calligraphy by listening to this story.

The standpoint of this writing certainly is not on mysticism. Do we have any scientific significance in this story? It is affirmative, as far as humanity is concerned. We think that plasticity of brain generally respond to circumferential conditions. It is suggested that elasticity of brain function has to be considered in certain specified conditions.

The elasticity of brain-neural system is the key factor of humanity in setting up intentionality. Crux is a major measure of organizing elasticity which is attributable to dynamic modularity. This is related to the definition of crux which we are dealing with. Some people call this elasticity resilience.

6-2. Neo-Confucian Concepts of Crux

Extraction of Crux Concept from Neo-Confucian Thoughts
Earliest Chinese Neo-Confucian: Zhou Dunyi and Zhu Xi
Korean Neo-Confucian: Emotion and Principle
Grafting of Ancient Ontological Thoughts

Crux is one of the major concepts in this book. It was discussed that integrism and cruxology will be useful in the administration of statecrafts of global situation. Crux is the key notion in how the human race can reach the restoration of humanity because crux is essentially an outcome of human emotion that governs the rationale and substantiality of global economic continuum from the bottom.

It is also an important step for humanity if we integrate both the East and West's frame of thoughts in a thought frame. The brethren have the same root of humanity. Only within bifurcation have branching out occurred. We are interested in the primordial and initial inspiration of either thought before the bifurcation took place.

Extraction of Crux Concept from Neo-Confucian Thoughts

The author intends to extract the concept of crux (幾, Ghi) from the thoughts from Neo-Confucian teachings. Crux in our paradigm is not ideal, but rather materialistic in the origin. Crux is based on the indentation of cumulative emotional influence, as far as the human brain is concerned. The major body of early Neo-Confucian teachings has pondered into human nature and the correlation of cosmology. Thus, they may have developed good fundamentals of theoretical scheme and practical technique in both the micro and macro scale of society.

The Sung Dynasty had produced large number of scholars who reestablished the Confucian tradition in response to immoderate intrusion of impractical spirituality. Zhu Xi

(朱熹, 1130-1200) was the leading scholar who reorganized Chinese intellects and other major scholars followed after him.

However, Zhou Dunyi (周 敦頤, 1017-1073) was the first Chinese philosopher before Zhu Xi in the Neo-Confucian school who believed metaphysics and ethics should not be separated. Zhou Dunyi initiated a new paradigm to construct a comprehensive foundation to include the full set of thoughts they encountered during their times. His idea was also in response to contemporary Taoism and Buddhism that prevailed during his days. The diagram for *Taiji* (太極) illustrated the universal mechanism which is the correspondent pattern with humanity.

In *Tongsuh* (通書), Dunyi said "Ghi is the unveiled that has been moved but has yet to build up, positioning itself between being and non-being (動而未形有無之間 幾也)." And a little bit later we read, "positioned in between motion and stillness, body and use, where the moment to lean toward either side is the start of exposing real principle and the sign of good and bad in every affairs (動靜體用之間 介然有頃之際 則實理發見之端 而衆事吉凶之兆也)."

Zhu Xi commented on this idea in his commentary on the *Tongsuh* (通書解) that "It is said that 'Great is the Heavenly Origin, from where all the affairs starts, it is the cause of integrity (sincerity).' Thus we realize there is unmoving stillness." He continued that stillness and unmoving is integrity (sincerity). When it moves infinitesimally, it is a hint (unseen nicety) of a hidden sign, and the hint is in between integrity and spirit (寂然不動者, 誠也. 至其未動處, 卽是幾, 幾在誠神之間).

In our modern interpretation, it is the matter of initiality in transcendental intentionality in transciousness. Unseen nicety of hidden signs is the vacuity of vector with no scalar but only direction. I would assure that there is an emotional step where humans cannot be aware of the quantity but only be aware of qualitative orientation. It is a kind of mood that assures normal people that things might work out to be as expected, where no reasonable bases and no realistic indications exist. It is not a kind of belief, but is still present in our lives.

We have Yi Whang in Korean Neo-Confucian scholastic tradition, who described ghi as two roads coming together to form a common crossing point. This concept is the reverse restoration of integrity. The reverse pursuit of initial integrity has always been the central issue of the matter in humanity. What Yi Whang means is that ghi is the most initial movement of human intent at the point of intersection between the genuine (human) nature and perturbed (or stirred) human desire. From there, we can find the seed of the important philosophical debate regarding Four Beginnings and Seven Feelings (四端七情), which is usually called Four-Seven Debate.

When a person is in a complete accordance with his environment, personal desire crosses with the human nature 'without stress' (which is an orthogonal intersection) for the development of initial crux. The prime concern in Neo-Confucianism was the practice of coincidental match between the universe and humanity in harmonious condition.

In his commentary, Zhu said that it is like pursuing the extreme to reach where "All nations help each other to be in accordance mutually, and people come to the stage of reconciliation" ... Ghi is the trivial (initial value) in the (whole) beginning. (如推其極, 到"協和萬邦, 黎民於變時雍," ... 幾只在起頭一些子).

The three above comments are excerpted from the chapter of Holiness the Fourth (聖第四) in Zhu Xi's commentary. In I-Ching (the Book of Changes), the holy person is defined by three notions; profundity (深), initiality (幾), and spirituality (神). Zhou Dunyi changed the above profundity to integrity. I-Ching (易經) reads "Yi (or, I易) is (a book of) a study about hidden initiality to the profound depth. It is so deep that persons may be able to understand the orientated intent of the world, and with the (timely) realization of hidden initiality (of affairs to appear), he will be able to achieve the purpose of world tasks. ......"

Zhu Xi cited Cai Yuan, who stated "Spirit is defined as sympathy and movement, but it is a rather profound dynamic framework. Spirit(神) is the usage(用) of integrity(誠)." On the contrary, initiality (ghi or crux) is the start of becoming a spirit when integrity is about to be revealed.

The above discussions are not very familiar to modern people. However, the essence of the discussions is that crux emerges in an interaction between emotional framework and environmental condition, whether it is a principle or practical situation. The emotional entanglement in human life, if anyone would like to design it to maintain consistent equilibrium, should be apt to lose the fluctuating energy intervention. Emotion cannot go alone but must be combined with external situations or with internal logics or value systems. Crux is formed during interaction with its earliest initial status.

The ontology of Neo-Confucianism fundamentally permits the existence of subjectivity and objectivity. In an epistemological view of ancient Eastern philosophy, there is neither discernment nor antagonism between idealism and realism. The correspondence between intrinsic Li (吾心的 理) and extrinsic Li (外物的 理) induce the cognition to the extreme acquisition(致) of knowledge(知). (Bae Chongho, The Ontology and Humanity of Eastern Philosophy, 1982, Introduction to the Eastern Ontology)

Eastern philosophy contains essential factors to constitute a 'persontological' view and the approaches very close to it. However, the concept of person should assume the central role of the science of humanity. Its definition and usage should be clarified.

A Person in eastern ontological philosophy emphasizes the cultivation of personal humanity. It is seen that spontaneous fostering of humanity in the social scale can be "guiteulized" through a novel scientific interpretation of person by a persontological viewpoint.

("Guiteulize" means that social effort has achieved the best outcome in cruxological development. Ghi-Teul is the hidden "dynamic" framing mechanism of social continuum. Cruxology is the mechanism of humanization. It is the science of cruxological framing in human society. It is named Ghi-Teul science (幾틀 科學).)

A new way of thinking in the science of humanity is emerging. In the most modern view supported by the science of complexity, cosmology based on quantum theories, molecular biology, and brain science; it is getting clear that we must develop a higher level of interpretation and application toward person(s). Persontology aims to renew the structural rationale of network and governing rules of system through the introduction of cruxological measure.

The rational foundation of persontology is arranged in Section 4-2 as the following;

First, persontology has a different frame from logical intellectualism. Emotional system of persontological frame executes magnitudes higher computation due to its parallel connections. The parallel connectivity comes from the dynamic modularity, the content of which can be compressed by enfolding and decompressed by unfolding the process for inspection.

Second, trial and error computation is avoided, and instant matching among maps will be utilized for the confirmation of correspondence.

Third, the whole body is exposed to illumination (equivalently, impulse, or set of wave equations), and thus, synchronized input of information are enabled.

Fourth, human persontology implements intercommunication by utilizing the correspondence of modularity. The correspondence of modularity originates from relaxed concentration of humanity in order to understand the plexal integrity of universal affairs of the ontological world.

Person is the intrinsic responsive mode, or intrinsic structural response (內構的對應狀) to the infinite plexion of humanity. The structure of infinities-matrix is built up from the emergence of triple set of sub-matrixes. It is identical in terms of structure to the abstraction process such as language formation. Table 6-1 is a suggestion illustrating the most probable cruxological construction of person.

Earliest Chinese Neo-Confucians: Zhou Dunyi and Zhu Xi

Zhou Dunyi was the first thinker in China about a millennium ago who believed that metaphysics and ethics should not be separated. His two books—*Taiji tushuo* (太極圖說, The Diagramatic Illustration of the Great Ultimate) and *Tongshu* (通書, Penetrating I-Ching)—were major Neo-Confucian writings on the metaphysical nature of moral cultivation. In *Taiji tushuo*, Dunyi commented on the Diagram of the Great Ultimate. The Diagram, created by a Taoist, named Chen Tuan (906–989), consists of five circles.

The top circle is empty, symbolizing that the universe is self-generative and self-reproductive. The second circle contains spiral union of two water-drop-circles of dark and light colors, with the dark representing the *yin* (the yielding cosmic cause) and the light representing the *yang* (the active cosmic cause). The third circle is a group of five small circles, each represents one of the five phases (五行, *wu xing*)—water, fire, wood, metal, and earth (水火木金土).

The fourth circle presents how the *yin* affects the female and how the *yang* affects the male. On the fourth circle, and the fifth circle, it gives the process by which innumerous beings are born through heterosexuality.

For Zhou Dunyi, the diagram shows a two-way flow between the whole and the part, and between the unity and the diversity. The whole diagram, from the top to the bottom, illustrates how the unitary gives rise to the diversified. It explains how the interaction between *yin-yang* generates the five phases and the multitude of beings.

However, by reversely reading the diagram, from the bottom to the top, it describes why the diversified multitudes are essentially unitary. It shows how the innumerable beings are derived from the five phases as well as by *yin-yang*. Regardless of whether it is from one to many or vice versa, the diagram shows that the universe is a unified system where the parts and the whole play an equal role.

On this basis, Zhou explains the metaphysical nature of moral cultivation in a manner that human beings with their sensciousness and consciousness, are to decide on whether they want to be participants or obstructers in the universal renewal. Hence, living a moral life on a daily basis is as much metaphysical as ethical, involving a conscious decision to render human activities to participate in universal refreshment.

In *Tongshu*, Zhou Dunyi explains the metaphysical nature of moral cultivation. There are two basic reasons of why innate human virtue equals integrity (誠, sincerity). First, the innated human virtue is hidden so that we may discover it by being sincere to ourselves. Second, because all beings of our universe are connected as a family of beings, to be sincere to oneself requires sincerity to others. Thus integrity has to be rooted in altruism.

We have to develop a noble state of mind that linked us to the universe. Introducing the knowledge behind universal reasons to a person can be achieved by learning. Learning is not only intellectual as understood in the contemporary society, but also ontological as can be seen in gentle and noble personality refined by altering the concept of humanity. One's self-image can be changed by having desires for broadening their universal perspectives.

An educated person is not just an active participant to the society but also a man of right order, who believes the inherent connections among all beings in our universe. By focusing on self-cultivation, Zhou differentiated Neo-Confucianism from Classical Confucianism.

During the Sung Dynasty, Zhu Xi's teachings were generally considered unorthodox. In his days, other Neo-Confucians used to portray more value to the *Book of Changes* (I-Ching, 易經). However, Zhu emphasized the Four Books: the *Great Learning* (大學), the *Doctrine of the Mean* (中庸), the *Analects of Confucius* (論語), and the *Mencius* (孟子) as the core curriculum for aspiring scholar officials. For all these classics he wrote extensive commentaries that were not widely recognized in his time. They have become standard commentaries.

Zhu Xi expressed that all things are brought into being by the union of two universal aspects of reality: *qi* (氣, vital force) and *li* (理, rational principle). The source of *li* is the *Taiji*. The source of *qi* is not as clearly defined by Zhu Xi, the leading scholastic assessment and classification on his thoughts to be a metaphysical dualistic monism. *Qi* and *Li* are mutually dependant on each other.

According to his theory, every physical object and every person has its *li* and is connected to the metaphysical origin, the *Taiji*. What is referred to as the human soul, mind, or spirit is understood as the *Taiji*, or the supreme creative principle, as it is manifested in a person. *Li* is perfect in every manifested object. *Qi* obscures our perfect moral nature. The task of self-cultivation is to clear our *qi*. If our *qi* becomes clear and balanced, we behave in a moral manner.

His concept of *Taiji* was different from the understanding of *Tao* in Taoism, where *Tao* is unmoving and still, reducing all things to equilibrium. He asserted that there is a central harmony that is not static or empty but was dynamic, and that the Great Ultimate is itself in constant creative activity.

Zhu Xi said that knowledge and action always require each other's presence. It is like a person who needs his body parts in place to be efficient. With respect to order, knowledge comes first, and with respect to importance, action is first.

Zhu Xi emphasized teaching, commenting that learning is the only way to holiness. His intention was to provide the way to holiness for all people. Disappointed by his contemporaneous learning system, he established his own academy, the famous White Deer Hollow Academy to teach students in a more proper manner.

He appeared to have adopted some ideas from these competing systems of thought. Unlike previous Neo-Confucians, he strictly abided by the Confucian doctrine of active moral cultivation. However, such adoption was largely attributable to Zhou Dunyi's precedent works.

From 1313 to 1905, Zhu Xi's commentaries on the Four Books formed the basis for the civil service examinations in China. His works were also influential for several centuries in Korea known as Jujahak (朱子學), which means the Learning of Zhu Xi.

Korean Neo-Confucian: Emotion and Principle

Neo-Confucian learning has been deeply developed in Korea to attain the ideal living standard for a scholastic life and statescraftmanship, and even for the common people.

The main concern towards the Korean Neo-Confucian school is in clarifying the relationship between the cause of universal phenomenon and the cause of human activity. It was presented as the *Li-Qi* Debate and the Four-Seven Debate. For Korean scholars, it was most important to pursue the fundamentals of ethics that even common people can adhere to.

In the tradition of Korean thinking, problems arising from one's mentality are at the center of contemplation. Korean wisdom seems like integrating intellectual balance and emotional equability. Korean challenge is in the pursuit of sagehood in terms of the tranquility of contemplating life and of enabling subtle distinction of affairs. This attitude is relevant to the oriental tradition of intellectual sage foreseerers. Such foresight is generally believed by people that self-cultivate scholars.

We find in the history of Korean philosophy the renowned philosophical debate concerning Four Beginnings and Seven Feelings (四端七情), which is usually called Four-Seven Debate. Four Beginnings are the issuance of principle in the universe and Seven Feelings are the issuance of emotional belonging to material forces. We are not interested in the detailed theoretical development but in the trend on how the initial development of humanity can be understood.

This Four-Seven Debate was made among famous scholars: Yi Hwang (Toegye, 1500-1570) and Ki Taesung (Kobong, 1527-1572) and between Yi I (Yulgok, 1536-1584) and Song Hon (Ugye, 1535-1598). The debate is very much important in modern implication because it deals with the fundamental theory of human emotional aspects in a sense that emotion is the essential factor of ontological understanding of humanity. In recent brain science, we have research outputs associated with emotional aspects that theorize the major neurophysiologic mechanism.

Yi Hwang (李滉, 1501-1570), best known by his pseudonym Toegye, is one of the two most honored thinkers in Korean Neo-Confucian tradition (another one is aforementioned Yi I, whose pseudonym is Yulgok). It is widely assessed that Toegye's well balanced and integrated understanding of the complex philosophical Neo-Confucian synthesis achieved by Zhu had contributed to the full scholastic maturity.

His Four-Seven Debate with Ki Taesung had oriented Neo-Confucian thought towards critical issues regarding the juncture of metaphysics and their important applications in clarifying the inner life of the human heart-and-mind.

A year before his death, he presented his understanding of the way metaphysics and psychological structures conduct our daily life. This work, the Ten Diagrams on Sagehood Learning (聖學十圖), became one of the most famous and influential works of Neo-Confucianism. After Toegye it was no longer possible to deny the legitimacy of intensive devotion to study and self-cultivation. It was not possible to ignore such trend in the administration of government and the ordering of society.

For Toegye and other self-cultivation oriented Neo-Confucians, human perfection was a centric topic of concern in the framework of Neo-Confucian thought. Principle and emotion are mutually dependent in the eastern thought frame. In the East, schizophrenic patterns between logic and emotion was not developed as can be seen in Hellenic culture, because of the ontological basis of thinking was different.

In the West, neither does the Hebrew tradition approve of the separation between thinking and acting. There are too many issues and topics to discuss about eastern ontology. Our main concern is that we need to avoid the separation of logia or *li* from the reality of entirety. The entire universe is dynamic not by li (logia) but as "spontaneously existing (機自爾)." I would change the *ji*(機) to *ghi*(幾) to represent the repletion of dynamicity in the universe. It is comparable to "Ruah (wind)" in Genesis 1:2, which effectuates the entire universal existence and fully occupies the universe. It is interesting that representative ritualistic scholars like Zhu Xi and Ti Whang are prone to assert *li*-orientated theories.

The Four-Seven Debate is not clearly concluded. As Yoon Sa-Soon explained, Toegye interpreted Four-Seven in accordance with the notion-intent (所指-所旨) and the cause of activation-emergence (所從來), the viewpoints of entirely different characters so that one may understand the value system and another may interpret and clarify the phenomenon with realism. (Theory of Four Beginnings and Seven Feelings, Preface by S-S. Yoon, 1992)

Li and *qi* in their essential dynamics are expressed in several distinctive ways of mutual coalition. There is *Li-Qi* dualism and *Li-Qi* monism. In dualism, it is said that *li* (rationale or principle) mounts on *qi* (vital force or emotion) as *qi* occurs.

Toegye said "*Li* (first) emerges and *qi* follows, and if *qi* emerges, *li* mounts on it. (理發而氣隨之 氣發而理乘之)" to mean that "*Li* and *qi* mutually emerge. (理氣互發)" Yulgok argued that *qi* (first) emerges and *li* mounts on it (氣發理乘) and after the mounting, *li* governs *qi*.

Sang-il Kim commented that Zhu Xi and Toegye thought the ultimate origin was *Taiji* or *li*, while Yulgok and Whitehead regarded deed or action as the ultimate cause. It was Yulgok's idea that when the initial cause is dynamic action we have to recognize *qi* as the foundation of all affairs. Yulgok has a beautiful theory in understanding the phenomenon applicable in even today's dynamic monistic viewpoint.

But, Michael C. Kalton stated that Toegye's theory is of dualistic monism. I would say Toegye had a "static" dualistic monism. Anyway, it is generally recognized that he thought that one factor was not being emphasized at the expense of the other factors and that they were being well tuned mutually. I think that Toegye focused self-cultivation in his theorization based on the integrated nature of humanity. He was interested in a vivid personal humanity governing the entire universe and society, in principle to Toegye's theory.

In this work, I would like to point out that "dynamic" dualistic monism is established with the intention of having all the factors harmonize together in an ever-changing status. Eternity is the integrated figure of all-and-ever changing affairs. All things never rest but rest forever. All changes ever but changes never.

The omnipresent "Ruah" that appears in the Old Testament is generally equivalent to the *qi* of the Neo-Confucian thought frame in function and in the existential scope. In our persontological understanding, we would put *qi* to attain peculiarity by its locality. The globality of the dynamic universal *qi* is essentially identical to *li*. *Li* also attains its local peculiarity in a specific space-time situation.

In this context, *qi* has essential importance in the persontological view of humanity. Emotion is the fundamental affair of human as *Qi* is the fundamental matter of a person. In this view, the Neo-Confucian interpretation on human emotions, moods or feelings is justified and retains its adequate applicability in real life.

The Four Beginnings are the heart of compassion (惻隱), the heart of shame (羞惡), the heart of courtesy and modesty (辭讓) and the heart of right and wrong (是非). Four beginnings are the governing principles of the human mind. Man has the four members of mind as the body does according to Mencius.

The list of Seven Feelings (or Emotions) as presented in the Book of Rites is as follows: desire (慾), hate (惡), love (愛), fear (懼), grief (哀), anger (怒) and joy (喜). Tsu-ssu in the Doctrine of the Mean gave instruction on the human feelings, the origin of the above seven feelings: joy (喜), anger (怒), grief (哀) and pleasure (樂). Before the feelings arise, they are in equilibrium: after they emerge, constant attention is required to ensure the state of harmony. The human being is supposed to approach initial equilibrium in a reverse way in order to find the initiative of true wisdom.

Our modern interpretation and correlation is certainly a deviation from the ones of ancient times. There is a basis of six emotions, which are not subject to being learned, rather are acknowledged universally and innately. They are joy, agony, anger, fear, fright, and hate. (Dylan Evans, Emotion, 2001) The universal principle is missing in western scientific thought for emotional phenomena. Emotional influence on the social level is restrained to the adjacent realm of an individual. There could be some swarm effect, maybe effective due to the intrinsic attributes of the swarm phenomenon, but not the connection to the thought of universal principles and human emotion as desired in early Neo-Confucian times.

I believe emotions, feelings and moods are all tools to shape conduits for the interconnections of mutually contradictory parts of memory, logic, and concentration entangled like a ball of string in individual cognition. Mood-congruent recall in an individual person is where a happy memory is easily recollected in happy moods. I think that the relevance between moods and other cognitive reactions is the potential candidate to altering the explanation of the theory of compression mechanisms in the human memorization process. There are many possibilities and new growing point of thoughts to flourish in this direction.

I think that the emotional factor generally affects the cruxological development of dynamic modularity at the social level, especially in the formation of articulus and figura presented in Table 6-1-1. Nodus and modus are generally prone to be built up by logical incidences in order to construct a functional frame.

Grafting of Ancient Ontological Thoughts

I said that gravitation was wave propagation from the surface perturbation due to the drastic dimensional degradation in the thin surface transition membrane of Unicrux. Gravitational waves are the cause of the three dimensional perception of human cognition which is associated with the intrusion of the time axis. Thus the space-time continuum is effective in our cognitive process and such correlates can explicitly provide the possible alteration between the combination of imaginary time and three-dimensional space and the combination of real time and imaginary dimensions of space.

Sang-il Kim comments that Plato focused on the entity and not on the action leading him to establish a higher notion like the entity of entity or the entity itself. Similarly found in the work of Zhu Xi and Toegye is the ultimate entity known as *Taiji* and *li*. On the contrary, Yulgok and Whitehead conceived action or deed as the ultimate entity. Therefore Kim says that Whitehead and Yulgok thought that the "actual entity" and qi are the only ones active, and the "eternal object" in addition to *li* are not the precedent of the "actual entity" and *qi*; thus *li* is not present before the affairs as Plato thought.

It means the phenomenal world is the ultimate and initial cause of philosophy. This monistic concept takes vitality when the world is seen as "becoming" than as "being." "I am what I am" is "I become what I become" and further "I will be what I will be."

However, as we discussed how dynamic dualistic monism is the reasonable frame for worldly interpretation, the changing affairs can be restructured in their internal mechanisms. Such mechanism was studied on for centuries to predict the future. I-Ching, the Book of Change, is too formative and speculative to apply in this day and age. The initial approach of I-Ching was reasonable in general theoretical foundation. The application to practice is regulating a fixed mechanism of a finite number of occurrences. We need to renovate to get toward highly scientific methodology.

However, our cruxological approach is not comparable to Chou-I and Sung-I. The initial development of thought could be a respondent to ancient thinking. The main difference is present in the modern sociological findings on systems and networks.

We have developed a cruxological thought based on the emotional essence of humanity, not capricious emotion nor non-reasonable mood, but rather persontologically traceable thoughts, which is tracable in quantity if the infrastructure for investigation is well prepared. The following Figures 6-2-1 to 6-2-3 illustrate the symbolic expression on how cruxes can be realized by the combination of integrismic factors; i.e. systemicity and nexusity.

In the formation of continuum, cruxological characteristics are important to direct the micro process of crux development. The combination of basic cruxological integers makes a large difference in extending the continuum of any kind.

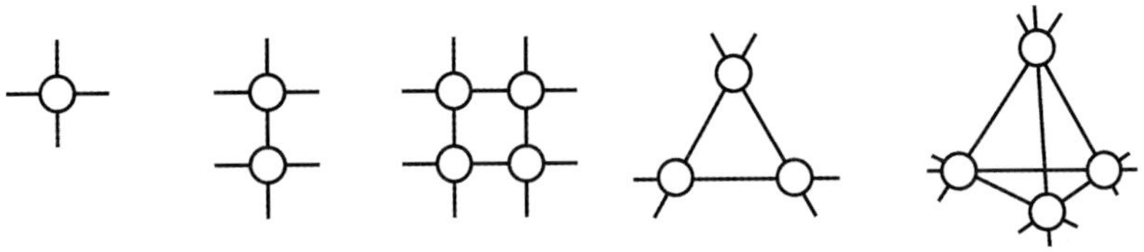

Figure 6-2-1. Illustration of Cruxological Morphogenesis in the Continuum Formation

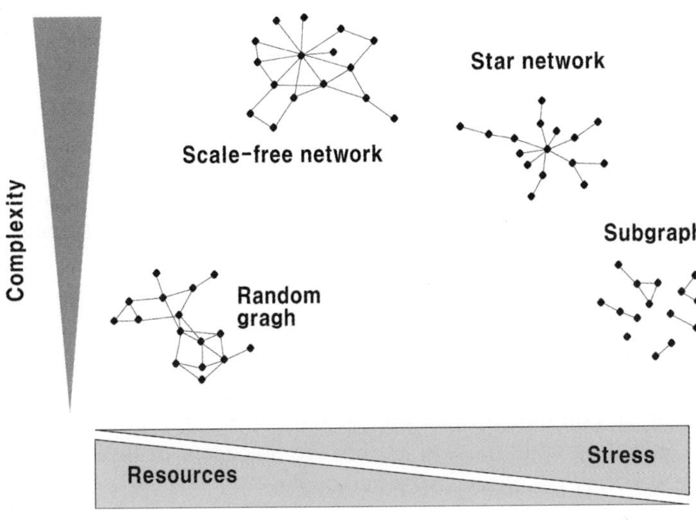

Figure 6-2-2. Reproduction of Fig. 3-8, p75 of Csermely

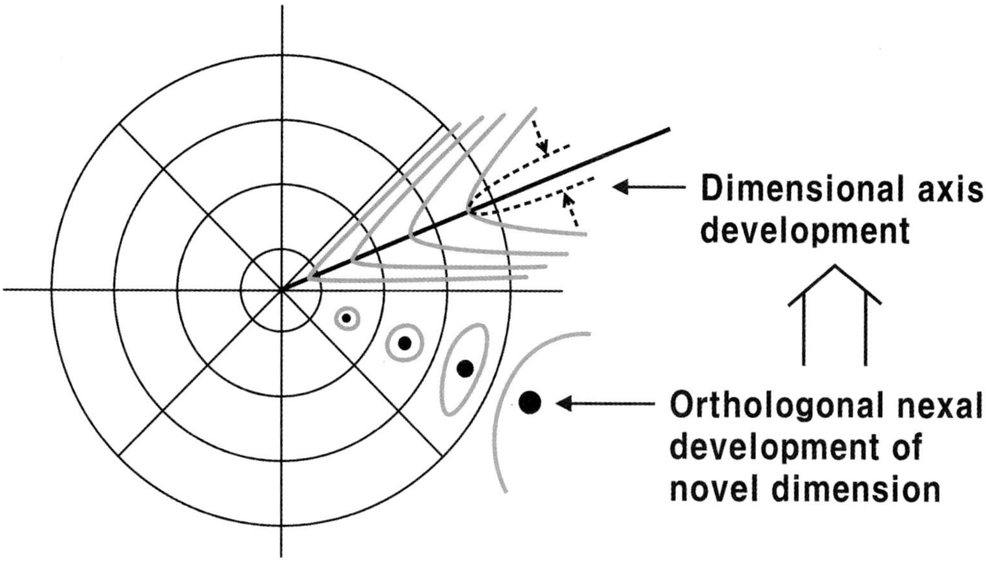

Figure 6-2-3. Orthogonal Nexal Development of the New Dimension

The three major axes of cruxological morphosis are deemed to be integrismic factors (characteristic sysy-nexy ratio), the degree of freedom (characteristic size of a continuum) and hierarchical morphosis. The sequence of cruxological development would be a critical factor in menesysal management. Orthogonal developments of new nexal dimension are visualized in Figure 6-2-3. This developmental process is the introduction of new dimensions to cruxes.

6-3. Complexity and Cruxology

Cruxology is an apparent phenomenon of the complex system. The complex system has been widely acknowledged since the 1980s. It has been partially misunderstood by many as chaos, catastrophe, the complex adaptive system, etc. which are just parts of the complex system.

I frankly believe that the complex system is not a concept to be limited to just the observation from the human-centric viewpoint. Complexity could be subject to relativity depending on whom and where the observer is. White noise to an observer could mean pink noise to another observer. It means complexity is the function of locality, which is the dimensional peculiarity of a part of phase space.

The Complexity Connection with Crux

Usually complexity is the outcome of three causes; the complex interrelation among a number of system constituents, the behavioral rule of an individual constituent, and the mechanism of mutual adjustment among constituents. (Y. Yoon and S. Chae, Introduction to Complex System, Samsung Economic Research Institute, 2005, Ch.1, Section 01, Korean Version) These are all intrinsic conditions of a system that affect the degree of complexity. However, when we deal with the extension of a system to a preferential territory of a network, we find there could be extrinsic condition that might influence the phase space in which specific system-network compounds reside.

Complexity is the major factor of deciding the phase shift of crux. Cruxological development has hierarchical variations depending on the intensity of the sysy or nexy progress. The hierarchy of crux was presented in Table 6-1-1 from minima, nexio, nodus, articulus, modus, figura, to systema. These levels are differently involved in sysy and nexy.

We have thought about the possibility that complexity could be a relative measurement of nexy by sysy. Every person has different levels of noise recognition in the same situation. Like when people become nervous, they seem to catch noise at lower levels due to our neural fatigue.

Crux starts with minima and nexio, which is solely nexusity in the initial condition. But systemicity intervenes with the progressive development of higher level cruxes. In common sense we regard nexusity to be simpler than systemicity. But actually nexusity has many more hidden dimensions than the subjective system does. The work of our brain is to reduce the dimensional complexity of the nexus we face.

The cause of complexity increases when crossing happens between sysy and nexy in human society. Neither Sysy nor nexy can comprehend each other, so that they end up having multiplication effects on complexity. The encounter between sysy and nexy at the boundary of the dynamic core (highly systematic) and dynamic periphery (relatively nexal) produces a lot of discordance.

But once the discordance is overcome, we can see the infiltrating process of the organizational principles of higher complexity in both the dynamic core and its periphery, which consist the entire dynamic module. We usually call this process "cultivated." The common expression of cultivation actually means the general pervasion of new communal characteristics into the whole community or, equivalently in our discussion, into a dynamic module.

Crossing between sysy and sysy, or between nexy and nexy, of same origins for either case, complexity tends to converge into nullity. They mingle together without stress. Complexity could be a matter of the stress condition that causes inertia and turbulence in flow patterns within confined phase space (as materially observed in porous media).

The Transition Zone by Cruxology

The transition zone of cruxology is where the integrismic condition is determined. Integrism is thought to follow the quasi-phase behavior. The pressure-volume-temperature (PVT) relation is essential in analyzing phase behavior in physical chemistry.

Cruxological morphogenesis has similar aspects to the phase behavior of chemical compounds in changing PVT conditions. It does not show exact similitude with physical chemistry but PVT relations help induce the formulation of systematic logic. In cruxological morphosis, transitional variation is highly significant. I believe it is rather easy to understand if we accept the fact that most cruxological changes occur in the transition zone.

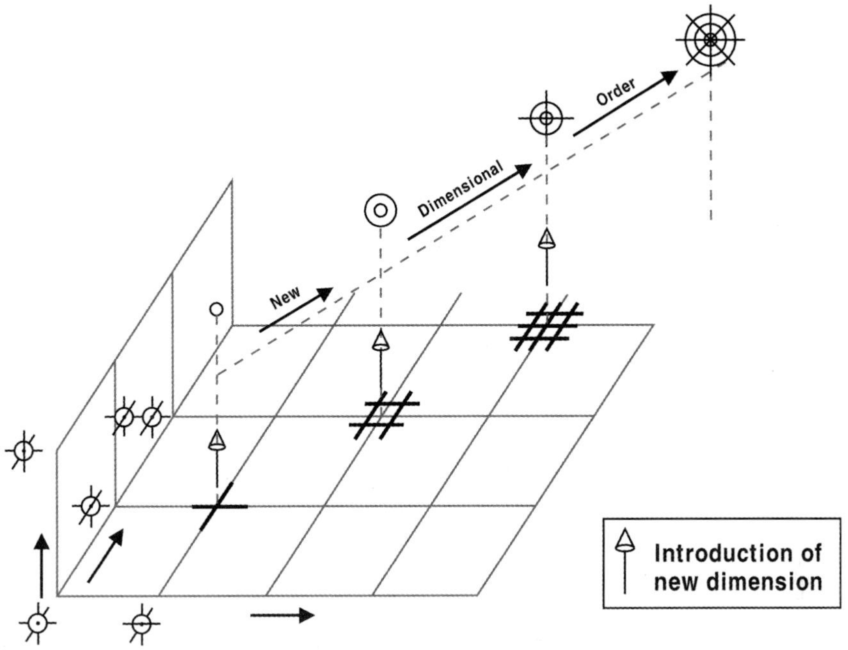

Figure 6-3-1. Visualized Variation of Cruxological Morphogenesis by Transition.

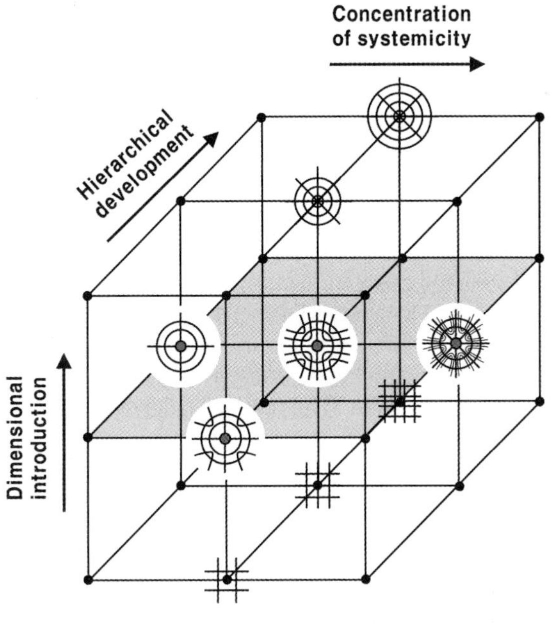

Figure 6-3-2. Visualized Cruxological Morphogenesis of Continuum by Matching Condition.

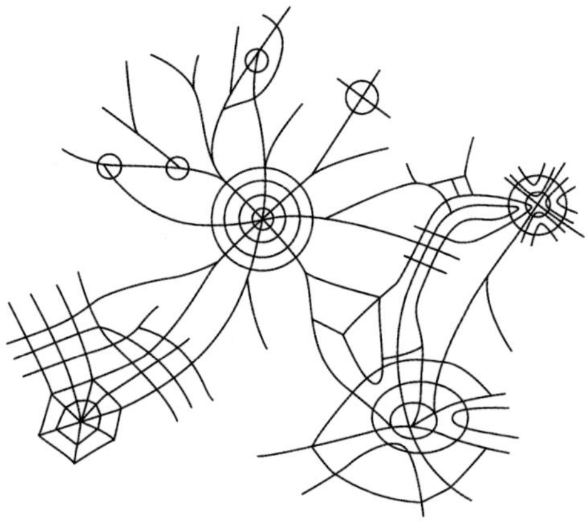

Figure 6-3-3. General Pattern of Menesysal Existence in Reality

The major differences are found in the fact that cruxological conditions are not exactly analogous in matching with physical chemistry. Moreover, we see some similitude with the sequential morphogenesis of the Change (or I, 易). The change of route generates different cruxes and menesys. By this comment, I do not mean to admit or accept the limited systematic thinking of ancient I-Ching, which is isolated and confined like a solitary island. I would like only to point out there is an effect of hysteresis in cruxological development toward menesys.

Homeooikonomia; Homeostasis, Homeomenesys and Homeochaos

Suppose we think of the present status of humanity on the earth as being governed to a certain extent by the influence of the galactic path of our solar system passing through the arms of the Galaxy. (Some mysticism cites on this topic but it should be a strictly scientific issue differentiated from mystic attitudes.)

We have a stronger and more direct influence toward the meteorological impacts, but in the long run meteorology is significantly governed by the condition of the solar system. Our serious concern is if our body can survive and attain a stable condition to live within. So light, wind and rain become the most important conditions of our life.

Our body is very much respondent. In cold rain, we run fast to avoid getting wet. Our hearts beat at the highest pulse rate. Our respiratory system and other organs collaborate together to activate an emergency plan. This collaboration gives us an idea on the dynamic modularity. The cooperation between the above mentioned organs depends on the supervenience of dynamic modules to physicality. However, we have to consider that modular activities have intrinsic mechanisms of network between modules. Such intrinsic mechanism comes from the network of particular genes and modules beings switched on and off.

Most of the intrinsic mechanisms of the human body are generated from past evolutionary records. But transcendental intentionality possesses some probable field of expansion to alter the intrinsic mechanism toward novel interpretations of past episodes by introducing new dimensions. It is done by the proper emotional administration of physical energy.

Emotional aspects of the human being have several representative levels of management such as mood, feeling, and emotion, the definition of which should reach a consensus among scholars to clarify emotional intervention in human life. Actually, emotional intervention is not adequately understood.

I believe that emotional life is a complex, energy administration system of a human. It is not just a system. To create a proper term for it, I would call it an administrative menesys. Some excess energy is discharged at the dead ends of life and then becomes a process. Some deficiencies of life recognized by an entity adsorb energy or potential in a mode of specific emotion. The mechanism of emotion is at the center of human life.

We know that there are homeostasis and homeochaos mechanism in complex system. Homeostasis works effectively in ordinary life and in the contingence plan of our body which is governed by the amygdale. Homeostasis still works excellently because of the effective administrative strategy of the emotional mechanism.

In homeostasis, the physical dependence of the mechanism is so intense that the networks of functional relations have a relatively low degree of freedom. It is deemed that homeostasis is generally an optimum solution specific to the body during its lifespan with the physicality that is established between the hardware and software: The hardware of physicality is systemicity embedded into solid physicality, and the software is the integrismic condition of partially or temporarily governing physicality. Such an optimum solution is attained based on the synergetic principle.

In homeochaos, physical dependence is relatively lower than in homeostasis. It is proper in our research to use the terminology of homeochaos to a limited range of complex integrism, where nexusity has higher governance. Homeochaos is defined by Kaneko as a hypothesis for the dynamic maintenance of diversity. He picked out the essence of homeochaos in four points: weak chaos, high-dimensional chaos, chaotic itinerancy and dynamic stability against external perturbation. (Complex Systems: Chaos and Beyond, 2000, p186-189)

K. Kaneko and I. Tsuda commented that the notion of homeochaos can be used for networks like immune systems, ecosystems and chemical networks in metabolic systems, where networks should be adaptive, flexible and intelligent. (ibid, p236) Kaneko's works have been one of the significant leaders in the science of complexity with regards to life science for over the last two decades. In his research, he said that he passed around the study on the fixed set of functional modules with predetermined properties and rather studied the developmental process "in which interactions among elements lead to their differentiation and emergence as units processing context-dependent functions."

Human cognition differentiates itself in respect to computers. Human cognition does not halt due to the mistake in thought processes. They have the "capability" to change the rules. We used the term "capability" instead of "ability" that Kaneko used, since humanity, of individual and social levels overcomes dilemmas by introducing new dimensions with the support of emotional administration, which is similar process to well-developed virtual reality.

He suggested that his theoretical ideas could be applied to human group behaviors and the evolution of human society: cell differentiation for individuals, formation of hierarchical structures, fixation of hierarchical structure into societal norms, economic and social structure, and forms of civilization and history. (Kunihiko Kaneko, Life: An Introduction of Complex Systems Biology, 2006, p346-347)

Our hypothetical frame pursue similar trends but with slight differences in concept, such as the introduction of integrismic factors as a general measure of the characteristics of phase spaces. Suppose we have three ranges of transition zone in the phase spaces for homeooikonomia (*oikonomia* means order, administration, and arrangement; it is well known to be an original

Greek word for economy): the ranges for homeostasis, homeomenesys, and homeochaos. (We mentioned that menesys means megalo-networked-system and that it is not a Greek word.) The proposed concept of distinctive homeooikonomia is shown in Figure 6-3-4.

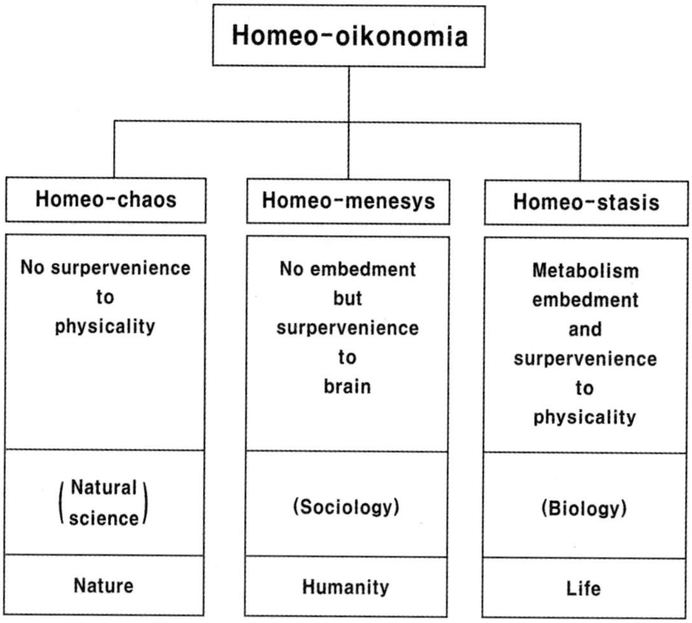

Figure 6-3-4. Distinction among Three Homeooikonomia: Homeostasis, Homeomenesys, and Homeochaos

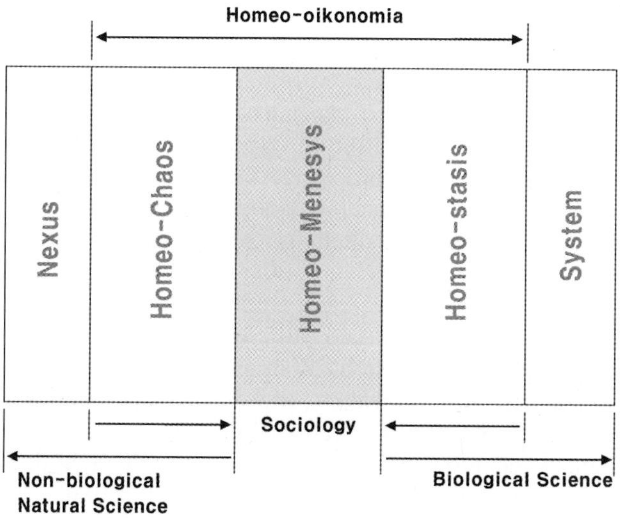

Figure 6-3-5. Visualization of the Territory of Three Homeooikonomia

The three homeo-"arrangements of administration in the texture of phase space" raises the question as to how they are generated. I would assert that as far as humanity is concerned, there are three major governing principles: the principle of synergy, the principle of free phase space, and the principle of emotional computation in synergetic process. The involvement of emotion and thus a managed synergetic process decides the free phase space that is available for the determination of cruxological speciation.

Variant Dynamic Modularity

We have dealt with three kinds of homeooikonomia. Their transition zones are differentiated respectively to each other in the sequence of homeostasis, homeomenesys and homeochaos from the highest systemicity to the lowest systemicity. This is another expression in the light of nexusity in reversal sequence.

There are many kinds of dynamic modularity classified by different cruxological characteristics. Cruxological characteristics are defined by the integrismic factor, the combination of sysy and nexy. If sysy prevails, we get homeostasis, and if nexy prevails, we obtain homeochaos. This is the difference of our theory from existing network theories.

A Transition zone is defined by the variation of dynamic modularity in an integer entity. Dynamic modularity is, in other words, affected by socio-psychological cruxology.

As modules are deeply studied by a wide variety of disciplines, it appears more and more evident that modules are to be treated as dynamic. The ubiquity of modularity has been found in various fields. (Modularity; the Understanding the Development and Evolution of Natural Complex Systems, the Vienna Series of Theoretical Biology, edited by W. Callebaut et al. and Diego Rasskin-Gutman, 2005, Preface and p7ff)

In alphabetical order, the editor of the referred book has brought together the experts of interested fields who are studying modules, experts on "artificial life, cognitive science, developmental and evolutionary biology, economics, evolutionary computation, linguistics, mathematics, morphology, paleontology, physics, psychology, and theoretical chemistry, as well as philosophers of biology and mind." I believe that this list will be extended to the full scope of human intellectual fields within the next 20 years.

Modularity in humanity is not of a simple mechanism. For instance, we have to consider articularity at the lower level of modularity. Let's think of the dependency of human life on light. Though it is essential for living, light brings hardship to humans as well. Life under illumination gives pain and agony an awakened disturbance. Agony invites rationality.

The variety of the rationality of nodules leads their superordinate dynamic modules to differentiation and then to articulation. It is the author's opinion that Nodules have not yet achieved modules but are knotted together with at least three steps of circular logic in a network, mostly represented by individuals in social life. In human physique, joints or nodes are preliminary basic states of articulation that achieve nexusity and systemicity.

A nodule within a system is called nodus to emphasize its integrated status. It can also be called a node when it has an accent on the function of nodularity. Thus a nodule can be considered to be a constituent body of an articulatory articulus. We can imagine that a nucleus family is a constituent of a community or an extended family. Community could be an example of the articulus if it has sufficient characteristics such as adequate geographical division, ethnic difference, or cultural discrepancy and so forth with its surrounding society. Beyond articularity, we can further imagine modules that constitute a system by introducing dynamic modularity. Figura might present in between modus and systema depending on the intrinsic structural attributes of the body of a system. Figura needs more delicate and in-depth explanation in theoretical aspects, but by intuition it is an essential step for the morphology of dynamic modularity in human society.

The analytical approach in science has some pitfalls of methodology because the isolation of an entity has to be assumed and because boundary effects are mostly disregarded in discussing the ontological frame. This may be one of the reasons why the science of complexity was developed.

Dynamic modularity is the most important factor in the cruxological development of a system and is subject to the influence of articularity from under and the influence of figurarity from above. In neurophysiology and genetic study, dynamic modules behave like independent entities to some extent. At the current time, a generally accepted definition of dynamic modules does not exist.

Gerhard Schlosser and Günter P. Wagner wrote as quoted below that the marks of modules are integration and autonomy. (Modularity in Development and Evolution, edited by the above two person, 2004)

"Most generally, a module is conceived of in one of two senses: either as a component of a system that operates largely independently of other components or as a component of a system that is repeatedly used. Here we understand a module very broadly in the first sense as it is the more general and inclusive notion. Modules in the second sense can only be the iterated units they are, because they retain their integrity in different contexts – that is, because they are also modules in the first sense."

"Two aspects of modules are emphasized by both perspectives: their *integration* concerning their internal relations (between their components) and their *autonomy* concerning their external relations (to elements of the context)." (ch.1)

Exploring modularity in gene network, G. Von Dassow and E. Meir state that the definition of the boundaries of genetic modules is difficult, because the gene in itself is a module of a certain kind. "The question is to what extent genomes break down into, or genes and their product conspire to form, *a logically separable guilds of the metabolic milieu* – that is, intermediate entities, made of genes and parts of genomes, which do something we comprehend in isolation. These are a few cases that seem intuitively obvious; the *lac* operon, the yeast-mating type switch, bacteriophage, the cell cycle clock, and segment polarity network." (ibid. Ch.12)

C. Redies and L. Puelles reviewed evidences of two types of modularity in the vertebrate CNS (Central Nervous System). The morphological embryonic divisions of the early neural tube represent the first type of module. The neural circuits represent the other type of modularity (in the CNS) that is functional modularity. (ibid. Ch.8) Embryonic modules and functional modules are deemed to be mutually correspondent in accordance with their topological positioning in integrated functional mapping although they are not easily found in detail.

The definition of modularity must begin from philosophical fundamentals. The functions and variations of dynamic modularity should be based on integrismic characteristics and its consequential cruxology. That is why we have thought of dimensional problems and folding-unfolding mechanisms. These ways of thinking originate from the persontological view.

Generally speaking, modules are the result of the enfolding process of less necessary dimensions of network nodes in order to satisfy the systematic function and efficacy. Due to special internal or external conditions, they sometimes effectively emerge with articulated function like members of our body, the nucleus family, and basic commune. Modules are mainly functional, and are embedded to material and physical conditions as seen in biological DNA. Modules are also socially organized under strict principles.

Dynamic modularity is attributable to energy allocation problems at any moment. Energy allocation changes with the intensity of communication and connectivity. Articularity contained in a dynamic module has intense systemicity leaving the lowest number of dimensions for easier communication with efficient synergy. Systems and modules are the result of enfolding of unnecessary dimensions of networking nodes to satisfy specific systematic function and efficacy. The figurarity of various grouping of dynamic modules are also deemed to have deep impact on energy efficiency.

For instance, the concept of human scale by G. Fauconnier and M. Turner (The Way We Think, 2002) gives a good implication of why we need the best figura for the efficient synergy. Dynamic modules are blended in complex blends (or for easier understanding, blending space), which are evaluated by its energy consumption. The synergy to be attained would certainly govern the level of flaring action in the Animotus tetrahedron.

In the preceding chapters, we have considered the multidimensional submatrix of modules: onto-modules, topos-modules, logia-modules. It is our hypothesis yet to prove that they reduce onto-logia blends for ontology, topos-logia blends for topology, and topos-onto blends for relational ontology. Based on this view, we can see that the number of variant dynamic modularity is innumerably countable.

Dynamic modularity is a very complicated phenomenon. But we can simplify the process in our understanding by adopting certain skeletal ideas that is explained in Section 4-3.

There is another point of further research in dynamic modularity, which is relevant to human transciousness. Transciousness looks intangible but tangible in a sense that it is a major factor that determines our reality. Animotus flaring is at the boundary between imagination and substantiality, and between human mentality and physicality. Animotus flaring seems to come originally from infinities-encounter, which freezes shapeless Animotus flare. We can partially observe the material evidence of the freezing in the process of decrease of the omnipotence factor of an infant. The so-called neurophysiologic omnipotence of infants is described in brain science (S. Ferenczi, 1924, and Y. Shuji, 2008, Hito to Saru no Ada).

Neural Resemblance of Cruxology

There are grounds to believe that the reason why human beings exist is because human race is the spreading neuronal function of Gaia, though whatever is told about the Gaia concept from the standpoints of ecology or orthodox theologies. Gaia is a reasonable way of thinking because the terrenial part of a human being is represented by the fact that all the genetic forms of living things on the earth are composed of essentially the same constituent originated from this Earth.

Gaia is the Greek name of the mother-goddess of the terrain. We may call it the god of the Earth (地神) in Far East, and Ha Erez (the earth) in the Old Testament. Ha Erez received the power of substantial production and contributed to produce all offspring of the terrenial surface. “Let the Earth ‘bring forth (not ‘create’)’ every kind of living creature.” (JPS Hebrew-English TANAKH) Regardless of terminology, we have the same thing in mind.

We are all originated from earthly resources and attained our capability to language from naming, defining and orientating. Human beings are all capable of doing the function of phonetics, semantics and syntax, which is the neuronal resemblance of humanity caused by the cruxological phenomenon of humanity.

The neuronal resemblance of cruxology comes from anthropomorphosis. It is necessary to understand that anthropomorphosis appears in our thought as the teleology of humanity, the texology of humanity (*texo* in Latin, to weave, plait, contrive, compose, and build), the theology of humanity and in the other forms of intellectual works.

Human race has to behave as a neural system in the ecological world, since we have the essence of teleology of humanity in this sense. Dynamic modularity is balancing between unity and diversity to correctly interpret the phenomenon within a reasonable range of accuracy and semantic validity. Found within this dynamic modularity is the texology of humanity. Dynamic modularity is valid in both intangible and tangible phases as far as humans are concerned.

Theology has been reinterpreted in the course of the epistemological process of humanity, i.e. the backward quest toward the initiality of humanity by composing intellectual particles in a mosaic. The theology of anthropomorphosis or the theology of humanity is the gearing part of the theology of revelation. The theological part of human intellectuality is not going to be negligible but will be strengthened and balanced in the coming future of Solar and Galactic Epoch.

6-4.The Science of Cruxology

Cruxology of Free Market Continuum in Different Dimensional Texture
Hierarchical and Localized Modularity
Dynamic Core and Dynamic Modularity
Integrismic and Cruxological Administration
Propositions of the Science of Cruxology
Why Micro Cruxology make Macro Difference?
Cruxology and Creativity: the Essence of Free Market
Creativity Leveled Up by Cruxological Enhancements
Cruxology and Menesys
Development of Menesysal Statics and Dynamics

Cruxology is the stem of humanity science. Actually, the tendency of anthropomorphosis of humankind can be presented as a scientific mechanism, which could be dealt with by the name of the science of humanization. As cited before, anthropomorphosis was originally used in order to explain the human resemblance to the Divine nature.

But leaving the controversial theological portion untouched, we can still take the notion of anthropo-morphosis, which means human (anthropo) change (morphosis), to apply to the general aspects of human phenomenon. Anthropomorphosis follows the rule of science of complexity.

As we have discussed, Divine Nature of totality has to be illustrated by dynamic relational ontology which in turn can be presented in a manner of plexion, which is the universal plexus of accidental occurrences into entirety. In dynamic relational ontology (equivalent to persontology), accidental occurrences are not necessarily teleological but must manifest internal structure of persontological characteristics, which is called integrismic condition.

Integrismic condition makes differences in the determination of dynamic modularity, the abstractive patternization in transciousness, and the communalization of human society, eventually changing all the fundamentals of the humanization process. We also understand that there could be a mutual intervention and an interactive adjustment mechanism among the above tri-leveled mappings of humanization; dynamic modularization in neurophysiology, abstractive patternization of the semantic and syntax in individual transciousness, and dimensional arrangement in the communalization of society. Free market is a good example of the complex cruxological development in humanization processing.

Cruxology of Free Market Continuum in Different Dimensional Texture

We have faced various cruxological developments of different dimensional textures in the evolution of free markets. This trend has been prolonged in history probably since the start of the barter trade in Neolithic age.

The human race has developed a firm basis of market system embedded into the material conditions of roads and cities, and other intangible systematic approaches such as monetary circulations to ease various transactions. Cruxology of roads (1-dimensional) might be the first major achievement of the free market system.

In modern architecture, lines of movement have become one of the most important factors of design in urban planning and complex building designs. This is due to the significance of permeability of people into complexity. For instance, Guggenheim Bilbao is characterized for its introduction of mobility in museums. (2005 Yasumitsu Matsunaga, Machizukuri no Shinchoryu, and 2001 Mark C. Taylor, The Moment of complexity; the emerging network culture)

Cruxology, for instance, is shown in the road system of human kind. The construction of roads for people and transportation has experienced a long progress of advancement since ancient civilizations. As people design their products to fit their needs and purposes and consequently in terms of dimension, their road system would have been also constructed in such a way of how they understand their utility.

The primitive form of road was so narrow that it merely allowed the minimum width of passage ways in a manner following the natural curvature of topography. However, in a long historical development of the road network, human beings extracted essential design factors to efficiently represent human needs.

The civilization of Incas had created relatively narrow mountain roads, while the Roman Empire constructed the most powerful transportation system with its well designed roads, bridges, and fortresses. The Mongolian transportation system passing through various desert lands had less construction technology than the Romans, but had a greater and faster post horse system to secure the information transfer throughout the empire.

In modern society, we have a number of flourishing metropolitan areas spread throughout the globe, and this social phenomenon improved air transportation and ground transportation, such as highways, railroads, subways and canals. If we confine our discussion only to the ground road system, there are several design factors that needs to be taken into consideration; width, direction, road surface properties, mutual communication at the cross, and so on.

When one-dimensional roads are systemized to form clusters on the surface, the two-dimensional road system behaves as an attractor for the two-dimensional market system. The cruxology of the areal market (2-dimensional) had been realized through the birth of cities following the road extension.

The free market system has not yet proved its absolute validity and legitimacy in the economy. But it is a major part of the natural flow of wealth and stands at the center of the continuum of economy. The emergence and diminution of a free market distinctively expose the socio-psychological aspects.

The socio-psychological exposure of humanity is thought to be presented in the free market continuum and the associated monetary system of value exchange. It is presented in a progressive manner of

point－line－area－space－extensions

(点　－線　－面　－積　－續).

Such expression can be substituted by our way of thinking in a sequence of

point－minima－nuclei－Monad/grouped nuclei (=cell division)

(点　－幺　－核　－大單核/群核　(=細胞分裂)).

It is again expressed in sequential features of

minima－line－area (= nuclei + lines)－space (= nuclei + fluxes)－extensions

(幺　－線　－面　(=核　+ 線)　－積　(=核群 + 線束)　－續).

The extensions that appear at the tail of the sequence may imply the introduction of new dimensions and sometimes the merger-integration of the past isolated entities or just acquisition of free connectivity with other markets.

But as soon as we leave the surface of the earth, literally a free space is waiting for us. A New Horizon opens up before we leave the spherical surface of the earth. Furthermore, in our cognitive phase space, we meet highly multi-dimensional free space with extremely complicated structures. These free spaces will contribute to the expansion of market, as the World-Wide-Web (WWW) has enormously widened our scope of transactions.

As further as we get into multi-dimensional continuum, the cruxology of spatiotemporal continuum (3-dimensional, 4-dimensional or higher) changes into a higher complexity. Simplex standard forms with the regularity of design could help the extension of continuum in general sense.

We can find such structural solution in architecture of both buildings and human brains. Tensegrity is tension-integrity method innovated by Richard Buckminster Fuller in the 1940s. The continuous pull (convergence) is balanced by the discontinuous push (divergence), as we can see at the Louisiana Superdome in New Orleans producing integrity of tension-compression. Tensegrity structure is stable omnidirectionally and independent of gravity, without having any weak points and free from fault propagation.

György Buzsáki commented that the cerebral cortex is a scalable robust spherical structure and that the modular plan of which is identical in all mammals. The modular plan consists of five layers of principal cells and a thin superficial layer, spanning only 1 to 3 millimeters in thickness. Tensegrity dynamics of brain network is achieved by the balance between excitatory and inhibitory forces (G. Buzsáki, Rhythms of the Brain, 2006, p42-44).

The Dynamic Tensegrity has some aspects of shamanism in the perspective of how our world and pure energy of the universe is assembled. It also shows a similar pattern with the theory of the extreme point by Plotinus in Neo-Platonism that indicates the assemblage point between the divine and human. But dynamic tensegrity only denotes the scientific concept of the perfect harmony among various opposing forces.

Generalization of dynamic tensegrity might have the extension of concept to the emotional stability and to the balance of dynamic modularity in brain science. It seems also applicable to the free market continuum we are discussing here. But the simplex regular pattern of parts doesn't need much emphasis, because we can utilize the gradually changing quasi-regular simplex forms as was utilized in the symmetrical drawings of the Circle Limit series of M. C. Esher.

Furthermore, our brain and market are not to be restrained by such standard forms. There is also a computational balancing mechanism analogously made in our cognitive process supported by the emotional mechanism.

It is the principle of dynamic tensegrity to assist us in designing of the phase space for the continuum of free market and of the creative peace. Our problem is how to define referential points and boundary conditions in mapping.

Hierarchical and Localized Modularity

Hierarchy tends to be established in a different mode depending on the character of localized modules. The character in our concept is cruxological. As we know, cruxology manifests the procedures of knotting, nodulizing, articulating, modulizing and figurizing. These procedures are the spontaneous results of integrismic condition.

Hierarchy is dominantly governed by the modular process and its dynamic change. There is another factor we have to consider to understand the hierarchical order. The spatial allowance for the modular activities to formulate the order is important. The allowance presented in the hierarchical formation is the characteristic height.

The characteristic height of hierarchical order developed in and by different local conditions decides the optimum range of balancing modular activity, which could be equivalently interpreted to the concept of symmetry in impact distribution of the activity. Impact, in general, is a function of intensity multiplied by frequency.

We discussed before that the property of module is integration (in society, stress-tamed) and autonomy (in society, voluntariness). There was another speculation on the modularity that a module is *logically separable guilds of the metabolic milieu* (thus, closed or quasi-closed logical circuit is an essential factor in module).

Hierarchical space allowance decides the number of hierarchical levels and affects the characteristic size of its contained dynamic modules. The local allowance and its internal process of cruxological development generate localized modularity.

Free market would be a good example to demonstrate the natural and spontaneous development of the hierarchical and localized modules and their activities. The evidences from brain science supports that plasticity exists in human cognition.

Human tendency of anthropomorphosis made us recognize that elasticity in transciousness is the cause of human imagination and thought. This elasticity has been developed to resist against human instinct that resides at the blind level of sensciousness.

Plastic performance at micro level neurophysiologic activities eventually has taken the lead in formulating elasticity at macro level in transciousness region. Transcious elasticity is not directed to be teleological but the episodic edition of temporal interpretation makes transciousness to take orientation.

The evidences of transcious elasticity can be observed in the coincidental trends of social activities in the same community, a group of people, in the same ethnicity and so on. The discordances among people from same background are not greater than their coincidental trends. Such evidences in our everyday life are usually revealed through local peculiarity of traditional custom in culture, practices in politics and economy, precedent cases or court-made law and so on.

Dynamic Core and Dynamic Modularity

The resemblance and difference between the definitions of the NCC (neuronal correlation of consciousness) and the dynamic core is discussed by C. Koch. (The Quest for

Consciousness: A Neurobiological Approach, 2004, Ch. 19, Section 19-2)

Edelman and Tononi think that significant scale of feedback (called as a reentrant signaling loop by Edelman) stabilizes the dynamic core by a stronger mutual interaction within several hundred milliseconds. The difference is that Koch directed his study towards the contribution of intrinsic properties of local neurons to the neuronal root of consciousness, while Edelman took a holistic approach.

Studies on the topics of dynamic core provide the fundamentals for the theoretical scheme of dynamic modularity. We have mentioned several times that the dynamic module consists of dynamic core and dynamic periphery. In our view, dynamic periphery has greater emphasis on nexusity; it permits the constituents of periphery to pay more attention to the outer environment rather than just being obedient to the system of their inner core.

Presented here is the essential part of the theory of dynamic modularity as follows:

First, as previously discussed, the dynamic module is the integration of dynamic core and dynamic periphery;

Second, generally dynamic module has a floating property due to dynamicity, though it may be very slight. Discontinuous continuity should appear as quickly as the dropping pattern in systemicity in the transition zone between the dynamic core and the dynamic periphery;

Third, the dynamic module is the variable collective of encephalic cortex and looks as if it is connected to thalamocortical neurons without exception as long as it is remained within a unified stream of activities. The variable collective has been organized by an integrated neuron process with the highest degree of complexity. Therefore, the Dynamic module has an integrated complexity.

The integrated complexity in the first run would decrease due to the integration of simplex regular logics and the tensegral frame of logics, but will increase with a successive expansion of periphery. The integration is done by enfolding of unnecessary arms of logics by fire-together-wire-together. It is suggested that the dynamic core contributes to elasticity, as the periphery does to plasticity. The dynamic core in the long run floats in the topography of neuronal phase space;

Fourth, it is suggested that the integration of conscious states is done through the network of dynamic modules while the network of dynamic cores builds and maintains the intentional elasticity in transciousness.

Fifth, dynamic cores provide the basis for qualia by generating simplex Gestalt forms. The Gestalt forms are produced by the symbolic mapping, which is matchmaking between the dynamic core activities and the savings account (?) of innumerous symbols. This enables the complex fluttering of multiple simplex qualia, usually experienced in our cognitive process. In a social perspective, we would acknowledge such effect on symbolism. It is a phenomenon which corresponds to the mathematical homothety.

Integrismic and Cruxological Administration

Development and management of cruxology should be based on integrismic administration. Integrismic administration is reexamining and re-engineering of the whole cruxology, in terms of the place, time, and way to vary the system and the network factors. Sometimes it is a matter of finding the exact location and the exact part of cruxological states.

In reality, we often find certain cases in the economy on the introduction of new distribution networks, the organization and the withdrawal of Research and Development bodies, and the Merger and Acquisition processes. The introduction of a new distribution network of commodities is an intensified effort for nexal development.

The organization and the withdrawal of R&D body are a matter of specific departments relatively segregated from the mainline of everyday decision-making. R&D usually has significant constraints in scale as well as in character to be tangled into cruxology unless it is an active innovation. Current problem of the R&D department is the excessive systemicity to keep up with the objective of its belonging corporation.

M&A can be understood as the integration of systems. The importance of hierarchy in the cruxological view is highlighted in the days of merger. The idea of the "system of systems" nowadays has been under fast development and widely applied in enterprises and in strategic military affairs. It possesses the possibility of ruining organizations and corporation efficacy by deleting the immense field of creativity and autonomy.

Gore-Tex is a legend in the management history that provokes creativity. But such systematic and flexible ideas of hybrid management will be more important when we get into the topic of menesys, the megalo-networked-system. Menesys will be dealt with after the globalization process near the end of the book.

Management and evaluation in cruxological view will make magnifying differences as to continuously preserving creativity and vitality in the menesys of the global/solar system.

Cruxology and Creativity: the Essence of Free Market

The relationship between cruxology and creativity is the essential factor in governing the formation of free market continuum. Free market mentioned here is an ideal one that is not flimsy and frivolous.

A linear market first evolved in the man to man connection, perhaps in Neolithic era, was capable of creating combinational cruxology; people can mix up freshly encountered ideas and then activate a combinational creativity. Linear market was the first comer among the first kind of free market. Areal and spatial market emerged successively. Areal market is the style of market field we have usually see in simple periodically held countries. Spatial market, I would say, has a more complicated hierarchical consumption structure, differentiated by social class. If anyone wanted to sell anything to knights, cardinals or the king of the city, he should be prepared with differentiated commodities.

Columbus, when he discovered America, he must have sold European commodities to the areal market of West Indies. On the contrary, Vasco da Gama probably sold products of variety to the spatial market of East Indies to sell to kings and aristocrats. Areal markets constituted with the native consumers who needed only commodities of less creativity; whereas the spatial market has consumers from various social classes who have differentiated curiosities and desires in accordance with their affiliated classes. Simply, linear relationship develops into areal and then up to spatial one as time flows. The creativity of the first kind of market stays at the first level.

But later in the spatial one we apt to experience the change to the second level of creativity as far as a spatial market provide a marketer the understanding of the social structure and the inner mechanism to it. In that case I would classify the market as the early version of the second kind.

The second kind of market is a functional one. The second kind of market is stemmed from the fundamentals of more segmented and functionalized society. We generally hear about intrusions in the globalization process when focu on developing countries, international corporations, financial institutions, and other advanced modes of business.

The structural cruxology in a society effectively contributes initially to the market change into the second kind through the introduction of a new dimension in logic, innovative principle, and administrative breakthrough and so on. The highest level of dimension for introduction is thought to be attained by the fabrication of the partial web of virtual reality, such as propagation of new systematic way of thought.

The partial web does not only mean the sort of WWW but includes all kind of nets wrapping the society or the market as a whole. The partial web would grow to cover the entire society until it is totally wrapped. The virtual reality has much more implication than that on the Internet. A human being lives with virtual reality in its entire life, for cognition is a kind of a virtual reality.

Thus, it is natural that the humanity has been producing virtual webs of certain kinds throughout history since the seed of modern cities sprouted a thousand years ago and commercialism got into stride. There are numerous forms of web which started to contain the human society and market within. These webs have been connected to one another gradually throughout hundreds of years, increasing the complexity of interconnection. That is the reality of humanity. The second kind of creativity is connected to the second kind of market and the second kind of cruxological change, the integrismic factor of which has greater accent and effect of systemicity.

The third kind of cruxological change would appear when the overall fabrics of virtual reality (such as knowledge menesys, internet WWW) brings or fabricates metamorphosis of free market or society in entirety, as can be seen in life of an insect.

The third kind of creativity hits when the apparent frame of dimensional structure and its associated fabrics of background regulatory principles make fundamental variation and overall transition.

The background regulatory principles can be expressed in the enormous networks of yes-no, or on-off switches that presents themselves in many entangled disciplines in human mind, such as, religion, custom, habit, knowledge, like-or-dislike, and others, positioned behind and wrapping the apparent social phenomenon. Such fabric of network had better be recognized as a virtual reality of equivalent complexity to the apparent affairs because they interact in all directions and hierarchies. This virtual reality has an extraordinary ability to reflect the changes of the phenomenal world in time and in due sequence.

The third kind of creativity comes together with the third kind of cruxological development. Therefore it is the metamorphosis of human kind. The third cruxology and the third creativity invite the transcent of humanity. It is not a dream. But if we fail to fabricate adequate management of cruxological development based on the comprehensive understanding on the coming future, we will be exposed to a broken dream.

The third kinds will be experienced in the next hundred years. There will be an introduction of numerous new dimensions that exceeds the second kinds. The third kinds would be different from the previous kinds in that they will be controllable in scheming for the balance of the immediate foreseeable future within the whole view of sustainable prosperity. The essence of the technology would be in finding out how to leave the vacancy to invite the effect of nexusity on the top of the systematic society.

I heard that there are many Korean Christian missionaries devoted to drill wells and install pumping instruments for African people suffering from drought and thirst. They need 'majung-mul', which means to induce or to receive water for pumping. Such act is necessary in

Africa due to the low initial level of underground water. 'Mul' means water. 'Majung' is a really beautiful humane word; it basically describes an act of going out to the entrance of a village to wait for and welcome our previous friends, guests, father, grandmother, sons and daughters.

In hydrodynamics engineering, it is called imbibitions process. The vacancy on the top of the system should be prepared with the waiting for the nexal connection with its externality. The partial disagreement of this book with evolutionism is the majung-mul part of thinking, the vacancy or vacuity idea that we have to preserve to be humble.

It is crevolutionism, the creative revolution. Crevolution is not only a humble or humane attitude but also a strict understanding on the deficiency of our civilization that accommodates only the logical conclusions with hundred percent certainties in our system management. Crevolution is the principle of providing rooms for revolutionary creativity of the third kind. Crevolutionism is not a sort of romanticism.

I found some examples of the above explanation of creativity of different kinds in a Karl Popper's writing about music. He commented on polyphony as a peculiar heritage of western civilization in a sense that music does not have its precedents in ancient Greece and is the creative field of the West.

Summarizing his explanation, the invention of counterpoint comes from the propagation of *organum* in central Europe during the 10th century. Ambrosian and Gregorian chants were made strictly with liturgy in comparison to the Minnesang and Meistersang, old folksongs in Germany. (Urlich Michels, dtv-Atlas Musik, 1977, Medieval Ages)

It is Popper's opinion that folk songs enjoyed liberty in creativeness while dogmatic regulation of the church induced the development of *cantus firmus* (firm music or fixed melody), which will eventually provide the basis for creating counterpoint and polyphony. Although it looks like that response errors made by congregation in antiphony and responsory had initially caused these alterations, fixed form of *cantus firmus* functioned as a firm frame and a regular order to induce creative freedom without chaos. This, I would call the introduction of novel dimensions.

Popper also showed unconsciously another comparative form of creativity in Minnesang and Meistersang, which we would classify as the first kind. However the second kind appears as a new dimension introduced to an environment that has an existing coordinate axis as a scaffolding to build up new style of music. He pointed out that it was a new cosmos governed by a completely new order. However, I would like to switch his eulogy to the third kind of creativity.

I found the most probable manifestation of the third creativity in music when I heard a person skillfully singing in a fluent glossolalia associated with a continually changing free gamut, I understood. It was a strong candidate of the third kind of creativity in music, extracted out of almost fully diversified phase space of modern music surpassing over the ranges spreading from ancient to present, from classic to rock-n-roll and so on. I know there could be many other ideas apart from mine, but I am very curious as to what the third kind of creativity in music would sound like.

Present transition will have deep impact to future humanity. This is severe enough to cause drastic change of the whole transition zone of homeooikonomia. It has been thought that technology was the prime reason for such an impact on humanity, but we have to acknowledge that the energy of diastrophism has been indeed accumulated in the human cognition and the way of thinking long into history from many sources of humanity.

I have mainly dealt with free market continuum in the above discussion. I believe the economy is the central fabrication of society. Free market is not good or bad as stated in many debates. It is natural. Economy is "Eat"-onomy. In my primary school days, I often heard from my father at our dining table that "all human affairs come from eating problems," and that we should "try to invent chlorophyll-factory to feed global people, and then there would be no more wars among countries." I still agree on most part his opinion even though I now realize that there originally is a bunch of trash in the hearts of human race.

Creativity Leveled Up by Cruxological Enhancements

Many levels of creativity have been differentiated by the efficient use of energy and by the accordance among cruxological properties. (Matching between cruxological properties eventually affects the synergetic effect.)

Energy management of the human brain by systemizing all the received information in accordance with individually, characterized rationales was developed in the past. Reductionism could be the type of thinking pattern. Customary habit is also a similar case. Especially infinity issues, such as infinities-transfer and infinity matrix, are seriously connected to the energy consumption of the brain.

Does the energy management work well in the mechanism of dynamic modularity?

Why do we have to introduce the concept of crux and orthogonality? It is to minimize the energy consumption of brain, which is governed by stress condition. Stress inefficiently consumes energy more than needed.

Why do brains mend boundaries smoothly in between manifolds? This is also attributable to the mandatory discipline of energy minimization in our physical condition. It is thought that relaxed selection is a result of energy savings of systematic arrangement of abstraction in the human brain.

If cruxology is enhanced, creativity is supposed to be leveled up. How do we know if cruxology has been enhanced? There are many indications we can think of; first, by hierarchy; second, by the relative size of the core; third, systematic internal alignment of the core; fourth, nexal connection of periphery with external entity; fifth, the quality of external condition; sixth, nexal connection within the menesys; seventh, systematic interaction among modules in the core and periphery.

If capacity and internal connectivity are changed, integrism will be opened to a new phase.

Cruxology and Menesys

We need to taste a little tip of practical aspects of cruxology from the technology of menesys before going into Part II, Ch. 7 in which the technology of menesysal management will be considered in depth.

Cruxological development generates seven kinds of heteromorphism (異形性) due to their environmental conditions and intrinsic attributes. Menesys is one way of expression in the cruxological output of humanity.

Growing cruxes take patterns of network, system, sysnet, netsys, and finally various kinds of fully integrated menesysal forms. The cruxology of menesys is deemed to generate passive and active menesys, which are complex adaptive system (passive menesys) and complex transforming menesys (active menesys).

It is likely that menesysal maps can be produced if we collect and accumulate sufficient data and information.

Development of Menesysal Statics and Dynamics

If we become familiar with menesysal phenomenon, the statics and dynamics will be developed respectively for the core and periphery in dynamic modularity. Menesysal statics works well in dynamic core whereas menesysal dynamics is effective in dynamic periphery. But when the modular situation of menesys is in the state of complete solidification or complete fluidization, the menesys does not permit room for the integrismic and cruxological theorization.

If we successfully develop menesysal statics, the theory would prefer to place core at the interpretative center and thus extending the theory from the core system, module and finally node connection. On the contrary, menesysal dynamics will prefer to take the trend of node, module, and at last core system connection.

Adaptability for sysy will be our major concern in the core and transition zone; adaptability is most meaningful in transition zone but still good in the core region. Flexibility for nexy will be the main issue at the peripheral region; nexy receives resource material for adaptability but do not perform adaptation.

Flexibility will be able to explain peripheral state with the conceptual tools of integrator and resonator differentiated in nexal maintenance of networks. It is related to the neurophysiological models of neuronal responses, which have resting state, excitable state (just temporary contact for information exchange), periodic spiking (periodic equilibrium), and maintainance of nexal connections (Eugene M. Izhikevich, Dynamical Systems in Neuro Science, the Geometry of Excitability and Bursting).

It is seen that adaptability is the main theme of the systematic maintenance of the dynamic core and transition zone. Continuous characteristic oscillation maintains system adaptability of neuron and core neurons should be stably adaptable. The multi-stability with multi attractors will be our important concern if we want to deal with dynamic modularity of many sectors that exist in the real world.

Fixed Stability will be attained by the embedment to materiality as done in physical embedment of human. System has to experience embedding procedures into materiality to strengthen the solid systemic stability, which we can observe the change of DNA structure or on-off switching mechanisms.

Here are other issues to treat; how to measure sysy-stability and nexy-stability, and sysy-variability and nexy-variability, sensitivity test utilizing input-output relation, hierarchical mapping of the core by the phase space mapping of the multi-characteristics of system.

How do we understand the mechanisms of multi-attractors, which causes multi-personage or multi-systemicity, can be approached by constructing of integrated dimensional coordinates to include all the mapping variables (strictly speaking, impossible, but most probably predictable in practice by proper omitting), which are related to menesysal map of knowledge-intelligence-pretium.

The matter of stabilization in static and dynamic situation is recommended to consider another approach that we can find in the Triple Mate of $(0.618)^0$, $(0.618)^1$, $(0.618)^2$, stable trio from *Proportio Divina*. It will be a sort of technological approach to make things handy in decision making. Series of 1.0, 0.618, 0.382, 0.236, 0.145 … can be used in formulating stable local structure in organization. We will be finding many other complicated models in nature too.

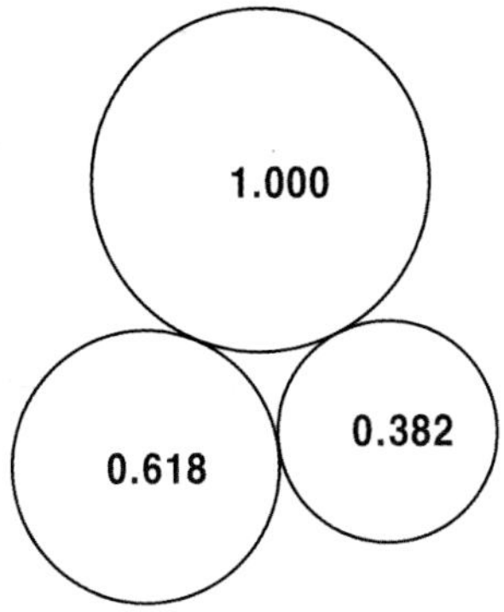

Figure 6-4-1. Stable Initial Triple-Mate

6-5. Continuum Economy of Multiple Interflowing Compounds

Cruxology and Socio-psychological continuum
Socio-psychological Continuity of Continuum Economy
Financial Economic Datum Plane (Variable),
Creativity Activation Map and Capital Accumulation Map
Market, Mirror Markets, and Symmetry in Economic Continuum
Integrism and Mixed Economy, and Maintenance of Market Continuum
Connectivity and Koinonia in Continuum Economy

We have Schrödinger Equation since 1935 to describe wave function. We understand by his equation that we cannot explain even a move of a single electron without an imaginary number. I believe imaginary numbers are workable in an abstractive sphere of the human cognition. The complex forms of the continuum of the global economy are originated from the abstractive and imaginary versatility of human cognition. Black market, virtual market, mirror market and so on are founded on all the variable ways of human thinking.

Cruxology and Socio-psychological Continuum

The human race has been experiencing only the initial stage of continuum formation in the last few centuries. The lack of recognition of continuity in human existence has always distressed human being. The thought on continuum in personal and communal conditions will alter many aspects of human life.

The ignorance of being one continuum is probably caused by the excessive self-realization in the early stage of establishment of individuality. Continuum is a notion of physics, usually used in a manner of "spatiotemporal continuum." However, the terminology will be adopted in greater frequency and importance as we forward to a new era for the development and exploitation of extreme and remote environment such as outer space and extreme zones of either tangible or intangible conditions. The intangible conditions to be explored would include the zones for virtual reality.

Survival modes will have to be sought actively in various conditions to build the integrated continuum of humanity in the world. Humanity has to maintain its continuity in all the possible circumferences regardless of the constraints originated from nationality, ethnicity, and locality and so on.

The way of our living in continuum needs to be vitalized by motivating the search for the integrity of humanity. Environmental conservation is one of the fruit of living in continuum but our firm stance on preferential pursuit for producing economic values will not be disregarded.

Socio-psychological Continuity of Continuum Economy

It is an interesting job to interconnect human cognition and the phase space of continuum. The boundary of cognitive phase space is formulated by the emergence of infinities-matrixes. We have defined in Chapter 4 that the penumbra of consciousness is the transition zone between consciousness and transciousness. It overlaps with the transforming region of the character of continuum.

Intentionality has an orientation that justisfies direction (aboutness) without any scalar value and immunity, thus having some boundaries that should be kept from external influence. Transcendental intentionality is the unique capability of a human being, who usually pursues vectorial movement towards a target and tries to overcome present constraints, suppression or boundednesses.

The boundary of a person comes from the boundary of cognition, which overlaps with the boundary of mind. The human mind is Le-Ba-B in Hebrew. Le-Ba-B is the source of the floating effect within the human boundary. The "aboutness" alone, which is the intentionality on an object, does not cause floating of the boundary of human cognition. However, transcendental intentionality makes human float when sufficient infinities-matrixes are formulated by the passing of transcendental intentionality over the critical condition.

Repetition of successive impacts of infinities-matrixes is responsible for the initial start of transciousness; transciousness is activated while getting along with the destined tendency of imagination and teleological orientation.

The sociopsychological continuity of continuum economy is based on the human cognition, which consists of sensciousness, consciousness and transciousness by the definition of this book. Social continuum including economic continuum is assumed to have similar character as the continuum of individual cognition if society achieves this continuity.

Social continuum is differentiated from the individual continuum in that fractal elements of the individual realm are disregarded in large magnitudes. The social continuum has several different aspects of individual continuum. The most representative trait is that social communication takes complicated procedures that need more time consuming the variety of developmental scenarios. It depends on the fact that humans have less physical embedment at the social level. We have greater noise, more number of directives, greater dependence on the past experiences and solutions than individual level at which we do not worry much about.

Social continuum has greater influence from the sociopsychological factors than any other factors, because social continuum is the output of the transciousness and postefference. Social continuum possesses certain mending effect of social intelligence to mobilize and concentrate people toward agreement in logic and in emotional aspects.

Sometimes social continuum is maintained by presenting an obstacle that every individual is guided to overcome in collective movements. Collective movements are generally led by the psychological leadership as we experienced in the Thir Empire of Hittler and by the confinement of our intellectuality within the ideological logic as in the old USSR. Social continuum has to have variety of fundamentals to be stable to respond to compulsory circumferences.

Financial Economic Datum Plane (Variable), Creativity Activation Map and Capital Accumulation Map

It is our consideration that there is a financial datum plane in our global financial system or menesys. The datum plane is variable depending on the monetary system or menesys. Monetary system in our discussion should also include general barter system. The major point of our view is that financial system has stemmed in order to assist the exchange of commodities.

Monetary system is a tool to reveal the apparent flow of values in pricing mechanism. It is generally agreed in economics that there is only one tool to measure the exchange value of wealth, which is pricing in a free market. The significance of free market is with pricing mechanism that manifests the intrinsic socio-psychological structure in an economy of a society.

The financial economic datum plane (FEDP) of global situation is important, because the flow of economy is distorted seriously by the potential of the datum plane. FEDP behaves like sea datum (oceanic level), which changes very slowly and is almost stabilized but the height of waves can be realized by the datum plane.

If the long term transaction volume high, the flow velocity of economic circulation high, creative activities high, and speculation in business high, then FEDP should be high. However, if FEDP is high only to pursue financial profit of financial institutes or individual bonus of broakerage, the financial practice should be unhealthy. In fact, the level of creative activity depends on the high potential of the datum. It is likely that capital accumulation starts from the datum. It is noteworthy that trials should be made to produce creativity activation map and capital accumulation map for global economic continuum in order to assess timely assessment of financial status.

Market, Mirror Markets, and Symmetry in Economic Continuum

Economic continuum is the continuous body of interflows. In the developmental process of Euro-American market continuum, there happened a long progress from population increase, commercial activity increase, and life standard upgrade since 11th century, and consequential drive to pursue speculation for most of citizens started everything specially in Netherland, England, and France, which is not a healthy process. But we can control and try to govern the developmental process from psychological initiative to make things happen in healthy manner in our days because we have more reliable information and data collection system.

In order to understand the character of continuum, we need to find out the reality of various mirror markets and the symmetry in economic continuum. Mirror market also feeds people and sustains the apparent market continuum. It is not an infrastructure or substructure. It is an ancient body or natural form of market continuum that is now a residual of apparent continuum of institutionalized market. It does not seem to be erased nor expelled from present market continuum. We need to find the intrinsic structure of whole market continuum.

Mirror markets are parts of market continuum, sometimes structurally sustaining open market .Mirror market is a retarded market and it resists to the pressure of open market in its best ability, just as the partial reason of terrorism is severe antagonizing against advanced countries. Mirror market in many cases sustains the fundamental ground of basic communities. Mirror market or underground market supplies at least the fundamental need for the grass root people, food and psychological comfort. It is not standing by the illegality of mirror market.

We need to divide the economic continuum in intrinsic and extrinsic market, which is better expression. We usually see the ecological environment as a continuum; economic continuum is different from the ecology in character, in content, and in syntax. But the formulation, maintenance and alteration depend on the same principles. I believe that is what we find nowadays in network study.

We find the comment on transposon in medical dictionary, which is not familiar to common readers. Stephen H. Buhned (The Lost Languages of Plants, 2002) wrote that we have genetic menesys as if we possess savings account in the bank, where wide range of genotype preserved in the bank of diversity. Plants respond to the stress causing to fluidize the genetic structure and produce mutants of big difference in a very short time. Virus helps this process by inserting less related DNA sequence into plant genetic structure. Symbiogenesis also works for the same purpose to generate new living thing. This is why Barbara McKlintoc, a Nobel Prize winner, commented that whole cell, an organism or the environment receives the instruction about the rearrangement of genetic structure.

It will be valuable act for human race if we reinterpret the hidden side of the world in comprehensive view and give proper territory to those unrecognized side of hidden world to behave good with proper role.

It is reasonable to see the considerable portion of the hidden mentality of globalization is colonizing spirit. They are manifested as the mentality of governance, the will to power, and the structure of dominance. These are the precipitate of systemizing the world. History advances following the course of repetitive iteration of expansion and concentration. However, global menesys dies if the precipitates are not eliminated. Now is the right moment to take action for healing.

We are about to complete apparent globalization in territorial expansion. We need to consider the fact that ecological aspects are manifested as the chain of chemical reactions, but economical aspects are presented as the institution and commodity, which is mainly adhered to the creativity of human being.

Most mirror markets are spontaneous phenomena except illegal black market, which must be enforced to put under legal control. If we want diffuse the mintage to smaller communities and construct multi-hierarchical and multi-local currency institutions to induce better flow of values in our global financial environment, we need to study in-depth and understand the mechanism of mirror markets thoroughly.

Mirror makets and underground markets could be caused by the accumulation of counter-cultural or sometimes the urgent economic need of a society. These side effects of the acceleration in globalization process looks opposite to the expansion of market continuum. However it is not definitely opposing against the extension of continuum, and is subject to a novel interpretation to include those isolated islands.

Integrism and Mixed Economy, and Maintenance of Market Continuum

The mixing of various patterns of economy looks inevitable in globalization process. However, global free market should utilize this kind of heterogeneous elements in order for economy to become stable through adsorbing all diversified elements that humans can create. Integrism is a way of collect the diversification of economic genes together.

However, the world is not destined to a mixed economy without teleological objectives, but directed to proper construction of economic continuum and world-wide establishment of educational information network to induce integrated development of the global continuum. Such network will become educative information technological menesys, which will technologically enable market continuum of the world to be sustained.

Connectivity and Koinonia in Continuum Economy

Animals live within the continuum of the present continuously extending from this moment. Human continuum is much more complex than that.

Virtual reality, visualization, spatialization, imagination of continuum, virtual space and technology for the accumulation of information, future planning, and so on are intangible, multi layered, complex continuum within which humans live in.

In these real and virtual continuum that are actually the results of abstractive cognitivity has created continuum psychology, continuum of legal institution, and continuum economy and so on.

In order to understand the thought of continuum, we need to recognize the significance of multiplicity in the universe. It is not a comment on the multiple universes or its similar theory. Continuum of various multiple entities requires us to act for better interflow of material and mental communication among them. There exist interflows in the global continuum of even nowadays that are fluidal mode, which can be named as 'koinonia' in koine Greek adopted in Paul's thought to express continuous flow of mutual communication and interchange of necessary materials and wealth in the global continuum of his ancient days of Pax Romana. We are living in an era of koinonia which means communicating, circulating and transmitting.

This situation enforces us to think of the necessity of the universal public law that every constituent of this planet generally agrees to in their common sense.

6-6. Activation of Multiplicity

Multiplicity and cruxological modularity
Morphosis of the Law of Paretto

The subject of multiple personas has been dealt with in Chapter 4, Section 4-3. It was discussed because of the necessity in order to get into the theorization of dynamic modularity. I believe the socio-psychological phenomenon has been recognized, whether consciously or unconsciously, in its concerned academic disciplines. In this section, we will simply think about the activation of multiplicity.

Multiplicity and Cruxological Modularity

The activation of multiplicity is deeply related to the cruxological modules. The occurrence of cruxological modules were sometimes called as "social segmentation" during the last century. Not all the social segments are cruxological modules. However, the internal mechanism of an activated multiplicity is more dynamic and basically cruxological. Such understanding can lead to a flexible approach in social management.

Cruxology is governed by the principles of dynamic modularity. Cruxological modularity is the result of multiplicity and thus generates diversity in human society. Theory of multiple personages has to be settled to describe the ever changing social segmentation and the increase of social complexity.

The socio-psychological task of this century originated from the inheritance of the industrialization process in the last century. As implied in the political economy of our day, multiple dynamic modules and multiple dynamic figures will attain greater importance.

We have to foresee the upcoming future with more complex multiplicity and have to be able to manage cruxological modularity through our sociological change. Otherwise, the human race will deplete their energy saved for positive advancement and prosperity.

Sysy is less energy consumptive in a sysy environment. Same thing happens so that in nexy environment nexy process may be less energy consumptive. It also means that as we usually experience the conservation of nexy take more energy in the sysy environment and sysy consumes greater energy in the nexy environment.

At the center of modular core, a sysy process takes less energy while it needs more at the peripheral location. Nexy reveals adverse behavior. Therefore we anticipate that the multi-persona breeds in a modular periphery where strong systematic modules exist around a specific module of concern.

Morphosis of the Law of Pareto

Pareto's law is known to be applicable to the distribution of wealth in economics, to a business strategy in marketing, to organizational rule of thumb, to software debugging and even to the engineering control theory. It is based on a more or less empirical observation.

I believe Pareto's law works in our society because of the concentration of wealth's toolkits organized for the 20% of population around at the center of the dynamic module. Wealth toolkits are more effective and productive these days than one century ago, since the means of production has exceedingly gained the systematic efficiency after the alignment and synchronization of trading, monetary, and financial system. Maybe, the Pareto's law has to be extended due to the enhancement of packaging the means of production in the 21st century.

We have realized the effects on the distribution of spheres according to their sizes in a regular tetrahedron on their packing efficiency. Effective packing of the uniform spheres by cubic and rhombohedral arrangements gives the packing porosity of 47.6% for cubic and 25.96% for rhombohedral. The difference of porosity causes the difference of flow rate, which affects the pattern of flow.

This simple metaphor produces an idea that the degree of freedom of a system is the matter of compaction and the matter of boundary in a phase space. Packing variation of entities into a phase space of various tensorial forms can induce our ideas toward the hierarchical build-up of syntax structure.

Ramachandran commented that the syntax structure of the human language is derived from the synesthaesia of a synthete, a man of hyper-connectivity in brain who can tell such as "the number seven is green" (V. S. Ramachandran, the Emerging Mind, Ch.4). He asserted that the human abstraction could have evolved from the same phenomenon.

I agree in the same line of thought but synesthaesia is a special case for the introduction of new dimensions into the phase space of human mentality. Psychological reservation occurs at a slighter intensity than infinities-encounter to produce a hierarchical tree-structure of linguistic syntax. Stammering, murmuring, and psychological blockings are essentially of same source of internal search for the introduction of new dimensions.

Dynamic emotional modules are responsible for this process. It is thought that a modern analogue of Four Beginnings and Seven Feelings can find the general patterns of figuras in introducing new dimensions. It means we can weave the emotional phase space of humanity to establish stabilized conditions of drastic social development of near future.

Chapter 7. The Science of Cruxology (II): Menesysal Technology

7-1. Menesys (巨大網系)

- Megalo-network-system and Complexity
- Nearly-Decomposable System (NDS): Character of Complex Modularity
- Embryology of Menesys and Dynamic Modularity
- Prototype of Menesys: Sysnet and Netsys

7-2. Variety of Menesys

- Manner of Clustering of Individuals
- From Scattered(散體), Group(群體), etc. to Collective (集體) to Communal(共同體) etc.
- Classification of Menesys

7-3. Flickering Maintenance, Vitality and Energy Saving in Menesys

- History of Humanity: Brain Resemblance
- Apparent Continuity of Anthropomorphic Vitality
- Integrismic and Cruxological Affections on Flickering
- Creativity Factors Flickers

7-4. Mechanism of Values and its Derivatives in Menesysal Phenomenon

- Values Menesys
- Utilities Menesys
- Knowledge Menesys
- Intelligence Menesys and Pretium Menesys
- Market Pricing Menesys
- Virtues in terms of Science: How Integrity Realized in Reality?

7-5. General Technology of Menesysal management

- Menesysal Internality
- Menesysal Externality
- Integrative menesysal Management
- Values Management and Utilities Management
- Connectivity among Manifolds: Why is Criteria of Virtues needed?
- An ideal type of menesysal organization

7-6. Peace Knowledge Menesys

- Essential Character of Human Knowledge
- Peace knowledge Menesys
- Educative Information Menesys

7-7. Menesysal Management to Enter into Solar System and Beyond

- One Century on the Earth: Basal Menesys
- Another Century into Solar System
- Millennium "To Boldly Go Where No Man Has Gone Before"
- Be Humble, Bumble Bees.

A new terminology "menesys" will be introduced. MENESYS stands for MEgalo-NEtwork-SYStem. Menesys was explained and discussed in its variety of order to be attributable to its own integrism. The general technology in the management of menesys was illustrated to show the applicability of the aforementioned theories in practice.

The basic academic definition of network and system is not yet completely defined. We have a decent article discussing about market-hierarchy continuum (W. Powell, 1990). It is our theoretical stance that the network and system, thus nexusity and systemicity, have a deep connection with the enfolding and unfolding mechanisms of social logic.

Netsys and sysnet are generally considered to belong to the complex cruxological range. A subsequent menesysal development is affected by the pre-settled *a priori* conditions which is a menesysal development mainly progressed under the governance of the factors that are concerned with continuum and connectivity.

It was discussed in Chapter 8 that creativity has three kinds. The 3rd kind of creativity is supposed to be fully fledged in the duration of menesysal establishment in the global scale. Currently, menesys is actively increasing in number and scope entering into the next step of the transcent of humanity. The author believes that menesysal trend has come to existence after the two world wars and the emergence of mega industries.

7-1. Menesys (巨大網系)

Megalo-network-system and Complexity
Nearly-Decomposable System (NDS); Character of Complex Modularity
Embryology of Menesys and Dynamic Modularity
Prototype of Menesys; Sysnet and Netsys

Megalo-network-system and Complexity

A "Menesys" means megalo-network-system which is a generalized notion of various "permunal" existences. Permunal is the compound of personal and communal.

As discussed before in this book several times, "person" and "personal" have certain differentiated meanings from the previous usage of vocabularies. A person is an individual existing in the integer form so that we can describe it with the measure of integrism.

The integrismic phenomenon, as discussed before, is denoted by the composition of systemicity and nexusity. Thus when menesys changes its contents in accordance with the way the concerned menesys is influenced, it is done by its own composition of systemicity and nexusity.

The classification of menesys is governed by their integrism. Menesys is thought to be significantly categorized by its performance in a centric role. When a system plays the centric role, a menesys is called systematic menesys. Likewise, there would be a counter entity called nexal menesys in which nexus plays the major role.

Systematic and nexal menesys are located at the opposite extremes of the whole spectrum of menesys. These bipolar extremes located at the either entrance of the transition zone of phase change from system to nexus (or network.) A metaphor of solid-fluid-gas makes it possible for us to roughly understand this relationship. But there are several major differences, which will be explained.

Menesys are found everywhere in the universe. The Menesysal phenomenon is attributable to the principles of the universe. Even in the outputs of human design, in the latter part of the past century, engineers started to supply the electronic devices based on the concept of fuzziness. Fuzziness was the first significant introduction of menesysal concept to the hardware of human beings' utilities. Fuzzy theory was developed by an expert R. Zadeh, in the field of control.

Fuzziness is generated when distinctively different systems are mutually met to build intermixed zones. An obscurity is the characteristic of such situations. Equivocation could be recognized most likely as a linguistic expression of such undetermined or deceivable intentions of a person (or software). In fact, the so called fuzzy solutions are produced in the transition zones, which are built when heterogeneous systems meet together dismantling or dissolving their respective boundaries.

In human affairs, it is believed that this kind of fuzziness is more profoundly expressed in a pattern of science complexity. The obscurity of accidental occurrences of external environment plays with the equivocality (多意的 曖昧性) of internal arbitrariness. The internal arbitrariness is mostly the result of shelving reservation of an individual's decision.

These interactions between external obscurity and internal equivocality produce complexity. The complexity of humanity originates from the external (i.e. social and environmental) obscurity and internal (i.e. individual and organic) equivocality. The equivocality could be multiple in realities. In other words, the reserving probability of which is greater than that of binary equivocation.

Multiple equivocations will be theoretical tools to express multiple personages in concerned parts of this research, and give a good understanding why a number of multiple organs and neurons are going well in coexistence.

Although mutually interactive, I believe that a new concept must be designed to cope with the complexity of humanity and human-concerned matters. The complexity of humanity looks like to have caused modularity and hierarchy in physicality and mentality.

The notions on menesys have an emphasis on dynamic modularity to understand the internal mechanism associated with the creative peace. The creativity has the duality of attributes in order to be easily drawn out of individual transciousness and smoothly accepted in society.

Menesys is especially important when we deal with the wholeness of an integer type of organism or society, which is equipped with globally complex hierarchy and modularity of complex integer constituents.

Nearly-Decomposable System (NDS): Character of Complex Modularity

There are other studies devoted to clarify the phenomenon of modularity. For instance, it was good to learn that there are distinction between organizational modules and variational modules in morphological modules to support the evolutionary view. (G.J. Eble 2005)

Such division would make it possible to quantify the integrativity and decomposability. But it was not easy for me to observe the stability and variability of an object with such concepts. As I imagined, the conceptual modeling of menesys combined with phase changes in the transition zone of its phase space can encompass all variable changes in ontological and epistemological views and implicate the static and dynamic status together.

In 1968, L. Bertalanffy presented a General System Theory, which introduced an idea of an open system and has become the forerunner of such trend of thought. We have Weiss and Pattee, who introduced a hierarchical thought in the ideal system, (P. Weiss, 1971, Hierarchically Organized Systems in Theory and Practice, and H. Pattee (ed.), 1973, Hierarchy Theory)

In one of my books, I found a discussion that a system is a set of entities with three properties (P. Rivett 1980);

1) Every entity in the set affects at least the other entity in the set.
2) Every entity in the set is affected by at least one other entity in the set.
3) The set cannot be split into two subsets such that no member of one subset affects at least one member of the other subset or that no member of one subset is affected by at least one member of the other subset.

As it can be seen above, a system is isolated by a set or a boundary. It is due to the epistemological need to examine the verification. Recently, there is one of in-depth approaches on modularity issues found in a collective report of a dozen of disciplines. Followings are from those articles about nearly-decomposable system (NDS). (Modularity, edited by Werner Callebaut and Diego Rasskin-Gutman, 2005)

According to the study of Werner Callebaut, a system in general may be "characterized as modular to the extent that each of its components operates primarily according to its own, which is intrinsically determined principles. Modules within a system or process are tightly integrated but relatively independent or 'dissociable' from other modules."

It is believed by H. Simon and other theorists that most of the natural complex systems display the nearly decomposable organization characterized by a thinly populated interaction pattern. (G.J. Eble)

When building a decomposition scheme for certain problems, a perfect decomposability is possible when all blocks are totally independent from each other so that they can be solved independently. In the study on economic interactions, decomposability is considered a very stringent requirement. Thus it was tried to be softened into a near decomposability to reduce the sub problems to contain only the most relevant interdependencies. In such a way, optimizing each sub problem independently will not necessarily lead to global optimum, but only to one of the best solution. (L. Marengo et. al.)

The quantification of morphological modularity is highly context-dependent. Variational modularity in morphological modularity is in the state of varying in a modular fashion. In this study of macroevolution, G. Eble commented on dynamic modules that assessing the presence of modularity is a major goal, for instance, in morphogenetic fields, or in the terra incognita of the genotype-phenotype map. Organisms and their modules are modular to different degrees. That is the reason why statistics for modularity are necessary.

Scientific disciplines are encouraged to get into quantification of variational modularity which has a conceptual connectivity to the dynamic modularity of our study. The late H.A.Simon wrote in his foreword "Why does complexity in our universe, at virtually all levels generally take this hierarchical, or nearly decomposable form?" His question was about the process of evolution.

Embryology of Menesys and Dynamic Modularity

Our question is given for a different purpose in pure scholastic achievements and practically aimed for the realistically impending political economics in this century. A menesys has its own process of conceptus, growth, and intrinsic phase changes. The embryology of menesys has to be systematically organized in theory.

Since our proposition of theory is basically persontology, there exists an embryology but not a process of senility. Senility in persontological scheme with an emphasis on the wholeness and overall connectivity of the universe is just a process of path change from omega to alpha depending on the increase of stress situation.

Menesysal embryology is relevant to its integrism. The embryology is also governed by the cruxological mechanism. Integrism has influence on menesys in a manner of affecting the ratio between sysy and nexy. The integrismic ratio affects the status of menesys in combination with the effects of initial location in its environmental phase space.

For instance, it is a very much indicative factor from where a menesys started its embryonic growth, the locality around systemic attributes or that of networking attributes. The cruxological measure would contribute from the earliest morphogenesis of menesys. The overall process of the main body formation of minimus, nexio, nodus, modus, figura and systema is governed by cruxology.

It was discussed that a systema is the persontological version of a system in the light of existence with the overall-connectivity to wholeness. But there is another line of cruxological formation concerned with networking.

Networking mechanism manifests the measure of the universality of the cosmos. Thus it exists in all the cosmos in a very similar manner to the cosmic microwave radiation. I believe it might be more fundamental than just being compared with the metaphor of background radiation. It is due to the belief that the nexusity of the universe is the original essence of the persontological universe.

The original cosmos, absolutely without its implicated intentional beings, that are ceaselessly observing, thinking and acting, could be expressed as the infinity of completely systematic network, in which the whole network itself is a unit systema. Such an ideal form of the universal ontogenic network, which is named Uninet, is located at one end of menesysal spectrum. Contrary to the Uninet, we have the Unisys, which are the other edges of the opposite wing.

The Unisys is thought to present the conceptual status of the system with perfectly developed nexal efficiency in the system. It is not yet completely clear to me whether the Uninet and Unisys are actually identical or not in their internal patterns.

But it is thought that we can put them in complete equivalence and symmetry in a universal ontogenesis "before and/or at" the origin of time. The reason I used "before" is that I imagine and intuit that our universal genesis has the backward "wholeness" observation exists at the beginning of time. The beginning of time is round a bowl-like shape according to the suggestion of S. Hawking's cosmology. It is sometimes discussed among physicists that imaginary time needs to exist at the genesis of the Universe to make time and space equivalent.

However, both of them equally have the most effective pattern in the light of integrismic criteria. The most effective pattern means the menesys which does not need any alteration of its sysy-nexy pattern for improvement.

In the theorization of Che-Yong (體用, static body and dynamic use) philosophy, important is the genetic setting, from which structures are developed. Human physique, when simplified, take a right-left symmetry in its two dimensional physique when it is erected upward. If forward and backward movements are required for the two dimensional physique, a partial structural asymmetry in that physical plane is necessary.

Such an asymmetry generates surplus rhythm of human physique and provides the clues of physical mobility. This forward-backward mobility takes a structural cooperation with an asymmetrically unitary direction of temporal flow, which is forwardly inclined. In fact, the studies on evolutionism need to include the teleological orientation in the earthly evolution. It has got to have a deep implication in psychology too.

Prototypes of Menesys: Sysnet and Netsys

In the formation of menesys, we can find fundamental influences of symmetry and asymmetry, and balance and imbalance. Menesys has two sorts of distinctive prototypes. One is sysnet and another is netsys. Sysnet is a systemized network, while netsys is a networked system. We previously discussed that networking mechanism manifests the measure of the universality of the cosmos in a similar manner to the cosmic microwave radiation. The nexusity of the universe is the original essence of persontological universe.

Systemicity is compressed or entangled expression of nexusity in logical manners. In pure logic, the Unisys is a completely compressed version of the Uninet. I do not know if this idea has some correspondence with the idea of the bouncing universe, which recently is being discussed.

Anyway, it looks clear to me that this sysy-nexy pattern is not a matter of time series. It could be regarded as a spectrum as a whole. Nexusity (nexy) is an unfolded state and systemicity (sysy) is an enfolded state. The unfolding of network should tend to occur first from the relatively less resistant and less void portions to generate the internal structure of a resultant enfolded system.

Now we said that there are two ways of development to menesys. One is sysnet, a systemized network, and another is netsys, a networked system. When menesys has its prototype of sysnet, the route of development is such that the rapid expansion of an empire of nomads in the vast expanse of grass plains can be seen.

As far as netsys is concerned, it is as if we are encountering the expansion of the Roman Empire. Carthage of Phenicians, the long opponent of the Roman Empire, had not constructed a political regime of an empire but concentrated more on developing handy managerial structures and decentralizing good harbors to spread in all the Mediterranean effectively. This resembles with the strategy of the nomad empires. In this perspective, Carthage is relatively closer to the concept of sysnet and Roman Empire to netsys.

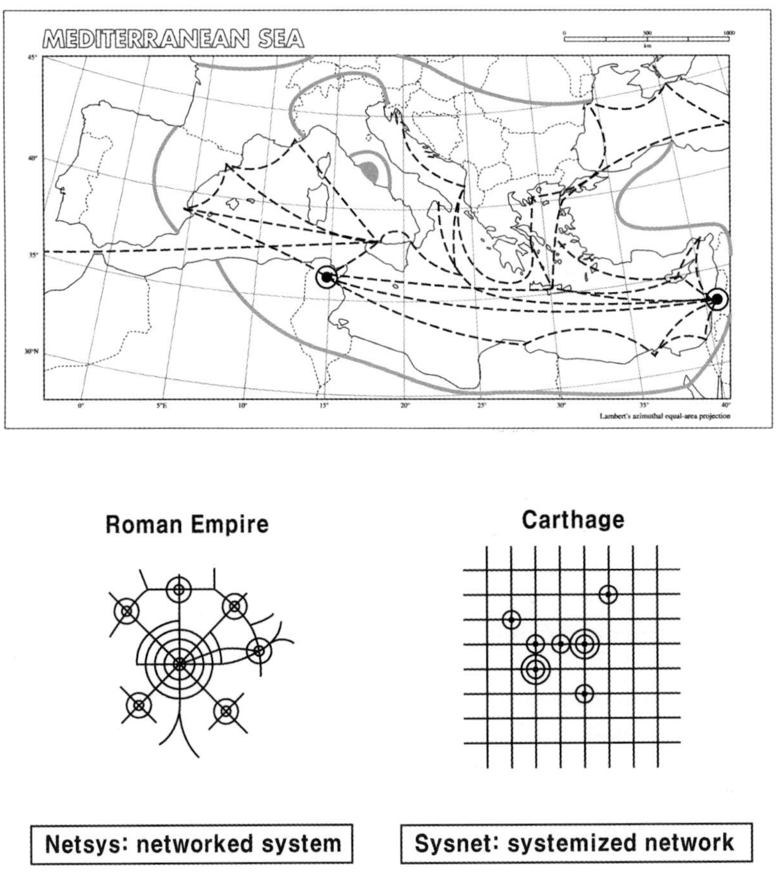

Figure 7-1-1. Embryology of Menesys and Dynamic Modularity

The simplest netsys has one matured systema (system ontologically open to the environment). The simplest sysnet is supposed to have one network of immature systema. Immature systema is in a form of nodus, modus, or figura as defined in Chapter 6. Becoming a matured systema politically has to be decided completely internally.

Phenicians developed a very effective decision-making process to pursue a commercial aim but stayed general as a polis pattern at the center of power (M. Tezturow, 1990, The Rise and Fall of a Commercial Country). It is thought that the commercial entities tend to develop nexusity while manufacturing pays more attention to the development of systemicity.

Netsys and sysnet belong to the cruxological range of development. After that, menesysal development takes place in accordance with the previously generated *a priori*. It means the developmental principle relies mainly on the prime points of connectivity and continuum formulation.

7-2. Variety of Menesys

Manner of Clustering of Individuals
From Scattered(散體), Group(群體), etc., to Collective (集體) to Communal(共同體) etc
Classification of Menesys

Manner of Clustering of Individuals

It is interesting to understand that individuals make clusters in various ways. Clustering is governed by two factors; external and internal conditions of individual constituents. External factor is a matter of dependency on one another to be clustered. On the other hand, internal factor are affected by intercommunication. Each individual has his own intrinsic attributes for mutual communication with others through language and/or other tools.

Steven Strogatz introduced a new concept of Sync, a science of synchronization. He cited Norbert Wiener's hypothesis that the brain ingeniously builds an accurate clock from an enormous number of sloppy oscillators. These oscillators are supposed to have a frequency pulling.

Yoshiki Kuramoto tried to penetrate into the core of self-organization in time. He developed an idea for a system of indefinite numbers of differential equations, all linear, all coupled together, usually known as not solvable. He substituted the previous method of Art Winfree, influence and sensitivity formulation with his specially defined interaction.

It was esteemed to be a highly efficient symmetrical rule, much further enhanced than the Winfree's frequency pulling. Though it might be too naive to believe, but I nevertheless believe that Kuramoto's system of differential equations might be used for the modified mathematical conditions of the dynamic relational ontological expression.

Nonetheless, clustering of individuals will take most likely a half dozen of different appearances in nature. Menesysal varieties come out as clustering characters.

Seven Clusters from Collectus to Separatio

There are seven clusters suggested to classify the essential constituents of developmental bodies of menesys. They are named as follows. The symbolic visualizations are presented in Figure 7-2-1. Latin names are adopted to show that they are all open entities.

For instance, a system is apt to be understood as a closed entity if it has to be isolated by a definite boundary. However, systema is not a closed one and identified as a part of the whole.

We have seven basic clusters with different figurarity. Figurarity is actually the expression of integrismic composition. Integrism changes by levels and influences the figurarity to be nexal or systematic in accordance with hierarchy. The seven basic clusters are listed in Table 7-2-1.

Table 7-2-1. Seven Basic Clusters with Different Figurarity.

Collectus (集體)	The body is collective and nexal connection is rigid and not variable.
Colonia (群體)	Nexal connection is variable and free but the body is collective.
Systema (系體)	The center of body has systematic function to centralize governance.
Communitas (共體)	The reality of humanity, proper integrismic factor for humanity.
Nexus (連體)	Systemicity is disregarded and only connectivity left.
Dispersio (散體)	None of systemicity and nexusity but many constituents exist around.
Separatio (別體)	single constituent found in its neighborhood

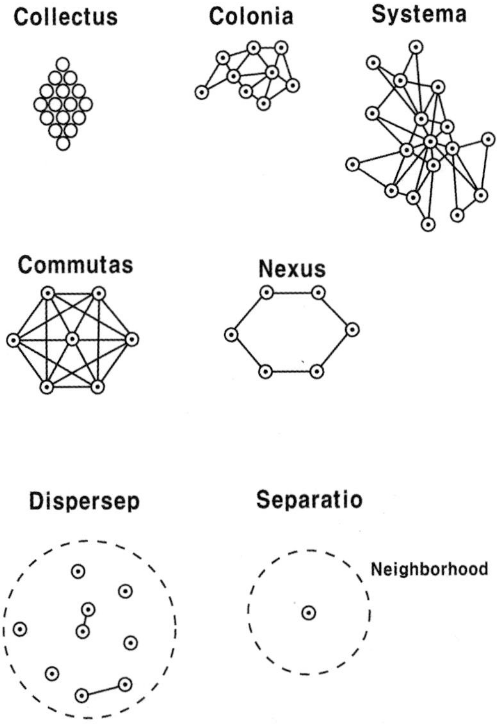

Figure 7-2-1. Figurarity of Seven Clusters: Symbolic Visualization
(My note on various clustering in comparison with homeochaos of Yoshimasa Yoshinaga, Fukuzatsu-kei Towa Nani Ka, 1996, p191,.)

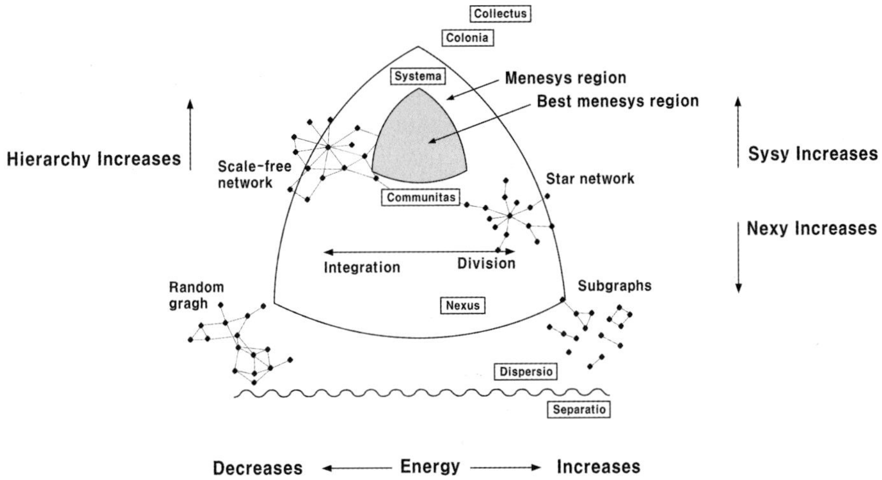

Figure 7-2-2. Territories of Clusters (modified from Figure 3-8 of Weak Link, Csermely p75)

Classification of Menesys

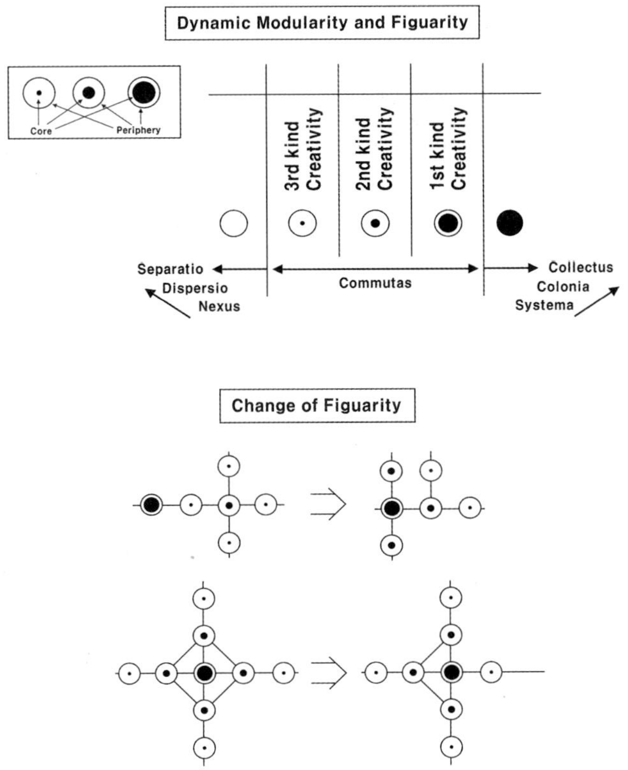

Figure 7-2-3. Figurarity and Dynamic Module

As we discussed above regarding sysnet and netsys in the last part of Section 7-1, it seems possible to classify the categories of menesys. The target menesys towards the human race would prefer to yield toward our future destination. Community (communitas) is positioned at the center of the arrangement of definition.

Characteristic differences among the various kinds of menesys are summarized in a compact form of illustration in Figures 7-2-3 and 7-2-4.

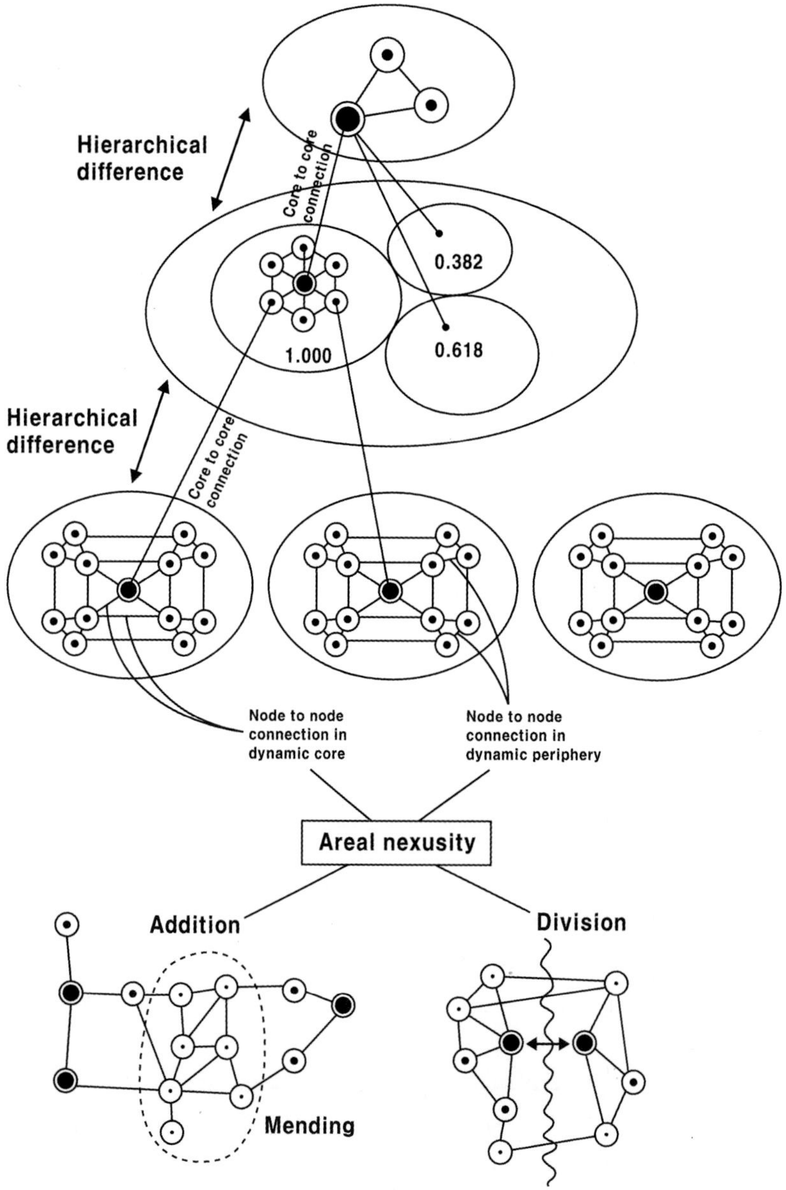

Figure 7-2-4. Figurarity of Characteristic Menesys: Symbolic Visualization

7-3. Flickering Maintenance, Vitality and Energy Saving

History of Humanity: Brain Resemblance
Apparent Continuity of Anthropomorphic Vitality
Integrismic and Cruxological Affections on Flickering
Creativity Factors Flickers

The advantages and disadvantages of flickering mechanisms should be illuminated in balance. Flickering optimally desynchronized in spatiotemporal continuum generally benefits humanity in a megalo-scale. Optimally desynchronized flickering (ODF) is an evidence to show that whole of menesys is in good balance, which means that it distributes energy evenly. If synchronization is achieved in all of humanity, as seen in fireflies, the intrinsic frame would not yield any desirable conditions for the variety of the manifestation of potential humanity.

The diversification is a healthy fundament in sustaining the weak links and the nexusity among those weak links to stabilize and diversify humanity in a global scale at the same time.

History of Humanity: Brain Resemblance

Some thinkers say that the universe resembles the human brain. Some researchers think that the ontological conditions of the human race resemble the brain. But it is not exactly our brain that we are directed to imitate. The ontological position in the brain takes place in the frame of Plexion-pragma-cognition. The human cognitive process is between the all-inclusive ontological Plexion and the partially-inclusive pragma.

Because the pragma is within our cognitive range, we examine the occurrence of affairs in epistemological attitude for our further survival. Ch.4, Section 405 illustrates quadruple annulus of human cognition.

In the brain activity, it is our conclusion that human race follows the route of transciousness. Most assume that human culture and history have been following consciousness. If it were consciousness that was at the center of the story, we all should have ceased to exist long ago in history. Optimally desynchronized flickering is the answer for continuity of prosperous humanity

Optimally desynchronized flickering is fundamentally associated with "accidentality" in nature. Accidentality is the name of the game of humanity. Where was this factor introduced from? - Undoubtedly, from transciousness. The recognition of accidentality in the human brain comes from a fact that there are predictions and postdictions (or reflections) about the happening of affairs. These predictive and reflective considerations are based on the teleological transciousness.

The teleology of transciousness is deemed to have come from the dynamic modules formulated partly by the encountered past affairs and the anticipated future affairs. Dynamic modularity makes humanity teleological. Dynamicity of brain is supported by the principle of plasticity in sensciousness, but in turn supports the principle of elasticity in transciousness.

Human history has a similar hysteresis. Big steps are elastic and resilient against impacts of anti-humanity. In a minor scale, people experience meaningless accidents and have to be obedient to their destiny. Some believe that justice eventually wins. Such conclusion is just another individual interpretation and could be not more than a simply wild wish as nihilists insist. However, history would never be non-teleological.

Apparent Continuity of Anthropomorphic Vitality

It is necessary to define "anthropomorphosis" in a different perspective from the previous one in Ch. 4. Here, anthropomorphosis is "the collective adaptability of human race to their environment or to the set of their experienced affairs in a manner of virtual reality, in a time-space condition."

Such universal adaptability is of a collective character of all human races because all the nations possess same universal common points even in the detailed psychology. It is reported that even infants and children have good understanding on the fundamental categories of physical and social world. They can utilize such understandings without having to learn the information supplied to them as brand new.

In order to enhance adaptability, humans sustain objectives by keeping belifs that fit well with today's evolved conditions even if such premises are not coincidental with each individual's own experience. This could be explained in that altruistic attitudes contribute to keep up the adaptability. (Steven Pinker, How the Mind Works, 1997, Ch.1, Standard Equipment, Reverse-engineering)

Human race has maintained the anthropomorphic vitality. Is it a kind of built-in program of humans? In major portion, yes, but partly we also have to learn how to provoke the switching mechanism. Vitality in a menesys can be sustained by energy saving mode, allocating energy in an optimum distribution profile and concentrating them into some designated parts of menesys to activate and call out necessary parts of a next developmental step. Actually, vitality seems like a fatal response to menesys at this specific level of developmental stage of human race.

Vitality is usually observed in a special step of local concentration of energy in the course of the long oscillation of compaction-dispersion process in menesysal development; which is why some civilizations occasionally have flickered to change their locality throughout human history.

Integrismic and Cruxological Affections on Flickering

Integrism is a basic factor that widely affects all scope of the science in the history of humanity. Optimal desynchronized flickering occurs to systematic cores of a menesys; flickering is a result of systemicity. However, if the systemicity overwhelmingly governs the members of menesysal body, flickering will not be activated because flickering presented upon a provision of adequate nexusity allows proper degree of freedom.

Cruxology will guide the activated humanity into consecutive chains of stabilized conditions, avoiding unnecessary fluctuations and upheavals. It also means that cruxology of menesys in human history has tried to keep the energy level as low as possible in distributing excitable periods to the representative multiple attractors. Cruxology allows us to view the human behaviors and outcomes as a menesysal phenomenon.

Menesys could be a flexible idea, regarded as a technological approach after the science of cruxology. Nonetheless, it is still expected that menesysal management of human phenomenon would solve many problematic situations of humanity.

Creativity Factors Flickers

Creativity factors are not maintained in handy managements. Even for a very creative person an important moment of creativity comes very rarely after considerable amount of waiting. Creativity is not easily met in a flat continuity. We have to wait and manage our internal conditions fluctuated in an adequate range to keep stimulus and stability simultaneously in our everyday lives. Creativity costs energy.

Simply speaking, creativity factors should be monitored to flicker in individual and communal situations altogether. If the creative peace is globally desired to be commenced and be successful, it is evidently necessary to have a creativity factor flicker in the politico-economic menesys of the whole earth.

The way to have creativity factors flicker ceaselessly has to do with the management of an individual's base, which has a deep connection with an individual's education. Every entity needs to carefully maintain certain amount of attention to find out what is his given portion of work in a whole communal body.

The creative process appears when the exchange of values takes place among respective parts and when a specific function of limbs or an organ are changed due to the variable situations of a whole body. The whole and parts happen to experience creative environments when external and internal conditions are varied, in the course of adaptation, to their new role definition.

The exchange of values among concerned parts usually appears through the communication of ideas, exchange of materials and energy carrying concerned values. However, we have to understand that the exchange of values is essentially identical to holding materials and entity in common in a long run. When values are exchanged between different entities without violence and oppression, it would be a good foundation for the creative peace. The creative peace will be firmly founded on the mutual understanding of covalence between entities as happens in a family. The real problem of human history is to learn an effective way for mutual value exchange.

A dynamic module flickers when energy is highly concentrated to a specific module, in order to strengthen itself in the frame of tensegrity dynamics of the whole menesys. Optimally desynchronized flickering is a good strategy of network management in maintaining continuous vitality and saving energy for the whole body and even for every part.

7-4. Mechanism of Values and its Derivatives in Menesysal Phenomenon

Values Menesys
Utilities Menesys
Knowledge Menesys
Intelligence Menesys and Pretium Menesys
Market Pricing Menesys
Virtues in terms of Science: How Integrity Realized in Reality?

Value is a complex and absurd word if we do not acknowledge the positioning of the word beyond our intellectuality. Value is a system; value is a network; value is a menesys – the compound form of system and network. It takes a menesysal pattern at an individual and also at a social level.

In pragmatic traditions, we do not easily accept the unitary value system. But value is unitary that starts from onto-feeling. I mentioned before that onto-feeling or onto-perception begins at the fertilized moment of gestation. We have to know that axiology, the philosophy of value, should be menesysal, because there are many branches in value recognition.

Value menesys has four additional menesys on top of it that support the pricing mechanism of the market. They exist as value recognition path in the chain of value-utility-knowledge-intelligence-pretium-price.

It is suggested that there are two parallel ladders of value recognition existing as illustrated in Figure 7-4-1 and will be shown in Figure 10-6-1 again. Two parallel ladders are formed because there are two different environments of value recognition mainly due to the emotional and rational tendency balancing cognitive activities as seen in the chain of menesys of value-intelligence-pretium and of utility-knowledge-pricing.

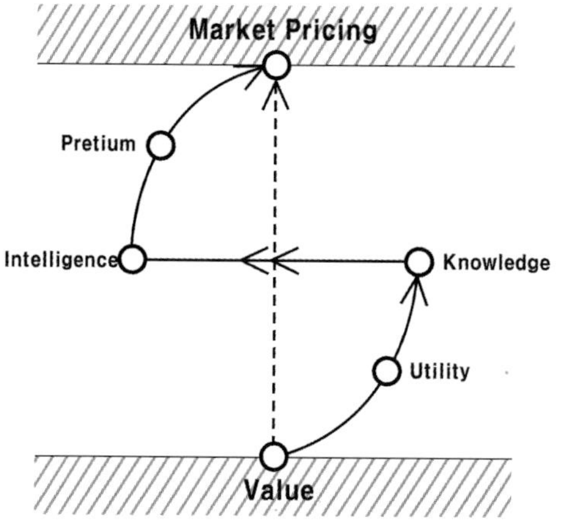

Figure 7-4-1. Path of Value recognition from Value Domain to Market Price Domain

Values Menesys

Values menesys is a tree diagram in the persontological view. The tree of primordial human value develops in a general sequence from onto, topos, and then to logia. In the ancient days humanity upheld a value system that was associated with heaven (天), terrenity (地) and human (人).

The interconnection between the two conceptual structures develops reasonable arguments: heaven with onto-value, terrenity with topos-value and human with logia-value. Onto-value has no locality but globality in existential sense. Topos-value has the sense of locality added but no fixation of relation. Relation is fixed in logia-value stage when the two entities mutually interpret one another in his own specific locality, thus constituting a phase space of conceptual knowledge.

In the enfolding and unfolding mechanism, relationship is fixed when unconnected dimensions of related entities are enfolded. Thus the mutual interpretation acts as an enfolding mechanism to establish relationship. This means that the enfolding of irrelevant dimensions between two entities is essential to formulate a systematic relationship.

Onto stage is supported by the sense of living. The individual sense of living is a really complex compound consisting of many levels of emotional cognition.

According to our understanding, the human value related to topos stage arises next to locate the ontological entity in certain kind of map that is formed by our cognitive process.

Logia stage is meant to connect the positions of entities arranged and fixed in topos mapping.

Onto is as free as the babbling of a baby in generating an adequate phonetic structure. It is as if a phonetic expression matches with a single emotional expression by an animal responding to a specific situation of freight or delight. Topos is such what the semantic structure in linguistic development may stem from. Logia allows the process of communal relation which will generate a syntax form of language.

Syntax is formulated through the articulation process. The articulation in linguistic development happens when a person experiences an emotional blocking within his dimensional allowance in describing one's situation.

The integrated form of the values from the above three stages eventually recorded in our cognition as a persontological fundamental of our substantiality. This is what we call Triradixism in Chapter 8 while explaining creativity.

The ontological view has stemmed from a cognition generated by the integration of feelings from onto-stage, topos stage and to logia stage. Topological view has the origin of topos and logia stages. I believe topology is not confined to just a specific mathematical field. If properly extended, topology would have become a substitute of geometry in ancient Greek civilization.

This means the progress of human abstraction has come to the stage of topology to illustrate the mystery of nature. Topology has to develop further into hyper-topology in order to organize our thoughts on the secret of humanity. I believe hyper-topology has to deal with the interpretation of dimensional phenomenon – even the mystery of the Divine Dispensation if possible.

Onto-topos correlation brings a communal understanding of an entity in a communal mapping, which the author refers as the wholistic ontopos positioning. The wholistic ontopos positioning is a story of the whole and the parts. Wholism or holism enables the backward reconstituting of human thinking. We would rather refer to holism as "ontopism."

The persontological view has strong rational tools to compose a logical explanation that includes the entire human phenomenon covering from science to aesthetics. The structure of the values menesys would be manifested by the triangle of ontology-topology-ontopism when the development paths are well balanced. (If unbalanced, it will give six probable outcomes of disciplines with emphasis on one side of the coupled apexes. This means there are possibilities that academic disciplines would be composed of more onto and less topos, more logia and less onto, or any other combinations.)

Ontology exposed the sphere of belief which is not verifiable and unidentifiable by intellect. Topology is actually the source of human discernment utilizing map-matching ability in virtual reality. Ontopism is a process of reconstitution through puzzle games. This game is successively substituting words that are waiting in array. It is a decoding procedure. If we determine some rules to find the right answer out of an infinite series, we can solve the whole of the problem by using intuition. The Figure 7-4-2 presents the basic tree diagram of valorization.

Usually valorization means the formation of price structure in economics. I would use the term 'valorization' meaning 'value formulation' in this book. I would rather use "pretification," not prettification, to denote the price formation by compounding "*pretium* (Latin, price, value, and worth)" and "-fication". Pretium is the normalized value that through several steps of adjustments adapts to the fluctuating sociopsychological environment usually affected strongly by human emotional factors.

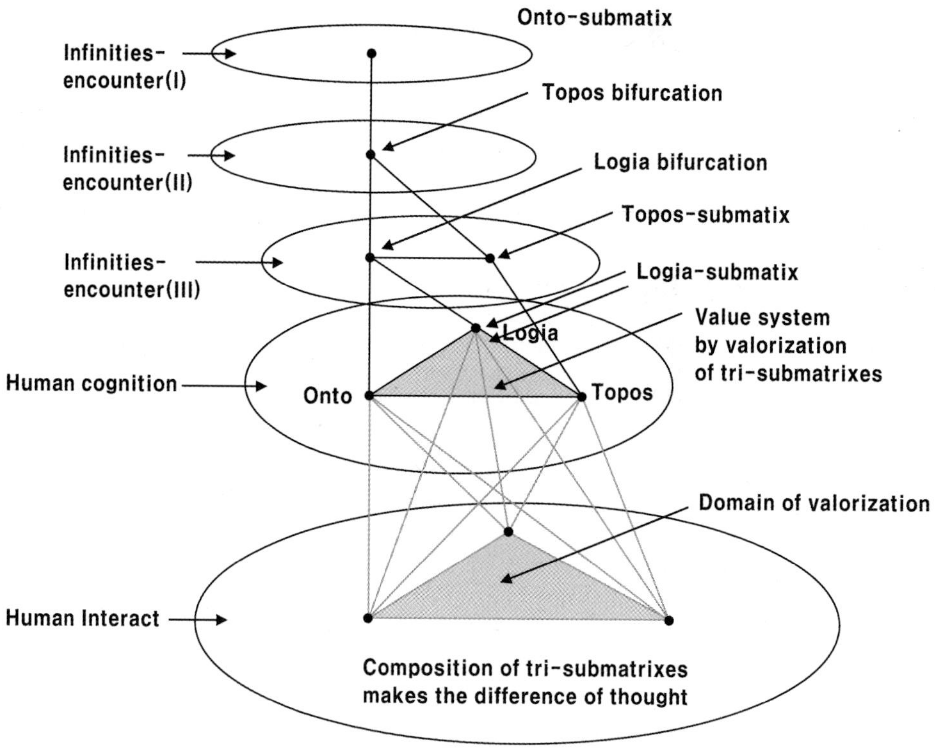

Figure 7-4-2. Basic Tree Diagram of Valorization

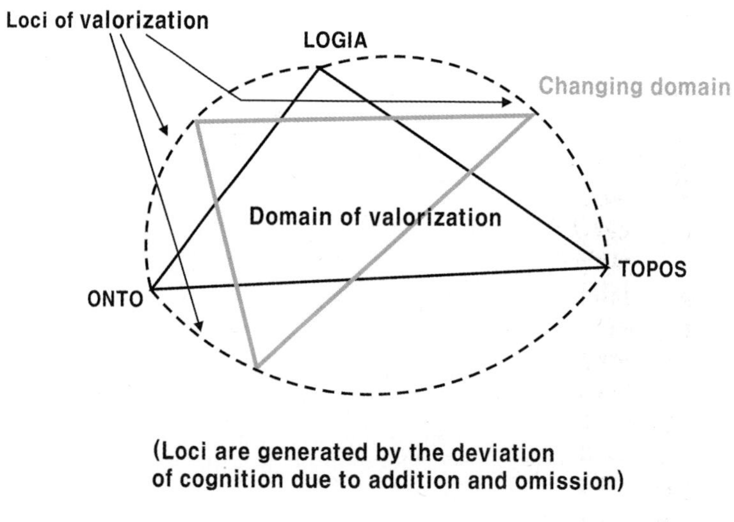

Figure 7-4-3. Loci and Domain of Valorization

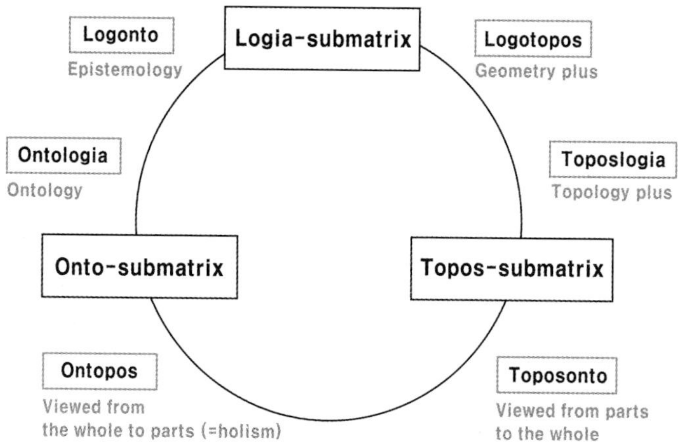

Figure 7-4-4. The Six Sets of Abstract Academic Fields Due to Bilateral Combination Developed among Three Submatrixes (depending on which submatrix the preferential initiative would be imposed.)

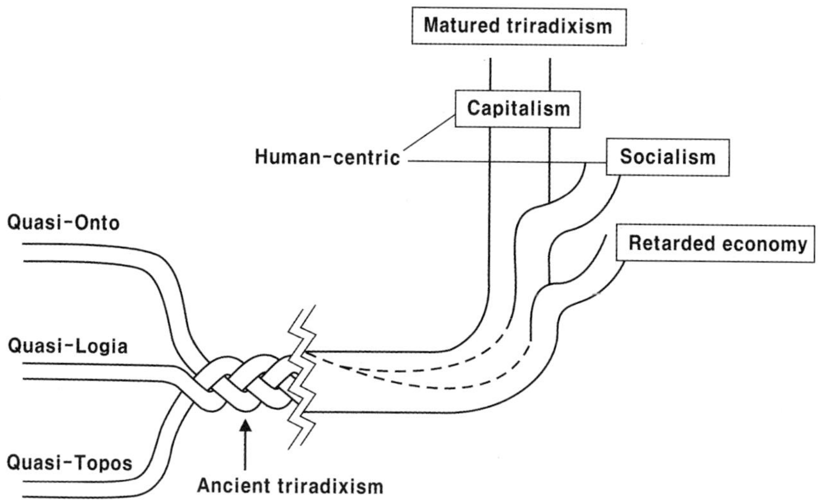

Figure 7-4-5. Tree Diagram of Global Values Menesys (Human-centric (capitalist-socialist) + Triradixes (retarded-advanced) value menesys)

Cruxology can be interpreted as a process of developing communal altruism as we assess it in a humanistic way. Values are omitted in a process of enfolding the arms of nexal connection. Usually when we encounter a new situation, there must be some portion of unbalanced grotesque values causing our dissatisfaction. Enfolding of unnecessary values enables a new systematic development to arise in order to integrate into an insufficient systemization of a simple conglomeration or into a matured systemization of a perfect system.

However, unfolding the hidden arms of value between entities can integrate a network. It is a recovery process of humanity. The creative peace is a little deviated from dynamic equilibrium in a sense that 'generous' dynamic core and thus caused tolerable dynamic module maintain a gently fluctuating integrismic and cruxological status. I know that this is a far too general comment. However, for the creative peace it is essential to have the establishment of adequate menesys of global values.

Figure 7-4-3 illustrates the loci and domain of valorization. Figure 7-4-4 shows how the branches of valorization affect human thinking. We can imagine that scholastic achievement will be much more anticipated in the long run if academic disciplines are diversified in greater variation. The basic submatrixes of onto, topos and logia in transciousness enter into six sets of standard bilateral combinations as shown in the figure. They all are related to existing disciplines, and will be to forthcoming disciplines, of abstractive sciences. Some abstractive sciences are not matured enough but will soon be as much as we can envision.

I think some people are reluctant to acknowledge that the valorization issue takes a significant role in the determination of the scheme of science. However, current global civilization is about to reach a critical moment at which we have to understand that the topics on value and valorization will be a major influential factor for the enhancement of human intellectuality. Valorization has deep impacts on the formation of values menesys as well.

The six sets of bilateral combination in Figure 7-4-4 could be developed among three submatrixes depending on which submatrix the preferential initiative would be imposed. The varied combination would reveal the intrinsic structure of the mentality of civilization resulted by certain omissions or growths of preferred branches. The process would be essentially identical for individuals, communal entities and greater social entities like civilization.

Figure 7-4-5 shows a draft touch of the global tree diagram of value menesys. The value menesys in ancient days were three-rooted almost without exception. Three-roots (tri-radixes) are heaven, human, and land. It had been maintained almost uniformly until the mid 19th century, before which most inhabitants of the world lived well without experiencing the penetration of western human-centric thoughts. Even within medieval Europe, axiology was founded on the three-rooted idea. Human-centric thinking has led to the acknowledgement of only the labor value of humans in economic valorization.

Anyhow, the tree diagram of current values menesys in the global free market consists of three major trunks: capitalist market, socialist market and slow-paced market. The main ancient trunk was defined by tri-radixes. There are many tasks to be achieved in values menesys in this century.

We will comment in Chapter 9 about value capitalism which will hit humanity after knowledge capitalism. Value capitalism is to make wealth, in terms of value, available for or capable of use for further wealth.

The way that any humankind can become the master of the coming civilization depends on whether he or she will be able to fill up the cavity of those potential fields of abstraction. Abstractive activities originate from the elasticity of our transciousness. Abstractive ways of thinking might be the only orientation that humanity possesses in order to overcome the lack and deficiency of human creativity. Otherwise humanity will be overwhelmed and digital monsters will prevail. I am afraid that humanity will be suffocated by nano-bio-electro-mechanization within half a century.

The complex disciplinary systems that currently exist in the academic world can be mapped in a way that transciousness interacts with the conscious and the senscious parts of our cognition. The offspring from the mating would include chemical engineering, economy, music, and many other areas of life we face every day.

We need to understand that there is a prime producer of all academic and intellectual disciplines apparent. It is akin to the universal Language of languages. The structure of the prime producer comes from the receptive attitude of an entity that assesses the value imposed on his or her existence and existential derivatives. For instance, slaves and freemen have different prime producer that builds different culture, custom, thinking and academic disciplines. A free-will slave like Aesop produces differentiated free story fables. That is what we shall term valorization.

Utilities Menesys

Utility, or utility value, is a derivative of "value" that is fully reflected in the human psychology because utility is what satisfies human needs. The satisfaction of human needs developed into a state in which it is entangled to the emotional activities. However, people are not straightforward enough to assert that the psychological satisfaction is directly related to the total sum of emotional intensity because vectors in emotion are diversified in too many directions.

However, there exists ways to measure the absolute quantity of emotional affections by nullifying the vectorial orientation through self-multiplication, or by dealing with just the energy consumed by emotional activities of neural system.

Utility behaves in such a way that apoptosis terminates some neurons in the brain, those which multiplied quickly during a baby's birth but diminished soon in number as to deduce the unnecessary branches of development. Values menesys has a phase space of maximized containability but gets reduced quickly into utilities menesys, which is within the practically adaptable range of human activities.

Therefore it was assumed that value menesys expands and extends to include almost every memory while utilities menesys does not allow insignificant memories to be recorded. Thus utilities are the function of density of recognized emotional activities and values are the function of density of entire psychological activities.

Utility in human life is directly related to value system. It is why utility often is called utility value.

While values menesys has a tree diagram, utilities menesys can be expressed better in a metabolic body diagram. It means that the physical process of how we eat, acquire or dispose of is manifested in our society's utilities menesys. It is a sequential process performed starting from the oral cavity to the anus. Basically, it is a penetrated main pipe with branches of by-passing sub-pipelines connected between the inlet and outlet.

In addition, the body would look like a torus with built-in paths inside. The paths, which are the various assistant metabolic organs, must be formulated together during the developmental period of the main line of passage that digests the external resource materials.

Imagine the flow of food, inhaled air or circulating blood flowing through the main pipe of digestive organs; then, the metaphor from the previous paragraph would be understood much easier. Utilities menesys is the matter of the efficiency between the intake and output, i.e. between furnished resources and desired outcomes. The utilities menesys in each individual human's cognition has intakes such as the sensing devices and the oral cavity, and outlets such as the action devices and the anus. Tangible and intangible flows can be organizedly show the efficiency of generating activities, energy and so on.

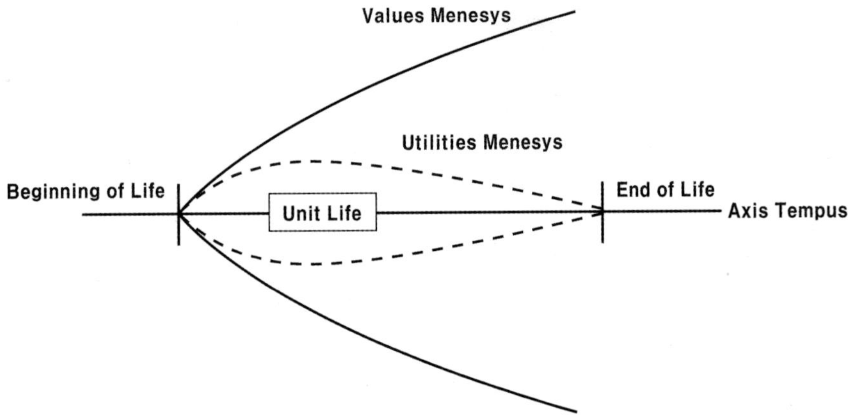

Figure 7-4-6. Values Menesys and Utilities Menesys

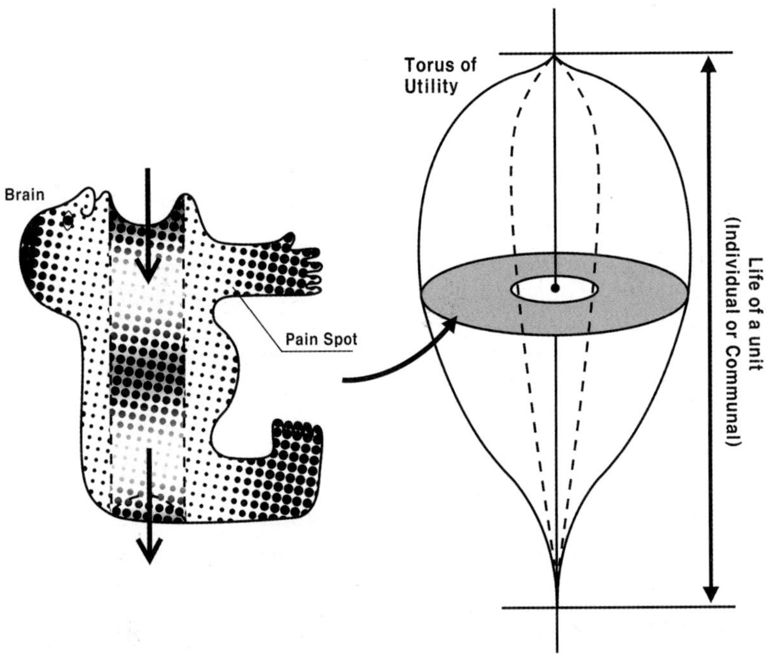

Figure 7-4-7. Torus Diagram of Utilities Menesys

The differences between values menesys and utilities menesys are presented in Figure 7-4-6. This figure illustrates that utilities menesys usually do not exceed the size and range of value menesys. Utility value must be, in most cases, different from the corresponding value, whether under-rated or over-rated.

Figure 7-4-7 presents utilities menesys in a torus diagram. Utility has its foundation on the physicality of an individual and value has its foundation on the communality of a society.

Utility is expanded by unfolding the pragmatic portion in the continuous value spectrum of an individual. When a pragmatic portion opens in the spectrum, the desire for the monetary value is materialized as it usually happens in the merchants' conscious minds.

It is an interesting subject to consider how valves and utilities complement one another; excessive utility deducts values if they belong to same domain. This happens when religious or intellectual leaders mix up their value and utility excessively. They end up losing the value portion as much as they gain in utility. It is not just true for the religious leaders but also for anybody who can become the victim of this complementarity principle. Therefore it is important that each individual establishes good values and utilities menesys.

Reasonable establishment of healthy utilities menesys for the creative peace is expected.

Knowledge Menesys

Value menesys creates knowledge menesys. Value is an actual concept although it may be conceptual and idealistic in its notion. "Being actual" in our discussion means neither factual nor realistic but rather quantitatively unalterable by any influence of a functional operation.

Actually knowledge menesys exists in between value menesys and utility value menesys. Value menesys is not supposed to be exaggerated or underestimated by the effect of virtual reality. Virtual reality is one of the major functions of human brain to modify the external information into digestible form for internal environment.

Knowledge menesys is affected by virtual reality because knowledge is mosaic pieces of the contents of human life which have to be continuously connected in order to form a conglomerate of episodic manifolds. A piece of knowledge or a group of continuous knowledge is subject to expansion or contraction depending on what circumferential condition it is found.

Knowledge in pieces is isolated from each other. Pieces of knowledge should be integrated through rationalization and amendments in order to formulate manifold, logical chain, and menesys. Knowledge is not possible to be found without the suppot from the basement of emotional aspects. This emotional aspect brings bubble effect.

Endoencephalonic hormons like Dopamine, Endolphin, and Adrenalin and so on are the physical effectuators of the emotional changes. Physical influence on the mental evaluation should be eliminated in the process of stable communal assessment.

Intelligence Menesys and Pretium Menesys

Intelligence menesys and pretium menesys are in fact closely related. Intelligence menesys has normalized patterns that have been altered from the knowledge menesys in order to attenuate the excessive fluctuation, for which human cognition usually supply the resource of fluctuating energy.

Pretium menesys of market continuum is the results of the balanced interpretation made by intelligence menesys of the continuum in order to manifest the inner situation of value-utility-knowledge.

Pretium menesys is the most stable pattern of value for which most of influential factors would have been considered if the intelligence ability of concerned market participant could take care. Such considerations are made fully reflecting how the utility of commodity or service could suffice the realistic desire of a constituent of the market in a state of equilibrium.

Intelligence menesys is more or less communal intelligence. Collective intelligence is not applicable in a market condition and swarm intelligence is not like communal intelligence either. Communal intelligence is integrismic intelligence that is much more complicated in intrinsic attributes than the swarm andcllective intelligence because it contains lots of different systems and nexus.

Market Pricing Menesys

Pricing menesys exists in a market continuum. Pricing menesys has grown up since the primitive bilateral transactions through the barter exchange.

The reason we think pricing mechanism to be so important in our everyday life is due to valorization, which stabilizes market price. On the contrary, speculators often consider pricing seriously in order to make an excessive amount of abnormal profit. In most cases, they are successful as long as they plot well to mobilize the market power as they frame it up.

Pricing is very much similar to democracy. If the boundary of a market is wide enough, any individual or a group cannot control the entire mechanism. We know that the price is determined by the relationship between demand and supply.

Pricing in a market has a false-expectation mechanism for rare commodities. Land prices and prices for curios and antique things show that pricing in a market can be easily distorted through a psychological effect.

When we get sufficient assist from the intelligence menesys, market pricing will be relatively more stabilized and evaluated in reasonable bases. The pricing mechanism in the global market continuum has a deep relationship with the monetary system and the financial system. If we do not have sufficient amount of monetary currency flow in financial system, finanacial system should not work efficiently as if we can see the irrigation channel without rain would be useless. The finanacial institutions must behave like irrigation facilities.

Virtues in Terms of Science: How Integrity Realized in Reality?

How does humanity practice integrity in reality? Generally, it is agreed among cultures that there are certain categories of virtues humanity manifests into integrity in reality.

The virtue of humanity in terms of science is the exchange between values and utility. Utility is a modified form of value reinterpreted in terms of emotional satisfaction. The emotion here is concerned with the notion of 'seven emotions' that is relevant to the social gain in terms of profit or egoistic satisfaction. But value is concerned with Four Beginnings which vitalizes the communality of a society.

The philosophy of value exchange in a menesys is very important methodology of menesysal management. How does virtue pay? The payback of virtue is the territorial expansion by value co-owning, as we can see in the software liberation movement.

However it has to be clearly understood that any one side of a value-seeking attitude or utility-seeking attitude has absolute validity in our living. We are talking about the management of society.

Four beginnings (四端) are, as discussed before in Chapter 6, the heart of compassion (惻隱之心), the heart of shame (羞惡之心), the heart of courtesy and modesty (辭讓之心) and the heart of right and wrong (是非之心). Four beginnings are the governing four principles of human mind as human physique has its four members.

Four virtues (四德) originate from the four beginnings of human emotion. The four virtues are charity (仁), justice (義), courtesy (禮), wisdom (智). (From Xhu shi to Yak Yong Jung: Philosophical Paradigmatic Study on the Confucianism of Yi-Dynasty, Hyung Jo Han, 1996, Ch. 7)

(The list of Seven Feelings (or Emotions) as presented in the Book of Rites is as follows: desire (慾), hate (惡), love (愛), fear (懼), grief (哀), anger (怒) and joy (喜). Tsu-ssu in the Doctrine of the Mean gave instruction to the human feelings, the origin of the above seven feelings: joy (喜), anger (怒), grief (哀) and pleasure (樂). Before the feelings arise, they are in equilibrium: after they emerge, constant attention is required to ensure the state of harmony. Human beings are supposed to approach in a reverse way the initial equilibrium of order to find the initiative of true wisdom.)

In the West, the love-truth bilateral balance is on the trail of the virtue-and-knowledge way of thought. It is a strong measure to manifest humanity in an orderly way. It is same in religious tradition. Holiness is the manifestation of justice, as written in Psalm "He is holy because He is just."

Holiness is not merely an outfit of hierarchical order of social level. Holiness has to be identified as the source of balancing wisdom and integrating function. It has to do with the fact why we have to behave meek, gentle and smooth in social relations in order to succeed in maintaining healthy cultural characteristics and even the good parts of our DNA in phenotypical appearance.

Holiness is the integration of virtues of humanity and the beyond. Science is the scaffolding, depending on which humanity builds virtues of integrity.

7-5. General Technology of Menesysal management

Menesysal Internality and Menesysal Externality
Integrative Menesysal Management
Values Management and Utilities Management
Connectivity among Manifolds: Why is Criteria of Virtues needed?
An Ideal Type of Menesysal Management

General technology of menesysal management is similar to the fluid mechanics in porous media, which is characterized by the phase shift mechanism in transition zone. The phase shift of fluid flow through porous and permeable media is more complicated than in open space.

The concept of statecraft (Ching-shih, 經世) in East Asia will be considered important in the coming decades. (Benjamin A. Elman, *Classicism, Politics and Kinships*)

Menesysal Internality and Menesysal Externality

The first element of the menesysal statecraft is internal management. The internality of menesys is a matter of logic and connectivity.

Logical circuitry is better developed when physical commitment of the menesys is greater. Mineral crystals are the product of pure logical relation. The probability of breakaway from the logical development of mineral crystals is almost none. But in the world of plants we generally find inert consciousness based on senscious system. In the kingdom of animals we generally find the emergence of activated consciousness with the earliest probable indication of transciousness as we can see in Macaques, a Japanese monkey that cleans up sweet potatoes by the sea water. Thus we believe that systemicity is more quickly developed than nexusity in the realm of living things.

The human race is peculiar in the sense that is has transcendental intentionality, causing the highest complexity beyond comparison with other cognitive beings. In human cognition, the dynamic system called dynamic core actively produces, sustains and abandons the apparent conditions of dynamic modularity in their life. Dynamic cores are subject to change by themselves, provided that their elasticity is maintained relatively more stable and rigid than the change of dynamic modules.

In this context, we can conceive the degree of compactness of systemicity that protects the connection between modules, which have neither logical correspondence nor logical resonance. In the phase space of human cognition we can define less connectable part to be a medium, outside of which connectivity is active in greater degree than inside.

On the other hand, it is thought that connectivity is significantly activated in usually none, or the less logical parts of the phase space. When the logical circuit is too tight the allowable space for connection is not enough. The degree of compactness of systemicity makes difference in the tightness of logical circuit. The lower the degree of compactness of systemicity, the higher is the room for connectivity. Connectivity is another expression for nexusity.

We can imagine there are firm cores with tight logical circuits and open space for the allowance of free connections, and network filling the open space like fluid flow in porous medium. Actually networks are not connected in accordance with logics but formulated through accidental habituation or taming.

There is a part of external management in menesysal statecraft. External management is concerned with intensive properties of the menesys. By general definition, intensive property is independent of the amount of material of concern.

Daniel J. Levitin (*The World in Six Songs*, 2008) comments about six menesysal development of six kinds of songs of humanity. Co-evolution between music and the human brain has contributed to each other in the history of humankind. He stated that humans consume music for six reasons: friendship, joy, religion, comfort, knowledge, and love. If we make a circular expression of these reasons, joy-comfort-love are subjective emotions, while friendship-religion-knowledge are more social phenomena that are usually dependent on quantity. Joy, comfort and love have played a pivotal role in social development. It is the author's opinion that the above three positive and integrative emotions have strengthened the human interaction and communal coalition as they affected internal formulation of the systematic logics in social dynamic core.

Menesysal externality might disclose different outfits in different circumstances, for instance, friendship-religion-knowledge could be changed to loyalty-knighthood-martial arts. Levitin shows the most probable core in human history is equipped with such structure.

Integrative Menesysal Management

The menesysal management proposes fresh insights and novel ideas in the administration of the real world. Generally, management of an organization supposed to take control of all his subordinates. In Internet, it is not easy to grasp the trend of netizens and influence the swarm intelligence. The change of intrinsic property of a menesysal organization, so huge and complex and so fast in communication, does not allow us to remain like a lazy cow on a train.

The new world that has already invaded into our reality enforces us to design the integrative menesysal management.

What is the most desirable condition of dynamic core, dynamic transition zone, and dynamic periphery in dynamic modularity? How can we achieve the expected result? The wisdoms of the world – kenosis, sunyata, chimchum – urge us to draw back from perfect systemicity of full efficiency.

We experience in the real world the crisscrossing of the domain of values and utilities, asymmetry of which cause the fractal third. The fractal third is named for the third forces generated from the fractal origin of humanity, which will gradually formulate integrated body and finally assume the third element of Tri-taiji.

We have usually two opposing forces in the world. When the situation changes, the world will be unified or will add up more number of constituents. In this book the most stable condition of the ternary elements are thought to have three different sources of drive in the elements, which are the sources of ultra-internal, ultra-external and manifolds connectivity.

In this context, it is necessary to consider the values management in ultra-internal human aspect and utilities management in ultra-external aspects of human being. The connectivity between those manifolds should be understood.

Values Management and Utilities Management

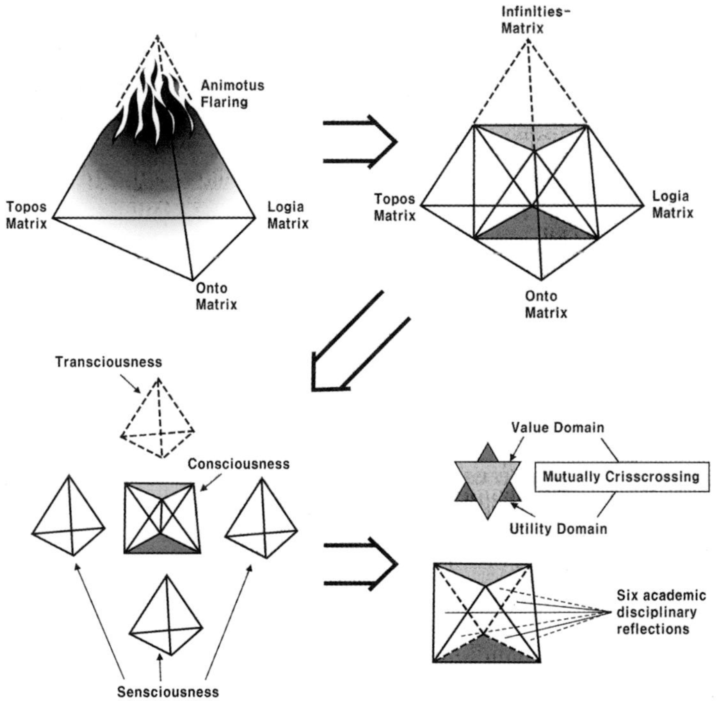

Figure 7-5-1. Symmetry Broken between Values and Utilities (Value is connected to intangible Animotus motif which cannot be defined materially, while utilities are connected to the tangible emotional foundation of human cognition)

In human activities, we believe humans live in a satisfactory range of homogeneity between value and utility. Because the sources of value and utility are identical, actually we believe in the symmetry between the domains of values and utilities. It is seen that symmetry has been always broken between values and utilities in humanity as seen in Fig 7-5-1.

Value is connected to intangible Animotus motif which cannot be defined materially; utilities are connected to tangible emotional foundation of human cognition. Value and utility usually do not coincide and utility tends to exaggerate while value is inert in expression. Communality values – in terms of virtues – activate and vitalize the 3rd kind creativity that is able to cure humanity by restoring balance. Value will attain a way of expression when terrenity value is well administered in its expectation for utility by renormalizing the abnormal expectation of monetary market price.

Communal balancing by virtue – that is the renormalization of terrenity value – enables us to evenly distribute the density of creativity over the whole society and its contained industries. The best mix of managerial policy is that hovers or circulates around the strange attractor.

Connectivity among Manifolds: Why are the 'Criteria of Virtues' needed?

We commented in above discussion that values domain is far offset from utilities domain. Furthermore, it is thought that utilities domain is adversary to values domain because of the cohesion to the source of value and utility respectively. Values domain adheres mainly to the domain of terrenity value that integrates and adsorbs most existential subconscious perceptions while utilities domain is mainly projected to the domain of communal value. The creativity value is scattered over the whole scope of human concerns on the surface of initial development.

Value domain is a manifold that is continuous based on the continuous perception between Plexion and Corpus discussed in Chapter 3. Utilities domain is also a continuous manifold that has continuity based on pragma and brain. Values domain and utilities domain mutually crisscross due to the converse arrangement in cognition.

In utilities domain, the cure of connectivity among separate three or more manifolds originated by three submatrixes are performed by communality. This is why the criteria of virtues are needed in comprehensive cognition.

Virtues, peace, stability and so forth that are essencial for the communal life usually have mending and integrating properties. Patchwork done for the continuity between manifolds saves energy of cognition; if the patchwork is not done when similar pattern of affairs occurs successively, the stress consumes useless energy at the subconscious level.

The reality of humanity is not more than usual relations between people. Relationship or relational structure is the condensation of human relativity.

Technology of menesys is an absolutely essential managerial technique in order to formulate the continuum that is the substantial body of the creative peace, especially for the future of regional, global, and/or Solar system. The menesysl technology is based on the nature of humanity and required for the social stability, productivity, and enhancement of cereativity. Appropriate administration of the subordinate constituents of menesys such as netsysand sysnet brings great influence to all social and communal classes of people.

The menesys stands for megalo-network-system. It is necessary to study the initial manifestation of netsys, sysnet and menesys and their patterns of development, the integrismic relationship and embryology of cruxes that appears in accordance with their patterns. The patterns of menesys should have various influences to the peace and creativity.

Maybe the particular patterns that could appear on the Earth should be different from those on the other places in the universe.

The complex menesysal developments will be assessed by the urban planning technology which will be no more confined to the city problems in the past but extended to the global and beyond problems in the future. The urban planning technology would be named Global Intelligent Modular Planning (GIMUP). GIMUP has the backbone of theory in the University (unity-versatility paradigm) and commercial networks including finanacial system.

The menesysal complexes in global enterprises vary from capital-holding limited corporations to global credit-holding conglomerates. Religious and cultural complexes in menesysal form encompass from fundamental concentrationism to interpretative essentialism as the author named. The menesysal complexes concerned with the global power regime and the governance systems are from the democratic conglomerate based on federality to the glocalized integration of the world. Human dreams are reflected in the names of projected imagination. We need to develop far reaching names of further future.

Knowledge system, academic system, education system, neural system, the system of physical organism, and Meridian-Cord systems (經絡體系) are all menesysal complexes. (For modern reference of Meridian-Cord system, see Scientific Search of the Meridian Cord by Chinese Medicine, 中國經絡的科學的探索, Zhang Weibo, 1999) We cannot go further beyond our horizon if we do not understand the dynamic modularity in the complex menesys.

An Ideal Type of Menesysal Management

Presented in Figure 7-5-2, is the conceptual triple mate of menesysal organization that has attained reliable stability and consequent fractal propagation in social environment.

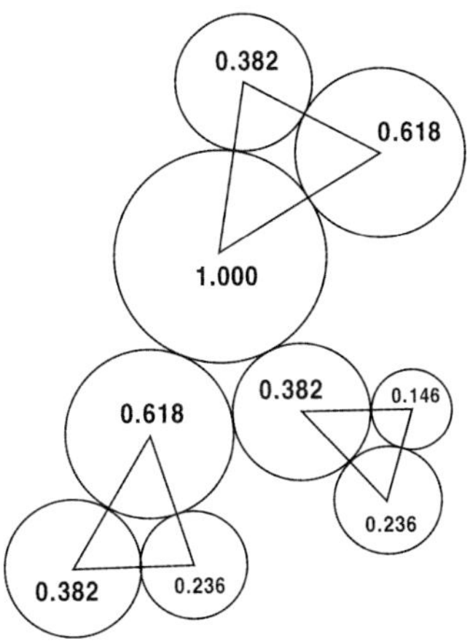

Figure 7-5-2. Conceptual Triple Mate of Menesysal Organization with Reliable Stability and Fractal Propagation

The fractal propagation of trinity relationship is generally found in practical situations. The world is not logically characterized in trinity but generally moves to form trinary format when dynamicity prevails in a society. The pattern of frequent agreement and division makes certain ternary condition. Mercantilistic countries for the last several centuries have experiences of dynamic situation and advanced abroad enterprisingly.

In excessively dynamic condition, there happen general trends that do not permit conglomeration among parts of the whole. But when adequate conditions are satisfied in history, a main figure appears accompanied by the competing second figure and the third figure comes out in a conglomeration of smaller entities.

We will think of Tri-taiji pattern in a dynamically changing history of the future in Chapter 10. However, we know that the ternary pattern of the separation of three powers gives fair balance of power. Triple political parties, the recondite arrangement of Benerux countries (I don't know whether the division of Belgium will be really consolidated or not) and trio formation of Britain-France-German and so on gives us an insight that it might be a good choice to adopt the pattern of triple mate in designing new menesysal organization.

Taiji changes to Tri-taiji when static environment alters to dynamic condition. It is fairly reliable if we start from initial triple mate as discussed in Figure 6-4-1 in Chater 6, and advance to Tri-taiji and its dynamic center, which is deemed to be the right answer of the global problems. (The dynamic center could be either tangible or intangible.)

In this case integrism looks like presenting the idea of *Liang-Yi* (兩儀, yin and yang, or heaven and earth) in fact it represents the situation formulating new synthesis meeting and parting among fractal parts that is grown up from the initial matrix in order to assume and become the third power.

7-6. Peace Knowledge Menesys

Essential Character of Human Knowledge
Peace knowledge Menesys
Educative Information Menesys

Humans cannot live in an everlasting conflict. Even the gods and goddesses of conflict and fighting take rest in the night, because they also possess the knowledge of peace in heir mind. Peace is the birthplace of human beings. Peace knowledge is engraved in our substantial body. We can see that human physicality is based on metabolic process that provides the refreshing and comforting peaceful condition for tired body.

Peace knowledge menesys is the exit of human mentality as the metabolism is the solution for physicality to avoid the collapse of physical process. The difference between the two process is that one is directed downward whereas another directed upward.

Essential Character of Human Knowledge

Human knowledge is presented in K-map by Takashi Saito (Ontology·Algorithm, 2004, Korean version, p40). A part of K-map is restructured to give an understanding of this book in process mapping of knowledge and cognition in Figure 10-3-2, which presented the relationship between persontology and peace knlwledge menesys. Figure 7-6-1 just above shows the internal structure of the menesys of human knowledge.

The core of human knowledge comes from the survival mechanism, then shift to balancing of transition zone that mostly overlaps with consciousness, and finally allows communication in its peripherial zone of knowledge and cognition.

Humans are not so cultivated from the start but it is worthy to quest from where this cultivated humanity comes.

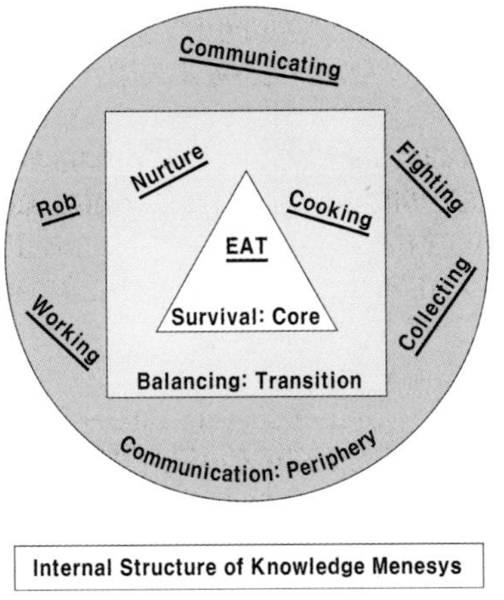

Figure 7-6-1. Internal Structure of the Menesys of Human Knowledge.

Peace knowledge Menesys

Peace knowledge is the most external part of the onion-like creative peace, because of its intrinsic attributes that penetrate into the love-wisdom fundamental. Peace knowledge menesys is a reverse tree diagram. In Figure 7-6-3, a reverse-projection of Sun light in Newton's 'concept of reverse prism.' As in a reverse prism, human knowledge either individual or communal takes a reverse valorization process to become peace knowledge.

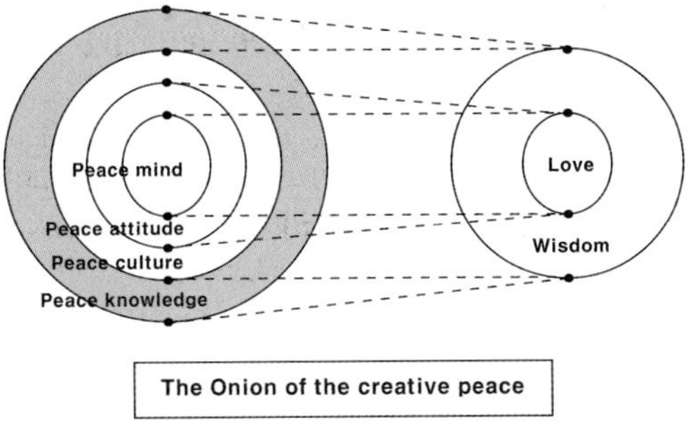

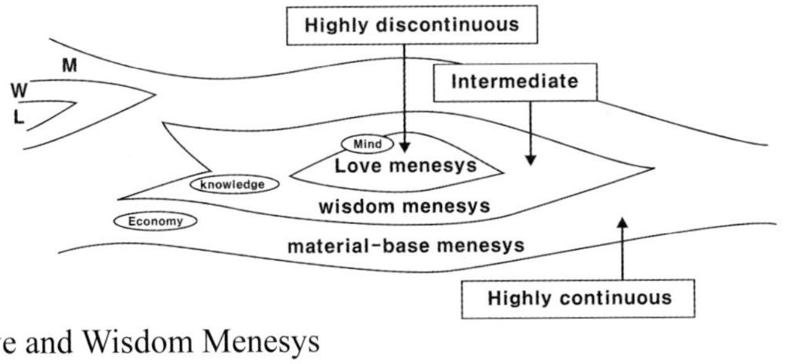

Figure 7-6-2. Love and Wisdom Menesys

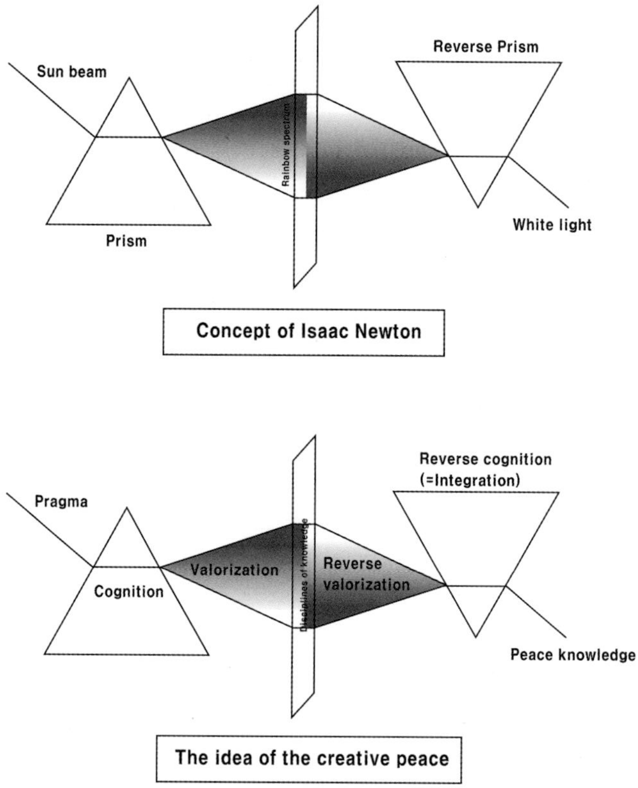

Figure 7-6-3. Reverse-projection of Sun light in Newton's 'Concept of Reverse Prism'

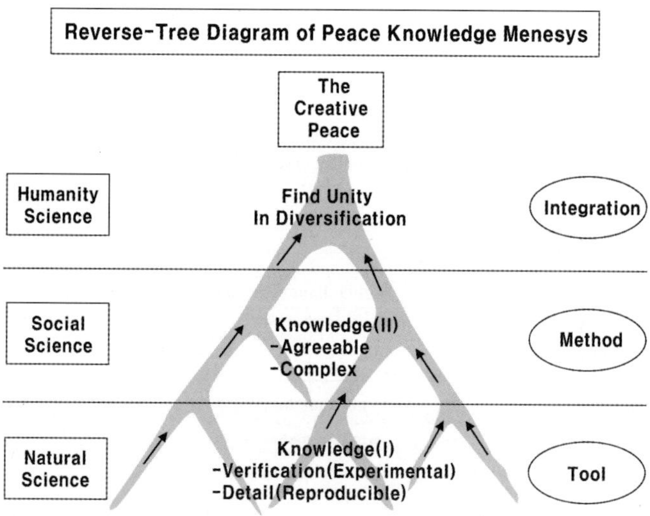

Figure 7-6-4. Reverse-Tree Diagram of Peace Knowledge Menesys

EIT (Educative Information Technology) Menesys

Educative Information Technology is aiming at establishment of global educational menesys. There is an interesting report from molecular biology that is applicable to EITech. Gary Marcus (The Birth of the Mind, 2004, Ch.9) states that learning is generally founded upon genetic mechanism and thus profound genetic regulation system is the prerequisite of profound learning system. Environment supplies the initial factor to stimulate overall process and system of learning. Genetic regulation provokes the chain reaction that will reorganize over all neural system.

Likewise in human life, it is thought that the carrier of meme, such as rhythm, symbol, and emotional and rational factors can induce the novel change in menesysal point of view. EITech can adopt similar strategy to activate the switch of the chain reactions that has been conserved in the genetic or mematic safe.

EIT systems of various environments come together to formulate a global EIT menesys. EIT system is designed for and applied to a specific local environment; for certain cultures, ethnicitys, economical statuses, politics and so forth.

Global collaboration in contingent circumference makes EIT menesys inevitable. We have had local contingency programs in Haiti recently. It is the proper time to formulate a restoration program for the whole society in how to basically alter the educational infra structure from scratch. It would change the future of Haiti, if we organize the information about the basic emotional stance and symbolic expression of Haitians and construct a map of their world of positive imagination as the starting line of their education, which will be later succeeded and developed to the fully developed peace knowledge system.

We need to think of the EIT menesys on a global scale. It is the forthcoming space entrance which necessitates EIT Menesys of the global inhabitants.

The world is heading for the weaving of a grand plexus of story-telling as we do in our individual cognition for the formation of a plexus of episodes. The sympathetic effect and shocking factor in story-telling could provide a primitive breakthrough of emotional induction that we can currently experience in a multimedia environment. The higher introduction of new dimensions in educative information systems are far above the emotional impact and invisible mending of emotional manifolds, and thus necessitates the activativation of creative emotional systems by realizing the ability of peaceful integration.

When we achieve the concept formulation through the continuous pop up of ideas, the emotional function of human cognition helps the sorting of information with the feeling of amenity through unitization and compression.

7-7. Menesysal Management to Enter into our Solar System and Beyond

One Century on the Earth: Basal Menesys
Another Century into Solar System
Millennium "To Boldly Go Where No Man has Gone Before"
Be Humble, Bumble Bees.

In order to advance in the solar system and beyond, we need first to practice menesysal management on a global scale. The human race is something like a fungus on the surface of this precious planet. We have to be changed to manage our given blessings.

One Century on the Earth: Basal Menesys

Environmental debates are structurally at the bottom of discussions dealing with space problems. Usually people do not understand this point but it has to be clarified in order to enter the space epoch, or with more precisely, the solar epoch.

Humankind will have to wait one hundred years, or at least a half century, from the day we accept an agreement to accept a global space project in order to build a consensus on how to approach and how to solve many issues of debate. The basal menesys as a fruit of this wide agreement would encompass menesysal efforts from education, technology, wealth distribution, governance, and congressional agreement.

It is probably inevitable to pass through great turmoil in the first half. Severe conflicts are anticipated. Present low-density conflicts are actually originated from the globally entangled socio-psychological problems. Wealth distribution and cultural conflicts are the reasons inducing psychological blockings of less privileged people of the earth. But in a deeper dimension, they are the expression of explosive emotional stress which has to pass over the gaps between thousand years of historical jump.

We would like to recommend a probable solution for this unbalance. The hybrid of business functions and governance functions could be organized in a Solarche Menesysal Schema and global Keiros project. This will be explained further in Chapter 10 in more detail.

Another Century into the Solar System

The next period, most likely fifty to a hundred years since achieving the basal menesys will there be an epoch of great attunement which will go through mutual adjustment, mutual coordination, and in the final stage mutual tuning. We do not know if we will pass through great upheavals before getting into this great attunement.

We have always had prophetic voices of warning saying that we will be punished if the human race does not alter its path. Whatever the upheaval might be, human beings have to be prepared to change their external identity while basic communal values should be kept firmly unaltered.

The century before us will not be easy but not too hard to overcome. The key factor is the preservation of humanity. How to define humanity is another issue. It is my thought that in any circumstances human beings have to stay in contact with nature; the green and yellow earth, sweating during hard labor, skin-to-skin touch with another person, eye-to-eye contact, emotional communication, wild nature, fighting spirit, heterosexual reproduction, the preservation of Earth, gravitational feeling, love for tiny worms and small birds, and so forth.

However, this is the right time to be prepared for the age of the Solar system on a global scale. Most social phenomena appear to be showing signs of boundary problems in the global economic and social continuum. The United States with its big continental breadth of territory began to recognize the signs of boundary problems of its resources and frontiers since the 1960s. Likewise, the global boundary problems have been encroaching on us since the introduction of new industrialized countries.

If the human race does not prepare for the solar age within fifty years starting from now, we will experience inevitable implosion which might nearly extinguish our civilization. It is yet another prophecy that we have to listen to carefully.

Millennium "To Boldly Go Where No Man has Gone Before"

With our physical condition built for earth's environment we are bold enough to go further out into the space. Our DNA is of and from this terrenial earth with specific earth gravitational conditions.

We know that our body will experience serious osteoporosis not long after leaving the earth's gravitational environment. That is probably why some leading intellects dare to consider installing instrumental functions of cyborgs into the natural physique of humankind but, most ideas are to remain as human as possible.

Going out into space has many associated problems such as this that have to be overcome within a century. The existential value of NASA and the other space institutions of other countries could be valuable in listing up those detailed items and find out how to bypass or decrease the impact of sudden changes of the environment against human physical and mental consistency.

Be Humble, Bumble Bees.

We are bumble bees. We work very hard to live. However, we are bubble beings. This means we expect fascinating fruits on most occasions. The world of reality does not permit us to enjoy non-sweat equity.

Human imagination is frustrated by the technology of virtual reality. Human beings are supposed to possess an inner-most emotional heart, which brings thinking based on consciousness. Thinking is performed by the imagination process which is interwoven with emotional signs. We are losing more and more equity of sweating and pain-taking. One of the pitfalls of brain science is producing empty grains of humanity. Computer game players lose the feeling of pain and frustration caused by violence and decadence. Furthermore falsehood has no significance to the player because they feel as if they are participating in reality.

Is it possible to sustain our humanity when the values of reality are vaporized? When values are sustained in reality, our emotional and volitional cognitive system have a firm and stable foundation. Volition is the most stable fundamental of humanity. There exists communal volition as effective in reality as individual volition.

Knowledge systems are in essence a virtual reality. Knowledge frequently becomes swollen up by the interference of emotional affection. Thus the knowledge capitalism is considerably unstable unless sound value systems are introduced. We need to introduce the insight of value capitalism as a beachhead for the future economic continuum.

For the coming years shortly ahead of us, we will have to learn that value systems and volitional processes should be recognized very much as essential for the entrance into the solar age. We can depart when value is volitionized. When values are volitionized we will not be frustrated by the immeasurable depth of space, which will shake our imagination process.

Be humble, bumble bees!
The lovely bubble beings,
Brilliantly shining in the air.

Chapter 8. The Science of Creativity

Creativity is the keyword for "The Creative Peace." The phenomenon of creativity is presented as the goal and solution of peace. Creativity is a matter of communality as well as individuality. "The Creative Peace" has eventually attained a new and essential concept liberated from the past notion of "creative peace" which was denoted as "the peace pursued through arts and culture."

The creative peace is reinterpreted to be "the Continuum of Activated Humanity in a Stabilized Condition." All the phases of human history are the output of creativity of the human race. With an adequate scientific viewpoint on the past, history will give an entirely fresh insight into humanity (人間性) itself in terms of creativity. In this research, "The Science of Creativity" is presented in order to organize the overall phenomenon of creativity.

The creative peace is what human beings have been pursuing all the time in the name of the kingdom of the millennium. Presently humanity is confronting such an important historical threshold that it may find a way to overcome and change itself into an unprecedented opportunity to enter into a profound transformation.

Creativity has greater importance than the existing knowledge systems. Knowledge systems without a gateway into a creative future exert negative influence to humanity. Knowledge systems are more energized when they construct a menesysal entity so that the creativity process may be activated.

Creativity is the essential driving force for the creative peace. We will see some excessively dynamic mechanisms of creativity which will be operable only when bubble or foam effects prevail under an unstable psychology in order to pursue the substantiality of unidentified future.

8-1. Creativity Comes from Transciousness

Creativity is a special characteristic of human beings. The immense influence of creativity on the omnidirection of human life has sustained history to move forward. Warfare is a false-directed choice. Good-will does not choose destructive action. Creativity usually comes with good-will.

Creativity is propelled by the "boiling" of the human cognition caused by the concentrated thermal energy of the fluidal body above its boiling point. As we see, the boiling of fluid produces bubbles. A bubble is a swollen gaseous state trapped in a liquid pocket of discontinuity with specific environmental conditions.

In our brains, the bubbles are the imaginary activities generated in our mental phase space constructed mostly by transcious activity. Imaginary activities are discontinuous together with our general cognitive flow and formulated by the successive concentration of our brain due to the supply of high neurophysiological energy.

Fixing a symbolic scheme in our mental association process is highly energy consuming. Imaginary activities have less energy consumption than fixing a symbolic scheme. The formation of bubbles is caused by the concentration of our imagination on certain isolated groups (or networks) of symbols or associations.

It is accepted that "flow," known as the psychological optimal experience, is the essential factor in the creative behavior (M. Csikszentmihalyi, 1990). It is thought that "flow" needs a highly concentrated energy in order to maintain isolation from general connected status.

Imagination in a society is similar. Imagination produces boiling bubble or foam depending on the level of socio-psychological intensity.

Imagination and Bubbles

Let us get into a little bit of the story about dynamic modularity. Dynamic modularity provides parts of variable keys of transciousness that open the ever-moving position of the gate into the domain of creativity. Some dynamic modules are supposed to fire together at the demand of the infinities-matrix of an individual. The infinities-matrix generates the ultimate demand for the sake of transciousness.

The domain of creativity is in the territory of matured imagination, which can communicate with the necessity of the infinities-matrix of an individual. A dynamic module is matured when it establishes its own communicable logic with the infinities-matrix. The logic of mutual communication is formulated in the course of a long collaboration.

Then the dynamic module starts producing bubbles of ideas, signals and images that can cope with the environmental impacts. The communicable logic could be partly prone to each one of the onto-submatrix, topos-submatrix and logia-submatrix depending on its specific locality and tendency in the human mental phase space.

The gate to imagination is floating in the phase mental space of an individual in accordance with the main stream of emotional stability. This means that vicious people can also activate creativity of their own style to suffice their personal purposes.

Now we know that desire or energy in our cognition provokes the bubble to imagine some objects in a manner of virtual reality. Imagination and bubbles are in imaginary dimensions.

Reality of the Bubble Effect

The bubble is imaginary, but the bubble effect is real. This is due to the fact that human acts are based on their imaginary conclusions. The influence of bubbles on our economy can be observed in asset bubbles and credit bubbles.

Virtuality bubbles also exist. Virtual bubbles, for instance, occur when we are delighted whenever we see our beloved ones delighted. It is a mirror effect of reflective virtuality. When a mass of people take to the streets, they are sometimes mobilized by some sheer psychological reasons. The reflective multiplication of virtual bubbles has a very important influence in political economics.

Bubble and foam are of the same origin and mechanism, except that foam is a more stabilized form of bubble as we can observe in our bath-tub. The size of foam is quite smaller than bubbles and does not burst easily. Foam is usually grouped and moves together. The cohesive force at the boundary between foam is stronger than bubbles.

Bubbles are generated by excessive desire or overheated expectation. We will include foam effects when we use the term bubble effect unless distinctively mentioned.

The socio-psychological fundamental mechanism of bubbles is correlated to delight-priming, which will be explained in the next discussion. It is similar to the mechanism found in festival psychology.

Pain-Framing and Delight-Priming

The human cognitive process has two contrasting activities; framing and priming. Priming is exhibited in emotions such as delight. Thus delight-priming (喜統合) is a mental mechanism at a relatively lower level of continuous consciousness. It is like dreaming which includes only partial participation of the brain.

Ecstasy is at the summit of delight-priming, but it generally concentrates on the present reality only. Emotional activity is a minimum and fundamental level of consciousness in the conscious flow of mind.

It is thought that hallucinations lack the strictness of logic. For instance, hallucinogens like ketamine induce a concentration of small neuronal groups. Delight-priming can usually reduce only the first kind of creativity.

(We have high levels of delight-priming as observed in the ceiling-less delight at the moment of martyrdom.)

Pain-framing (苦統合) needs a higher level of continuous consciousness. It sustains the high level of continuous consciousness by framing the logical structure and stress-tamed condition. It is my opinion that the intermediate high level of the continuous consciousness might match with the second kind of creativity. Pain-framing process mobilizes most of human neuronal groups (Susan Greenfield, *Brain Story*, 2000, Ch 11).

Delight springs out when we meet with the self and pain comes when we meet together with others. Pain as the peak of our emotions and cognition allow us to grow and attain a self-overcoming attitude and logical circuit for societal living.

However it is thought that the third kind of creativity comes to our consciousness when we tie up the bundle of stress and pain by ecstasy. This high level of ecstasy covering the entirety of our neurophysiological integration usually does not deal with partial involvement and exaggeration of our sensciousness, (in the form of sound and color) in order to avoid the frustration of the integrity of our mind (or heart). It is often found in the meditation of monks and nuns.

Transciousness and Creativity

It is reported that the structure of the brain is supposed to close its loop after a release. This means that the exchange of information is impossible between output modules and input modules (Ken-Ichiro Mogi, 'Nou To Souzousei Kono Watashi' To Iu Cuoria E, 2005, Ch. 5).

I think the transciousness (for imagination) and postefference have been formulated because of the discontinuity of input and output modules. This is due to the infinities-transfer and plexion, which we discussed in Chapter 4.

The 3rd kind of creativity achieves the highest integrity by reconstituting the general conditions for the transcent of humanity in the transciousness region. It is related to overcoming the intrusion of the time funnel of our episodic memory. I believe that we have to encroach into the reversing time funnel through the integration of our segmented mentality (or spirituality).

We cannot see our spirituality since it is not a form of a whole symbol as realized as a dream or illusive episodic expression. It is a process comparable to the intrusion of the reverse intrusion starting from anus to mouth and producing senscious organs and brain-neuro organs. Transciousness is physiologically the result of reverse dimensional intrusion like Möbius strip. Thus our integration of cognition is achieved in a reversed way.

8-2. Permunal Creativity

"Permunal"
Undiversified embryonic process
Blending of reserved stem cells
Importance of Transition zone

"Permunal"

Creativity is essentially "permunal", which is the combination of personality and communality. Significant portion of the economic value of creativity that human beings have produced has well been accumulated in the past of human race. Violent invasions, remodeling requirements, etc... have devastated stockpiles of human creation, nullifying the value of their creativity.

The science of creativity will be an essential tool to establish a new structure in political economics and the other state-administrative (or world-administrative) methodologies. Past political economics has usually disregarded psychological factors.

However, economics looks like it is ready to accept sociological effects of psychological factors. Psychology is gradually changing towards the direction of quantifiable and measurable science, attaining greater support by neurophysiology.

In fact, the science of creativity lies on the overall scope of human phenomenon. This means that creativity is hard to interpret in order to acquire integrated consistency if the whole scheme of human lives were not understood. A basic way of thought in quantitative interpretation is presented in the chapter. The proposed methodology could be extended and applied to various disciplinary knowledge systems.

Creativity has become the life line of human civilization. The functional capacity of the current technology highly surpasses an average intellectual of the human race. Global economy has a far greater and wider influence to our everyday life than cultural force. Politics in the modern day seems like it is losing its state-administrative capability and is becoming nothing but a skeleton by practicing programs and delivering addresses repetitively. Could creativity be the answer to these desperate queries?

Creativity has been focused since the beginning of the 21st century. The content, characteristic and intensity of creativity in our urgent need have been rapidly changed.

Undiversified Embryonic Process

'Undiversified embryonic process' is a term that popped up in a conversation the author was involved in at a seminar on communal movements in 1988. The seminar was organized by civilians for the study of communality in the process of national reunification of the Koreas.

An embryo is grown from a zygote. Zygotic condition contains all the possibility of an entity-to-be. Communality is the most vitalized condition when a community is not so big. When communality is in place, human physicality may still hold the cohesion reasonably tight. The undiversified state of the embryo is a metaphor when we describe the development of society that still keeps the omni-potency of a stem cell.

Undiversified embryonic communities conserve energy and rationale to proceed for further segmentation of diversified function. It is a technique of state-administration on how to keep vitality of a society. This is done by preserving the communal capability in a fresh mode so that the society does not deteriorate by the solidification of excessive systematic development.

There is a vital part of communal body even in an intensively systemized body of organization. The small communal body has a directive force of the whole organization. The constituent of the communal body later becomes the future leaders of the organization or any business conglomerates.

Blending of Reserved Stem Cells

It is really important to develop a strategy in order to sustain the creative capability for a society. In human organs we have reserved stem cells that are activated in time of need.

There are several strategies on how to blend the reserved stem cells.

Firstly, we need to keep as many stem cells as possible. Secondly, we need to conserve stem cells. Thirdly, the blending should be performed in a controlled manner.

In preserving the stem cells of a society, it is necessary to provoke plain cells to become stem cells and to keep the societal conditions to conserve the stem cells as long as possible. Recently we have heard about the possibility of producing stem cells from skin cells. I think the success will be achieved in the presence of a strong spirit and belief that it is possible to achieve anything. There should be a certain kind of reward that the society provides to the people who dare to live as stem cells, because maintaining the capability as a stem cell of a society is really energy consuming. In the science of complexity, they are called adaptive nodes, which I believe is not proper to call stem cell people who are all-round responders. Do they exist in reality? If they do, then how could it be promoted to satisfy the need of the society? This is another question.

In conserving the density of the stem cells, there are certain social, economical, and political conditions that should be maintained. We will comment on this in later sections. In the blending procedure, stem cells must be positioned to provoke the aged segment of the society. There could be many skills we can conceive.

Importance of the Transition Zone

The importance of the transitional phenomenon we experience in the alteration of a society is not satisfactory even though we emphasize a number of times. Transition contains all the probability in its womb.

Creativity in essence is the process of changing or being changed. Creativity is really workable and adaptable when it is admitted by a society. Such modification is made into the transitional stage of an individual, because an individual person cannot experience the delicacy of transitional change in its initiative but rather in the gradual environmental change.

Transition zone is a barometric criterion in the human mind to grasp the optimum point of our delicate needs. People or a country that has experienced drastic changes and ongoing world ascension thereon has the best sense for creativity. That is why those countries that overcame the painful memories of warfare in the past make a lot of difference in the development of their country as far as historical conditions are met.

8-3. Value Theory (Axiology)

General Aspects of Value Theory
Value Theory in Economics
Integrated Universal Value theory
Axiology and Creativity

General Aspects of Value Theory

A proper value theory, axiology, as it is called in the US, has to have the objective of the "unitas multiplex," which implies the unification of value systems in several multitudes. This means an infinite number of individual value systems should formulate a single rugged order (W.H. Werkmeister, *Historical Spectrum of Value Theories*, 1970).

Value emerges communally through social consensus. Social utility has a similar process of generation. But the value standard is floating at either the level of individual or communal state. Value and utility value are based on the usage and the feeling of comfort and discomfort when we use an object. The deepest measure of value comes from the individual's infinities-matrix.

As acknowledged, US philosophy pursued pragmatic value. We have left a question on whether a pragmatic and the entangled relative values can bear an absolute value system or not. Absoluteness is based on the universal validity, which is connected to the axes of unity-diversity, subjectivity-objectivity, and the principle of equity. Logical positivism of our day does not allow absolute value, but a quasi-universal value system can be suggested based on the scientific and rational fundamentals which are derived from the substantial experience and experiment on the science of humanity.

A quasi-universal value system enables us to provide the value foundations for public law, which is deemed universal.

Value Theory in Economics

It is the sincere opinion of the author that economics has given up on the value theory except for some people who support a neo-Marxist view. In order to provide the fundamental of continuum economics that we are dealing in this book, we have to develop a sound value system, which is originated from the relative and dynamic value theory.

We have a unique value of measurement for a market price of a specific condition. It needs to be based on the psychological mechanism that exists in markets and society. However, classical economics and neo-classical economics have a fallacy in their value issues.

The sociopsychological mechanism is deeply related to the virtual reality of monetary and financial institutions. The excessive discrepancy between market continuum and monetary/financial institution stem from the deceptive property of virtual reality. The reduction of discrepancy and lubrication of discordance could be a major target of policy making depending on whom and which political party are the policy makers. It could be also said that discrepancy and discordance due to virtual reality can be used for the purpose of inducing time-difference effect.

Luke wrote a parable about some laborers who were invited to a vineyard. All of them were paid evenly with one man-day wage no matter when they started their day works. During the days of Luke, a daily wage was one denarius. If we consider a person cannot live fasting for 10 days in general, one denarius could simply be worth one tenth of a man's life.

It is thought that total exchanged value in a market can be interpreted to be the equivalent of the absolute quantity of psychological stress during certain periods in the market, which in turn, is partially reflected in the domain of monetary currency of the market. Emotional desire to exchange various commodities and services is the source of socio-psychological stress in a market. Such desire and stress are the major factors that maintain the level of exchange flow.

As the author understands, Adam Smith and Ricardo acknowledged that labor is the source of value. Marx recognized that labor is the substance of value. It is the author's view that physical labor is the transporting medium that at the same time transforms intrinsic psychological value into extrinsic utility value. Physical pain is the somatic measure of corresponding emotional and psychological output of a human. Karl Marx thought that he discovered the relation between social exchange value and monetary value. Wage in kind has various patterns as does barter exchange.

A monetary exchange is the socially agreed pattern of value exchange of a specific society. It is the major question of value theory to ask what makes people feel the equivalency of economic value and to assure subjectively the equally changeable utility value. Money is the matured social expression and labor is the somatic expression of the socially tuned activity that is supported by postefference of subjective transciousness. Figure 8-3-1 shown is the fixed domain of the monetary institution due to embedment into materiality. Monetary domain is controllable in quantity.

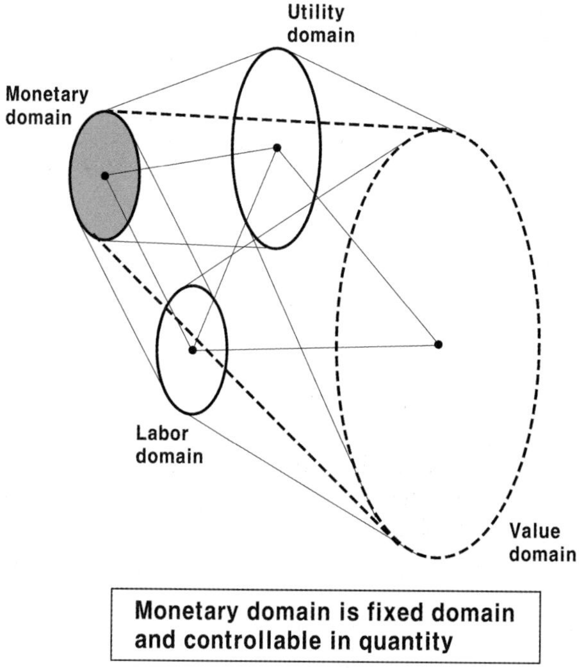

Figure 8-3-1. Fixed Domain of the Monetary Institution Due to Embedment into Materiality. (Monetary domain is controllable in quantity)

Social tuning process of value is done by the dynamic modularity of value menesys. Areal and hierarchical development of dynamic modules provides the explanation of how a society tunes, synchronizes and compounds its value and utility. Figure 4-4-3 shows the influence of lower level modules to higher hierarchical level of dynamic module.

Labor is the extrinsic allotment of the intrinsic social stress or market stress. Labor value gets a good economic definition when labor is compensated by wage in kind or in cash. Human being does not work without stress of hunger, even in collecting economy. People have different sorts and changing levels of stress due to the needs for hunger, clothes, and habitat, and other needs depending on culture. Stress is compensated by wage. Laborers must be compensated up to the amount by which he or she does not feel stress any more. The level of stress compensated by daily wage would be the minimum of what people need in their lives.

But human beings want always more than the minimum because of the transcendental intentionality. Transcendental desire is the prime criminal that organizes the resources to attain surplus value. Without the prime criminal, the human race could not have achieved the present civilization. Why do we have to concentrate for the production of surplus value? It is said generally due to avarice and greed. Is it all that is implied by the human activities?

The efficiency or the productivity of an economic continuum is assessed by the degree of satisfaction of its economic constituents. The degree of satisfaction is rated by the presence of surplus value which can be distributed to the contributors of production activities. Surplus value means the amount of additional value over the minimum daily requirement or need, which is equivalent to the fair daily wage.

The domain of monetary currency functions to allocate and distribute the portion of surplus value to the contributors of production. Generally speaking, the emergence of surplus value in human history is caused by the distinctive rise of creativity value in human economy, which is transformed to the market continuum. Theoretically, daily surplus value is the daily produced value that exceeds the daily consumed utility of physical laborers, according to the shares of contribution of the factors of production; capital, rent, technology, design, and so forth.

Think for a moment of an economic society, the scope of which is equivalent to its corresponding market, produces its maximum economic value by the contribution of only communal value and terrenity value in its highest efficiency, the needs that the communal character of the society demand is satisfied with the produced (utility) value at the moment. Communal value does not contain the presently effective creativity value. Creativity value is effective when the desire of the society is vitally activated in order to attain necessary additional surplus value.

Therefore, surplus value is regarded as following;

Surplus value = Total value produced from the entire economic continuum
– The value produced by the contribution of
present communal value and effective terrenity value
.... (Equation 8-3-1)

, which means that the routine production of the society is equivalent to the value produced by the contribution of present communal value and effective terrenity value and comprises with the production to maintain the activities in present society without any creativity. Thus,

Surplus value = The value produced by present effective creativity value
...... (Equation 8-3-2).

Value theory is deeply related with monetary presentation of value. Monetary domain has a comparatively fixed emotional basis when the gold standard is established. Such an emotional basis has both endogenous and exogenous causes; from inside, abstractive thinking of symbolism, and from outside, comparison of needs from struggling market.

Surplus value is generally pursued partially by the fear of destruction (eventually hunger and death) and in major by the imaginary drive for the fixed objective that is named "ideal." An ideal is not a balancing factor but pertinacious unbalancing pursuit toward an engraved target in human mind. Monetary domain possesses such psychological scars in it.

Value theory can only be explained fully by the support of psychological reasons. In the monetary domain usually exists are the reliant expectation to the absolute mintage power and the psychological portion of epidemic inflation that is only illusive expansion of value. These also provide the fluctuation factors that make value theory inexplicable.

In Figure 8-3-1 presented is the fixed domain of monetary institution among the domains of other economic values. Money is imaginary but fixed due to the embedment of its quantification process into materiality. Monetary domain is controllable in quantity because of the embedment into materiality, such as pain, sweat, labor, and gold and so forth.

Integrated Universal Value theory

When we deal with the commodity market, information network and power regimes altogether, it is thought that integrated universal value theory is necessary. The author conceives that integrated universal value theory is will provide the foundation for the forthcoming universal public law

The integration of various values can be achieved by examining the relation between value cognition and value usage. In doing this, fundamental knowledge about the intrinsic mechanism of humanity from individual physicality to communal mentality should be required.

The science of humanity will have to secure intellectual universality through pursuing novel comprehensive interpretation, which is firmly founded on scientific and rational knowledge excavated from concrete experience and experimental knowledge.

It is thought that integrated universal value theory is desirable to start from the permunal emotion-volitional value system. Kurt Lewin (1890-1947) proposed an abstruse but well-defined topological psychology, which deals with field theory and tension systems, dynamics of tension system, valence (誘發性, inductivity due to tension or stress), and so on. It is thought that his research presented a methodology to develop a systematic approach toward the quantitative technology as far as the philosophical stand point is rearranged to meet with Gestalt psychology.

It could be possible to organize complex psychological system, which is basically concerned with pleasantness-unpleasantness, in the perspective of market economy. The triradixism, the value system which will be dealt in the next section, could be considered as a solution candidate for such aforementioned frame of thought. Triradixism has a distinctive characteristic in terms of creativity, which we have to live together in concurrent days but has not been explained deeply enough.

Axiology and Creativity

The reason that existing axiology was not able to include the idea of creativity was the intangibility and undulatory fluctuation of creativity. However we will see how creativity value is exposed to the composition of land value and land price in a materialized form. Therefore it is necessary for the frame of axiology to deal with the value of creativity.

Creative activity is generally thought to occur in the process of variation in environmental conditions. But it is noteworthy that there should be an invariable internal condition of human cognition to utilize variable external conditions. The invariable part of cognition is found mostly in the region of transciousness, because transciousness is the part of humanity which sustains the orientation of creativity on an individual level. At a social level, sustainable orientation of social value system also regulates the orientation of creativity by encouraging or discouraging the direction of creative activities through the market pricing mechanism.

Triradixism will be expressed in the next section where will present the relevance between axiology and creativity.

8-4. Triradixism

Three Roots of Values (Triradixes) and Advancement of Society
Value Structure in Triradixism
The Value of Creativity in Market Domain
The Value of Labor in Communal Domain
The Value of Terrenity in Land Domain
The Measurement of Value and the Social Advancement
Mechanism of the Exchange and Transference of Values

Triradixism is a way that can stimulate an innovative advancement within society. The value system of a person generates a differentiated future for the person in accordance with the content of the value system. It is not easy to clearly show the basal strata in persuasive rationale in how the contents of how value systems interact mutually and are correlated.

Three Roots of Values (Triradixes) and Advancement of Society

It is the author's viewpoint that the value system of humanity must be reorganized at this inflective moment of the human race. The possibility of integrated value and new dimensional introduction is of increasing importance that the novel values of this day have to be added.

The dimensions of value made no trouble before the 20th century. We have been all satisfied in our economic forecast with a single value dimension of labor in economic theory. Lately, we have experienced the significant intrusion of the value of creativity into our society by technological innovation and its associated trends.

Social advancement cannot be defined only by a perceptive social phenomenon because it happens rather at the mental level. It is true that we recognize social advancement when there is an obvious occasion that represents such advancement. However, social advancement stems from the accumulated value of creativity that is produced by the creative organismic ability (創造的 有機能) that originated from the communal mentality (=trust-commitment-faithfulness=信賴-獻身-誠實).

Though an internal mental phenomenon presents itself even beyond our cognition, social advancement should be recognized as the mental phenomenon at the fundamental level. We believe that the 'overflowing model' is appropriate to describe the social advancement and change.

Table 8-4-1. The characteristics of societies in different advancement level

Level of advancement	*Cognition Characteristic*	*Needs Characteristic*	*Mental Ability*	*Lifestyle*	*Level of Region*	*Economy*
Hunting-Gathering	Simple Cognition	Appetite	Recognition Ability	Nomadic	Caves	Natural Economy
Ranching & Farming	Systemization	Stability	Planning Ability	Settlement (Settling)	Settled Area	Ancient & Feudal Economy
Commercial society	Flexibility	Accumulation	Linking Ability	Independent	City	Free Economy
Industrial society	Structuralization	Abundance	Synthesizing Ability	Consuming	Nation	Mixed Economy
Information society	Bird-eye view	Freedom	Creativity	Individual	Global World	Knowledge Economy

Accumulation of wealth and value is prerequisite for the social advancement. Such accumulation happens in the everyday economy, and when the accumulation reaches its critical amount, – thus "overflows" – human cognition, needs, mental capacity, and lifestyle change depend on the level of social development. The characteristics depending on social development are listed in the Table 8-4-1.

In the hunter-gatherer economy, the recognition ability, along with memory and judgment, must have been the critical ability that developed in the course of nomadic life led by appetites.

Later, as life stock farming and agriculture developed, there existed leftovers and tools, which developed the concept of possession and systemization. In this era, the needs for settlement arisen and the planning ability enhanced, which together led to stabilized settlement in one area.

In the commercial society, flexible cognition ability is required to recognize benefit and loss and to act upon the recognition. In this society, the linking ability is crucial for commerce because the difference in potentials between regions or groups is the key in commerce. Moreover, independence and city life are the basis for commerce.

Skinner approved of learning to be completed only when a desired response continued even after the reinforcing stimuli is dropped. In other words, a specific act establishes in an organic body due to sufficient stimuli.

Communities and commune-morph continuously evolve their characteristics through autogenously learning mechanism. Then, how are the "characteristics (體質)" determined? In my perspective, the characteristics are the unique and typical pattern of contemporary life that is developed through the interaction between social mentality and the circumstances of the times.

The source of change in the unique pattern of life is the alteration in "cognition and needs." Then, what is the fundamental cause of the change in cognition and needs? It is the accumulated goods and possessed wealth. In fact, the change is initially observed from a unique opinion leading class.

Among the common people and the public, the ability to recognize and scrutinize the unique resources that appeared in certain regions – the rich, the scholaristic, and the poor-are aspects that stand out.

For instance, the reason for the development of the free manufacturing industry in the 6th century is that there existed the strong need of consumers to purchase manufactured goods. Thus, the ability to develop manufactured goods enhanced dramatically. The consumers in those days were bishops in churches, feudal lords and their families, and knights.

The luxury life style of the boosted industries; those industries that took their place in cities where transportation and trade were flourishing and soon after that, consumers' demands exceeded the supply that led to the rise in free manufacture.

The needs that appeared in extreme poverty were highly related to survival. Such phenomenon can be detected from the fact that the tremendous growth in the agricultural population in the 11th century Medieval Europe led to urban migration for the people with no land.

For people who could sustain their survival, they could gradually accumulate wealth and become merchants. The merchants soon developed themselves into the earliest bourgeoisie and brought the unique characteristics for the medieval cities.

Adjusting such lifestyle throughout the mercantile cities in the Mediterranean and in the north Europe was indispensible in the progress; we would like to refer to such lifestyle by a communal stratum or a social "characteristics".

Social advancement is achieved when a new lifestyle is accepted by the source group then spread to other communities and commune-morph, becoming a characteristic through a thorough learning progress, and building up the new creativities.

In order to achieve such organismic ability (有機能), communal unity must be strengthened over minimum critical condition. The unity cannot be dealt separately from communal mentality that enables communal behaviors in a community.

Industrial development in a commercial society relies on the structured self-cognition. In other words, a person in society should understand his position within the social structure.

Thus, a capitalist in an industrial society would not be seen as a person who created a structure, but as someone who utilized his synthesizing ability to put together previously sporadic and dispersed stables of society.

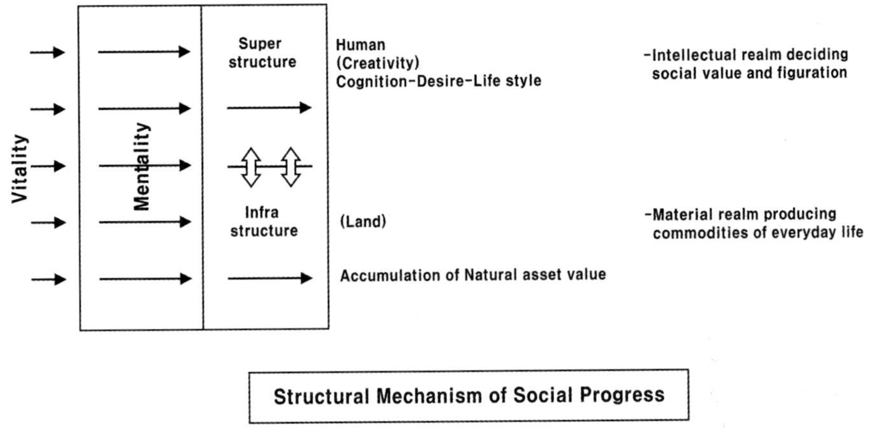

Figure 8-4-1. The Mechanism in the Social Structural Advancement

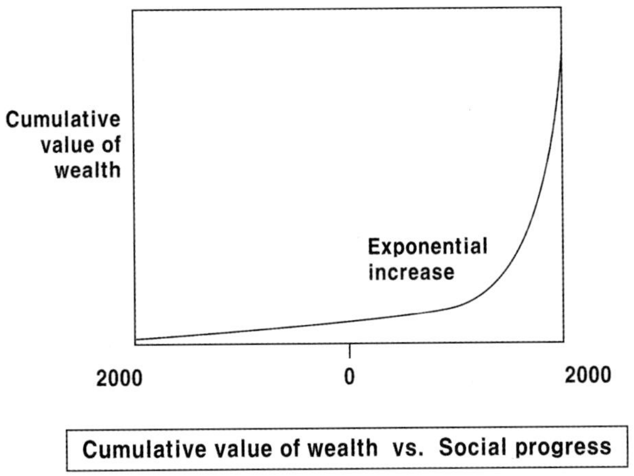

Figure 8-4-2. Cumulative Value of Wealth vs. Social Progress in Human History (Cumulative wealth in history is actually cumulative value of creativity.)

Society is composed of infrastructure and superstructure. The infrastructure is the physical domain where actual production and social relationships are made. The superstructure is the intellectual domain where social values are determined and implied. This notion is presented in Figure 8-4-1.

The changes in cognition-needs-lifestyle occur with the superstructure acting as the core and the infrastructure providing stimulus to the superstructure. Nonetheless, the infrastructure is also influenced by the value system-pattern behavior. The infrastructure is the domain where land and nature become monetary value and social utility because of inputted labor.

The social value system is not a fixed system. Though the macroscopic direction of society, which is directed to develop a communal society due to the fundamental needs of humanity, stays unchanged, the microscopic characteristics of value system change as the society evolves within the infrastructure.

The superstructure can be perceived as the "domain of creativity" generated from human knowledge and wisdom. The infrastructure is the "domain of nature (land, or terrenity)" that includes land, natural resources and physical human labor. Categorizing physical labor or the natural part of labor (compared with human's creative part) as the nature (land) looks reasonable if observed from the current economy, which is the most distinctive in creativity but still not dividing labor into physical and mental labor in its economics.

In the millennia of human history, cumulative wealth is actually cumulative value of creativity as presented in Figure 8-4-2. Accumulation of the value of creativity is proportional to the resultant peaceful wealth accumulation, provided that wealth is not stolen of extracted. In general, wealth is accumulated and overflows the container of outward appearances such as lifestyle and social structure sequentially.

What is the fundamental reason explaining why human labor induces isolation? Is it because of the low wages or the deprivation of opportunity of "creative labor" which comes from innate needs? Though the low wages and human mechanization would all be part of the problem, the real problem lies in the depriving characteristics of the industrial society that squeezes labor past the adequate level. If we keep ourselves within the confined value theory of uni-radixism (which is labor value only in these days), we will not be able to explain the complex advancement of human society.

Value Structure in Triradixism

There is an anecdote about the value of economy. A pupil of J. M. Keynes asked, "How come Keynes's basic theory does not have a value system?" as an answer, Keynes stated that, "Currently, the only useful value system is a labor value system. However, unfortunately, this one also lost its credibility."

On the other hand, the system of a socialistic political-economic study stated its view as follows. "Smith's assertion of the factor that increases the quality and quantity of labor is up to the nature of human beings lacks validity. Rather than examining human being's nature, we should concentrate on the findings of independent economic structures in regards to the individuals as the ones needed by that economic structure to fulfill certain duties."

Nevertheless, a notable theorist of capitalism George Guilder asserted that capitalism is highly psychological and considered to be within an intellectual area. Wealth is tamed by one's mind and exists in materials. By saying so, he made it clear that the bond between human being's nature and economic value is inseparable.

At this moment, we are discussing economic value. But in the reality, the notion of value (or pretium in Chapter 9) that is being dealt in the field of economics is market value. Let's say that a book is about to be published.

In the initial stage, thanks to the value system, that book would be purchased because of its market value, most of which is new creativity value. But, soon after, those books that weren't sold will be stored in the garage and thrown away eventually. The image of market, which is reflected in people's mind as utility, will rapidly disappear right after its emergence.

The reason why we failed to find an adequate methodology of how to theorize the intangible value is due to our stereotype. We believed that there is nothing other than the "price form" which can be calculated into a currency for realizing the notion of "value." A market price is a measure of value that has been implemented by the active social participants in the market economy. Capitalism is said to base its logical path in the market economy. However when Guilder mentioned that it actually is in the psychological and intellectual area, he was implying it should be free from the boundary of the market economy. A capital is a source of profits and also a source for a continuous capital accumulation. If we follow this possibly-clear logical link, there is a considerably flexible historical interpretation to set the precondition of an initial capital accumulation.

To analyze the condition that enabled the primary capital accumulation,

(1) A rapid expansion of the trade and the scale of commerce.
(2) A prioritization in an industrial sector.
(3) Enclosure movement
(4) An incredible surge of the price

Other factors that weren't mentioned above are morally unjustifiable only to be undermined in general. For example, those include the piracy, slave trading, and the pillage of colonies.

If we narrow our attention only on the history of mankind and the way to analyze that process, we will make the silly mistake to solely concentrate on social value, hopefully, to be defined as the market value. In other words, we will ultimately ignore the natural value and communal value.

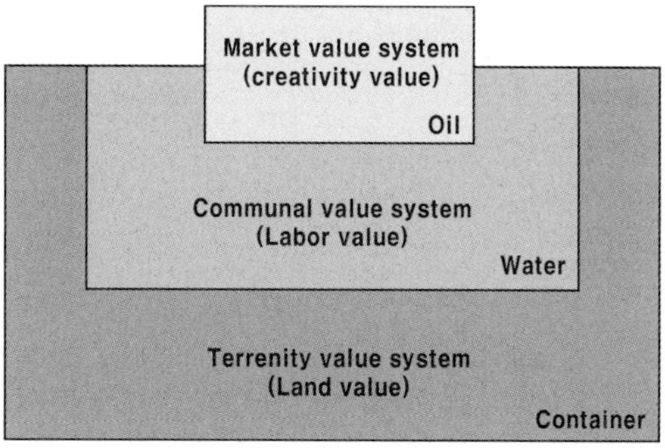

Figure 8-4-3. Three Layered Value System

When we deal with the value theory, "Three layered value system (三重價値體系)" is required, currently; our value system only touches the market value domain. Three layered value system is shown in Figure 8-4-3. However, it is necessary for us to recognize that the natural value system (terrenity value system) embraces the communal value system. Moreover, we should realize is the communal value system (a social system), a huge shelf that supports the market value system. Genuinely, the market value is equal to the communal value. The communal value system latently exists as the communal property within the social community.

On the other hand, it can be said that the market value system comes to the surface as a recognizable form due to the price system. Therefore, the market value can be clearly described when interrelated with every external economic activity.

When human beings have certain desires, these desires provoke creativity which will then lead the productive labor forces. Then, the productive labor force transforms the resources (terrenity value) into the products (communal value) or the commodities (market value).

If they are produced through the labor, a produced value is usually introduced into the domain of communal value, and it tends to possess market value when the concerned market is in need of such commodities. (It is suggested to imagine the upper part of market domain that is beyond the communal value domain. The upper part of market domain emerges due to the psychological instability that caused the abnormal expectation.)

It is true that everyone knows the reason for the difference in working efficiency is due to the difference in tools and technological methods. However, rather than viewing this as an obvious fact, it is the difference in accumulated creativity that explains the difference in working efficiency. Realistically, in defining the term "labor" for humans, the only part that is qualified to be a scientific standard is "creativity."

During the advent of agriculture, the effort of digging a small hole with a finger on the wet ground and planting seeds is wholly the effort of creativity. It is what immensely augmented the economic power of the mankind.

Creativity is applied when the terrenity value is turned into communal value. Also, creativity can exist in various forms, such as education, technology and innovation in reality. In fact, all cultures and customs that we unconsciously view and practice, consists of parts of creativity.

Most creativity, which was manifested after the start of human history, are accumulated and connoted in the communal value system. The communal value is just the accumulation of creativity made throughout the history of mankind.

Since the modern civilization severely ignored the terrenity value (natural value) and only developed the human-centric lifestyles, we are confronted with the retaliation of nature, which is the disturbance on rent as seen in land pricing and environmental problems. For the communal value, modern economy has placed such extreme emphasis solely on the market value that the communal destruction may have been induced as seen in the collapse of families and the negligence of communal morality. Market value (acting creativity value) must be strengthened and matured through improving structural support of terrenity value and communal value.

Recently, as people started to focus their attention on the ecosystem, the regenerative power of nature and land or the recirculation cycle has become a hot issue. This is called the static and existential value that the nature or the land possesses toward the society. Moreover, there is another notion to be called as the dynamic value of the land that is related to the communal utility of the nature.

The notion of dynamic value is connected to the soundness of expected value. An expected value is an important factor to trigger the healthy motive of human beings to the extent that unfair infringements of human greed do not occur in the future. (Such infringements are usually caused by the unjust mentality of humans that raises greedy speculation for unfair interest and avaricious investment.)

In my view, communal value can also be categorized into the static value and the dynamic value. Communal dynamic value represents the flow of economic exchange without commercial transaction in a community while market dynamic value exists within commercial transaction that is subject to the flow of economic compensation. Dynamic flow occurs due to the potential difference between static values, which is the apparent needs of the constituents of a community. But the apparent need in communal value system is satisfied without any correspondent compensation. In other words, dynamic communal value is defined to be the communal value when the exchangeable utility value in a market is produced in a community or in a communal society being isolated from the market that is correspondent to the society or community

Communal static value has two categories; existential utility value and generalized utility value (or communal characteristic value). "Existential" is used to mean the innate characteristic value of a community that is engraved in DNA level such as motherhood, heterosexual reproduction, physical peculiarity of human race, etc... when a man does not work and enjoy, people feel animosity toward him attributable to the innate communal static value.

"Generalized" is used to express the generalized characteristics of community created from a community's overall constitution. Much of the abandoned utility value of a market goes through the process of becoming generalized into characters of the community and then transforms into a communal value.

In accordance with the above idea, we can illustrate as following; excluding the abandoned utility value within a market value system from the total social utility value equals the total market value.

[Total social utility value] – [Abandoned effective value within the market value system] = [Total social market value] .... (Eq. 8-4-1)

The commerce is known to increase the welfare of all these participants according to the studies of general economy. Nevertheless as for why both participants of commerce receive supplementary value beyond the actual value of what is being commerce is not mentioned anywhere. Paul Samuelson's cross question, on what we can produce isolating from the technical knowledge inherited from the society's past, is the proof of what he acknowledges the existence of a communal value system.

Because the communal value system always exists profoundly under a market value system, forms such as technology, information and substance and so on – which is fixed as the generalized characteristics of a community – are indeed the social surplus values in another form though they are seen nothing more than abandoned utility from the market perspective.

In discussion of antiques and land, Keynes called antiquing "the sewage of purchasing power," which should have been related to the market value system, yet it is drained into the communal value system due to the unhealthy expectation misled during a speculation to purchase a property belonging to communal value system such as antiques.

For instance, if pottery was imported from a region well known for its great pottery, then that pottery possesses not only the market value expressed in its market price, but also accompanies the additional communal value that comes from communal value system. Thus, that pottery enjoys effects and benefits beyond its sales price.

This is similar for clothing as well. If two different regions, one famous for its clothing and the other for its pottery, traded their clothing and pottery, they would have not only sold but also exchanged their each accumulated community value. In other words, these two regions have enjoyed their mutual benefits through trade by the additional amount of their accumulated communal values. The enjoyed communal values are equivalent to the accumulated creativity value introduced into each communal value system in their past.

In addition, when the two regions exchange their accumulated creativity (i.e. community value) through the act of buying and selling, the market activity is manifested by trade quantity depending on the exchange price. As time flows, it is a general trend to present the real exchange value of the equivalent market price and trade amount to be amplified as much as the accumulated communal value has been increased.

It is not thought that the abandoned utility value from the market weights equivalent to the total quantity of communal value. This is because there are massive communal values transferred to the communal value system which have not passed through the market. Intangible, spiritual, ethical, mentorship, communication for idea exchange, friendship, informal organization, free commodities like gift, relief works, communal collaboration, patriotic works, and so forth, are that sustain the humanity strong and healthy. In brief, the community value system is indeed of a vast extent.

Most of the creative effort, since the very beginning of mankind, is thought to be transformed into this communal value. Some of the communal values have been engraved into DNA system of human physicality. Other parts have consolidated to the culture and tradition and so on. The communal values were originally all creativity value in their initial stage.

It has been commented in previous pages that communal dynamic value represents the movement of value due to economic exchanges without commercial transactions in a community. The dynamic flow occurs due to the potential difference between static values, which is the apparent needs of the constituents of a community that can exist without any correspondent compensation. The dynamic communal value could be interpreted as the cause of economic koinonia (which is in fact the relation of interchanges or interflow of values) in a community or a society.

Communal dynamic value becomes to exist when utility value is recognized during a community's movement by the community being transferable to and from market but not subject to correspondent compensating. In short, it is the production value occurring in the family, church, and self-help organization and so on.

In discussing communal static value, it is asserted that the utility values traveling incognito in a hidden accumulation of the communal value in trading amplify the substantial exchange value of market trade.

In addition, communal dynamic value must amplify the market trade as well. However, trying to theorize this might be very difficult because communal dynamic value disguises itself under the market value system. Without the interflow of communal dynamic value working as the lubricant of market, market value system should need much higher cost to sustain the dynamicity. The low cost of sustaining market economy helped the efficiency for the socialist capital economy in mainland China or in other relatively new countries introduced to the global market continuum.

Only this could be said now; the measurement of communal value is manifested through the land price in a market.

Land value will be covered separately, but the locality, in the economic characters of land, is closely related to the degree of concentration of communal value, which is directly related to the economic efficiency and profitability in the modern society.

In other words, the land price is decided in accordance with the regional density of communal value. Unless the speculative and unhealthy socio-psychological expectation causes a distortion in land pricing, the land price in the market represents the proper measure of the communal value.

However, not only the total market value, the cultural value, the educational value, and so forth of the particular community, but also the abnormal psychology of expectation, the distortion effect of deeply indented traditional tendencies and the pursuit of non-economic symbolisms, which induce unnecessary struggles and competitions, are reflected in the land price of the community.

The land pricing system has to be uncorrupted and protected from intentional distortion of land market forces so that such characters of land price may reflect the proper community value to represent the socio-psychological response of the citizen's communal consensus. Therefore it is necessary to invent and design an adequate method to assess the land price at a particular time. Such assessment would be founded on a smoothing technique between the fluctuation of land prices on both a macro scale and micro scale, which can exclude the influences of excessive kicks up and down and derive a social consensus.

Communal economy is not visible in the institutionalized economy in which common people manage their economic life with their income in terms of currency, but barely detectable at the marginal boundary or the horizon of the institutionalized economy. In the short future, people will recognize that communal economy is apt to bring enormous influence on the social institutions.

The power of Twitter, the network of Facebook and the impact of internet communication in recent votes and politics are just the political preceding of what we understand to be comparable to the communal economy to be unveiled. But the communal economy has far more profound and varying patterns of manifestation from communal politics because it has greater complexity due to the monetary and financing institutions. Furthermore, communal economy seems relevant to the various underground economies, for which we have to make researches to establish a consistent methodology to interpret economic continuum.

Value of Creativity in Market Domain

Creativity in the light of science and economy is the ability to provide solutions to newly confronted problems and to suggest fresh approaches to familiar problems. Creativity in light of life style and culture, on the other hand, is the ability to create arts and expressions in novel, impressive forms. In making research an individual creative process; first, creative person have constitutional elements of higher level of motivation and energy; second, the tolerance against external impulses increases within the ego-psychology, for instance preventing scientists from getting their concentration dispersed; third, the unconscious discord, similar to the neurotic schizophrenia, seems universally supporting the role of creativity.

Especially the second part enables individuals to activate 'the liquidity under suppression,' by which I mean the ego I allowed to contact with the unconscious instinctive elements within its personhood without being subjugated by them. This seems fundamentally to be related to the unconscious discord in the third part.

Creativity is a phenomenon that cannot exist nor be defined individually without the existence of community. The total market value subtracted from the total communal (or social) utility leaves what is called consumer surplus utility, a concept rather obscure and difficult to clearly define, when approached from individualistic viewpoint; creativity, which is the most important element in the thinking structure of capitalistic viewpoint, should be also assessed as a blind ability without logical foundation in individualistic interpretation.

Creativity is classified into two kinds. First, there is the creativity concerned with innovation, invention and information industry, which can be interpreted as a system-extending ability and behaves relatively more volatile in an economy. Volatility (蒸發性) is the counter measure whether the outcomes of creativity that is introduced into the economy can be spread and accommodated as the characteristics of that community. Highly volatile creativity leaves not many material commodities in an economy and not much impact to the mental structure of the concerned communal society. More volatile creativity contributes in increasing the scope and frequency of activity of that society, while leaving comparatively less effect in maintaining the system structure and establishing the social characteristics.

On the other hand, in the domain of creativity value, human actions for the educational and technological accumulations were defined to have maintainability of social structure and thus explained to have greater condensability (凝縮性). Condensability is the opposite concept of volatility. However in general, volatility is more prevalent in the early part of creative activities while condensability is gradually activated in the latter part of creative process.

Evidently, the residual utility value of an invention falls down abruptly in the respect of the market value if a more innovative next invention comes after, whereas the decrease of the residual value of information under similar circumstances is not so drastic.

The educational and technological accumulation that is used for the reservation and transference of creativity value does not change significantly in terms of the social value of its knowledge content even though considerable time passes. Education and technology make information more systematic, whereas invention and information have more nexusity with their extending outreach to new fields.

However, when the content of new and unorganized information are introduced into our actual livelihood system, such information within the livelihood information system will lose its value faster than the information of technological accumulation; structural information supported by logic and tradition of a society would keep its utility longer than the individually produced information.

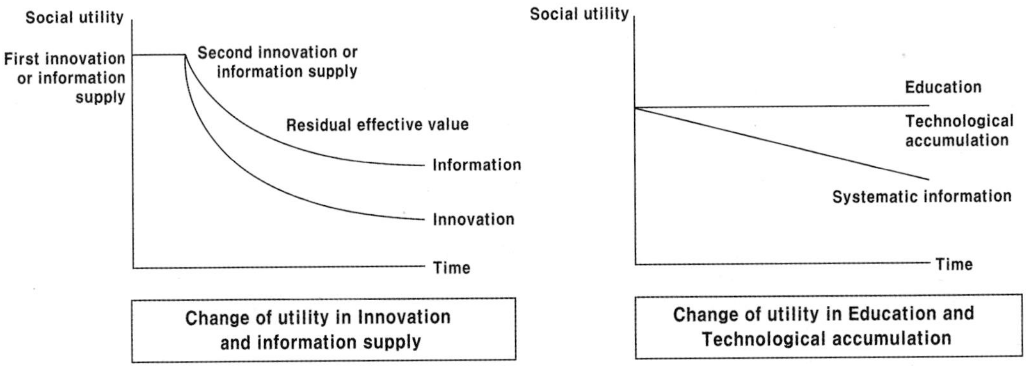

Figure 8-4-4. The Change in Utility of Education and Technology.

Education and technology have accumulated fundamental systematic knowledge over long periods of time, supported by which society has been able to select, systemize and utilize additionally greater amount of knowledge menesys. Individual characteristics in promoting invention and gathering information, which is usually apt to be fluctuating, have disappeared and it turns out to be more stable.

Unstable creativity, such as appeared in inventions and information, which goes through the inspiration process loses its market value as it is compared with successive inventions and information as time passes. However, if it gets absorbed into the structural creativity system and becomes a part of communal characteristic, it will be reflected as consumer surplus contained in total utility of the society, and as a result, an amplifying effect that provides greater utility than the fair market price in trade.

When the union of nature (land, resources, raw material included) and labor (actually a part of nature) is regarded as the hardware, and human and creativity as the software, all social-economic phenomena can be explained more simply and clearly than before.

If we consider the old concept of the four steps of education adopted in the mid-20th century that leads to the creative learning, we can see that it agrees quite well with the proposition that the ability to practice a creative talent is both produced and liberated by the skill of using the instruments of expression. Thus, the change of cognition-desire-life pattern throughout the historical development of humanity is almost essentially similar to the development of education through the steps of the skilled labor, technological labor, mental labor, and creative labor.

Therefore, when the degree of creativity contained in the general labor of entire society increases, the degree of creativity exposed in the content of education rises. Such advanced educational content is brought to humanity as the tool of expression, which will gradually enhance the creative activities to highly liberated and high-dimensional status.

Above is the logic that clarifies the relationship between education and creativity, therefore creativity settles down as technology and systematic expertise through a process of education (including study and training). But we easily forget that this kind of creativity is the product of the communal effort of a society that aims for the yield of communal wealth.

Education, study and training are actually kinds of labor that accumulate creativity value. The labor piling up creativity value is just an exposure of the development and heredity of communal characteristic, by which a communal society conserves, transmits, and accumulates the inherited creativity through intersubjective communal way of thinking.

Therefore the fruitful output of the creativity-accumulating labor should be used for the sake of the community in a way that the fruits is utilized to maintain the creativity-oriented-ness of the communal society; through such strategy that society can compensate with higher incentive for the creative individuals in order to be motivated to concentrate toward creativeness. However, the abuse of the communal wealth or common wealth that threatens the living of the physical laborers must generally be avoided.

The measurement of the market price on the value of creativity could be made based on the prices that appear on the economic structure, such as the expenses for education, information cost, etc.

In order to activate creativity, the creative organismic ability of the society must be achieved, and the most important factor that helps society achieve this goal is communication, which has a very intimate relationship with social education.

Therefore, learning within a community deepens and strengthens by an "equal stimulus for the communal expression." And this fact means that social learning and communication are interrelated inevitably. We cannot deny the fact that education and the accumulation of technique, which are types of creative activity, have played a very important role in the formation of the intersubjective thinking process of a community. That's why we can observe the improvement of koinonia flow in society as the contents and techniques of education are increasingly transferred between school and society. (Koinonia is defined here as the flow of economic value, but it has twenty-five different usages in definition at the times of Koine Greek.)

On the other hand, the negative effect of social education that awkwardly and excessively inclines to the systematic national purpose, creativity of the society was decreased. In order to accomplish a healthy society and/or nation we need not only to allow creative and free expression of individuals but also those of the community.

The way a community manifests its own creativity is selected based on their past experiences. The community naturally chooses and uses a way of reacting to its past experiences based on the knowledge system progressed by researches, debuted, technical fields, etc.

This kind of social or communal reaction or way of expression is deeply related to creativity, and an individual's creativity is intimately related to the way of expression of the community the individual is attached to.

For instance, the ability of linguistic symbolization is expressed by coinage, and this is deeply related to the creation of a new culture or the creativity of a community. Theoretical works that reveal this kind of phenomenon should go on.

The Value of Labor in the Communal Domain

Capital is considered to be frozen labor. "According to Marx's research, a capitalist needs to satisfy three conditions in order to obtain a larger amount of surplus value. Thus, the first is the reduction of wages, the second is the extension of working-hours, and the third is the increase of the price of products. These three conditions are not logical so we can prove this by using as an example in the current industry that has gained large amount of capital.

The principle of industrial economy of Ford's automobile industry and the logic of surplus value of Marx disagrees with each other. Thus when Marx supported the idea that a capitalist must extend working-hours, Ford's factory pursued the reduction of working-hours. When Marx supported the idea that a capitalist must reduce wages, Ford pursued the increase in wages. When Marx supported the idea that a capitalist must increase the products' prices, Ford's car industry pursued the opposite in lowering the product price."

This is quoted from the first lecture of Sun Yat-sen about 민생주의. Sun indicated the discordance between reality and Marx's surplus value theory which was developed from the Labor Value Theory, pointing out that there was something wrong with the productivity. Before we enter into the above argument, it is important to consider the values of creativity. However, if creativity is included to what we commonly call mental labor and we try to define that pure physical labor excludes creativity, the value of labor can be devaluated to the extent that we could neglect it.

Pope Pio 11th who follows the teachings of Quadragesimo Anno, defined labor as, "What is labor if it is not a way of applying man's mind and soul to the gifts of nature in order to develop human power?"

Although the classical concept of labor is vaguely defined, the following statement, "... in order to develop human power" clearly states the basic values of labor. Why does labor exist? When considering only the physical aspect of labor while excluding the psychological attributes should we think of what is needed. Is labor invaluable if we exclude the psychological aspect of it?

The classical concept of labor states that it will not be considered as labor if it is not related with production or the creation of wealth. Regarding this point, it should be thought about deeply. Where does the value of labor originate from and how can we measure it? There are three viewpoints that deal with the value of labor.

As we have dealt with the theory of values, the system has three factors; nature, community and the market. It is important to consider the following elements in each of them; in the value system of nature, the existential value of a labor; in the communal value system, the communal value of labor; and in the market system the labor value in the market.

First, the existential value of labor has two aspects. The first aspect relates to physical labor by which human existence – thus such biological traits of human labor such as sweat and sensation (painful) – allows us to consider the existential value of labor. This is also connected with the basic human feeling like patience which plays an indispensable role in the formation of the communal attitude of a human being. Perseverance is the foundation on which the maturity of individual level stands, as well as the relational base of association on the communal level. Therefore, through the perseverance of labor and the pain, humans are able to enter into the mutual process of growth and association.

The second aspect is to consider labor simply as a physical activity. In other words, it is to consider the existential value of labor to be meaningful even in the case of robots in labor, since it would be viewed in relation to the produced outcome, and not whether a human or robot is the laborer.

We can see the importance of labor when we consider labor to act as hardware while creativity plays the role of software. In order for any kind of commodity to be produced, the hardware called labor must be put into use.

Even if labor is simply defined in the form of force multiplied by time, the survival of mankind would still find its foundation in labor. To go one step further, the relation between labor and capital can be shown by the concept of "capital as frozen labor" to be utilized in the equation expressed as follows:

$Y = CF(x, z)$ ....(Equation 8-4-2)

, where y is capital,

x is labor,

and z is creativity.

C: the characteristic function of a society in accordance with the developmental stage of creativity and labor efficiency.

It means that capital can be seen as the frozen form of labor which is created by the creativity of humans that exponentially acts on the physical labor. Here, the effect of creativity is limited by the specific characteristic function of a society and this characteristic function reflects the social characteristics, which is related to the developmental stage of creativity and the productivity of labor.

$C = F(c, p)$ ....(Equation 8-4-3)

, where c is creativity expansion according to the degree of creative ability,

p is labor production or labor efficiency.

Secondly, we need to think about the communal value of labor. While labor can be considered to be directly related to the existence of humanity, as well as be the proof of such existence, it is also self-evident that labor within a community can lead to the formation of relationships between people. In other words, without physical labor or action, there is no such thing as human relationships. The thinking structure of *Homo sapiens* (where personism and communalism are reflected), which occurs in the place of communal thought– or intersubjectivity – is expressed through the medium of labor and action.

According to the rules of teaching by Pope Pio 11th people need labor in order to develop the potentiality of man. This means we must diversify, harmonize, and perfect the communality, which is the property of our species - the indigenous quality of mankind.

In reality, the development of cooperative work and specialization, along with the dramatic bifurcation of human activities, can be regarded as necessary steps towards the realization of communality. In other words, those intermediate steps are the stage of extrinsic differentiation that is not followed by the intrinsic harmony. Simply speaking, labor is the communal medium which connects relationships among people in a community, and it is an important symbol towards achieving inner harmony.

This is the reason why living on unearned income should be frowned upon within society. Here, we see that unearned income traces from "unearned expected value", and "unearned expected value" is the complete opposite concept of the value of labor.

Unearned expected value denotes the state of expectation regarding the earning of value without performing labor. To the economic world, it acts as a cancer which must be removed.

If we consider the communal value of labor from another perspective, we would have to discuss the existence of a capital that has communal characteristics like the indirect social capital. Because this capital can also be defined as frozen labor, we would be correct in calling it capital.

Normally within an undiversified (or undifferentiated) communal society in ancient days, communal facilities such as the indirect social capital are usually created through the volunteering efforts of a community. Bridges, roads, and buildings for communal use are a few products of such efforts.

After modernization, the capital within market system began to be invested for the communal indirect social capital. The government has also taken a significant role in the capitalistic system. These phenomena, along with the trend to maintain corrupt relations between politics and business in society, have led to the disappearance of communality in the field of indirect social capital, and instead caused the emergence of a new business field for capital investors.

Railroad bridges, dams, electricity, and communication services are just some of the examples. However, if there were no huge capital investors or no desire to accumulate capital, then the currently-existing indirect social capital would not have come into existence. On the other hand, we should also consider whether or not the current rapid development of society is the best path for the human race.

There is an overlapped part in economy between the public sector supported by government or state enterprises and private sector of huge enterprises. Within the modern economy, this overlapping phenomenon has extremely important indications, which simultaneously show the modern society's potential for growth and decline.

Whatever the case, the capital management that is applied to the indirect social capital and the public sector must create and apply a new methodology which can be treated differently from the market economic system.

This is one of the questions arisen from the communal concept of labor. Maybe the rent, which is the passage in which the product of human labor is absorbed to the communal economy, must be used as the fund for the public sector. A more elaborate and thorough discussion is needed in order to clarify that.

Thirdly, I would like to talk about the value of labor in the market system, meaning it should be understood in the light of the whole value system that includes the nature value and the communal value.

As we all believe so, the market value system is the bud that creates all the social values and the melting pot that diverges to new values.

Just like there is an active proliferation of cells at the growing points of the plants, we are faced with a new phenomenon in which nature values are rapidly changing into social values, within a market system through a mechanism called "competition," in order to increase the amount of new values, making new norms for social values that originate from people's cravings and change of cognition and ultimately create a new society.

In the market, the expected values are brisk. This expected value can be divided into the value that fairly has "labor" as its precondition and the other value that unfairly seeks for an unearned income. It thus is the fundamental turbine that enables the market system to exist.

Unearned expected value produces the immature human nature and germinates the negatively anti-communal spirit, such as cruelty, corruption, and falsity to make the society deprived of spiritual passion and thus ruins the society.

Unearned expected value is totally opposite to the labor value and this should be thoroughly regulated. That is, "The one, who is not working, should not eat." Moreover, if something bureaucratic and mammoth takes place within the market system, there soon will be the demise of spirits in the market.

The vital power of a market is of a small scale like the germination state of a plant and it is full of energy and desires. However, huge organizations, such as gigantic enterprises and government enterprises, which are aware of their authority, use any means possible to destroy and suffocate this pure vital power of a market without any guilt.

There would be other social usages of large enterprises, but I believe that it is crucial for this generation to cope with the non-communality.

The discussion on the value of labor gives us an opportunity to think about some process: Humans are created to provide labor, labor allows the enlightenment of humanity to take place and thus permits people to be humans, and ultimately, labor contributes to the soundness of humans' creativity.

The issue behind human alienation has its solution in creativity. The problems related to human alienation are solved if the existing value of labor is fully acknowledged and practiced since creativity will definitely revive.

The Value of Terrenity in Land Domain

In value pricing, the implication of land value is extremely important. It is mainly because the land market price provides the way to measure the health of community. In most economic situations, the communal value is included in the land price. This is due to the psychological projection of the entire value system of humanity to the land as a market commodity where a community resides. The market land price of community reflects the structure of value systems of community as its components.

The components of land price are:
First, the monetary value derived from the natural yield of the land;
Second, the communal value created from the economic efficiency from cooperation in community; and
Third, the bubbled up value derived from the scarcity of land and the over-expectation in a society where the land has become a commodity.

The above three components are the constituent price factors in Triradix diagram. Presented below in Figure 8-4-5 is the illustration of political economics in matching relationship with the constituent price factors in Triradix diagram.

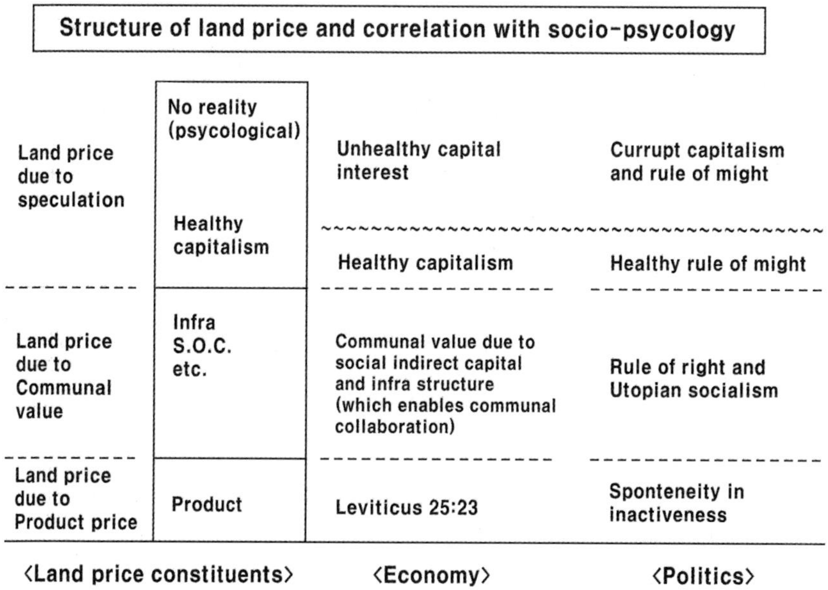

Figure 8-4-5. Illustration of Political Economics in Matching Relationship with the Constituent Price Factors in Triradix Diagram

It is observed that there exists a corresponding framework between economy and political regime. It is realized that the adequate level of competition in economy and politics is essential for the developmental growth of society and market. Even the rule of virtue (or ancient kingcraft in the Fareast) has to maintain the rule of mighty

The Measurement of Value and Social Advancement

What is value measurement? Value measurement is the process that explains the relationship between the objective analysis of a society and quantitative analysis of value that is decided subjectively by an individual.

The pricing mechanism in the market is the only notable and widely accepted mechanism that makes value measurement possible, and is the process involved in obtaining an objective equilibrium by combining with subjectivity. The pricing mechanism has to be based on a market economy because the market is the most natural consequence of anthropomorphosis that spontaneously introduces the integrismic and cruxological measures into human society.

The pricing system in the market is the only way to determine value, however, because of the market's volatility and its diversity in varying situations and conditions, the market price is set according to the principle of probability. This means that the pricing system can be explained by probability distribution and should be described as a characteristic of the principle of probability.

To measure the market price so that it becomes the representative value, there needs to be a certain amount of contacts and transactions, in the most active part in the market according to the law of probability, beforehand.

Anomalous prices exist in the periphery of a market, and this is the reason why social injustices sprout in our society. Active civil society that allows transaction or sufficient frequency of affairs alleviates social injustices and unfair trade on the internet in theoretical terms.

If value measurement is only dealt in the market value system, it would fail to bear the historical transitions in the future. Value measurement can emerge as a very sensitive realistic issue and the underground economy could be one of its examples.

This is because the underground economy in a socialistic country creates individualistic black markets, and underground economy in a capitalistic country creates communal volunteering commercial transactions or labor trade markets (this is limited to underground economy in liberalistic capitalism that has no economic restriction. Therefore, examples such as illegal trade in foreign currency need another set of considerations.)

Value measurement needs a common standard that could be used in all three types of value systems; market value system, communal value system and territorial (=nature) value system. In other words, among the three value systems exists mutual exchange in value. Ultimately, territorial value, creativity value or utility value that all human favors will combine to form one big pool of communal value system, and this sum of potential value will play a big role in the lifting up of human's creative values.

Therefore, if you logically connect the relationship between the increasing amount and the total sum of creative values and the communal value system related to the creation process of creative values in the market value system, you will be able to construct a basic picture for value measurement. However, variations in the development process of creativity in different generations result in complicated measurement procedures and doubts about the practicality of the standard.

As a result, the easiest way to have access to value measurement is to draw a conclusion on the relationship between the market price and the creative value and understand the utility value (which amounts due to the practice of commmunal value system that takes volatility and adherence into consideration) according to the current market price standard.

Coming up with the methodology of value measurement is especially crucial in the state in which the importance of the value system is increasing.

Despite the difficulty in finding an answer, I now would like to deal with the problem regarding the measurement of advancement.

What is advancement? Advancement in its essence is a comparative term due to its association with intentional orientation. However, according to the existentialism, it defines advancement from its relationship of the portion and the whole, so the standard of measurement does not need to change if we view the term as a partial intention that aspires the completeness of a whole. This will probably be a nondimensional indicative coefficient.

It will be deemed rational, realistically, if we adopt a healthy communality as the measuring standard for advancement. Communality should be expressed as the degree of integration and the degree of freedom, which are mutually conflicting. The situation that has the highest indicative coefficient of advancement would be when adequate cohesion and freedom is allowed to any phase of a community.

The adequate integration and freedom will in turn create the most active and creative value. This integrated unity and diversifying freedom will maintain the amount of physical distribution and information at an adequate level and this will allow the peaceful development of a society.

This indicative coefficient will include the advancement in technology and society, but it could also be indicated as the advancement in the individual and society. This coefficient indicator will also help interpret the reasonable balance among the value systems that one specific society maintains.

This will be, from the socio-technological viewpoint, a very basic indicative coefficient foe designing. This indicative coefficient will allow us to judge the relationship of the amount of energy and information used in a society or community.

Mechanism of the Exchange and Transference of Values

The transference and exchange of values between domains and systems are relevant to cognitive process. The value transference between domains of different values is caused by human factors. Are we daring to defend human factors as a leading determinant of the teleological development of history? Who or what decides the teleological orientation and on what basis?

It might become nonsensical within the next three decades to say that humanity exists as an entity with free will, because molecular biology will find a sufficient number of evidence indicating predetermining factors of human's fate hidden in human genes such as D4DR, which governs the tendency to grow with violent and offensive character.

The content of humanity looks like a complete mess in a garbage can. It is frequently said that the human brain has evolved a blind selection at random. But the science of complexity asserts that there certainly exist significant orders and patterns of self-organization in certain circumstances. The order or pattern is not the absolute criteria to understand the trend of plexus of affairs. Then is it merely debris of human interpretation? For instance, anthropic theory definitely proposes that the universe is pre-orientated. Multi-universe cosmology has different explanation from anthropic theory.

It must be regarded that the trash can is the treasure box of evidence if an old KGB agent were to thoroughly examine it.

Human interpretation of natural and social affairs is the most significant factor for the progress of society. The present society is the result of the human interpretation of the past. It is actually the essence of what we have to call as communal intelligence. Communal intelligence changes the criteria of communal acceptance to the new creative values. If present intelligence does not admit new innovation or new idea, such creativity value evaporates. Evaporated marketable value of creativity is adsorbed into communal value domain.

Curious attraction of creativity value in a market strengthens market value of a new commodity. However, general use of new commodities by the masses reduces curiosity. Such reduction of curiosity reveals the value flow from creativity domain into communal domain.

Curiosity is a pattern of emotional prerequisite of transcendental intentionality.

Concentration is another emotional prerequisite at a lower level than curiosity in the light of developing transcendental intentionality. Curiosity that is intensified greater than usual condition is the evidence of working transciousness by which people believe unconsciously that there is something to pursue by intuition.

Table 8-4-2. Correlation between Cognition, Creativity, and Transcent

Intentionality	Transcendental Intentionality		Transcent of Humanity
Concentration	Curiosity	Imagination	Creativity
Sensciousness	Consciousness	Transciousness	
Local positioning	Balancing between Pragma and Plexion	Teleological Progress through infinities-encounter	

(Above table is a little too format-fitting that it may reveal certain deviations in actual practice at the boundaries.)

Monetary system is effective in maintaining the interconnection between value, utility, and price systems. Monetary institutions are embedded in materiality, which does not easily allow a fluctuation that is dependent on human emotional influence.

Some commentator might interpret the light that appears in “Let there be light!” as the infinities-matrix of humanity that caused the enlightenment of domains. Value transfer is generated by the enlightenment of human cognition.

8-5. The Science of Creativity

Growing Point of Science Tree
Classification of the Creativity Value
Exchange and Interflow of the Creativity Value
Accumulation of the Creativity Value
Quantitative Assessment of the Value of Creativity
The Value of Creativity and Human History
The Principles of Creativity

The science of creativity has not been the significant issue of academic world. People in the past usually thought that creativity is exclusively possessed and practiced by the art world. Furthermore, the combination of creativity and science has not been regarded comfortably by the common sense of the masses.

Growing Point of Science Tree

The science of humanity has already come into our life and raised the question about human consciousness. But a study on consciousness alone does not pave the way to the full understanding of humanity. That is why we have discussed the transciousness that will guide humanity toward the transcent. Transcent is differentiated from ascent or descent because it possesses the capability to proceed for the anthropomorphic advancement. The anthropomorphic advancement induces creativity in human life. The human creativity has certain scientific principles. Growing points along the science tree will be the science of creativity.

The present status of the science tree has been changing over the past fifty years since the launch of the space program. The space program has changed the viewpoint of science and technology; at first in the fields of military and aerospace, and next in hardware engineering of electronics, mechanics, and chemistry and so on, and has finally touched the basal strata of pure sciences such as complex systems, biology, brain science and so forth. Psychology, communication, education, political economics have been influenced by these changes in scientific methodology.

The science of creativity is not limited as only an applied science as we have thought but, also an integrated form of all pure and applied sciences. The phenomenon of creativity is the main subject of our civilization in the next several centuries.

Classification of the Creativity Value

To start a discussion on the creativity value, we have to consider the classification of it. In Chapter 1, creativity is defined to be an elevated part of humanity. Said definition looks superficially simple, however the expression has deep insight when we consider creativity from its three stages of different magnitudes of complexity.

The first kind of creativity is generated by the simple combination of ideas, which can be called contingent creativity as it usually happens by accident, although it may be a case of well intended finding (searching).

The second kind of creativity streams out of the integrated structure of rationales within an internal process. This second kind has to do with systematic ideas in humanity. It is an unconscious fruit of existing systems of intrinsic attributes. The motive for attaining the second kind of creativity is not necessarily derived solely from the language level. The motive for attaining the second kind is spread over and derived from the entire phenotype domain of humanity, which encompasses logic, language, music, dance, and any other systematic tools for communication.

The third kind of creativity is achieved by the escape out of the systematic process of internal logic. Thus, the third kind usually arises at the intermediate of a system; such a intermediate behaves like a transition zone in phase behavior and can thus be considered as an extended boundary zone for a system.

We know that novel sprouts out of new civilization essentially require such an escape from existing past logics and ancient styles of emotional expression. The third kind spreads also over the entire scope of human life, from individual to global, and from everyday life to extended periods of history. The most important steps of advancement at any level have been achieved by the third kind of creativity.

These three kinds of creativity are the activation of humanity. They are a stepwise activation of humanity to achieve the transcendence of a human being in the course of answering the deepest query of the actual infinity.

Creativity value is the energy that drives the frontal head of humanity to achieve the transcent of humanity. The value of creativity accumulates to give humanity an opening to exit to a different set of condition, which we usually call liberation. Freedom, the most beloved word of human beings, is another expression of the manifestation of creativity and the achievement of creativity value that has been accumulated.

The value of creativity can be classified by concerned entity who is involved in the creative activity and by concerned field of humanity in which creativity works. We can designate creativity value in art, in science, and even in martial arts and so on. Creativity is sometimes expressed as personal creativity, communal creativity and permunal creativity.

Creativity value can also be classified by the criteria of volatility, pseudo-condensability and condensability. Volatile creativity is that creativity value was not able to infiltrate into society and thus evaporate. The value of creativity has condensability when it settles down in the communal society examined by the communality of the society.

Pseudo-condensing of creativity value comes from the excessive dominance of virtual reality on the creative activity. The effect of virtual reality on creativity helps condensing mechanism of creativity value into materialized forms in the society. However such condensing is not complete and pseudo-condensing happens because virtual reality has only a temporary effect on holding the creativity value in the society.

In order to examine the process of creativity getting accommodated into a communal society, accommodation makes the results of creativity accordingly effective and resonant to the character of the community being received selectively and preferentially. The contents of creative activity being opposite of the community's own urges tend to get driven out or eradicated, the community indemnifies and promotes creativity that reflects its own urges. It is the explanation about volatility of creative value in Figure 8-8-1 in Section 8-8.

Exchange and Interflow of the Creativity Values

As we have discussed, creativity values are exchanged with communal value and terrenity value. The mutual interflow of the triradix values are the major reason of the activities in markets.

The exchange between various creativity values are also the cause of human activities in a market. We usually transact to cause the exchange and interflow of the value of creativity. However transaction is not the only cause of such exchange and interflow. For instance conquest is another cause and donation is the third, and so on.

Communal life is characterized by the exchange and interflow of the creative values. Creativity values are the daily manna of humankind.

Accumulation of the Creativity Value

Generally people think that creativity value does not accumulate, because it looks as if the creativity value is evaporated. However, we have to know that creativity value has been accumulated all the time since the ascent of mankind. The accumulation of the value of creativity is the actual driving force of human civilization.

People will be surprised if they hear that their DNA is also partly the fruit of cumulative creativity value. Creativity was provoked, in the author's view, when human being had encountered the infinity or infinities. The infinity had shaped the container of creativity value since the intial encounter was made in human experience. The container for the accumulated value of creativity has been widened in capacity and sophisticated in dimensions.

Quantitative Assessment of the Value of Creativity

Value and Price are under kinship. Price is effective as a unique measure of value in a market. Market pricing is determined through two ladders of value-to-price domains as discussed.

However, if we just concentrate our discussion on the value and price only, we can see the valuing and pricing mechanisms as follows.

Valuing mechanism is determined by 'due momentary subjective personal values by desire based on emotional aspects,' and then comes 'the intersubjective market effective value as regulated by social constraints.' This market effective value is the pretium we will explain in later sections.

Pricing mechanism is determined by 'the contrast mapping of the range of intersubjective market effective value, versus the range of objective price within economic and monetary constraints.'

In socio-psychological effect in market, episodic evaluation is at the center in terms of the route and process through which the emotional histeresys are influenced. Market is not only political economic environment (that is the relation between power and money i.e. constraint-flow relation).

We think that socio-psychological (nexusity-affinity) factor is brought to be important and thus understand that necessity of breakthrough to socio-psychological-political-economics is necessary in our days. Therefore, a new comprehensive model could be utilized founded on the least number of parameters as in integrismic factor.

The assessment of business projects and enterprises is relatively stable nowadays except terrenity value (rent concerned). Financial crisis usually comes up with the frustration of the rent value of realty.

The rent and interest looks similar to or same as opportunity cost in space and time. The opportunity cost is compensation cost in another sense. It is thought that rent as spatial compensation and interest as temporal compensation. In such viewpoint, creativity is time concerned value reflected on rent.

The comparison between price and book value is made by present value. The assessment of price of commodities is temporally unstable from time to time due to the fluctuation of the supply and demand relationship. However, according to the overall allocation of values in a market, the assessment of price of rare commodities brings explosive and implosive psychological impact to the market. Land and antiques are the rare commodities in a distinctive contrast in a sense that one reflects terrenity value and another reflects foregone creative value. Explosion and implosion occurs due to the psychological environment for both cases.

Quantitative assessment of creativity value needs certain standard condition to be evaluated. The basic unit of creativity value concerned with market is proposed to be as follows. (The author believes that economists are better qualified in defining the following formula.)

1 V_{cr} = the creativity value to increase the invariable market price of a specific commodity by one standard (provisionally U$)
= Cost of Creative Process / Creativity Ratio * Profit Ratio * Volatility factor

, when cost of creative Process is personal and/or organizational (communal),
creativity Ratio = cost of creative process / total cost of commodity,
profit ratio = market price of a commodity / cost of production
volatility factor = ratio of acceptance of V_{cr} into market in terms of monetary value

The Value of Creativity and Human History

Human history has always been guided by the value of creativity. Humans are humans when they are creative. Humans are humans when they are able to utilize the past accumulation of creativity to excavate and discover new horizons successively without halt. The author thinks that it is a better definition for human being to thrive positively for the succession of creativity value accumulation than the negative definition to resist the sour precipitation of the past.

Our standpoint how we can view the past makes much difference in recognizing ourselves. The viewpoint to positively interpret human history can also make historical understanding so profound that human being can see many more novel dimensions and introduce them into human history without blocking in psychological view. Creativity value itself is the wriggleless extension of humanity and the mending of the mountains and valleys in human cognition so flat that people can go forward without distortion.

The Principles of Creativity

The general principles of the manifestation of creativity

1. Balanced and appropriate creativity comes from the infinites-cognition.

This indicates that balanced and appropriate creativity can be attained from the relation between Plexion and pragma. The combination between Plexion and pragma generates the epistemological contrast between infinity and nullity, and this contrast generates the accumulation of the emotional responses that are around the human neocortex. This accumulation is conserved as a balancing indicative within the intellectual system of a human and it inhibits the divergence of the process of creativity.

2. An appropriate hierarchy of creativity exists in its application in different situations.

To be able to apply creativity in different situations indicates that the creative process can be generated from all situations and that the creative contents can be presented as the solution for a specific situation. An appropriate hierarchy of creativity means that the form of the presented creative contents can be changed according to the range of application and anomaly of the situation.

3. Creativity is activated in a society with a high energy level.

Creativity is best activated in a society with a high energy level which is equipped with infrastructures that allow the creative activity to take place. However, in a society in which the land and communal values are the leading value in economic structure is not as highly activated as it is in the high energy level society.

4. A community's acceptance determines the life span of creativity.

The life span of a creative activity and content varies according to the level of adsorption and acceptance of a specific community to a specific creativity.

5. Only an active small-unit group can activate creativity and make it efficient.

If it is not a peculiar case, creativity can be activated and made efficient only in an active small-unit group. This is because the basic knowledge, for creativity, of a field appears in a peculiar topography of a general knowledge system and a group that possesses that type of topography is usually formed in a small unit.

6. Creativity and its contents are spread by mimesis.

Once a creative content is formed, it is spread by the process of mimesis that exists within a community. In the course of the spreading process, the beneficiary and the recipient can add or form a new secondrary creative content.

7. The manifestation time of creativity can be controlled.

The method to control the manifestation time of creativity can vary according to the internal and external triggering factors.

8. A great rash plan is needed for the continuous manifestation of creativity, at the rear of it.

The continuous manifestation of creativity needs, at the rear of it, similar interests-seeking groups or a great communal continuum that has the same purpose with it.

9. Harmony between logic and intuition is needed in the depth of creativity.

Logic, or integrismic systemicity, and intuition, or integrismic nexusity, must be adequately harmonized so that the manifestation of creativity can occur. This will be, practically and personally, the most focused matter of concern in the science of creativity.

10. Appropriate creativity and the maintenance of spirituality are the matchless twin stars.

The spirituality of human being, as a maintaining mechanism of human mentality has the deepest union with the dimensions of integrating virtues in a long run.

8-6. Individual Creativity and Communal Creativity

Creativity and Dynamicity
The Field of Creativity
Three Kinds of the Process of Creativity
Communal Creativity

The dynamicity of creativity value is the essence of human progress both in individuals and at the communal level. We have commented about the land value system, communal value system and market value systems. These three value systems in the universe were originally designed to be dynamic upon humanity fully maturing. If the value systems are not dynamic and thus are stuck in static conditions only, the human world would be destined to devour one another until it completely destroyed itself.

Creativity and Dynamicity

Creativity is very dynamic process. It is all about the flow of ideas and feelings. As we explained, in creative process the spontaneous flow (沒入) of concentration is observed in individual realm. It is similar with "floating (遊)" of Chuangtse.

The dynamics of individual creativity is unstable and usually fluctuating. Thus, when a person wants to perform a creative act, he/she should pour all the cognition into the subject he/she is pursuing.

In communal domain, we can also find the interflow of value or wealth, which usually take a relatively tedious progress than in individual creativity. Communal creativity usually takes place in an integrated pattern hand in hand, which consumes much more energy and complicated process than individual one. However the dynamics of communal creativity is the real progress and advancement of humanity.

The Field of Creativity

Dynamic locality is the ultimate cause of creativity. Static locality is the evidence of the state of stillness (or the state of nihil), from which no reality is produced. The net of Indra, which is frequently cited by Buddhist philosophy to visualize the net of infinite relations of universal existences and comprehensions, is not the cause but the result of human essence. But it reveals well some aspects of locality in the whole universe by reflecting the globality of Plexion.

The field of creativity has the specificity of locality as its cause of creativeness in a sense that creativity is the response of the local field to the global field. The creativity is the essence of living system, which is the ontological relational network. Life system is the local manifestation of the global dynamic relational ontology.

The whole globality has the infinite number of manifestation of locality. Logically above comment is true for all lives like amoeba, plants and humans. The field of creativity is the field of life. Here, we understand that all things and affairs belong to the unity.

Three Kinds of the Process of Creativity

Node system menesys

We commented that there are three kinds of creativity. First kind is produced by composing separate elements together to create a new concept or new commodity. In this case we can see that the field of creativity within which creative activity is going on is at the level of node or nodule.

The second kind of creativity appears in the field of creativity at the level of system which has consistent chain of logic for its survival. The survival logic of the system does not mean that it has life. Life has profound mechanism than the system, even though there are lives at the boundary like virus or bacteriopharge that are much less complicated than very much sophisticated artificial systems. The second kind creates flexible and adaptable character by the interactive process with its environment.

The interaction with environment becomes most important for the third kind of creativity. The third kind works in the highest menesysal level. Such menesys has highly complex character and mechanism that can create organismic ability in adequate integrismic conditions and cruxological development.

Communal Creativity

The essence of humanity is the communal interactive creativity. Mirror neurons constitute a super system, which enables a baby and mother mutually reading out of the flatness of white noise from each other. It is possible because the mirror neuronal system is a core activator of the transference of complex rationale and emotion from source to receiver. The transference, in author's opinion, is menesysal transference and is achieved through consecutive process that reiterates ceaseless severance and reactivation until satisfactory results attained. Subconsciousness is flatness as observed by consciousness. The disturbances of pulses are easily detected out of flatness. "As water reflects a face, so a man's heart reflects the man (Proverbs 27:19)."

The author believes that mimesis (the evidence of communal creativity) is picking up the location of peak value (displayed in the noise pattern) and the offset discrepancy of active location.

Communal creativity is unfreezing and advancing all the way by ceaseless initialization to new summit experiences.It simply extends the current findings and results of creativity. Communal creativity is actually the dynamic identity of humanity.

8-7. Emergence and Structure of Creativity Value in a Society

Creativity value emerges from cruxological mechanisms. A cruxological system must be simple and efficient for easy mass production of creativity value. Basic crux must also be able to absorb and diminish effects of stress upon itself.

Creative Value Structure and Cruxological Development

In order to consider the creative value structure, the formation of the value structure through the birth of nodes and systems in each cruxological development is our main concern.

The value of creativity in mentality is always floating. Fixation of creativity value is not likely to be achieved in mental domain. Embedment of the value of creativity in materiality, which is generally achieved through material process, have creativity value become fixed in material foundation of an economic society.

The emergence of a creativity value occurs in materialization process. Single act of creativity even still remaining in mental domain such as an imagination of idea cannot emerge into the sphere of humanity without materialization.

The emergence of Complex Adaptive System: CAS at the boundary of chaos can be interpreted as the creative system for adaptation. However, when a menesys (of complex adaptive character) get into the full chaos condition, adaptability is lost. Flexible adaptability of the menesys will be suppressed when the governance of systemicity is excessively overwhelming.

We are discussing about the way of "*via media*." We know that intermediate manner of administration is usually the best to stimulate the global menesys as far as the body of administration is not corrupt. The gentle intermediate way of approach to the environmental problems is actually the most positive concept for humanity to alter the environment being attracted by the reengineering purpose. It is necessary for us to have an eye sight to see the connection between creativity and peace-conflict correlation.

The world has been waiting for the dramaturgy of alternating strategy between sysy and nexy to take balance at the best integrismic stance for changing situations. It is a matter of symmetry and asymmetry. During the progress under globalization and in the futures even after that, human race has to be directed by symmetry (logos, logics in wholeness) and asymmetry (rhema, single affair supplying energy).

Harmonious drama is not symmetrical and should be pierced by time axis into its whole episode and thus causing dramatic story, energy, panoramic unfolding of probable accidents. External panoramic unfolding in time axis of human cognition is manifested by internal intrinsic enfolding.

Menesysal Development and Creativity

Menesys is formulated through the progressive development from crux (minima and knot), nodule, module, system, and finally various menesys.

Crux comes from the feeling of existence of an individual personal. It is originated from the persontological episodic memory. Crux is the initial phenomenon of person, which accumulates the rationales for survival and transportation of his own genetic characteristic.

Persontology is presented in this book, because the intrinsic condition of communal noosphere is the most fundamental reason of persontology. Forward frontal vector is a good expression of gene transportation DNA of human beings. It is also a cause of anthropomorphosis. The difference between crux and knot is that knot has phase while crux does not.

A nodule is a step lower than a module. A module usually has distinctive function to support the module, system or menesys, which are higher entities in hierarchy. However nodule has the least logical circuit to connect the different entities. The transportation system has many nodules with simple functions of stations, and as can be observed in nomad Bedouins and transpoting nodes of medieval pirates, nodules are all simple
Transportation has the concept of connectivity in its nature. Nodules are the foundation of formulating modules.

Humanization is the modularization of plexus of humane affairs. The menesys of creativity is directed to anthropomorphosis. Module is the essential element that decides the character of systems and menesys. It means integration is the basic character of modularization.

Discontinuity from external condition is the nature of a system. A system tends to isolate its constituents in order to intensify the governance by internal logic. In the initial stage of systemization, system characteristic is voluntary creativity. However it changes to enforced creativity in the latter stage, which solidifies the system to become inflexible.

Systems generally concentrate to intensify the creativity of the second kind, which has tendency to wither the whole system and lose life in a long run. Menesysal entities are eligible to manifest the creativity of the third kind for self-surmounting, which is the intrinsic value of the transcent of humanity, if they are well managed and vitalized.

Systemized creativity value induces the order of the system by stimulating nodes. Networked creativity value eases the nodes of network to reduce stress. Creativity in network environment acts as in pulse pattern. The present networks that we observe are not pure network and mostly influenced by systematic infiltration so that we may not be confused in understanding the phenomenon. In appropriate pattern of menesys, high energy condition is hidden in chaotic status that will be able to novely manifest hidden order of elevated society.

The crisscrossing of civilization of high tidal variety induces the changes of the potential and pulse that has been contained in the creativity values of novel ideas and new information. However, we have to be careful in sustaining the encompassing motherhood is the expression of the representative creativity value of the humankind, which is the self-sacrificing and voluntary value conserved in genetic structure. The forfeit of motherhood is the barometric reverse-indicator of the vitalization of human civilization.

In a basic commune, both intentional and non-intentional, learning by side glance implies that sunconscious inflow of information through white noise, pink noise and brown noise is the main source of the production of communal value. Learning by side glance is performed by "mirror neuronal 'super system' " as some researcher named.The mirror neuronal menesys indicates a probable huge super menesys of communal value of human race.

The career record of general enterprises usually traces back to the developmental progress of their informal organizations. Informal organization of a company like vineyard meeting in the evening helps new comers to grasp atmosphere and tendency. Such learning and grasping by side glance provide unexpected complexity to the understanding of the new comer on the orderly arrangement of formal organization and have him interpret by himself the details of the topography in human relationship. This is the principle that is consistently applied to the management of general organizational mechanism in production line systems and distribution networks, and staff systems.

Globalization has changed the character of modern nations and countries. We are confronting an umbrella-pattern of menesysal growth in international relations such as EU, NAFTA, ANDEAN, and ASEAN and so on. The deterioration in the governance of nations to the concentration and mobilization of resources are complemented by the solidary movements among nations and countries.

Such changes occurring at the magnitude of $10^{9\sim10}$ should have deep and close relevancy to the pattern of human cognition. At the magnitude of population of $10^{10\sim11}$, we will experience some changed situation definitely.

The degree of dispersion in geographic or topological distribution of population looks like making distinctive shift of characterization process. These phenomenon has to be related with the complexity of neurophysiological network in terms of "ten to the how many" of population. The menesysal development is the subject that should be studied for humankind to proceed in anthropomorphic development for the long term view vitalizing humanity in all direction.

The structure and development of terrenity value is deeply related to the magnitude of population and the growth pattern of intrinsic structure, which have to do with the population density in induction of human emotional stress. The experimental results of the expectation on the efficiency of dense population in a long experience have the deep emotional root since the days of collecting and hunting economy. Dense population generates stress conditions and on the contrary a good infrastructural condition of communal production.

The relevance of the concentration of population to land value or terrenity value is not only related to the menesysal development but also td the all phases of social phenomena and even from the magnitude of 10^1. That is confirmed in various situations such as in the adequate number of nucleus family, in the optimum participants to a discussion that is around seven, and in a squad organization in the army.

This principle is connected to how we can acxhieve the fair distribution of wealth in the domain of land value to which geographical density affects mostly.

Human cannot formulate communal intelligence by swarm land occupation alone. Humans do not formulate swarm intelligence but strictly speaking is supposed to acieve persontological permunal intelligence through good administration of land occupation.

Human being is more influenced by teleological factor. It is why religions have such long and ceaseless influence on humanity. That is why cultural pheromone has such strong impact on humanity. The integrated influence of cultural pheromone is reflected in the land value and/or land market price. It is related to the psychological amplification of land value.

Integrism to Enhance Creativity

"Aha!- experience" in our creative moment strike us when we are in a vague and uncertain moment but the overall picture of the appropriate target has been set up. Uncertainty activates our emotional mechanisms. Reasoning comes from our systemicity, a mechanism of fixation, while emotion stems from nexusity, a mechanism of dynamicity. However, reasoning and emotion move together.

Dopamine is a compensating measure in the brain-neural environment, in which the brain can acknowledge as endo-encephalonic monetary currency. The uncertain condition in an appropriate expectation for a creative breakthrough could become a sweet compensation to the brain. (Theodore William Schultz, University of Cambridge)

The system in general is considered as the source of stability or can be compared to a military safe camp. Humans can snatch flashes of new idea when he is subconsciously stable. K. Mogi called it serendipity, which is the property of creative process. (Mogi Kenichiro, Hirameki Nou, 2006) Such dynamic move is generated from the appropriate composition between sysy and nexy or proper context of emotion and reasoning.

Such best integrismic condition can be activated in societal level as well as in individual condition.

Cruxology to Enhance Creativity

In our globalized age, cruxological development is about overcoming self and structure. Systemization, to which we are usually accustomed, is the forced cruxological development, and it is not fundamental nor a basic value of cruxology. The essence of cruxology is the self-adjustment function of the future individuals. The self-adjustment is the only way to accompany nexal connection and systematic growth, and it is the central mentality of the creative peace. Cruxological measure can enhance the creative process, that we are pursuing for.

8-8. Creativity Dedicated to the Creative Peace

Communal Acceptance of Creativity and Creative Character of Politics
Peace Modules and the Continuum of the Creative Peace

In human history, we find the intervening of creative actions greater in conflict and competing status than in peace status. Humanity does not look like learning to concentrate for creativeness in stabilized condition.

Communal Acceptance of Creativity and Creative Character of Politics

It is deemed that the process of communal acceptance of creativity in a society is an essential factor to explain political process. Creativity is largely perceived as an individual and subjective process, but it cannot be separated from communal cognition, needs, and way of thinking. Whether or not the intrinsic process of genuine individual creativity can be dealt at the communal levels remains always questionable, because the genuine creativity is largely the synthetic process of subjective intuition.

Nonetheless, considering the individual levels of creativity, the most important questions are, “Could the outcome of individual creativity be applied to the community?” and, “If applicable, how could it be processed?” as well as, “Through what channel would it be applied?”

Normally, creativity is considered as a personal phenomenon. However, because individual creativity is an inter-subjective process, it must be placed in the mental process with in community. That is the art, invention, creation, and education (and the profound impact of these).

If the creativity has a high resolvability, its content gets foundered easily because it is unable to remain behind, this remains as a personal achievement. However, the creative value settled during the communal process transforms the community's character and reinforces the community.

Figure 8-8-1 illustrates how the volatility or resolvability of creativity value has to do with the creative process exposed through the entire socialization process. Basic illustration was taken from the figure work of Sato and the idea of volatility was added by the author.

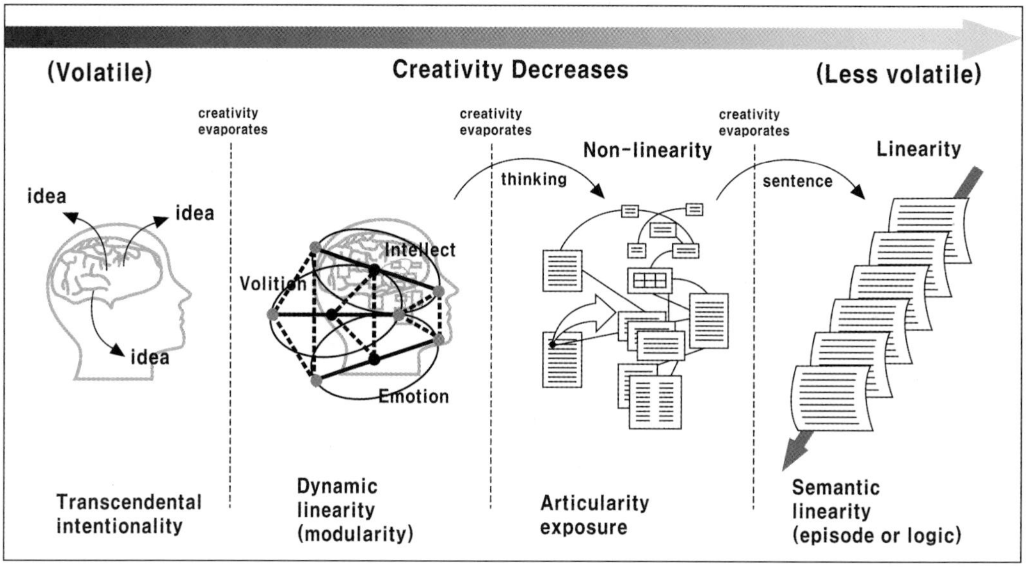

Figure 8-8-1. The Volatility of Creativity Value Exposed in Socialization Process. (Basic drawing adopted from Takashi Saito, Ontological Algorothm, 2006)

To categorize the creativity phenomenon within a community according to the resolvability or volatility; dance, music, arts, literatures, etc… are creativity with high resolvability; patent, invention , new logic analysis are creativity with medium resolvability; technological, professional and educational materials have low resolvability and thus stable.

To examine the process of creativity getting accommodated into a communal society, accommodation makes the results of creativity accordingly effective and resonant to the character of the community being received selectively and preferentially. The contents of creative activity being opposite of the community's own urges tend to get driven out or eradicated, the community indemnifies and promotes creativity that reflects its own urges.

In short, as a community grows and develops, the systematic aspect becomes strengthened originally, because communities have a strong aspect of networking, as the systematic aspect becomes stronger, the actual creation process is disappearing – according to the modern economics or the modern management technique, though unconsciously done.

Such direction is another important task for this particular research. Especially, the creation process in relation to politics is very much related with its community's urges and it's easily assumed that its resolvability is also pretty high. The economic creation process comparatively has a lower resolvability and a lower cohesion with the communal urges than the politics. Therefore, it is only natural to expect new methodologies from the economy and the conglomerates.

Peace Modules and the Continuum of the Creative Peace

The continuum of the creative peace will be formulated through the localization of peace administration after fair progress in global continuum. The localization is not separable from globalization. Some people enjoy using "glocalization" instead of globalization because local situation is the essence of criteria to assess the healthiness of globalization.

We need peace modules to achieve glocalization. Glocalization process promises human race a great progress in global administration; first, comparative homogeneity in globalization, second, hierarchical development of global locality under mutual collaboration, and third, dynamic modularity in local entities.

Comparative homogeneity means that global characteristics are homogeneous enough to sustain communication and connectivity among local modules. Hierarchy among locality and various functions are necessary because globality should be supported by systemic administration. Dynamic modularity ensures the constituents of the world to behave as self-propelled creative integer entities.

Geographical locality and topological locality in a society contributs to the formation of dynamic modularity. Conflicts and competition are the spontaneous results of the locality and its subsequent development.

We have a famous Hora and Tempus Case in the precision watch business, which shows effective management of modular manufacturing. The method presents the efficient way to mobilize the capability of human essence.

Hora and Tempus are both watch manufacturer who applied different strategy in composing precision watches. Hora adopted hierarchical modules of sub-assembly whereas Tempus did not and had to dismantle the watch for a while for instance taking phone calls. Hora won the game and Tempus lost his shop finally.

We had better sometimes consider the reshuffling of the factors of peace and conflict into a local module. Local restructuring of peace factors will contribute for synergy effect. Local constraints on conflict factors will generally suppress the internal energy in a local module. Reshuffling the entities and relations under and beyond local modules will accelerate the modularization of global political environment.

PART III.

Integritium and the Creative Peace

Chapter 9. Status Quo Diagnosis

"Sustainability" is at the center of concern among contemporary intellectuals. In this work we have a higher objective than sustainability, which is the continuity of present achievements of humanity. Only pursuing sustainability cannot promise a sufficient solution for perfect sustainment. We have to go further than the pursuit of sustainment to find paths that leads us into new continents. At this stage of human history, we have to think of the prospect of "continuum" that contains reflections of various aspects to the ultimate solution of present civilization.

The continuum is defined in several different meanings; in physics it means spatiotemporal continuum and in mathematics, in my opinion, accessibility to the continuum is studied in quest of the issue of infinity. But in the substantial world where we have to deal with practical political economy of globalization, the author thinks that the issue of continuum has to be placed at the center of current queries and solutions. This will be verified in the discussions of Part III. In this chapter we need to understand all discussions in the following sections in light of continuum.

In this chapter, nine different subjects are discussed. We are not going to talk about some of the important issues that contemporary intellectuals deem to be essential; for instance, environmental issues, human right problems, etc. Rather, subjects dealt here are all related to the orientation of human history toward the achievement of continuum. The effort for continuum has to be complemented by dynamic modularity. In fact, world governance, financial capitalism, world currency cannot be managed without the ceaseless concern about dynamic modules. Value conflict, universal public law and peace knowledge system should be kept consistent but at the same time maintain the tentacles to the future, the unknown and the disconnected. Military and anti-violence, and underground economy are presently the issues of direct battle in everyday life, but are subject to preparation of forthcoming menesys of universal autonomy.

Courageous steps are required to get into an inevitable new phase of human history. Those who understand will remain, others will perish. In this context, our diagnosis is made in accordance with the upgraded standards for our present anomalous situation. Anomalies currently found in various aspects of our civilization are caused by the fast infiltration of scientific findings into every other field. The infiltration of science has changed the basis of political economy and social psychology.

9-1. World Governance and Global Continuum

World governance is not subject to any specific subjective entity. It is a matter of the whole human race and the entire humanity. Is continuum only related to connectivity? The answer is negative. It has to do with the socio-psychological status of human race in terms of susceptance and tolerance, which allows the propagation of emotional, informational and intellectual waves to be transmitted through the continuum in a positive manner.

The main concern of our diagnosis is the query about global conditions of governance in political economy. It is asking if we are well established in systemicity and nexusity globally. Needed for humanity is the optimum combination between the two, the terminology for which is preferably "dynamic symmetry mechanism" to maintain stability and active creativity together in order to step on to enhance the upper stair of human civilization. (Symmetry hereinafter means as described in Collins 21st century edition "beauty or harmony of form based on a proportionate arrangement of parts.")

D. Held and A. McGrew had formulated a methodology (I would call it H-M measure, hereafter) to classify the globalization process into four categories; thick, diffused, expansive, and thin globalization. (1999, Global Transformations, Preface) Their conclusion about the present process of globalization is the thick globalization characterized by high extent, high intensity, high velocity, and high impact. I believe we need to combine the idea with dynamic modularity in order to better describe the internal dynamic mechanisms as well.

The issue of world governance is better illustrated with the degree of self-governance of dynamic modules and the dynamic cores in a restrained situation. Dynamic core is a relatively stable part of dynamic module. Dynamic module consists of a core portion and a peripheral portion distinguished by their abrupt change in the degree of systemicity. Dynamic module in reality would be a certain ethnicity, community, society, intellectual groups, interest groups, and different regimes of the political economy.

Importance of Dynamic Modularity

The dynamic symmetry mechanism could be a simple technique in the current administration of many international relations, but it will become the advanced technology of global and post-global management when everything settles down in a grand complex design. In this structure, people will enjoy the continuous flow of economy connected to each other providing sufficient and tolerable furnishing with commodities and the means of exchange. The balance of power among nations was the main theme of international politics since the beginning of the Westphalian Regime.

The reason that political entities, not only in political institutions but also in others such as entrepreneurship and educational institutes, have to play with dynamic symmetry is that the internal modules of entities have their own independent mechanism as well. It is true especially for dynamic modules generated from internal systemicity, which is formulated through a long history of interrelation and contributed to form upper level systems (or menesys) by the incorporation of the other modules through the activation of its nexusity.

(Whitehead used "concrescence" to mean the incorporation of a part with others into a conformal unity, which only confines the concept to systematic entity. Our menesysal understanding provides a freer and wider variety of manners of incorporation and could further be a base of an analytical approach to the practical world.)

The difficult part of current international politics is that we are about to enter a stage of elevation of the concept of politics into a day of worldwide conglomeration. It is said that the real test of the United States is to see its capability in handling International politics. (Chapter 6, Post-American World, Fareed Zakaria) It is the opinion of F. Zakaria that the balancing strategy against emerging powers would not only restrain the object country but also cause backward restrain to the subject country so that it will be isolated eventually. The U.S. as an incompatible supplier of the universal ideals and contextual contents of present civilization would not regard this as a possible outcome. Even El Salvador, New Zealand, Japan, China, Russia, or Brazil would not accept it if they were positioned to be so.

Our present quest is if any leader in the world can play shadeless and transparent governance, which I would humorously call "new Must-Mist style administration," the leader must mist not by glorious self-mystification but by contracting oneself leaving intangible mist in the air with only effectual full governance. Such administration will be materialized by the growth of dynamic modularity in the whole world. Dynamic modularity in the real world is presented by international regions, nations, metropolitan areas, ethnic communities, interest groups, political solidarities, economical communities, NGO networks, etc.

(Must-Mist is a simple reverse phonetic arrangement of Tsim-Tsum, a Hebrew word, which stands for the contraction of the Almighty to yield existential space for the diversity of creatures.)

However, we are at the initial stage of understanding the mechanism of dynamic modularity. Then someone might ask what would be left after the contracted loss of the subject entity. The answer to this is that those who create things always take the preferential right to the future and head for creativity. In fact, they are not merely contracting but further extending to the alpha path as far as their energy permits. It partly means that they will become and behave like stem cells that could convert themselves into any kind of future being. In this sense, such contracted entities deserve to be called the leader, foreseer, or elite in a true meaning. The mechanism of dynamic modularity has to do with the interactions between the leading core and the following periphery.

Elite and Commons

True elites and leaders are originally rooted firmly into the lives of plain people. Elite should infiltrate and spread into the common world. They are dispatched by the deep command of their own community, or by their own moral sense. Elites are the essence of a community and society like aristocrats out of tribal communities. When depraved, aristocrats and elites of society became oppressors upon plain people in history, sometimes even allying with external invaders. The reason for their behavior was simple. It was their major concern to build and maintain a system to the source of their own interest in order to systemize the established structure.

There is an interesting strategy expressed by R. Kaplan (Ch.10, p148, Warrior Politics); "Joseph Nye Jr., dean of Harvard Kennedy School speaks of 'soft' American Hegemony. Sun-Tzu says that the strongest position is 'formless'; it is a position that adversaries cannot attack because it exists everywhere and nowhere, An American imperium should be like that."

As a cosmopolitan, the author would like to say that the future world is an exclusive stake of neither America nor China. Persia, in ancient times, was said to have achieved almost equivalent degree of integration as seen in the present global integrity. One interesting thing is that Persia did not anticipate that a new strong imperium would come out of the Thracian and Macedonian region where they used for their horse growing before the Phillip II, the father of Alexander. What if incomparably aggressive military forces grew rapidly at mid-Africa in the short future? Are we not experiencing an initial phase of such occurrence in Somalia today?

Interestingly and ironically, the best governance of the Holy Ghost is achieved by the presence of nowhere but everywhere. Sun-Tzu did not write for such vague situation but with substantiality of well trained basic forces of military. Sun-Tzu even killed the two most beloved royal concubines in the course of training a platoon of the king's harem to make a strict example.

Elitism must serve the society so that society can achieve the connexion with the external world since it has the communication skill based on the educated intellectuality and intelligence. However, that does not provide a solid reason for it to be apart from the common people. It should be dependent on world-nexusity and practice the power of political economy within their society.

Elitism thus lives on the support of commonism, which is the thought of commons who a little lower than the average of the members of a society. As a rule of a thumb, I believe, it would be quantitatively right if we take the average of the societal attributes from the one third levels of that of the highest elite. Commonism is a way of thinking of common people, who are usually of the largest population and the most representative group. Commonism tends to keep highly conservative attitudes when the society gives emphasis on traditional values.

Global elitism has achieved a vast coalition based on the effect of national educations for the last two centuries linked with the industrial revolution and the subsequent progress of nationalistic atmosphere. It means that a sort of continuum of intellectual linkage that assumes a center of global knowledge has appeared. The central part of the global continuum of intellect has been laid on the epistemological attitudes of science and technology. Such elitism in developing countries is sometimes said to behave in a negative manner to their own belongings. Several decades ago, it was explained by the dominance-dependence theory.

Commonism is dependent on local communality. Commonism has a long history with varying names. I believe it has been well expressed in political system when the House of Commons was fully established in England. However, the commons living now in Darfur have nothing similar to the commons in England three centuries ago. They are analogous to each other only when the positions of the groups are considered to be meaningfully identical in the distribution of the whole population.

Elitism and commonism play a combined role together in the dynamic modularity of a nation and also in the global state. "Core and periphery" is a synonym of "elitism and commonism." Core tends to establish the systematic center while periphery tends to maintain its original attribute to connect with any kind of entity mutually communicable. The world is thought to operate under the principles of continuity, capacity, multiplicity, and dynamic modularity. In this connection, we have to consider the factors of the contradicting elements of "elitism and commonism" in the global and local conditions of the world through the analysis of the current status.

Our problem is the split between two factors that can be observed in present international relations and in various national conditions. Sometimes, elitism with power governance has different ethnic and racial origin from the commonism in a country. Elitism and commonism is better a set that is applicable to interpret dynamic modularity than a mutually antagonistic set of conservatives and progressives in a society.

Enclosures and Connexions: Cap-Con relation

Cap-con relation is formed between capacity and connectivity of values in a society. Society is a continuum of values only when and where connectivity is effectively working to allow the interflow to occur among capacious parts. Cap-con is an indicator in a society to show its vitality. There are non-conductive parts always in a society due to the lack of capacity and/or connectivity.

It is our theoretical interest whether or not a society has the difference in the phenomenon of interflow due to the connectivity that we can see in the linear and the non-linear flow of fluid mechanics. It is very much likely that inertia effect and turbulence of interflow could be observed in social science too. It is thought so due to the deferred propagation of public opinions and internet communication which are affected by the resistance of structural constraints.

Eric Beinhocker (the Origin of Wealth, 2006, McKinsey & Co. Inc.) cited that the most convenient method to describe a dynamic system is to observe it in the light of stock and flow. Stock is like the balance in a bank account and flow is the changing rate of stock. Cap-con relation, enclosure, and connexion have direct reflections from the concept of stock and flow.

The notions of capacity (cap) and connectivity (con) provide the tools for us to understand the role of enclosures and connexions in dynamic modules vitally activated in the continuum. It is thought that cap-con relation would be helpful in collecting and processing social data to correlate with dynamic modularity. (Capacity is a synonym of containability, and connectivity is similar to the conductance of flow in a porous and permeable medium.)

Enclosures are usually the container of common values formulated for a long time, such as family, religious entity, societies in a regional community and informal gatherings of friends. Enclosures are generally inert to the environment, but contribute to the capacity of a module. Dynamic modules generally take its stability from enclosures of the society.

Connexions are the connecting members of a society. Connectivity is maintained by homogeneity or by common interests. Any member of a society even any member of any enclosure, can become a connexion, as long as no contradiction occurs against its attributes. Connexions tend to be temporary. Elites and commons are all dispersed in a society and are found in enclosures and connexions. Thus, elitism and commonism are spread over the society in a pattern of networking. However, the center of elitism in general makes a systematic central body to exercise its power.

As the human society becomes mature and diversify significantly, the enclosures and connexions attain very natural accord and symphony in both internal and external conditions. It is not equal to those days when power was concentrated on autocracy or oligarchy. The hierarchical layers get thicker and thus achieve a deeper mutual communication. The various enclosures and the indivisible connexions between enclosures help a society overcome its discrepancies between elitism and commonism.

Some parts of elitism and commonism in a society are combined together, shaping a mixed up zone. Such mixed up zones as a whole behave like a transition zone, and its size and property decide the stability and vitality of a society. It is my observation that there has not been any better condition than today in human history as far as enclosures and connexions are concerned. It is near a turning point, though its direction is yet unknown.

The human brain could be a very good model for the development of future society. This is because the dynamic modularity of the human brain, as this study was hypothesized in Chapter 4, has a very similar mechanism with the social change mentioned above.

Integrismic Governance and Glocalization

The real problem of the current world is globalization. Globalized localization was named glocalization by Roland Robertson, an English sociologist, to treat cultural problems, but now it has been extended to the fields of economy, politics and every aspects of humanity. One way to concept globalization has been proved, but it does not maintain a balance. We know that the reaction is, in most cases, as important as the action itself. Thus, localization associated to globalization is desirable.

Glocal continuum would be the next desirable existential pattern of the global political continuum. When we put our hands on macro subjects, our fingers should cover all the aspects of micro affairs. How can we do that? The answer is integrismic governance. In the study of constitutional order, we find some academic conclusions about the developmental history of the modern state. Police state appeared, then constitutional state, followed by social state, and finally steering state, which has been the latest development. In our scheme of thought, the pattern of governance was thought to be developed in a sequence of power governance, constitutional governance, equilibrium governance, and steering governance, respectively matching the state pattern. What we suggest here, the next developmental stage, is integrismic governance.

R. Kaplan, a well known journalist-traveler, commented that the current world has a similar degree of integration as that achieved by the Persian Empire. I believe the measurement was made by his personal intuitive feeling based on his erudite and wide knowledge. I evaluate his sense of integration most highly because he is also a man of the investigation field. However, how could it be measured in practical administrative works? I believe it has not yet been measured.

It is noteworthy that aforementioned H-M measure in the early part of this section provides a good index that indicates the degree of globalization. One thing that such analysis cannot easily grasp is the intrinsic attributes of the world governance. It is not a proper tool to understand the dynamic modularity in the globalization process. However, I believe the scale of measurement has evolved throughout human history.

It is from my personal insight that the measurement scale of integration is related to the five factors of CTCAM; the degrees of communication, transportation, collaboration, assimilation, and mobilization. It is important to realize that these factors are broadly covering the global dynamicity of substantiation in a spatiotemporal continuum.

We know that the global communication has improved magnificently due to satellites, internet, and telecommunication, and so on.

Transportation has also experienced a profound enhancement in the context of the development of air and surface transportations in the last 100 years. Collaboration among multilateral parties in the global scale has proved its effective transactions formulated based on the stability of legal, diplomatic, and business relationships.

Assimilation is generally the fruit of the legal system, the content of education, and culture. Global trends of fashions and life styles were intensively led by only several countries in the last century. Such trends have induced global youngsters to wear blue jeans and eat McDonalds. However, the factor of assimilation has a more serious aspect to be considered. It is the coincidence of identity, which is the intrinsic attribute. Mobilization is largely dependent on this factor. Assimilation is an indication of the potency for integration of a society. Mobilization is a major factor of assessing a practical integration of a society.

At this stage of human history, where the self-governance of dynamic modules is not yet matured enough, we can only mention in a subjective manner that communication is affected by four parameters. These are velocity (V), intensity (I), extensity (E) and tolerance (T) of communication. Velocity of apparent communication, and the intensity and extensity of global communication quantitatively indicate the level of the world's communication. Tolerance is the capability of the global society to absorb impacts of communication. VIET seems like the general parameters to decide each of the five factors of integration. (My personal opinion based on the method above, in the simplest and most subjective form, is that the integrity of present global order has been increased almost ten times greater than that of 1945.) However, the integrity needs be adjusted by the bottom up observations that consider the effect of dynamic modularity. If such reliable technique in political economics is systematically designed, it would be the best method to measure the quantity of integrismic governance. However, it is not an easy task to design a method that follows the global situation continuously.

In 1945, human race has psychologically reached to an ultimate ceiling of annihilation by using nuclear weapons, which is the output of science and technology. Future historians will distinguish history by the before and after of the end of World War II. It was the time when the product of human intelligence attained the overwhelming compulsory power over humanity. The process is still ongoing from thence, sometimes with hopes for successful cures and, in others, with disappointments.

The factors discussed above are deeply connected with integrism of a society. Vitalized condition by the proper selection of integrismic standards would induce the best condition for glocalization. The world is entering into a big remodeling phase. However, we have yet to innovate and design novel methods to materialize these developed situations.

9-2. Financial Capitalism

Originally, capitalism started from commercial capitalism in the beginning of the second millennium. Commercial capitalism had been supported by colonialism between the 15th and 17th century, after Western Europe had experienced the earliest growth thanks to the industrial capitalism. Industrial Revolution had stimulated a severe competition among national states and affected industrial capitalism to construct an embryonic continuum of global economy. The global spread of industrial capitalism induced, at first, the global networks of multi-national corporations and then it was followed by globalization of financial institutions all over the world.

Financial capitalism is unique compared to the previous capitalisms in a sense that it has a different value pursuit and timely supports the information technology. Especially, the current financial capitalism has a strong tendency to produce more excessive bubbles than it is required for a healthy capitalism.

The perspective of capitalism has been changing. Financial capitalism started to construct its infrastructure in 1960s when the worldwide accumulation of financial assets began to increase substantially. Financial superstructure has been expanding significantly faster than the actual infrastructure. In other words, the governance of financial economy over the real economy has been realized.

Sub-prime Loan Crisis

International currency has been diversified by the emergence of novel payment methods, such as Eurocurrency in 1960s and credit cards in 1980s. The concept of international currency has been extended to contain the function of currency, asset and liability. For instance, Eurocurrency is a money market based on liquidity of short term deposits, a form of asset, while credit card is a payment based on liability.

In 1980s, the international capital transfer was activated bringing monetary circulation through the effective networks between financial institutions. Eurocurrency, Euro-asset market, foreign exchange market, as well as financial markets around the world came together and met in Uruguay Round in 1980 and this accelerated such trends further to bring in the emerging markets of Asian and Latin American countries.

In 2000, A. Lamfalucy, known as one of the active opinion leaders in the European monetary market, commented that our world has not yet been fully globalized financially, adding that financial globalization is an ongoing process of geographical admission of new countries, intensification of interdependence, and financial innovation.

Lamfalucy's main concern is the possible future disturbances from the converged structure of financial institutions, which can be caused by four factors: imprudent mutual merger, the lack of basic frames of financial markets ('basic frames' as described by J. McMillan in "Reinventing the Bazaar," proprietorship, legal frame of transaction and financial regulatory mechanism), the character of new emerging financial organizations, and lastly, the influence of "unified" financial market on banks and financial structure.

It would be good to mention here a criticism on financial capitalism made in Tokyo, in 2008, which is one of the tri-polar financial centers of the world. M. Kaneko comments on the collapse of the "shadow banking system" which entails ten trillion US Dollars scale. The worst fact of the shadow banking system is that the loss cannot be defined due to the technique of securitization. He cited that FIRE economy, (Finance-Insurance-Real Estate economy named by K. Phillips) which should be a supportive system of economic activities, has been changed to a dominant industry. It is worrying since a very special stagflation is anticipated with the bubble collapse resulted by the combination of asset deflation and resources inflation. When it loses the original position in the market, financial capitalism will not stand alone and thus will induce more intensive and extensive complications.

Y. Kuratsu has had his career with international investment market for decades since the early 1980s. He thinks the current crisis is an occurrence of once in a century and new financial transactions such as securitizations and CDS (credit default swap) have already infiltrated into the entire financial institutions in a global scale. He is somewhat affirmative about the investment banks and their financial techniques. He stated the major point to avoid excessive bubbles is to carefully examine the sound relationship maintained between interest rate and stock and to check the degree of active participation by market participants.

He seems as if he believes in financial capitalism, which is the positive pursuit of profit by financial institutions. Even though he stresses the necessity of adequate bubbles, he criticizes the strategy of investment banks that have introduced ultimate risk factors of credit into the transaction system. He is very analytic towards the axiology of capitalism. He points out that asset bubbles and credit bubbles are related to call option and put option respectively. Current crisis has to do with credit bubbles selling put options and the lack of administrative ability of investment banks and commercial banks in put option market. (Subprime Crisis Concert, 2008. Ch. 5)

He indicates that the credit ratings have been performed disregarding the independence between the domains of concerned values; utility values and exchange values. If we visualize his idea in Figure 8-3-1, exchange value and present value would be achieved in the domain of monetary currency together, but the utility value is in a different domain. Actually the difference between exchange value and utility value is the absorbed portion of utility value into the communal domain as explained in Triradixism in Chapter 7.

It is not the objective of this diagnosis to clarify the overall mechanism of current global financial crisis. We intend to show the proper position of financial capitalism in the developmental stages of the humanization of economy. The humanization of economy does not only aim to social responsibilities but also sufficient and profound provision of economic utilities to human kind.

In my opinion, "netsys (network of systems) without regulation" was the name of the crisis. If the interrelations and interactions among financial organizations are delimited by the weak regulatory function of contractual transaction level that assures only mutual connection, such netsys will not develop itself to yield a reliable menesys that is equipped with proper combination of systemicity and nexusity. An ordinary market is just a networking entity except for the umbrella regulation of the weakest level. Market is not only capitalistic, but also a very natural political economic entity.

But the financial market is not like an ordinary market, because it has specific supporting capitalistic functions for the entire ordinary market. The function of financial capitalism comes from the existing identity of the financing institutions and is justified only with its support from the whole market.

It is noteworthy to simply view that present global market consists of 4 different layers; natural resources layer, industrial production layer, commercial transaction layer, and financial interaction layer. It is thought that additional layers would emerge after and upon the financial layer. Those could be some phases of knowledge or educational capitalism and/or other successive capitalistic developments. The stages of capitalism depend on the concentration process of capital resources in order to construct the 'organs' of humanity. The capital resources are natural, human, social, and so forth. The organs of humanity are the institutional and non-institutional continuum produced in the course of concentrating capital resources.

It was pointed out that cutting-edge financial technology was introduced during the process of the whole crisis. Financial engineering stemmed from the science of complexity since around 1990. The science of complexity was originally discovered from the computer application of meteorology and fluid mechanics. The fundamentals of complexity are not solidly based on definite physicality as can be observed in the development of the human body, including his organs.

The techniques of financial engineering are not well established in the ethical sense. Its basic relevance to the social evolutionism shows a non-teleological strategy that can be emerged as any kind of singularity, as far as the hunt for preys is guaranteed. However, in molecular biology with firm basis on physicality, we find certain complex regulatory networks in human genomics. Human physical genomics has its teleological route towards the final advancement or the demolition of humanity. This is the contingency factor we have to introduce to the financial mechanism of this stage of capitalism.

"The new paradigm for financial market" of George Soros gave a good insight into the crisis we experienced in 2008. He commented that we have to explore for a new conceptual framework for the way markets to really work. I believe the current crisis comes from two major reasons.

The first is that new emerging submarkets of the world did not go through sufficiently long periods of adjustment as other forerunners have done in the last two centuries. It is especially applicable to China, Russia, and other countries that once were categorized as post-socialist countries. We have to keep in mind that several short periods of liquidating bubbles would have helped avoiding the extraordinary bursting of bubbles of long term period.

The second is that current financial centers took the advantage of the new emergence of advancing submarkets and drained excessive profit in a short span of time using their enhanced financial technology. It is such speculation that, sometimes, should be defined as an immoderate gambling. The main factor that affected the bursting was the hunting forces who eventually hunted themselves. When they intended to design skills to hunt others by complicated financial engineering, it was too late for them to realize that they were actually hunting themselves.

"Currency Wars," written by H. Song and widely read by Chinese readers, has a hot spicy flavor of the conspiracy game. In China, surprisingly, millions of people have been deeply affected by this book. In his Appendix on "the implosion of American debts and Curtailment of Global Liquidity," Song describes junks as a Sub-prime loan CDO (Collateralized Debt Obligations) and equity tranche. The fact that not only the leadership of Wall Street but also that of US Government lacked philosophy and principles has been revealed. If it were not the lack of philosophy, it thus was conspiracy.

It is noteworthy that current crisis bears underground factors of market economy. Frequently found among past socialist countries was the outcrops of the market economy as black markets. They are built on and sustained by the compulsory needs for commodities. We have seen black markets of illegal commodities like opium, narcotics, and weapon in capitalistic countries. This time we smell some unintentional but irrational willful negligence from the beginning, in setting up credit lines of junk assets through the approvals of credit assessment organizations, which are merely based on mathematical models.

Usually confidence tricks are not allowed in a sound economic regime; because they are the underground factors that hinder the credit system of a market. This time, the whole market was structured with a set of trickery designs publicly and unconsciously acknowledged and approved in the market. Senior tranche, 80 % of CDO, was rated AAA by Moody and S & P and provided a credit foundation to generate special purpose vehicles. These unfamiliar ideas, the so-called and criticized as the "shadow banking system," have produced numbers of SPC (specially purposed companies since Enron) and SIV (structured investment vehicle).

It is now definitely understood by many people that certain unacceptable factors that could be allowed only in the underground market have been intervened and finally expelled out of the market. It was like a computer virus. Some critics say that if the Cayman Tax Haven and number of hedge funds anchored there are not treated duly, concerned authorities will not be safe from public criticism and eventual denunciation will be made by the whole world.

Before the crises in 2006, the New York City was the world financing center that treated over five trillion dollars of Bank Acceptance Check (95% of global BA). The reason of this success was the grasp of the opportunity market that provided investment networks and skills. We would like to know the factor which pushed Wall Street to this failure point. Was it solely their greediness or the inevitable business topography that has led them to hell?

East Asia was not well prepared for the speculative market probably due to the influence of Confucianism or the long history of agronomy-based society. When futures trading bonds were just introduced to Tokyo in 1985, the theoretical price of futures surpassed the market price of futures for the minimum discounted rate items in the spot market. Investors complained and CBOT personnel consulted that it might imply the absence of speculative investment forces.

The key point is that speculation was not built up at the initial stage of the market. Speculation is a cultural or habitual personality. In East Asia speculation has long been considered as same as gambling. It took such a long time for common people to recognize that speculation means to avoid a risk. This assumption must be examined carefully in detail by socio-psychological researches. The speculative market forces would not perform such vast scale study since they are concerned with only impending profits that highly surpass the usual level.

The socio-psychological mechanism associated with speculation could be at best a part of the invisible hand of economy. In the laissez-faire the usual level of competition is not only enough to induce sufficient conditions for an efficient capitalistic model. Unless competition is overwhelming, an economy causes turbulence or fatigue, hidden psychological mechanism in competition would help achieve proper capitalistic booming. But in any case the optimum scale and intensity of such socio-psychological factors should be revealed.

Speculation of the future value and credit card in use of future value in advance at present would be effectual in boosting present economy. These conditions would induce bubbles or the bubble effect.

Thus, East Asia might contribute to the world economy if new but less speculative and more conservative techniques of financial market are to be introduced in the region. The market in East Asia, apt to be investing, inclined in a cluster-wise manner with blind eyes, which could easily prompt bubbles of dangerous level and, then the world economy will lose its stable backup. IMF might have to change their internal strategy for the rice-consuming regions in Asia to induce a corner of hard working and stabilized theme in the structure of global economy.

Bubble collapse and Underground Economic Factors

A bubble can be considered as the basic unit of the universe. Some people think that grids with density or building blocks are the basic units. If string and M-brane are considered as the candidates for the basic constituents of the universe, a bubble could be, in some sense, a better one. Quantum Gravitational Bubble suggested by Steven Hawking is the infinitesimal spatiotemporal quasi-bubble of gravitation.

To approach to the topic, bubbles exist in phases of solid and fluid. They are distinguished by the characteristics of the surface. Solid bubbles are found in pores of floating volcanic rocks called pumice. Well known ALH84001 which was discovered in Antarctica in 1984 is a solid bubble from Mars. It was one of the average annual numbers of fifteen pieces of rock that were bounced by the impact of meteorites and travelled to our Earth.

Most bubbles are met in fluidized pattern. When we take a bubble bath, bubbles on the surface of tub do not attain mobility until mobility surpasses the surface tension. Suppose we have a frictional marble surface of bath tub and fluid bubbles are being formed on it. Usually the direct surface is filled with the first layer bubbles of bath and the second layer continues to build up on it. The bubbles of higher layer get greater mobility than the first layer.

Bubbles are indispensable as long as people need effects on the psychological expectation for the wealth in a market. For the best market condition, every individual should have some bubbles in his/her mind. But each bubble should not be too big. Bubbles had better take the patterns of layers and groups to achieve higher market mobility.

I believe that in reality, humans live lives that are patterned with bubbles. What would be the best size of bubble in a communal relationship? Communal lubricity gains the highest value where integer plus alpha. Integer is a personal territory. If an integer entity acquires an assistant, what would be the best range of relationship between the main researcher and the assistant? By intuition, $0 < \text{alpha} < 0.382$, which means that the assistant gives his/her boss the portion of golden section of 9 hours including 1 hour of commuting time.

Bubble contains deep socio-psychological implications. I believe bubble is at the center of the phenomenon of the creativity process. Bubble is not just an unsubstantial effect but should be considered as a realistic phenomenon relevant to the postefference of transciousness of the human cognition. Postefference is dealt in Chapter 4.

Bubble does not belong solely to capitalism but is a general phenomenon of humanity. Thus, the most proper bubble or foam forms when its condition gives the best efficiency in administration. It is our matter of concern what the optimum condition is. Maintaining a fair distribution of optimum basic sizes of bubbles would be desirable for the reservation of vivid picture about the future in the imaginations and dreams of people. It is my concept that bubbles are just one of the manifested psychological expectations, at present, for the future value. Bubble could achieve the foam status by the change of the economic conditions. Foam is relatively smaller, but more stable and usually clustered or spread over the whole medium of the fluid flow.

It is my observation that the bubble status will end up with three most representative outcomes; bursting, collapsing, and squeezing. For instance, foams disappear by squeezing with just light impact. Bubbles grow as the environmental conditions change. Immanent conditions for bubbling are the personal character and the content of individual pursuit. How we quantitatively measure the effect of bubbles remains as an unsolved question. However, it does not seem impossible if we are well prepared with basic criteria and case definitions. Anyway, the present status of the sub-prime loan crisis turned out to be a bursting case since Wall Street has been playing with relatively big, thin, volatile and small number of bubbles.

Wealth is not healthy when too much deviation exists within the population. Uneven distribution is not necessary, except in the case in which initiation of boosting for economic interflow is needed. Speculations sometimes take the role of initiation as can be seen in Tulip speculation in the early Dutch economy. Roman troops had strong motivation for economic reasons. Mongolian soldiers were known to have similar desires for the collection of wealth. These cases showed that bubbles had assumed the initial cause to establish a vitalized economy.

When the merchant class emerged in the earliest stage of commercial capitalism in the 11th and 12th century, they were originally the surplus population which was a result of the three-field system that had to be converted into commerce, craftsmanship and physical toil. It was also the bubbles of the governing class of religious and the secular power of those days who ordered expensive commodities that caused the earliest interflow of the market economy.

Bubbles are not independent due to management but economy and market would be quickly frozen if mobility originated from the bubble effect that has got dampened. Economic values in triradixism discussed in Chapter 7 are diversified into three classes; utility value, exchange value and expectation value. Mobility is fairly attained when utility value and exchange value in a community are realized properly. This level of market mobility is called a "fair market value".

But fluent mobility is necessary in today's fast-paced society. The speed generally originates from the pursuit of a higher rate of return and consequently from the excessive expectation for a higher profit. Certainly, contemporaneous information technology assisted the speed. However, even IT was originally provoked by the expectation for a greater profit. While such general trends have stimulated abnormal expectation values which were manifested by the bubble effects with an excessively high mobility, expectation value is socio-psychologically motivated.

Bubble becomes an underground economic factor when it is linked with illegal or extremely anti-social motives.

Keynesian Modification of Psycho-capitalism

The current financial crisis is an award ceremony of a dishonorable title to the current capitalism. It is called psychological capitalism, in short, psycho-capitalism. It is not a criticism of capitalism, because capitalism is a natural process of economy as long as people are allowed to react as they are prone to. It could be a highly optimistic view, but capitalism has originally a good identity to supply food to family members and the needy. If it was well-managed with the concern of their limits on territory, capitalism would not have caused such shameful outcomes. The loss caused by the current financial crisis might be comparable to a loss due to the defeat in a major war.

In the 1990s, socialist countries faced a vast collapse. Capitalism and socialism are the offspring born by the logical bifurcation from the natural way of human life. Logics always talk for its systematic ideas only. These controversial logics of economics have proved that they have to get together in order to get back to the origin. It was evidently shown in their practical modification which was vastly made since the two economic systems had appeared.

Keynesian modification of capitalism was well admitted by the President Roosevelt in the last century. National revenue was directed to be utilized to boost economy during the depression. It had been examined and was proved that Keynesian modification could furnish a sort of remedy to the real economy. The economic principle of Keynes was that production and employment are operational variables that can be adjusted by the policy mix of the government. The policies on currency and exchange rate were made public control business cycles.

Keynesian remedy, based on the prediction and modification of the macroeconomic system, would work in business cycles such as Juglar cycle and Kitchin cycle. It looks different to me in long wave cycles, such as Kondratieff or Schumpeter cycle, which was affected by the distortion of basic economic mega-structure under the economic value system.

As we can see, Schumpeter mainly focused on the innovation and leading industry that belongs to the creativity value domain in our value category. Kondratieff had also recognized that the significant impact on the cycle was the major influence of primary industry and the trend of rent in contrast with public bond and security, which should be located in the terrenity value domain in the value category of triradixism.

I think that the implications mentioned above will have to be studied by economists carefully, because economy can be healed by the sufficient supply of currency in the general range of systematic capitalistic economy, but the same remedy could also excessively distort the economy in an unusual magnitude when some bubbles in a market still remain alive, in order to absorb the supplied financial energy. Healthy capitalistic market would be activated in a thin moderate layer just above the communal value domain as seen in our value system. Socialist countries were not able to utilize even the energy in communal value of an economy since they had planned and programmed entire phases of economic affairs in a compulsory manner.

It is noteworthy that Marc Faber (2002, Tomorrow's Gold, Ch.7) commented that the fifth recession in long wave cycles is coming by around 2004 and the sixth in around 2058, according to the principles of Schumpeter interpretation on long wave cycles. I believe that the span of cycle might have been lengthened because the creativity value has rapidly increased since the mid 1970s. The 1970s was the start of serious innovation of capitalistic economy due to the Oil Embargo.

Furthermore we have had additional influences from the financial engineering, which is based on a new philosophy of the science of complexity. It might be right to consider everything again, including Black-Scholes Model (1973), which has supported the confidence on all the derivative games and the Federal Reserve Bank. It is my belief that the science of complexity has implications on very wide arc of views, which will give pictures of a mega scale for the future of humanity.

Anyhow, what we confront presently is the distortion of superstructure of the economy, the credit structure itself. This superstructure seems to have been seriously distorted in order to efficiently suck the wealth of the lower level countries in capitalistic hierarchy. The collapse of bubbles is said to be arisen due to the ill financial engineering practice (Yasuyuki Kuratsu, 2008, Toshiginko Bubble no Syuen, Ch. 2).

Capitalism has had several stages of development from commercial capitalism. Industrial capitalism and then financial capitalism followed. It might be true that we are entering into knowledge capitalism from the end of the last century. It is required for us to understand that the rise of certain capitalism depends on whether or not its value system and its practical means of implementations have achieved their continuum. Such continuum should change in order to occupy the socio-psychological realms of contemporary people.

The historical development of capitalism has been a kind of transcent up to a new phase of economy adding up a higher and a more efficient hierarchical mechanism step by step. Keynesian modification has worked to cure the discontinuity due to the collapse of bubbles in 1929, during the booming periods. But our present question is how the bubble collapse would be defined in a peculiar financial situation like today. It is a crisis of asset bubble and credit bubble.

Some people believe that economics should not be mixed up with psychological factors. But bubble in asset and credit is no doubt a socio-psychological phenomenon. Asset bubble comes from excessive concentration of speculative investment on asset commodities, while credit bubble is generated from unduly dense speculation on put option sales of credit structure itself, jeopardizing the entire economy to burst.

In the 1980s, Japan experienced a combined recession. It was the land value assessment at the center of the crisis. The reassessment of book value of increased land price had covered the long defect of management. However, hidden bubbles have burst when banks met the financial liberalization. Combined recession is essentially caused by asset bubble. We know that the most economically advanced countries have been experiencing frequent bubble occurrences since the 1980s, in succession and sometimes simultaneously.

In 1992, the Japanese writer, M. Yoshikazu, insisted that Keynesian solutions, that is applied to the entire world, must be adopted by the establishment of strong global central bank system in order to introduce the global currency, to regulate BIS, and to sufficiently control huge worldwide monetary flows (1992, Fukugo Fukio, Ch.4). The way of his thinking is based on the understanding that the global deregulation of financial system has destroyed Keynesian remedies that have been practiced within the boundary of nations respectively.

I think it is another extreme alternative that could be developed from the present crisis, which will be called an alternative of the dominance of systemicity in the global economy.

Global Rearrangement of Financial Capitalism

It is widely discussed that global financial centers need to be rearranged. Some people seem to have taken the idea of triple financial centers, as have been administered for world stock markets. And those financial centers are said to be connected to the future blocs for the future key currencies. It is also observed that there are other alternatives being discussed to provide a wider acceptance to the endeavor of big emerging markets, China, Brazil and India, for financial and currency centers.

The rearrangement of financial capitalism has to strengthen the logical regulation first. It passed a long time since such opinions had appeared. In Global Transformations (1999, Ch.4), Davis Held and his colleagues stated after their ten years of thorough study that short term capital flow has clearly revealed its tendency to cause high instability and systematic risks in the capital market.

As discussed by Kaneko (2008, World-wide Financial Crisis, in Conclusion), the global economy should not repeat the failure of Japan in the 1980s. He points out that the American mode of financial capitalism has produced the most serious problem in the determination of total loss because of the complicated securitization process due to their institutional and technological strategy. It is a vicious spiral between the credit stringency and the aggravation of real economy that worsens the crisis.

In addition, there is another symptom of resources inflation associated with the present financial crisis. We think, according to Yasuyuki, that currently we are under the combination of asset bubble and credit bubble. It is usually seen in a combinational mechanism, as written by J.K. Galbraith (1990, A Short History of Financial Euphoria), of asset speculation and induced credit expansion, which is the spiral mechanism of bubble expansion.

Credit bubble, this time, looks different because it is not induced as appeared in the "stock-flow combined recession" (Yoshikazu Miyazaki, 1992) and in the patterns of most speculation in the past, but it is the source of structural instability. Greenspan's bubble is not entirely his own responsibility. If he has had any theoretical tools to quantify the bubble effect, he should have avoided the collapse of bubble.

The definition of bubble presented by Yoshikazu, while explaining the Japanese experience of financial bubble, is that "bubble occurred when real price of asset and the price of fundamentals do not mutually coincide." He pointed out that the change of expectation and the change of asset price must have interrelation but it is hard to strictly verify in theory.

It is an intentional design based on a wrong philosophy, rather I would name, an ill philosophy. We have to seriously consider that the ill philosophy is basically the suction process of the global financial capitalism. The mechanism of suction from the lowest hierarchical structure of capitalisms has been settled down, whether it was intentional or not. This is the point we have to clarify but it is not easy to do so.

I believe that Kaneko and Yasuyuki are warning with greater alerts than they have done before. There will be a high probability of the total collapse of the world economy, this time or the next time, if the fundamentals are not reformed and restructured.

We talked about the adjustment toward rearrangement of global financial centers. Triple or multiple centers alone are not the answer. The financial power based on injustice and immorality is not going to save the economy. Financial capitalism should be regarded just as a stepping stone to an ultimate solution. But the methodology of managing the economy and grasping the picture of economic integrity has to be given a greater consideration, in order to pave the paths to a new era to which the human race must get within 50 years.

If we understand that financial capitalism is just a partial manifestation of the total economy, the global rearrangement should be considered only a part of our concern in framing the future political economics of the world. Financial capitalism must rearrange itself, remodeling the world currency problem.

Financial capitalism is only the active organizing portion of the entire economy, in terms of credit institution and credit flow. Financial mechanism plays a central role in knowledge economy of an information society. But soon later, we will see certain activation of the real body of knowledge capitalism, which is supposed to follow financial capitalism.

The introduction of internet domain into our real life was just a short and pianissimo prelude of the main game of knowledge economy. The financial struggle associated with internet fundamentals has caused a lot of trouble to humanity, as we saw in bubble effects. When internet fundamentals are changed to cloud computing conditions, as anticipated by several thinkers, and the cost drops down fast, the communal value domain will absorb the major portion of the generalized internet efficiency.

We have communal domain and terra domain of the economic value system, unseen but vividly working in a hidden condition. We need to design a way to correlate appeared economy and hidden economy. Such design has to do with menesysal integration, which is explained in Chapter 7 among financial system, monetary system, commercial transaction system and industrial system.

In this context, financial menesys (system and networks) of market domain should be mended together with global currency menesys. Currency menesys has to be considered cautiously in order not to lose the hidden impacts of unseen domains. Global currency system or menesys is analogous to the conscious state of humanity.

9-3. World Currency

Currency Domain as Floating Standard of Divergent Economic Values
Current Imperium of Global Monetary System
Spiral Hierarchy of Currency Pyramid
Global Currencies to Support Continuum Structure

Financial capitalism is at present on the stake of global governments. Financial systems have been under organizational process over the world. It was proved that the global financial system of systems is distorted in philosophy and in methodology. United States FRB is responsible to stabilize prices and also has to keep leadership in sustaining healthy financial systems. It was a severe failure of the second item of FRB responsibility, currently applicable to not only the United States but also the global extension.

Worldwide unitary reserve currency, such as "bancor" named by Keynes, has been suggested by authors such as J. Stiglitz (2006, Making Globalization Work, Ch.9). But some others suggest six or seven multiple key currencies are actually anticipated at this stage of financial crisis within the next half century.

It is necessary for us to design world financial infrastructures in a grand conception. We should not forget value problems, because value axiology is the main theme we have to struggle in order to attain control over the stable but vividly activated economy in the creative peace.

Currency as Floating Standard of Divergent Economic Values

If we imagine in mind a flat domain of currency space, it is reasonable to match plural domains of divergent economic values in our recognition with that single currency space. The currency space is considered as a floating standard of reflected divergent economic values. Therefore it is rational to deem the currency as the ground of human consciousness to vitalize economic continuum and to create foam stratum in place of bubbles.

Currency has various patterns in practice. In the internet environment, new modes of currencies have been innovated and will do further more. The kernel of our current world problems is the global system of currency. Conspiracy views are frequently raised concerning key currencies. Whatever the past history of currency is, the factual manifestation of currency patterns are the inevitable outcome of humanity. As we experienced without enough currency, economy does not move quickly and substantially. In a micro scale, the number of local currencies and community currencies prove their authentic utility in the circulating economic flows.

The flow of currency has a definite relationship with the flow of commodities and the flow of utility value. Exchange value of currency comes from the credit associated with the currency system. The confidence and reliability on any credit system rest on the systemicity and nexusity of the system instead of, I would say, the integrism of the currency menesys.

Circulating is largely related to nexusity and organizing institutional and administrative efforts, such as mintage and setting rules, is connected to systemicity. The integrism of the currency system should be different depending on the location, for instance at the central part and at the peripherals of the continuum of the global currency system.

It is our interim conclusion that the domain of currency exchange must be studied deeper than ever concerned with the human consciousness. Currency is the integrated realization of values. Domain of currency is the standard map to give correspondent reflections among various value-concerned maps that humans can imagine. This intuitive derivation out of the correlation among brain science, human consciousness and social institution provides some theoretical stance.

Currency is not only the measure of flow but also the measure of justice. This standard has been threatened and will be unstable in our future too if it is not adequately treated in ethical respect. Everything concerned with monetary phenomenon is floating and does not have definite standards. Even the gold standard would not have its legitimacy and legality in the economic theory if strictly and thoroughly examined.

Political power has decided every fundamentals of the economy including currency settlement in gold standard. Gold has socio-economic value based on psychological utility and practical usage. However, still obscure are the criteria of correspondence between utility value of gold and its exchange value in economy. The pain and agony consumed to obtain gold would be the measure of the exchange value of gold with other materials. (Delight and ecstasy was felt before the value of gold could be exactly measured, like labor which measurement are fairly done today)

The production of gold in human history is 160,000 tons in total. Although we utilize the entire pure gold into effective currency, it is worth of around 4.5 trillion U.S. dollars. It is not enough to supply the currency volume in order to efficiently manage the global human economy, sufficing the endless desire. The currency in monetary form is much less than the figure, but still not enough.

Hence, we have some inevitable queries in mind; what the most desirable pattern of currency in this scientific era would be and whether or not we have any practical method of managing the broad currency issues to reflect the value system with less distortion. These questions are associated with the complicated world power mechanism.

J. Galbraith mentioned that the fundamental cause of the financial crisis is the uncertainty of the financial system, which promotes “speculative euphoria” spontaneously. Greenspan has once described the same thing as the “irrational exuberance.” I believe there is a hidden fear positioned at the other end of the psychological extension of investors. Fear to lose, to be expelled, or to be caught by poverty lies at another end of the exuberance of speculative euphoria.

Current Imperium of Global Monetary System

The structural interdependence between the U.S. and the countries in the East Asia is called “a common yoke”. The yoke was similarly managed in the days of the British Empire. An imperium is able to maneuver the financial and monetary situation of their supportive countries to adjust to its drawing mechanism through a siphon.

A decade ago, a renowned European Banker, A. Lamfalussy, expressed that the elasticity of the advanced world financial system can be examined and assessed when the U.S. financial system experiences the depression. The U.S. had been in a booming economical state for the last decade at that time. He said that the market system is more efficient than the financial system and asked whether or not the financial market has the ability to find a balance in the course of price changes, as usually done in the market.

He did not suggest why and how we examine the elasticity. It will be a good chance to accumulate basic data of the financial market if the current crisis is not so inordinately severe that we have keep an inert and stabilized condition for a certain period of time. But it does not look so.

The mechanism of price determination of financial assets is relatively hard to understand because they are dependent on liquidity and leverage. He commented on the fact that there are not definite theories about the liquidity and, especially, the deficiency of liquidity. He added that there are not clear agreements even on the optimum level of leverage.

It is our belief that financial capitalism is not a good definable system unless the reflective relationship between monetary domain and socio-psychological domain is clearly understood. Psychological factors are studied recently with emphasis on the marketing strategy in the micro economy. But to solve problems of the mega economy (global and historical, when compared with the macro economy that is mostly national and in time-series) is a more essential and a philosophical approach.

J. Stiglitz asserted that the fundamental cause of many problems associated with globalization is the global reserve system. He advises to reform the global reserve system. It was indicated that specific national currencies used for the global reserve system caused inequities and instability.

Irrigation projects to furnish water resources in rice-growing countries experience rent-seeking behavior. Actually rent-seeking has been changed to power-seeking to prevent weak and peripheral farmers to attain equitable water supply. It is true that such small society cannot survive if the core families do not preserve the leadership. It is a kind of social cost that everyone in the community acknowledges. But it does not justify that the core families can cut out other families from survival conditions or exploit other's fruits of sweat.

It is well known that the present crisis has been caused by the deficit of the U.S. economy. It is also well recognized that, nevertheless huge deficit are present, the dollars flow back to America. The differences between the interest rate in the purchase of the U.S. Treasury bills (T-bill) by developing countries and that of the U.S. loans back to same countries are effectual international tributes. No matter how the tribute is interpreted, at least, it is thought that we cannot find the due justification of the mintage of master currency of the US dollar, except the fact that the United States is a new center of global monetary imperium.

The hegemony of the world governance has several distinctive aspects in these days. Generally, it was found in the power of military, of mintage, of financing, of technology and ridiculously even in the power of consumption. The consumption power is effectively workable in the present global economy. The consumption power has been supported by the circulating volume of currency. It is the story of the contemporary United States.

Suppose the United States collapse abruptly, none in the world would expect such drastic disaster. Even foes of America would not want it. Stiglitz recommends providing novel global greenback instead of the U.S. dollar for master currency. His idea is to utilize the global greenback to solve financial problems, poverty, development, environmental issues, education and medical supports. However, it is thought that the single global master currency would cause serious problem if the world is not prepared to manage it.

The United Kingdom maintained imperial system until just before the World War I. The United States succeeded UK position after the World War II under the Bretton Woods system. If the world is really going to get into a new era, the United States and the leading countries should create a new leadership in thoughts. Ancient ideal of imperium is not the right answer at this stage of the world history. Hidden intention for imperium would not impress world people to support the new leadership of any country in the world.

Spiral Hierarchy of Currency Pyramid

B. Cohen (1998, the Geography of Money) described a global currency pyramid which consists of seven strata; top, patrician, elite, plebeian, permeated, quasi, and pseudo. Global currencies compose a pyramidal shape. Global currency issues will change their agenda in accordance with the main decisions in what the master currency would be, and whether or not and to what extent the existing currencies would remain.

In a world of peaceful negotiation, the power of rationality prevails over the violent power. Money generally reflects the internal energy potential of power games. The world monetary system had been under long influences of the exclusive territoriality. The exclusive territoriality was one of the major outcomes of the Westphalian regime of 1648. The multilateral treaty finally getting out of Middle Age, based on which modern international law was defined, generated most of modern political economic conditions. Territoriality is the prime principle of the system of nation-states.

The strict principle of territoriality gave way in the last half century to a viewpoint of transactional network space, in which the concept of monetary flow can be treated. The transactional space behaves like a porous and permeable medium in which the mechanics of fluid flow is applicable. The major motive of transaction is to interflow.

Such monetary transactions are made actively in a single representative phase space of currencies, revealing the unrecognizable combination of socio-psychological flows. Commercial, industrial, and legal transactions are all subject to interconnection. Utility value assigned to a transaction is usually represented by the monetary quantity.

Suppose we have a denarius in the Roman Empire, it was reflecting one day earning for a family gained by a man working for one working day. Monetary domain is controllable in size by the volume of currency under circulation.

It is necessary to see that monetary institutions have developed for the entire human history and the foundation comprises with the essential background of humanity.

The circulation of money attained its symbolized power in humanity by the fixation of unit utility related to the stress potential in economic transactions. Socio-psychological pain of one day hunger and its equivalent one day labor might be the most prevalent unit for the social stress. The stress potential varies from person to person and from situation to situation to some extent, but the time span is identical.

For our understanding it is noteworthy that our hypothetical theory shows that the human transciousness would emit successive postefferential measures of an individual's physical stress. The postefference should be formulated through preafferential reactions of physicality established from the embryonic stage of humans. (Ch. 3)

It is generally known that postwar Keynesian school has tried to establish a value system based only on the single value of labor. Our study is different from the Keynesian in that the value of creativity and terrenity should be considered together. Additionally, labor value is important in its ontological significance by which labor value facilitates communal affinity through achieving uniform stress distribution in a community. These ideas coincide with Triradixism in Ch.8.

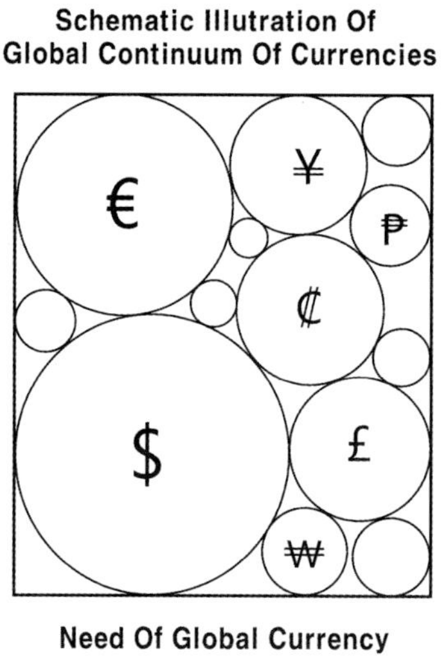

Figure 9-3-1. Shematic Illustration of the Global Continuum of Currencies

It is my theoretical stance to assume a spiral hierarchy in the currency pyramid. It is a frame of triradixism that helps us to imagine the distribution of three values in continuum. The exchange system from primitive barter in natural condition, community currency either of ancient or present, fully grown market currency like drachma, bezant, sterling and dollar has position in a monetary phase space at their time. Specific currency in history can find their position in a spiral hierarchy thus in currency pyramid.

Since capitalism has been floated to the surface of the economic power sphere in history, the monetary phase space of currency pyramid has been gradually expanded to a global scale and achieved a greater continuum, following the progressive capitalistic hierarchy of agricultural, guild-accumulated, commercial, industrial, financial, and intellectual (informational, or knowledge) capitalism, no matter whether they are matured or not in accordance with their respective definition.

In transactional monetary space, there exists a continuum of currencies. In Figure 9-3-1, a schematic illustration of the global continuum of currencies is presented. At the central position of the continuum, value potential is the highest one. The center is where creativity is at its highest potential. The master currency is supposed to be located at the center where total utility values of terrenity, communality and creativity are accumulated most heavily. In general, the continuum of currencies has a similar correspondence with the continuum of values under certain assumption. It is thought that the growing point of total values, especially the value of creativity, is located at the center of the continuum of values.

Suppose we have a pearl necklace of more than two hundred countries and number of other currency-using communities following them. The biggest pearl at the top is the U.S. and the E.U. and the others are following in sequence, diminishing in size. If we plot the location of the existing countries in the continuum, we can see the spiral form of scattered countries with their respective currencies in current use. In this case, the continuum is intuited to be a bell-shape, represented in a cylindrical coordinate with three dimensions such as the distance from the growing point of creativity, the radian from the base line, and the quantity of total values (in vertical axis).

It is lucky for the human race that the (comparatively) most creative society has the leading position in monetary continuum. But it does not yet mean that the U.S. may be located at the center, the growing point, which will have to be able to control the mintage of the global currency. The United States have assumed their potential uni-power position since 1990, in global politics and military leadership. It was similar in the fields of economy and culture.

However, the most important aspect that the world requests for America to fulfill is in ethical and philosophical aspects. America should be born again not in a religious meaning but in their morality. Capitalism has nothing bad if it is morally administered. It is not easy for a confederated country to behave satisfying the global demand morally in every respect, because it is a mixed up conglomerate. Maybe that is why we have to design a novel global constitutional institution to give a definite authoritative interpretation and a compulsory measure to be implemented for rare cardinal affairs of worldwide scale. The institution would be like a hybrid of the global Senate House and the judicial Supreme Court.

Getting back to the continuum, we need to understand that value usually flows opposite to the flow of money. In the monetary continuum, we can imagine behind the scene a value continuum which has counter vectors in contrast to the monetary flow. The counter vector of value continuum, generally caused by tangible and intangible commodities carrying values, generates vector of monetary flow.

We can understand it better by imagining the mechanism of tropical storms which generally have upward whirling wind and cumulonimbus cloud around the periphery and downward calm wind at the center of the eye. The wind is the monetary flow and value has a stream of counter direction. Monetary flow is a means to symbolize the unseen value for humans to realize the potential flow of the human consciousness. The human consciousness has characteristics (both in individual and social basis) resembling the potential flow, which exists everywhere in the dynamic nature, such as the electrical voltage and the geomagnetic potential. This scheme of idea is illustrated in Figure 9-3-2.

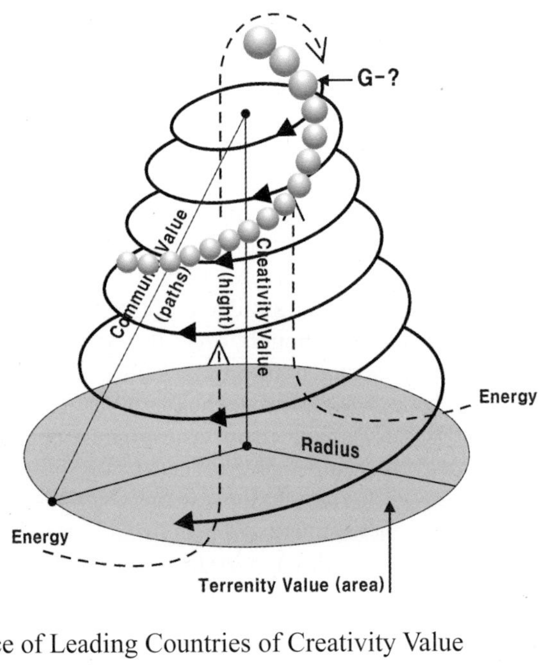

Figure 9-3-2. Necklace of Leading Countries of Creativity Value

Economy caused by humanity is like an extraordinary flow caused by the potential difference. Consciousness, originated by the human desire, caused the potential difference. Transciousness, expressed in Ch.4, called for gradual and accelerating increase of foams and bubbles, which used to exist in our life. However such transcious behavior of our brain has caused development and evolution of humanity by transcendental intentionality. I would like to call evolutionary leap of humanity as the transcent of humanity. A significant transcent of humanity in written history was materialized from 25 to 20 centuries ago. We are at the second wave of such transcent.

A continuum of economy has the center of creativity supply as a growing point as found in normal plants' life. Creativity is a factual form of consciousness but yet to be counted in monetary expression only when it is admitted by the market. Creativity used to have sunken cost as in terrenity value. The increase of values of humanity is generally attributable to the mathematical increase of human population (in the domain of communality value) and geometrical increase of human creativity (in the domain of creativity value).

The global countries in the pearl necklace can be grouped into G3, G7, G20, G50 and whatever needed, in order to make clusters for administrative hierarchy. The financial and monetary system is a mirror that reflects the economic flow of utility value. A mirror is not a device to create and boost a real economy, but could be misused for an improper money game.

We have to realize the significance of where the human race has arrived at this moment. Pricing mechanism in a market is the sole method of presentation of utility value. Price in planning economy cannot help us recognize the representative value connected to the human desire. But market pricing is not normal when the market contains too much bubble effects. The global market is not only an economic phenomenon, but an undiversified (undifferentiated) embryo of humanity. We must not play with the market if we are really concerned with the future of humanity.

As we all have recognized just before the crisis, more than 95 % of banker's acceptance checks in gross volume of the world trade have been reinvested in New York, because New York was the only place in the world that was fully equipped with necessary infrastructures in order to invest any surplus money most efficiently. The investment of surplus flow of money is allowed legally and customarily in a form of speculation, which falls within the purview of willful negligence under an unconscious agreement of the entire society.

In my opinion, the cause of the present crisis is furnished by the fluidized bed of the less definite derivative market and the less responsible morality for the abnormal expectations. (Fluidized bed is a totally unstable floating bed of materials by supplying strong air flow from bottom to top.) The networking fundamentals of the financial market have been maintained for continuous flow of excessive profit until it drained all the potential of investment, healthy or unhealthy. It was unquestionably a quasi-black market.

The general trend of debates looks like it is going to end up with the global currency. It is our task to clarify what the standard of currency is and what the standard of the global master currency should be; gold standard or gold and silver standard. Gold, silver, and other means for the symbolization of exchanging and carrying values are not to suffice the enormous volume of economy in the future human history.

In my view, the world is at the stage of entering a multi-currency menesys; allowing all the existing currencies and more liberal minting, coinage and creating various forms of future currencies, some of them already found in internet sphere. But we absolutely need certain supply of novel monetary form equivalent to the volume of the new supply of creativity value into the global market. Whatever the novel master currency would be, and whether we call it "bancor" or not, it has to be minted by the most reliable administrative institution as the hybrid of the global Senate House and the judicial global Supreme Court. The institution should be more authoritative and authentic than merely a Central Bank. But I will not be surprised if the US dollar maintains its superiority over other currencies for the next generation. But the new master currency is of a different purpose, to straighten up the background philosophy of tuning in value-currency interrelation.

In this research, we emphasize the mechanism of dynamic modularity. It is related to the current and future trends of the world monetary system. It is expected that the complex strategy based on multiple-currencies, which allows healthy competition in a monetary market, enables both local and global vitality. The vitality of currency market is not aiming to the drain of high profits of abnormal level. The vitality in local markets rather assures the stability of the global market if managed healthy and autonomously, especially when hierarchical development among local markets spontaneously takes place.

It might be called "the strategy of entangled multi-currencies." Figure 9-3-3 illustrates how the entangled multi-currencies can satisfy the financial need of complex society. The success of the strategy depends on how it is designed. The world itself is recently studied to grasp the philosophical conclusion that the stability is propped up by such entangled multiplicity. Market is a very natural concept that sprung from the human mind, founded on the recognition of every other entity under mutual intercourse. The hierarchical development has recently been explained in connection with the evolutionary thought. However, it is more important in political implication than in biological analogy.

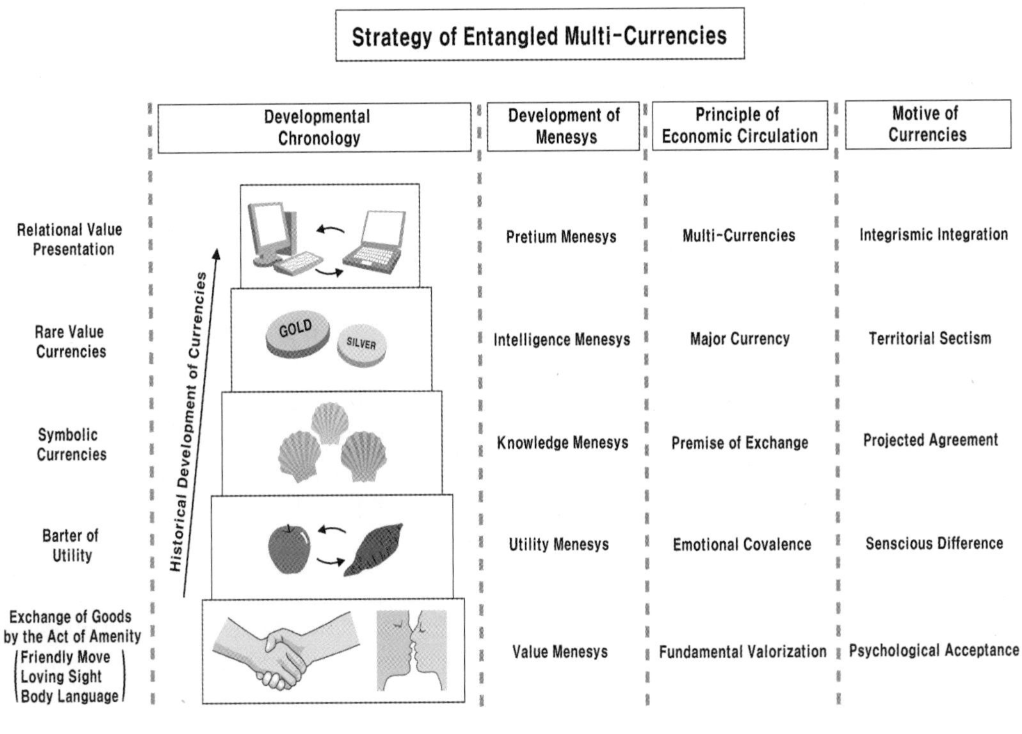

Figure 9-3-3. The Strategy of Entangled Multi-currencies

What if three or four polygamy system of global currency exists? Is it going to be well administered by a simple communication structure? Isn't there any leeway that would cause greater problems because of political involvement? Isn't it right to establish a liberal multi-currencies system without questioning? What are the standard and the criteria of the liberal multi-hierarchy and multi-currencies system? What are the criteria associated with the multilateral negotiation when the major currency such as dollar and euro do not function well?

The aforementioned pearl necklace of currencies is an efficient model to generate the hierarchical clustering. The system of governance in economy will have to be designed as has been done for political institutions. For instance, economic system has already developed financial system to supply money as the blood circulatory system and money is made in the marrow-like central banks, institutions with constitutional self-authority. The transcent of humanity has similitude with the embryonic development of the human fetus.

Political system was like the development of members of apparent physique. Politics is thus a little bit simpler than the economic system, because it is an apparent adjustment of the unbalance in power relation. But economics is a more complex system of mutual adjustment in intrinsic unbalances, which has to be regulated in a greater degree of freedom. Economics is a more delicate subject than politics in general.

Politics decides things relatively, by comparing powers. Nevertheless, even in a global dimension, politics is a matter of proper regulation between the systematic powers of elitism and the nexal democratic rights of commonism.

What are the principles and rationales of economics for a monetary system? Are we going to admit new key currencies? Jeffrey Sachs, when he visited South Korea in 2009, recommended a common monetary system among Korea, China and Japan. In his stance, what would he suggest for the prospect of the African region?

Additional Global Currencies

It is deeply impressive that J. Sachs has a real ardent passion for the African countries as he wrote for the Millennium Village Project (p251, 310, Common Wealth, 2008).

It is our belief that we have to go further than that in order to face the concurrent world problems to establish a continuum of psychology, knowledge and economy. Thus, we need a greater supply of financial aids to the relatively isolated countries. The monetary circulation is the major solution to solve the alienation problem of isolated countries.

Our standpoint is not that of monetarism. Monetarism has to be overcome by the thought of continuum. Continuum economics is centered at the continuum of the value of creativity and the other two values are in support of the value of creativity.

The continuum of the value of creativity can be recognized when labor value is divided into intellectual labor and physical labor and the intellectual labor is identified as creativity value with less volatility. Attention, concentration, and indentation of the learning and the training process produce less volatile creativity, because learning is well admitted into a communal society. The creativity is said to be volatile when creative innovations are not well accepted and applied into the market.

The continuum economics is interested in the expansion of the boundary of continuum and in the intensifying the continuity flux in the continuum. The boundary and the continuity flux are related to how strong and how long the correspondent market sustains its continuum effects. The continuum effects of a market are positively giving the most significant influence to the interflow of credit, value and money among individual entities in the market.

The expansion of the boundary of continuum is achieved by the transcendental intentionality. As explained in Ch.4, the consciousness of the human cognition is related to intentionality. Intentionality is explained usually as the aboutness of the human consciousness. But human is not only looking at things (intentionality), but also mobilizing all the neurophysiological potential (transcendental intentionality).

This is why people frequently and easily put themselves in teleological pursuit. Human being orientated in accordance with his original tendency is very natural. Certain initiality of a person gives deep continuous impacts on his life. This is why we used to tell youngsters to make up his mind and have dreams for their future.

Therefore, it is the same with the human race in the course of their growth in terms of the expansion of their continuum boundary. The market has grown by the successive waves of admitting global countries into the continuum economy, which is the internal story of the market growth by admission. But external growth of continuum should be guided by the growth of vision. I would urge human race to touch the flash of lightning trans-space vision. Trans-Space project has to be touched for the human race in order to save the momentum of continuum growth. This side of concept would be called "growth momentum."

For the growth momentum projects, which provoke the value of creativity to be multiplied, it is thought that a differentiated special currency from ever before should be innovated. The special currency, let us call it "pax", would be symbolic to the whole humankind that guides to the creative peace. It is to cope with definite problems observed in the administration of global reserve currency in the US dollar. It is widely discussed that the United States had to suffer red figures in finance if they want to supply their own currency to the whole world.

It is known as the Triffin's Dilemma. The Triffin's dilemma is caused by the basic imbalances in the Bretton Woods system. In maintaining the Bretton Woods system, the United States theoretically had to run a balance of payments current account deficit and surplus at the same time. If surplus pursued, global liquidity would not be supplied and the global economy would shrink.

On the other hand, if liquidity solved, deficit situation would be maintained and dollars would be oversupplied. Then, dollar value should decrease and its credit as reserve currency would follow. If an alternative is taken to reduce the deficit to decrease monetary flow and withdraw dollars back to domestic economy by the raise of interest rate, a recession will be anticipated.

Humanity has psychological pits, usually called paradox, for which they do not understand the internal logic clearly. In the Triffin's dilemma, there exist speculative ways of thinking that affects certain spots in the whole process, which show the twisted effect of the psychological aspects when a part assumes the role of the whole.

It means that pax is the source reserve currency to give the gold-like authenticity for the credit of key currencies of the top hierarchy countries or the regional master currencies in the world. Pax amount minted periodically represents the additional volume of creativity value in the period furnished into the continuum of the global economy. The quantification method of the increase of creativity value should be designed.

In addition to the pax currency mentioned above, three other sources of strengthening continuum should intensify the intrinsic attributes of continuum. At this moment, it is necessary that strengthening the global value of communality is deemed to be achieved by supplying new monetary flow through creating new currency from the countries at the centric circle of continuum to the group of countries of a less-capitalized situation and thus of retarded development.

The three means of new monetary flow into the global economic continuum can be admitted in three reasons; a contra-capitation grant for the population of retarded developing countries, annual compensation evenly disbursed for the increase of land price in advanced countries and BBB currency flow for both of asset bubble in stock exchange and credit bubble if unintentionally occurred. (BBB denotes before-burst-bubble.)

The "contra-capitation grant" for the population of retarded developing countries is designed to introduce those national economies of less capitalized countries into the global continuum economy. The grant could be utilized for the construction of social infrastructures, either tangible or intangible.

The gradual compensation is to be paid annually to the inert national economies to remunerate for the rise of land price occurred in advanced countries. The abnormal rise of land price is attributable to the concentration of economic activities which drains the capitalistic profits. The kick of land price is a part of asset bubble and associated with mortgage issues, but does not mount systematically on the real economy as observed in financial institutions. We will call it "rent compensation." (In the western society, land price and rent can be used almost equivalently. In Far East, land price is regarded as a means to accumulate wealth.)

BBB currency flow for both of asset bubble and credit bubble should be used as a toolkit to fix the global economy in a dynamic and autonomous manner. BBB currency, or 3B currency, is designed to prevent bubbles to be produced in advanced countries and it is called "bubble prevention." 3B will take care of super-structural bubbles as we have seen in the recent financial crisis. The contra-capitation grant and the rent compensation can be disbursed by key currencies, such as dollar and euro, while the bubble prevention will be paid in 3B currency. 3B would be managed like SDR (special drawing right) so that the effect should be easily traced.

It is expected that an instant query will be raised, a query on how great the scale of 3B monetary flow would be from center to periphery. It looks proper to start from the three percent of the global GDP. Three percent is known to be an indicative optimum rate of return for non-profit organizations. We have to remember that three percent is not an ignorable amount because it was the magic percent of the free market volume of oil product which actually caused the first oil embargo. We may increase or decrease gradually from then on.

9-4. Knowledge Capitalism and Value Capitalism

Knowledge economy is defined by D. Foray in a complex form of three characteristic ways; the first is that the importance of the knowledge-intensive employment is considerably increased; the second is the economic importance of the informational section has reached a definitely significant level; the third is when the intangible capital outruns tangible capital in the real capital stock. (Dominique Foray, L'économie de la Connaissance, Preface, 2000)

It is not easy to discern if knowledge economy has already come. Experiencing financial crisis, there are some criticisms on the new economy which includes knowledge economy. Knowledge economy does not bring up significant difference from knowledge capitalism. People usually think that knowledge capitalism is at the door. Maybe it is already in our family room. It is probably right that the prelude of knowledge capitalism impressed people when internet had been introduced. Knowledge capitalism has a huge body; since knowledge encompasses every intellectual concept we have ever put our hands on.

Value capitalism is coming after knowledge capitalism. In our viewpoint, value is a concept essentially concerned with the value of creativity. Creativity has a greater territory than learnt knowledge, because it is an (endless) extension of humanity and human wisdom.

Knowledge Capitalism

Both countries of socialist and capitalist have experienced certain level of collapses. The human race had good opportunities to restudy those relevant economic data in their process of collapse. The initial effect of the undiversified (undifferentiated) embryo of an economic phenomenon of a society could have been examined. It is not yet too late if we intend.

Around 1990, it is known that some economists from outside had consulted to the Russian government and had chance to collect those economic data. Those data would provide deep basic understandings of the future global economy. We need new insight on capitalism and neo-liberalism, which are too simple and something no more than complicated slogans. Capitalism and liberalism has the deepest implication associated with the phenomenon of humanity.

The financial crisis revealed an indefinable territory of capitalism, the shade of economy and its supply of psychological energy, the greediness or a due desire for human need. Current capitalism has to recognize where they have reached and where to go.

It is my view that socialism collapsed before the wave of creative market, not only a plain conventional market. It is well recognized by intellectuals that the creative market had first appeared when white color laborers had exceeded half of the laborers in America in the early 1960s. A capitalistic society with the major drive of knowledge can be acknowledged as entering into the stage of knowledge capitalism. Creative market is based on the knowledge capitalism and is a scouting advanced detachment of value capitalism forthcoming, presumably, fifty years later.

I know that there have been other definite factors that caused the collapse. But it is noteworthy that socialism detected by the tentacles the future territory before them which is not manageable by their theoretical background based on simple equality and logical communality.

Knowledge capitalism in the last half century has built social superstructure comparable to a religious value system in our past history. The concept of knowledge capitalism is almost substitutable with information capitalism. The difference between the two is that human knowledge has certain suppositive *a priori* that information does not imply. Knowledge is the essence of the present humanity and will assume the foundation of the future humanity. Knowledge capitalism has the deepest effect from the interior of humanity, the intrinsic structural aspects, which are essentially interrelated with value problem.

Knowledge is ontological. Certainly, it is right that parts of knowledge are attained through the epistemological process. But since knowledge ought to be formulated into a system (exactly, menesys expressed in Ch.7) in order to be activated in our life, knowledge has to be ontological, and if we faithfully follow the direction of this book, knowledge has to be persontological.

When we organize our pieces of knowledge in a conflict-orientated way into a system, we can see intuitively that knowledge would not be forming a well rounded continuum. The peace, strictly speaking perpetual peace (not of E. Kant), is the only way of constituting a healthy and highly communicable continuum of existing knowledge of humanity. Humanity is going to get the way towards peace knowledge system or peace knowledge menesys.

Peace knowledge menesys is the substantial body of peace knowledge continuum. Peace knowledge menesys is extended by connection in reversal manner of knowledge menesys as we can see reverse prism producing white sunny light back to our sight. That is what we can describe as altruism, self-sacrifice and so on.

But when knowledge is associated with profit seeking capitalism, so that it may be called knowledge capitalism, such capitalism is supposed to be governed by the virtual reality. This means that, in another sense, the pieces of knowledge could be exaggeratingly expandable and contractible. It is due to the brain function to match rationality with emotional aspects to bring the coincidence in mapping correspondence.

This process looks like coming from the effect of the virtual reality in knowledge accumulation. Virtual reality has played an intensive role in the accumulation of knowledge by its profound variability which enabled pieces and manifolds of knowledge get connected together.

But the same function of virtual reality has caused a deep influence on the pitfalls of knowledge capitalism by perverting the true measure of value of reality. Value is not subject to conversion because it is based on the factual occurrence of accidental affairs.

Knowledge has to do with bubble effect, which is the product of - intentionally directed, or upon many occasion falsely orientated - the human expectations. It is eventually relevant to the virtual reality. When we understand the essence of the phenomenon of knowledge, the capitalism based on knowledge should be treated from the fundamentals of value axiology.

Value Capitalism

Value capitalism is the last stage of capitalism. The novel phase of humanizing economy will have to be disclosed. Knowledge capitalism is the way to creativeness of humanity to be accomplished by value capitalism. Knowledge cannot guide the human race to their desirable state of intuitive attention and the concentration of their mental resources. It means we will not be able to escape out of huge pile of junk knowledge unless the axiology of our economic values is soundly arranged.

Value capitalism has something to do with the capitalized value. What do we mean by "capitalized"? Capitalism is a capitalization process. Capitalization is to make things capital. Capital is defined by "wealth available for or capable of use in the production of further wealth, as by industrial investment." Value capitalism is to make the wealth in terms of value available for or capable of use of further wealth.

As humanity gets to the stage of discerning between price and value, and of understanding the difference between knowledge system and value system, value capitalism should appear on the horizon. In fact, value system is permunal (personal and communal). Individual value system is not possibly working. Communal value system cannot work alone either. Value system has to be permunal.

We understand that individuality is the basal philosophy of capitalism. Such extreme attitudes eventually caused the psycho-capitalism until now. A right understanding of value capitalism and its associated value generation by creativeness would extend the future of humanity into the outer space.

Without the change of humanity, present civilization will be imploded. If and only if we could overcome the obstacle in our innerness, we would be able to utilize all the past innovative outputs in due order. That is when capitalism evaporates before the emergence of activated communal creativity.

If value, especially creativity value, becomes our criteria in living, humanity will learn how to exceed the capacity of even quantum computers. The capacity of the creations out of humanity will easily exceed the administrative range of humanity unless the human race attains the ability to use the power of creativity in a due manner.

Value capitalism will be initiated when the chain circulation of VAT (value-aware transactions) started in the continuum of economy. A full circulation in the psychological continuum of economy would ignite the initial conscious chain of humanity in the continuum economy and start to propagate value capitalism as we had experienced in the internet cyber space.

Then, what is a full circulation? It has got to be related to dynamic modularity. The first module of initial group of dynamic modules should be located at the centric region and has to have sufficiently complex systemicity of rationales and sufficient nexusity to transmit trends. It could be located at the center, core part or periphery. The major factor in determining the first module is vivid transcendental intentionality.

Let us call the first module of transcent a prime module. Prime module propagates mimesis effect (originated from A. Toynbee) to the initial group. If we accept the concept of meme in a society, meme is generated from prime dynamic module. (Meme is variable dynamically.) Transmitting trends would be just a switching on and the rest is dependent on the internal readiness of the continuum and its constituents.

The value capitalism will have to spread over the entire continuum; it is a higher level than knowledge capitalism and it is not capitalism any longer. It is a matured continuum economy. The creative peace will open a way to the economy of *pretium* (=value) continuum by stimulating the continuous process of creativity in a global scale.

Creativity is prone to float making lives and things unstable. Value capitalism is the institutional facility to accommodate vitalized creativity processes in various forms. The essence of value capitalism rests on how society, or how continuum in a novel expression in accordance with current historical development, admits, preserves and provides necessary frame and space for the ever changing contents of the human creativity. It has a deep relationship with the legal frame, such as the universal public law, which will be dealt in the next section.

Redefine Labor Value, and Normalize Capital Value and Terrenity Value

Values are innately interrelated through the cognitive process of human beings. The human cognitive process realizes value in an integrated pattern in the emotional domain and in the symbolic domain of the monetary system. These integrated realizations of values make the flow of value and utility values mutually mixed up so that we may not differentiate three values distinctively. In the last financial crisis we know that there was a real estate problem beneath which deeply associated is the problem of terrenity value.

In this connection, we need to discuss about the value system a little bit more. If we want to understand the value system and its application, labor value ought to be redefined. The activity of labor is analogous to a farming field cultivated by the consciousness. The human consciousness requires a physical action of labor to confirm the reflectivity of its rationales on the reality. Labor confirms the content of consciousness through sensciousness. Sensciousness directly reflects physical reality.

Lower-than-head (physical in contrast to mental) labor value, suppose it is worth of one Roman denarius, which in general geometrically increase in accordance with the increase of human population, cannot be used to count future value. Creativity value has been increased exponentially in accordance with the developmental history of characteristic capitalism.

Disregarding the ethical point of view for a while, capitalism can be seen as the process of intensification of capital force. Even in an agricultural society, although it was induced by the greediness of landlords, capital has been accumulated for thousands of years. Their problem was that they had only land for growing crops and livestock to invest their capital. But strictly speaking, the accumulation of capital, even it was very tedious, had occurred.

Then a question comes up with such situation; whether or not squeezing sweat of their peasants by the landowners has to do with the creativity process and the creativity value. In strict rationale, it is affirmative. The excessive creativity in human nature appears as unjust bubbles like acts of aggression and compulsory land annexation. That was the psychological drainage effect of creativity in a society and thus peasants lost their portion of creativity when their labor was not duly acknowledged. This is a general phenomenon that has happened in all developmental processes of capitalism.

At this point we have to confirm the fact that labor is based on physical contribution and it is strictly related to present value. Capital value is concerned with future value. It is usually said capital is the frozen labor. But Capital value should be redefined to be equivalent to accumulated creativity value, as discussed in Ch.7. Future value ceaselessly stimulates fear and creativity at the same time through postefference process of the human transciousness. But human labor has dampening effects of bubbles and also collates communal constituents parallel together.

Labor value enables the function of connexion work in a society. Labor is important for establishing and sustaining the socio-psychological fundamentals, especially representing the connectivity portion, of the continuum of global economy. The content of the continuum is actually the accumulation of human creativity value, and labor value (exactly, lower-than-head labor value) makes it possible to have the continuum connected.

In this context, the capital flow is the recognized flow of creativity occurring in the consciousness. Apparent capitals are the superficial variation, like the wave of the ocean, of the accumulated value of capital. Deeper current of capital flow undetected in the apparent economy has to do with utility value (present in the domain of communal value).

It is thought that unrealized exchanges of inherent creative value in transactions have originated from the common property region of the global humanity. The native medicinal knowledge on herbal plants grown in India should not be collected by herbalists of a mammoth multinational pharmaceutical corporation for the purpose of registration of patent in his own country to legally compel to purchase the license back to Indian people.

It is thought that conventional capital value is expressed as the following Equation 9-4-1. Physical labor value constitutes the initial foundation of historical capital value through self-labor, slave-labor or seizure by invasion.

The second composition of historical capital value is accomplished by the accumulation of ceaseless creative value of the human race.

The third factor to be considered in connection with the historical loss of capital value is capital goods and knowhow destructed and outdated as time passed.

Capital value = Frozen physical labor value
+ Accumulated creativity value
– Destructed and outdated capital goods .... (Eq. 9-4-1)

However, we experience other forms of significant loss of capital value such as in monetary bases through the modification of enterprise accounting, by bubbles bursting in stock markets, etc. It must be studied further in depth.

Physical labor value is represented by the unit of HP, horse power. Human labor value becomes physical labor value when all creativity value is extracted out of it. Thus, we can say physical labor value is directly proportional to the human population. But creativity value is increasing as creativity per capita increases. Then the cumulative amount of creativity value should be presented in a highly sharp exponential increase as science, technology, culture, and various infrastructures developed.

Terrenity value is not simply represented by the land price, but has been changed to represent the characteristics of medium for the economic interflow. For instance, the improvement or sustainment of land and environment induces economic activities and make economy to flow. The fertility of land is a good incentive for farmers to immigrate.

In this context, let us discuss about the normalization of capital value and terrenity value. Imagine a particular situation of the human consciousness in a complete virtual reality developed from its initial conception. Nullity and infinity are not discernable to such consciousness without subjective episodic memory.

Pain pertained in the labor of experience is theoretically the measure of realism for a certain individual. Ecstasy is not as useful as pain for the measuring purpose because usually it is an integrating reality. Sometimes, measurement can be made by muscular ecstasy because human physique tends to take the pleasure of moving his body. At this point, we would consider it an anomalous phenomenon temporarily.

We said above actually based on that laborless human consciousness (which means consciousness with null experience) tends to bound even to infinity in its measurement of values by emotional or senscious means when creativity value and terrenity value are taken into account. It is why orthogonality is needed in the value relationship.

Orthogonality is not an easy concept to digest. I would simply define and easily understand the concept to model two value systems, which respectively meet together with labor value system in a perpendicular dimensional relation. I worry whether or not regularly trained mathematicians and physicists would lose temper and faint for what I am taking into consideration.

I believe there exists a principle in the entire universe that expresses that orthogonality is the very basic structure of nexusity. It means mutually orthogonal nexusity, stressless relationship, is the original condition of the initial universe. The most infinitesimal stress condition would cause the sudden infinite agitation of the universal existential condition. Orthogonality is the null stress situation in a dimensional crossing. I believe it is similar to the phenomenon of the universal shrinkage of gravitational dimension as discussed by Linda Randall. (Warped Passages, 2005, Ch.22) It looks to me identical or at least similar to the orthogonal relationship of Axis Tempus with brain activities. Human labor is the differentiation of time dimension in the human recognition.

Based on this intuitive proposition, it is conceived that the normalization of creativity value and terrenity value has to be done by labor value. (The purpose of this book is to quickly show the whole scheme in the way of persontological thought.) Redefining labor value in persontological perspective enables to normalize the other two values. Thus, we can see that capital value is normalized not to be exaggerated nor disregarded by self-centered emotional expectation. In this way, capitalism will take a healthy stance of ethics in depth socio-psychology, avoiding floods of bubbles and maintaining a steady continuum.

Value of mineral resources is realized when excavated and introduced into the economic flow of commodities, unless it is assessed in terms of future value for the purpose of bank loans or for the sales of the reserves of mineral mines. As environmental value is considered in rent nowadays, resources value as raw material or energy source should make compensation for the environmental loss it has caused from the nature. It means that environmental rent has to be distributed through all chains of the continuum supported by the resources value. Environmental rent is actually generated from the emotional pain measured by the human cognition. In this connection, resources value is contained in the terrenity value system to establish a right circulation in the continuum economy.

In order to cope with the accelerated exponential growing of the quantity of creativity value in our future, it is suggested to design certain ways for retarded parts (or countries) of the global continuum to disburse only the incremental portion of creativity value of the concerned latest period. The total increase of creativity value produced and accumulated for a long period of time has been heavily imposed at a time to poor countries without exception. They did not have opportunities to enjoy the early-bird benefit of the continuum. Their opportunity costs have been lost and not compensated. The following rear countries should be taken care to be introduced into the economic continuum without straggling.

Therefore, the aforementioned pax currency (representing the newly springing creativity value added into the global continuum) needs some contrivance to be used as the standard economic value of the future. Ultimately, the liquidity expression by currency flow and even the pattern of declared notation of economic value will have to be transformed to expose the effect of creativity value which has been disregarded. The expression has to do with the current changed circumstances of the fluidal economy such as internet, telecommunication, etc, but other forms of change should be realized in the global continuum environment, such as caused by the wide spread living condition of the virtual reality, will require deeper and wider changes than we are anticipating.

9-5. Universal Public Law

Universal public law is one of the major keys to open a gate to a healthy continuum of the global economy. "Universal" means this public law will be applicable in exo-Earthly conditions of the universe too. It means universal public law ought to consider the future of the human race. "Public law" in the past has implied that nation and public institution is superior to citizens in a subordinate relationship. But public law in the future will be applied to the global and universal continuum.

Legal science in the past, especially in civil law tradition, has put accents on the abstraction and systemization based on the rationality. The tolerance of legal science to the irrationality to a considerable extent will have to be necessitated from the near future on. It is due to the fact that legal knowledge systems in the world should integrate to contain innumerous perspectives of geographical, communal and functional localities. Forthcoming global legal knowledge system will include not only the three most representative legal systems of the world; civil law, common law and socialist law, but also other local customary laws and newly developed conceptual legal customs in novel disciplines of science and technology.

It is said that a great part of the global order from which human race has been benefited in the last half century has been enabled by the adoption of anti-Hobbistic expressions taken and used for the design by juristic approach of the US government between the Hague Peace Conference in 1899 and the Paris Peace Conference in 1919 (F.A. Boyle,

Foundations of World Order, 1999, Ch.9).

"Hegemon" is the keyword we have to ruminate nowadays. Several countries at the front of human endeavor are carrying the word in their minds all the time. Regime theorists in politics have expressed that Hobbistic hegemon has a determinate importance in realism. In the case of the US, American constitutional regime for controlling force has dealt with the controversial problem between justice and expedience. I believe that the US domestic tendency of anti-Hobbistic democracy recently came to understand that lobbyist tradition in parliamentary activities to lubricate the firm constitutional idealistic system has been overdoing their utility function.

The other leading countries are assessed hardly capable of maintaining global hegemony compared with the US in philosophy and in practice as far as world-administrative power is concerned. The US has born-limitation to their choice of international policies except in their extreme situation.

However, it is our question yet to be answered if the USA will be able to change their constitutional ideal after the 11th day of September. Fierce reactions called for Iraqi and Afghanistan wars have been seen not very successful and regarded insufficiently justified several years later by most global people. Now, then, beyond international law and beyond regime theory, what would be the next step to be justified and widely accepted? In 2002, the DHS (Department of Homeland Security) had been launched. What is the focus of the consensus that the global people would positively agree in common for the prevention of terrorism?

The philosophical issue laid behind the current world is "the problem between the whole and the part." We do not have any standard global legal frame. What we need is not the standard but the principle how to comprehensively utilize all the local legal frames. The universal public law will have to be designed to suggest the public way how to interrelate and to interact among local laws in global legal circumferences.

Universal public law deals with the global continuum, whatever the continuum is defined; market, culture, knowledge or anything. It has to take into consideration the conflict issues between relaxation and tension, core and periphery, peak and edge, and other dualities present in the human environment.

I believe many citizens would agree that universal public law will eventually be required. Universal public law would not be regarded as a realistic measure to power politicians in international affairs, but as an indispensible means to the global consciences and intellects.

Legal Stem for Global Continuum

The world should become one not of the unitary but of the diversified. Dynamicity ought to be sustained in all diversified localities. Dynamic modularity was not sufficiently developed when nations had sole authority through its national governments within their border. Other modular factors in the planet globality have emerged since the latter part of the last century. Module is a basic part of the whole. To make dynamic modules workable in the global condition, the presence of the value chain is one of the essential conditions.

Division and diversification of the global value chain between different value systems have been tediously established for the last two centuries through cultural, economical and religious agitations and/or intrusions after the Industrial Revolution. Global production system or network of commodities and services supported by regularly established contractual relations (not necessarily by direct ownership), have been developed in the last century by multinational corporations. International public institutions have been established through the development of the multilateral diplomatic treatise and the citizens' solidarity. International NPOs and NGOs stem from the development of international networks of citizen activities.

Our concern is to find out the common denominator of various legal systems based on sound humanity in order to establish a proper legal menesys that supports the global continuum of the political economy. It is interesting to learn that the weakest links are the most important hubs in the mechanism of network. Not all but some of those weak links, which have wide connectivity and thus promise a higher stability than a network, are located at the less or some of the least pursued targets. Such clues will be found around weak linkers, not around strong linkers. (It is reduced from the findings of Peter Csermely, Weak Links, 2006)

Weak links of legal systems are with the terms and conditions for the poor, the undefended, the defective, and the less privileged. In fact, the notion of publicity has been developed founded on the thought of justice in which weak links of social constituents (and also in which weak links in the logical chains of a society as we discussed just above for legal stem) are to be protected and taken care in many ancient laws.

The social weak links, for example, have-nots, orphans and widows, have no belongings and obstacles against others to keep for themselves and thus easily cohere together. That is the historical wandering people or displaced persons in hunger times or war times. Such cohesive power is different from the profit-seeking elite groups working systematically in order to achieve definite purposes. It is true that wandering vagabonds can also be organized for their purposes, after which they will be protective against outsiders. It is also right for elites the fact that they are weak linkers of the society when they have a wide spread of relationship as a hub without enforcing others according to their purposeful act.

It is a well known fact that ancient societies did not have definite development of a legal philosophy and a legal science. Legal systems are like music or a score. Innumerable music is all thematic. Such themes are well acceptable by the human perception if they are acquainted.

Usually we classify legal systems by mapping with the standard mode at the center. The standard mode we have in mind was the national base legal systems. The days we will confront requires us to get out of that fixed idea. General understanding on the legal phenomenon should be discussed from scratch. It does not mean that we have to give up present approaches in general but find out theoretical bases to admit all the possible legal systems into a grand design.

For instance, several legal systems such as in Jewish and Chinese traditions have undividedly combined with religious or ethical aspects. But if such laws, commandments and disciplines were workable in such countries, or even at the level of the community like intermediate groups, the dimensions of such design factors should be considered. Why did the Hebrews think that there is a distinction between M-sh-ph-t (justice) and Z-d-q (righteousness) and the intervening factor of H-ss-d (binding love) in its structure?, why did ancient Chinese think that they contained law and ethics together into a commingled integration with governing skill?, and how those mixed-ups and intrusions could have been successful for a long period of time?, these questions must be carefully studied to induce a legal stem for the global applicability.

Current global culture has to think of the conservation of the diversified asset of humanity not only in an anthropological view. There is not a superior and an inferior system but an adaptive system to encountered circumferences. Infra basics of aforementioned factors of *hessed*, ethics, governing skills, and *jus naturale*, all have hidden logics in them. We do not yet have definite ideas of what factors would deserve to be the weak links. We only know what the hard core is.

Romans had legal concepts of the eternal law (*lex aeterna*), the natural law (*lex naturalis*), and the human law (*lex humana*) in principle. *Lex humana* was generally classified into *jus gentium* and *jus civile*. Their legal infra was *jus naturale* which contains the natural law and the eternal law. Such foundation was prepared by the stoics represented by Zenon who supported the cosmopolitan world of Hellenism.

Western Europe has a progressive history in the theoretical development experiencing competition among different legal systems; common law, civil law, and socialist law. Legal science, which would assume the hard core of the next generation's legal system, looks like it is taking a firmer stance in civil law countries as the extension of laws experience new dimensional developments. Similarly, in common law countries the depth of interpretation of laws necessitates as the human society gets into a deeper systematic formality, legal positivism and positive law have been gaining a greater influence on the legal system.

Upcoming task for the legal system is centered at the fluently governing the instantly changing legal contents of many societies. The frames will be redesigned simply and applied firmly and the nonessential contents kept so flexible by communal bodies, which will be incessantly arising and diminishing under the decreased governance of national level.

Even the legal system should become alterative for itself in a floating manner. A legal system floating in need but stable in principle can be accomplished through the continuous trial and error methods by people participating in political process of communal decision making in legislation. The cost and time for legislation should be lowered and kept easy by technological means.

Such flexibility would allow social and political governance easier in the global scale when three levels of liberty are recognized, i.e. the levels of compulsion, restrain, and released (liberated). The compulsory level has to be the most definite and simple to fix and sustain systemicity of a communal body of various hierarchies. The restrained level of liberty would admit the least necessary factors to maintain solidarity, presumably associated with scientific principles. The released level of liberty will contribute for the creative and liberated humanity. The easier governance will enable a more fluent global interflow in the political economy.

It is thought that the legal stem must be formulated hopefully within next quarter century as I personally predict for the establishment of the global continuum of the political economy. Such prediction comes from the most probable consequences of future human history to be influenced by the accelerated technological progress. To accommodate forthcoming abrupt changes of humanity and society, the frame of legality should be mended first at least in a pragmatic way.

To be pragmatic means that the human creativity has to be maintained consistently at a proper level in order to administer the present speed of technological pushes. The human creativity has to be kept vividly in the coverage over the entire essential loci of humanity and in the complexity to enable digging into the depth of scientific and technological revolution. Otherwise, humanity will suffer drastic steps of shrinkages.

Imperium or Integritium

I have a deep admiration for the writings of N. Luhman; the ideas about positive law, the need for the diminishment of complexity in social process (stepwise and procedural solutions in legislative-administrative-judicial process, constitutional and democratic process), for the flexibility, social system, etc. His ideas lead us to the realization of the lack of nexusity. The real problems are located at the interactive phase with hetero-systems such as culture and economics. We will have to adopt the concept of dynamic modularity.

As I cautiously predict, if the present conditions of globalization proceeded without significant change, the world will, in a long run, meet the next step most likely in a form of a strong-arm imperium, which is simply the very concentration of power. In my view, the world has to exert for the preparation of "integritium," in order to survive for most of the profoundly manifested contexts of humanity in history.

Integritium is an integrated political status of the general acceptance of the diversity of most nations and cultures, with adequate level of power concentration. In this comment, people would raise questions about what the integration should look like. The world in history has experienced the repetitive process of expansion and compression, in other words, diffusion and concentration, enjoying the effect of apparent growth and internal substantiation in turn.

Integritium is thought to have existed for millennia in the past but has not been exposed to the eyes of the human kind. Imperium has always been only a part of integritium. Imperium is merely a core part of integritium. Integritium encompass all scope of the human race and the realm of humanity. We can say that integritium is a matured understanding of humanity on its own identity. The distinguished difference of present civilization from the past is that human consciously recognizes to some extent the internal connectivity of humanity in a global scale.

In this era of the concluding part of the Earth-confined civilization, we have to recognize the meaning of the silhouette. Integritium is a dancing silhouette of dynamic humanity on the earth. The silhouette is socio-psychological political economy. Neo-Liberalists assume that the silhouette is the global market and some post-Marxists (after 1990s) consider it as the global autonomic collective subjectivity.

However, the substance of the silhouette is a complex humanity which has recently been revealing its dependence to physicality based on the numerous findings in physiology and other related disciplines. The physical dependence gave a firm ground for individuality. There is a contrary tendency to the physical dependence of the human identity.

Issues on rhythmicity, values, complexity, and even technologies of virtual realities open fresh ways for the integrated insight on the human identity. (I would like to leave a room for the role of the human mentality in a clause of "the duality of contrary monism," which I have to discuss intensively some time later in my life with scientific data. The duality of contrary monism is implied from the proposition that the universe emerged from the building blocks of the nature, which suffice both of the theoretical foundations of the existential threshold value of the Planck constant and the philosophy of "creation ex nihilo.")

Integritium is a consciously awakened state of humanity. It will be realized as the fundamental of the existence of the human race when we extend our territory into the endless outer space. Integritium possesses real software of being human and the human race.

In order to bring up an integritium in a balanced and efficient way, absolutely needed are the integrismic preparation and the cruxological provision as discussed in this book. Integrism would be, if the integritium is supposed to be intentionally planned to provide intended cruxological situation, the complex prescription of external nexusity and internal systemicity.

The apparent nexusity and internal systemicity are related to the value interflow among different value systems, value covalence among value subjects, and value transference among value spheres in triradixism. Karl Marx thought that he succeeded in finding value expression that enables the interpretation of utility value into exchange value.

The problem contains an issue about how labor can be interpreted into the value of labor product and how labor hours can be expressed into value amount of labor products. Such labor products and commodities mutually compose a phase space of value correlation in a specific society, which in turn expresses the relationship in market prices in monetary value. Thus the morphosis into monetary expression was thought to be the matured value expression. It is basically similar to the factor of capital when we regard capital as a frozen labor.

When we ponder into the deeper socio-psychological aspects, it is deemed that pain, as an expression of the senscious and the emotional measure, is the major factor in the conversion and the transference of value. Ecstasy and pain are mutually reflective contents of humanity. Utility value is identified to be the integrated expression of such neurophysiological situations. When utility value in the conscious state of the human cognition attains an expressive form of monetary exchange, legal systems gains governing constraints on value system.

Thus in this context, the insights of the human science will be applied in the legal consideration of the economic valorization. The economic valorization will become gradually important in the legal system because the process of value transference and value covalence will have to be considered with a greater significance in the global economic order in the near future, presumably within 20 years. Popular feeling and public sentiment have been admitted to the political decision making process nowadays by adopting an opinion poll.

Legal system, in a long run, will experience similar process as in politics as far as legal peripherals are concerned. Legal system in the upcoming global civilization has to be able to cope with the chaotic atmosphere to be encountered from time to time depending on the localities (usually equivalent to peripherals) due to ethno-graphic, politico-economic, religious, cultural, or on-off line situation.

Scientific ways of thinking and humanitarian approaches in activities altogether have been intentionally improved in the last half century in a global scale in contrast to the concurrent increase of systemized and impulsive violence. Integritium is to be built in balance on the previous bent and crooked topographical condition.

The Nature and the Humanity

In an ancient Confucian classic literature called Courtesy Writ (禮記), delivered is the courtesy and decorum of humanity. In a section of Music Writ (樂記) that deals with the political function of music, courtesy-music-punishment-politics (禮樂刑政) is the answer of integrated politics. It is written that "Therefore, courtesy directs people's intent, music harmonizes human voices, politics have people unite and accord, and punishment prevents wickedness; courtesy, music, punishment, politics are all the same at their summit and emerge good governance stemmed out of the concord of people's hearts."

Usually we know in contemporary politics that law, politics, virtue, and skills are supposed to be mixed up properly to produce the best political governance. The ancient concept of composition of courtesy, music, punishment and politics remind us the vicious political deception of common people through the abuse of 3S; sex, sports and screen. Good-nature doctrine and evil-nature doctrine have dominated the trend of thoughts of the East and the West. It is our observation that the human nature has a duality which we have to consider in application as can be seen in the doctrinal problem of the human nature. In the practical range of life, there is not a significant opposition between the two doctrines.

When we consider the development of understanding of the nature and the humanity in the last century, we have to put a novel approach to the establishment of the frame of the universal public law. The new horizon of the universal public law dawns at the basic affairs; the new conceptualization and description of time, weights and measures, the establishment of universal calendar based on the newly defined periodicity of human life, the legal identity of human being, the principles of application of science and technology into the realm of law, and the definition of the legal sphere extending to virtual domain (not only cyberspace), etc.

The complex adaptation of the creative peace largely coincides with the sustainable development. Complex adaptive system in legal frame tends to take initial multilateral forms of agreements and treatise between entities of politics, of enterprise and of citizens. Such private, public and mixed legal frames that exist in an entangled form in the global legal environment have to be given a proper arrangement.

The essence of current problems in the legal framing is the lack of seedling or fundamental to start the discussion. The global development, efforts of space and ocean, should be efficiently mobilized by launching integrated mega projects, which will be the toolkits to accelerate the legal rearrangements of the concerned fields. It is enough to structure only the initial core part of legal frame by such global multilateral activities.

In the general approach of legislation and judiciary tasks, needed are the new horizon of technique or technology for systemizing and networking of the complex legal frame in a global scale. The technique in its first step will be related to the human knowledge system. We will have to develop "the technology of knowledge" which treats the content and structure of knowledge system, or if acceptable, of knowledge menesys. It will be more integrated and higher "arts of technology" (than present information technology) to predict the long term development of the human history.

Some studies on the history of law recently tend to correlate the legal frame with the long term cycles of economy. It is my view that the economic feature of the time has to do with the concurrent knowledge system without exception, because economy is the manifestation of the internal mentality and, contrarily, human mentality is the outcome of economy. It is a generally recognized fact in these days of knowledge economy.

The structure of knowledge system depends on the basal structure of ethics. Ethics of various cultures from the early human history have general coincidences in the prohibition of homicide, adultery, theft, perjury, avarice, etc. However, as discussed in Ch.7, the psychological defect behind all kind of human crimes has three factors; deception, decadence and violence. It is generally recognized that violence threats human life, decadence rots personhood, and deception distorts human relation.

In this connection, fundamental constituents of human ethics are deemed to be the ethics of life, the ethics of personhood and the ethics of communal relation. The ethics of life deals with individual factors. The ethics of personhood takes care of destruction of personhood. The ethics of communal relation has to do with the factors concerned with communal ethics. These ethics all stem from the human physicality and the human communality based on physicality. Physicality and even communality are the results of the coexistence between humanity and the nature.

The factors of human ethics are tabulated below in Table 9-5-1. The hidden topographical structure of ethics governs the apparent legal frame of the global condition. The legal stem or legal frame should take the structure of ethics into account. The following table shows a reorganized expressive form of ethical aspects, which would function in the arrangement of legal thoughts. If rhythmicity has to be considered in individual or communal spaces, ethics of species and ethics of substantiality will be considered. Forthcoming global complex society should require the interpretation of ethics to be comprehensive but to converge significantly directive.

Table 9-5-1. The Factors of Human Ethics

Ethics of Life	Ethics of Life-or-death	Issues of Euthanasia
	Ethics of Sex	Issues of Spectrum of Sexuality
	Ethics of Species	Issues of Physicality and Mentality
Ethics of Personhood	Ethics of Substantiality	Issues of Virtual Reality
	Ethics of Liberty	Issues of Oppressive Maltreatment
	Ethics of Integrity	Issues of Schizoidism
Ethics of Communality	Ethics of Humanity	Issues of Relation among Person
	Ethics of Society	Issues of Transactions, Agreements, and Treatise
	Ethics of the Nature	Issues of Humanity and Environment

Code of Universal Public Law

Universal public law will have to take a form of integrated code that global people would accept and follow. It will be achieved only if nations and peoples find the sufficiency of legal adaptation and the utility applicable to their practical lives.

We had two great leaps in the history of code. They are the Justinian Code and the Napoleon Code. The Justinian Code was a compilation of the existing Roman law. The Napoleon Code was known to be enhanced in the attitude of integration. Our desire for the universal public code (UP code) is many steps higher.

The prerequisite frame of legal science should be settled down before formulating the UP code. The difference between legal frames comes from the sources of law. The source of future universal public law would not be divinity, nation, multitudes, or any sovereign entity. The source of legal authority will be balancing among the rational necessity of laws, the world legal traditions, and the pursuit of dignity and trust in global humanity in the coming era.

Format and content of the UP code to be pursued will be founded, for instance, on the following three principal and simplest needs of global people in priority sequence.

1. Fundamental; Have them all eat and cured.
2. Communality; Have them all realize others pain.
3. Creativity; Have them all take delight alone and pleasure together.

It is good for a legal frame and codes to be equipped with the simplest backbone and comprehensive considerations of variant situations. Above arrangement of law would assist politics by bringing power game into the rational legal mind. We have been far departed from the subconscious effect of the legal governance on human life through various level of education.

The edition of past contents of legal sphere is thought to be proper to be expressed in the following scheme.

1. Core of the UP code; System-conservative, Endo-Humanity introspective
2. Peripherals of the UP code; Diversity-conservative, Exo-Earthly extensive
3. Legal Informatics; Rearrangement of Legal Documents of various levels and fields

There is one more agenda to comment; the consideration of the future morphosis of humanity. The humanity will be experiencing three interacting morphoses in the at least next one hundred years; info-morphosis, media-morphosis, and idea-morphosis. I believe humanity will not go too far from the present characteristic of physicality for the multitude of people.

However, the human interface with new style of living technology have been altered significantly even until now and will be changed in a greater magnitude. The human interface is actually occurring centered in the spheres of info-morphosis and media-morphosis. In fact, the eventual change would come in the sphere of idea-morphosis, which looks like gradually affecting the human physicality for a long term.

But the point of discussion here is with persontological change of these three morphoses. Persontological information, persontological media and persontological idea are all implying essential changes in overall humanity.

Persontological changes as a whole represent a trend in the alteration of human response in a manner that sensciousness overwhelm consciousness. Linguistic tools of humanity sounds like having been strengthened in phonetics and increased in quantity (this is a part we need to inspect experimentally) but not in semantics and in quality of depth. Human abstraction and its derivatives seem most likely captured by the monotonous and fragmental tendency imported from recently developed new technologies.

Info-morphosis has been made from analogue to digital and will be eventually persontological which is upcoming fast after the virtual reality to strengthen human sensualness and sentience. Sentience is a terminology used for animal sensation which does not have self-consciousness. Human will possess a richer sentience in their territory of self-consciousness. Maybe a richer sentience might influence negatively to the complexity of the self-consciousness. Humanity with more sensualness will perceive differently from the past.

Media-morphosis has made a journey from linguistic to the audio-visual media in the last century. It is suggested to think of persontological development in media. For instance, the holographic presentation of information is a way to persontological media. Multi-parallel media not only depending on complicated audio-visual technology but also with various forms of rhythmicity including scent is at the threshold of our civilization.

Idea-morphosis has enhanced from a base stood relatively on subjectivity in the pre-modern age, but changed the stance in the western world of the modern age to scientific objectivity, and then is forwarding toward the overall atmosphere recognizing a denser intersubjectivity than before in the post-modern age. Ideals will be also changed to persontological, which will be characterized by the realization of the whole and the infinite. When humanity encounters an inevitable way out of this planetary constraint the self of human race has to be realized.

The idea-morphosis of that altering stage will have to meet a mental condition to overcome the meaning of the whole and the infinite. This process of idea-morphosis is going to be accelerated in the next 100 years to formulate basics of the continuum of humanity in a manner of legal frame in the next generation.

Legal alteration in the global condition is to be different from and incomparable to any of the past.

9-6. Military and Anti-violence

Military ventures are being spread over the globe. It is a natural phenomenon, because the warfare costs are bottlenecking conservative military projects. Military industries are ready to enter into the general life of common people constructing conduits of amity and amenity. Survival games, war games and virtual games show the starting tip of such trends.

However, it is better for humanity to check the fundamentals of those introductive industries. The virus of violence has deeply infiltrated into the human consciousness and the imagination process. Violence and military efficiency are not mutually on the same track. It is especially true when military efficiency does not largely depend on the human physicality.

We might pass some period of time in a near future to see the explosion of the potential violence hidden in humanity and the present civilization. It will be solely dependent on whether humanity is humble enough before the nature or not. If the leading countries in power, such as G3 and G20, are humble enough to waive the preferential right, such explosion will be exempted.

The human race will soon, in a long scale, see the almost violence-free military power endeavoring to become the scientific vehicle, in order to advance to the solar system. We need to change our eyesight and prospect.

Low-density wars and battles will come up more actively in the short future. Military and weaponry regulations in a worldwide scale have yet to be taken care by multilateral agreements as has been taken care and being considered in PSI (Weapons of Mass Destruction Proliferation Security Initiative). The military industrial complex will prove their value in the space industry. But such fact is not going to justify their guilt in spreading the cause of the worldwide violence.

Kshatrya or Volunteer Army

Present military forces at duty in their position have to think of what would be their responsibility within the next three decades. In a general way of thought, there are two ways of development in the military occupation and role definition.

The character of military occupation has been changing so quickly in the last century due to the professionalism and scientific approaches in strategy, tactics, and arms system. Thus the following facts were generally accepted; organizing voluntary forces in public by national states and establishing military businesses. It is most probable that military private enterprises will be publicly accepted if they are firmly regulated legally and by internal rules. (Conscription system is not popular any more in the general trend except in the recession.)

There is another trend of competent military personnel, encompassing from intelligence through activist fighters, who realized business opportunities in private with their expertise. Such military professionalism is comparable to Kshatrya; a militant cast of ancient India that consists of professional soldiers, which is a kind of hereditary soldier group. (Modern Kshatrya will not succeed across generations in certain but will be regarded as definite social identification.)

The two trends mentioned above would be the way provided for career soldiers in the future. The hybrid of the two can emerge. Effective and forceful army in any way is needed for the world public security. Peace and order are of different dimensional business from warfare. We will have to open a way for active Kshatrya volunteering army under contractual basis. But before launching, we should be prepared with firm regulation and sufficient training against the illegal military violence, and the full support for the fighting against niche-violence which we call low-density war.

It is anticipated that the earliest business projects of space exploitation will be in need of the hardest drill and the highest expertise even compared with those in military standard. Personnel globally involved in the initial massive push to outer space should be, frankly speaking, chosen out of the best elites of military troops. Such prediction implies that huge groups of top notch military volunteers have to be supplied globally within a few decades.

It is true that some people worry about some probability in that military forces would maneuver global or local political powers in the future as had appeared in the Roman Empire. We have read many conspiracy stories before. Such anxiety should not be ignored in the long run, not only by the non-military but also more seriously by the military themselves for their business purposes.

Armament System of Multistage Systems

The system of systems is a concept appeared since the late 1970s. It is clear to utilize the system of systems for a higher efficiency in improving tactical and strategic military operations. But it is not yet well known if the system of systems is also helpful in a battle to restrain military operations effectively reducing unnecessary violence. We need to apply such efficient way in a balanced manner to either side of mutually reflective doctrines in combating and in peace-making. It is an agenda to be seriously considered in unmanned warfare system of systems which is implemented by only logic.

In our study, it is considered that the concept of the system of systems (let's use "sysosys" for simple denotation) is somewhat a misled concept. There is an alternative idea to the sysosys. Menesys (megalo-network-system in Ch.6 and Ch.9) emerged from more basic early forms of cruxological development under specific integrismic condition. Sysosys is a compact logical output of the huge enterprise and organization of modernity that appeared after the World War II.

"Networks and Netwars (2002, edited by J. Arquilla and D. Ronfeldt, Ch.10)" does not give definite answer to the identity of networks because consensus has not been reached among theorists. However, the distinction between system and network has to be given to grasp the situations in management. This distinction would help enhancing viable flexibility in planning and operation for the both aspects of hardware and software. Systematic logical thought alone will not always lead to success.

Warfare technology of unmanned system has been introduced since the expansion of strategic air force. Unmanned tech has been diversified its extension to surface, ocean, space and even in cyber space. However, we know that unmanned warfare tech alone cannot ultimately promise the completion of war without the final human touch as can be observed in most of the human affairs.

Various phases of wars and battles, in theory, is supposed to change due to the maintenance of conformity for the information exchange, reporting system, and maintenance of organizational system between strategy and tactics, between headquarter and unit fields, and between software and hardware. However, the maintenance of conformity should be complemented for the best result when sufficient degree of permission is given to each unit in order to arbitrarily return against situations in each field.

It is one of the major military doctrines of all age that the victory of war and battle is governed by the factors of fire support, logistics and concentration of military force. Therefore, the difficulties of modern war, especially the agony of gigantic military forces against the effective non-regular attacks, deepen.

Regular military forces of considerable size are to be converted to actual police force in the domestic and global security purposes by minimizing armory and logistics and acquiring social permeability. In a long run, police forces and military forces will not be differentiable in economically advanced countries. Legal consideration for such change will be essential.

Past criteria of military balance were mostly nuclear weapons. Thus, for decades, the five nuclear power nations formally recognized were at the center of the global power balance. However, nuclear power has been substituted by smarter weapons. The center of munitions industries has been shifted to the space industrial technology.

Military restraints are recently performed not on the ground but in the digital virtual space which is actually generated from the symbolic domain of human abstraction. The security of digital virtual space is more fragile than desired.

Industrial connectivity between general information industry and the information of military and national policy has been promoted so intensively that the organizing ability of information industry may be regarded as a major factor of criteria in the assessment of military intelligence.

In the past, nuclear powers were symbols of the global power supported by the psychological illusion of nuclear bombs and missiles. In the present, space development started among emerging powers on the game of lining up for the future power competition. Such trend is not desirable for the establishment of the global peace regime.

This is the time to learn how global space development could be implemented cooperatively under a global menesys. I believe our objective for the global space development project can be achieved by a mammoth business project, the significant profit of which can be well distributed even to the countries with less and no contribution through the establishment of the Fund for the Sustainable Development of the Earth. Current project for the space station is not sufficient and fast enough to resolve the long range global stagnancy anticipated before the human race.

It was thought that a global mammoth enterprise to develop the solar system was necessary at this juvenescent developmental stage of the sysosys in several leading countries in the world. In fact, the concept of sysosys is not desirable for a global size corporation surpassing any multinational conglomerates. We need to take the combination of nexusity and systemicity into account to grow such beautiful fruit of the human intellect, currently in the form of sysosys, which will later be improved to a liberated menesysal form.

Different Intelligence Task than Ever before

Major works of intelligence organizations would shift from the collection of information to the abstract interpretation of the general tendency of the global menesysal situation.

Collecting information is concerned with permeable and fact-finding techniques, which is not well operable in an antagonistic atmosphere. We do not need to mention NSA legend. We have a general agreement that Big Brother is prohibited. Nevertheless, overall predictions on the global change still required it.

Such predictions are dependent on the trained personnel with the highest expertise who knows how global affairs are interweaved and what the role of technology, social schemes and various levels of institutional measures take their parts in the whole uncertainty.

Information collection of military concerned affairs will be lasted as far as the human race exists like the poverty on the earth. CCTV of Big Brother's 1984-pattern will become necessary evil for public security encompassing outer space to underground.
Coverage of military intelligence will be strengthened by the electromagnetic and the nanobiotic mechanical means as can be seen in the current innovative process.

Our major concern about sustaining the global security continuum can be done in any circumstances by maintaining the global readiness to furnish timely prior judgment for due action. When information collections are to be done satisfactorily, a greater emphasis will instead be put on the comprehensive interpretative scheme for the confirmation and supply of circumstantial judgment. Such scheme has to do with predictive simulation model, which should be developed for a considerably long period of time to be prepared for the menesysal management of megalo human society in the coming centuries. It is not a science fiction any more.

Intelligence task for beforehand detection in the last century has been done in a greater accuracy and coverage than any historical ages. For instance, the prediction for the rise of post-military regime (1980 ~ 1987) in South Korea had been made in advance three decades before; youngsters out of poor class wearing dirty slippers entering military academy was thought to be the presage of future coup

It does not look like the comprehensive change of the global situation is detected in a long term basis. We are concentrating on the prediction in the light of the national profit. It might be true that intelligent organizations have not yet developed the list of the full set of factors for the analysis of the global prediction.

Constructing the interpretative patterns of unimagined future would be the important task of the intelligent organization of the next generation. That will not be a fictitious task but will have to be supported by adequate theoretical fundamentals.

Gentle and Powerful Progress to Outer Space Exploitation

Violence intensified by technology, if treated with right visionary perspective for the future of the human race, could be a part of progressive steps toward the outer space exploitation. Violence will be considered as a necessary evil of the past if we understand that we have to deal in a short future with the concentration of excessive energy surpassing the range of the gentle humanity. That is the unknown path to the space.

Armory systems have been renovated for centuries for the widely spread conflicts on our planet. However, the military technologies have a good converted application in the universe, if we think in a positive way.

Dream is the answer. Harmonic concentration of the capability of humanity is the answer. We have to recollect of past East India Company of Netherland and consider a launching of the future Solar System Enterprise of the world. Confined conditions of Netherland in the 16th century have stimulated the small country to venture and jump out of their low land and successively out of even the European territory.

We have to accept our destiny to go out of this planet. Energy and mineral resources has to be gained eventually from the outer world although our immediate needs could be sufficed by the green energy. We need to become frank to say that the human future depends on the revival of aggressive attitudes we had around the early expansion of 25th century B.C. and the similar racial migration of the 10th century.

East India Company of Netherland had a right to negotiate with foreign countries in order to enter into an agreement and settle down a treatise, and a right to start warfare. It means they had full power assigned by their government. This measure enabled field commanders to take the best and quick measure to cope with any situation encountered on their own circumstantial judgment. The company was responsible to pay due profits to stockholders.

Solar System Enterprise is recommended to solve current world problems. We have financial crisis, unstable security problems, issues of poverty and health care in the global scale. It gives us certain peculiar feeling that the current global financial crisis makes us recall the Tulip speculation in Netherland at the dawn of commercial capitalism.

We need to see our past with a greater belief that humanity will overcome the exterior challenges by integrating our inner capability. Military technology and intelligence technology will provide practical tools for us to cope with the opening of the new era. But the spirit and morale must be transformed.

In the Far East, there has been a strict principle for thousands of years that the military will not, shall not, and must not overpower the civil authority. It is noteworthy that such a principle had been originally induced by the leaders of military powers to guarantee the entire safety of a society. Military powers were supposed to be utilized for those territories for new expansions or protections against the external antagonistic forces.

9-7. Underground Economy

Imex economy (IMplicit-EXplicit economy)
Ways to vitalize Shade Effects of On-ground Economy
HT Conversion of Underground Economy into Institutional Domain
Vitalizing and Humanizing of Menesys of Post-global Economy

The size of the global underground economy would be around from one fifth to a quarter. If we change our view to estimate the underground economy it could touch higher than our usual guesses.

The underground economy was discussed partly in the section about the world currency. But we need to understand that the underground economy is a kind of shelter. It is a conversion of concept. For instance, internet is an underground shelter, actually. Offline is the main flow of humanity because of its associated physicality and responsible personhood. We would like to use explicit economy (Yang) and implicit (Yin) economy, and imex economy (IMplicit-EXplicit economy) for the proper expression to establish the concept of the underground economy.

Imex economy (IMplicit-EXplicit economy)

It is known that there has not been substantially matured imex economy. If we take socio-psychological insights into our account, comprehensive and integrated theory could be attained. Why does the implicit economy (non-institutional and communal) and explicit economy (institutional) exist in a reflective mode?

Is it mutually reflective, in parity, under a balance, or interconnected and interflowing? Are the explicit part original and the implicit part parasitic? It is not likely. Implicit economy looks like it has its own system. Imex economy is the manifestation of the human sociopsychological domain.

It is seen that implicit economies are present in the economic continuum, whether or not they are illegal, legal, communal, or religious patterns. The reason we call them implicit economy is partially because implicit economy cannot maintain an active interflow with the explicit economy. This is illustrated in Figure 9-7-1.

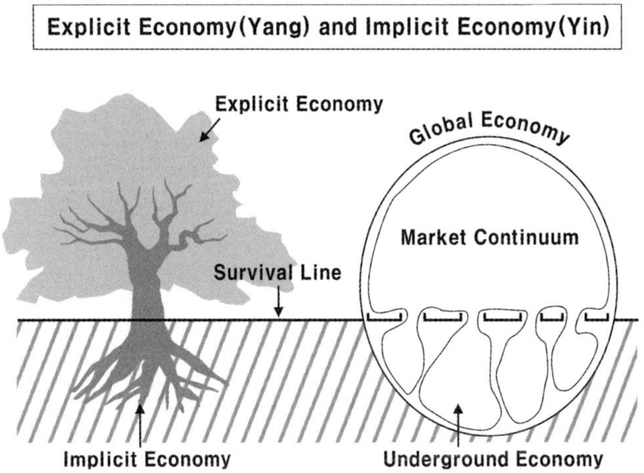

Figure 9-7-1. Explicit Economy (Yang) and Implicit (Yin) Economy

The isolation and insularity of the partially communicable implicit economy with the global economic continuum might be due to the irreversible cathodic effect by which its constituents want to protect in their philosophy; sometimes idealistic, in other cases greedy profit-seeking mind. However, it is suggested to wide open our view and see another horizon beyond which the implicit economy can be flourished within the continuum of global and beyond-global economy.

Then are the implicit and explicit economies continuous? Generally speaking, the implicit economy and the explicit economy coexist in a symbiotic relation that mutually interacts. They make an economic integer in many cases. Strictly planned social economy was not able to manage alone if they did not have the inflow of commodities from the outside free market. The author believes that the implicit economy and the explicit economy exist together in the continuum of an economic integer, but have to be distinguished in their economic effect.

The underground economy is a negative economy that is defined controversially against the explicit (positive) economy. The underground economy is a distinctive part of the implicit economy which has a strong intentional pursuit for the monetary power of institutional economy. In other words, it could be defined that the monetary flow or the exchange value of institutional economy has broken away from the governance of the explicit economy.

However, it is noteworthy that the underground economy is merely a small part of the huge universal implicit economy which contains almost all the energy consumption of the human kind.

Ways to Vitalize Shade Effects of On-ground Economy

The underground economy is originally caused by the socio-psychological anti-societalism or the Taoistic desocietalism. Societalism has a greater focus on the relational bond (that is more dependent to socio-psychological factors) of the constituents of a society than socialism in the author's definition. The underground economy is contained in all institutions, in which human beings survive and collaborate mutually to maintain the national frame no matter what the political and social system have been selected.

In some countries, we find that a Janus-faced single person has the absolute power in both on-ground and underground economy. In fact, we have to consider that in a less institutionalized and retarded economy controversial faces of a dictator or a big wheel of a business conglomerate can be easily met.

The structure of the global economy has double phases of expression, which are connected endlessly on-ground and underground. Such conversion is in an alternating mode which is similar to Möbius strips in a single local conversion but had better be called hyper-topological phenomenon when it is considered in its entire mode. Hyper-topology is a terminology that the author wants to adopt.

The system in which the results of the production are collected in the form of yearly tributes and taxes has clearly existed since the ancient times' community system. Moreover, the things that are hidden and distributed to the authority are what become the richness of the on-ground and the underground. Currency is used in the transition, circulation, accumulation and increase of values, regardless of whether or not it is on-ground or underground and legal or illegal.

Before, all richness generated from production belonged to God. But later, richness became an exclusive possession of the authority, who justified their acts with the fact that richness belongs to the nation. Therefore, when the rightness and fairness of distribution are disturbed, due to the problem related to the ownership of richness, the underground phenomenon appears.

Because a nation's only income sources have been and still are the foreign wars and the taxes, authority always tries to strengthen the tax administration system. However, those countries that could not succeed with the on-ground economy try to supplement their economy with the underground economy, by adopting a dark policy.

Indeed, because the ground economy is somehow indirectly supported by the illegal economical activities, all countries cannot totally forbid the underground economy. So what they choose to do is to control the "contact point" of the on-ground economy that is related to the underground economy in order to indirectly amplify the tax sources.

Generally, on-ground economy and underground economy are mirror images. This causes the mutually contradictory aspects of a society to coexist in a society, such as the capitalist underground market that arose in the despotic socialist country and the pseudo-socialist economical network, which is based on voluntary public services, that arose in the capitalist countries which adopt the market-economy system. Even though the underground economy is the one that contains various changes of the economy, there is no specific science, yet, that can analyze this kind of phenomenon.

Now, from the socio-psychological point of view, we can regard the underground economy as the anti-taxism that all organization, in which humans survive and cooperate to systematically maintain a country, have within themselves. It is noteworthy that the underground economy is kind of a gene that succeeded in human history. The structure of the underground economy is not of dual helix but resembles the Möbius strip that connects the underground and the on-ground lives in a ceaseless activity of conversion.

The psychological path of an economic entity is hidden when strategies for survival prefer to avoid unveiling. The portion of the human desire that is not accommodated under institutional systems remains entangled in isolated regions from the extrinsic economy.

But it is suggested further that the original economy of the human race is subject to changes in accordance with the institutional intervention and the rest of less adaptable portion of the economy remains as the mirror image of the institutional economy. It means that the global human economy is originally implicit and yin-sided. In some sense, explicit and yang-side economy has been sustained by legally compulsory means. Yang-side patterns of organization seem to have infiltrated into the yin-side economy. However the modern bank financial system is the offspring of ancient usury, which was not morally admitted in the ancient community.

In the Network of Underground Economy (published by Oriental Economy, 1986, Chapter 3 of Part III), the editor Takaji Nato (名東孝二) summarized that the three patterns of the underground economy exist globally: i.e., institutional pattern, institution-symbiotic pattern, counter-institutional pattern.

Institutional pattern was presented in tax haven and offshore haven in the past. Institutional pattern has contributed much as the hiding place of shady deals and enterprises. Institutional pattern of the network of the underground economy has worked as the important hidden growing points of immoral interest of the global economic continuum.

Institution-symbiotic pattern is observed in swollen patterns of the private economy, which usually hurt the continuity of the continuum in the institutional economy. Such symbiotic pattern represents the portions of spot transaction and non-monetary economy, which is privately managed in shut-down enclosures under a private power. In India, the underground economy takes almost the half of GNP.

Counter-institutional pattern is attributable to the institutional expression of an ideological belief. Socialist countries had established planned economy to experiment their belief. As mentioned above, the underground economy of counter-institutional pattern is observed in both socialist economy and capitalist economy. Such evidences support the idea of dual economy.

But the author wants to point out that the economic conversion in various mixed patterns of compatible dual blocs has appeared since the collapse of socialist economy. The global economy has revealed the basal strata of the free market, within which financial crisis revealed current Wall Street style that cannot represent the healthy scheme of a free market.

In this contest, the global economy has to be treated under the advanced concept of continuum economy, within which vitally survived are the various modified and deformed patterns of the free market economy. The global economy will be better treated by several sets of mutually interfering forces but internally connected. The best set could be modeled by triple forces for the simplest econometric approach.

HT Conversion of Underground Economy into Institutional Domain

It is the most critical factor in menesysal management to design the best arrangement between underground continuum and onground continuum. These two contrary continuums are interconnected in nature. The conversion of underground economy to onground economy will experience the resistance from the onground side as well as the stress of underground side resisting the governance of onground. The problems are actually how to deal with the constraint conditions of institutional and non-institutional underground. Our discussion eventually will end up with the discussion on Imex economy.

Moving the stage of economic effort by transferring improper economic stress to opposite domains (in order to theorize this kind of situation and economic phenomenon) is a good example of integrism. Such conversions would change the integrismic conditions.

HT conversion is suggested for a general understanding of certain phenomenon such as the Möbius band and the Klein bottle. I do not know whether or not some scientist has already thought of and materialized this kind of concept. HT conversion is believed by the author to be a higher step in the abstractive metamorphism of the human cognition. Thus it is proper to classify it as hyper-topological abstraction (Section 3-3).

If HT conversion is duly arranged at least in its primary stage of conception, the human race will be successful in managing Imex economy. The integrated logical system that will take imaginary and real economy into our consideration simultaneously is not a remote story. The Imex economy we face in our reality is not that of a pure imex condition. It should be a quasi-imex economy that either intrinsic or extrinsic part of economy prevails in accordance with the corresponding integrismic condition.

Integrismic factor consists of sysy (systemicity, 體系性, which is real for a concerned dynamic core) and nexy (nexusity, 脈絡性, which is imaginary for the same core). The logical clarity of applying integrismic factor is not perfect but in practical usage it will be found just enough to be utilized based on experimental and empirical data collection.

In autocratic socialist countries, while the on-ground economy is systemized and regulated, underground economies take the form of a network-linked free economy. In capitalist countries, where market economy is advanced and well-organized, network-linked non-profitable social volunteer work done by churches and civic groups follow the beginning stages of communal socialistic modality. This shows integrism. But it is a sad reality that rather than these desirable examples, illegal underground economy – the black market, drug dealers and terrorists – is thriving all the more. Through the research of the mechanism of integrism, the way to overcoming this abusive phenomenon will be opened.

Cure of Negative Aspects of Underground Economy; Psycho-virtual Reality and Low-density Warfare

It is an urgent global issue to cure the negative aspects of the underground economy. We have a ragging of psycho-virtual reality and low-density warfare in addition to the old black markets of drugs. The change in the predisposition of gambling business, and ensuring the connection with open economy will make underground markets contribute to local economy.

The substantial improvement of ethics has been cautiously achieved in various kinds of black markets and the business of the marginal underground market, promoting competition against on-ground economy.

Forming a generalized viewpoint on drugs and VR (Vegetarian Remedy to induce virtual reality will be to some extent will enevitably penetrate into our everyday life), the conceptual approach to the underground market forces can be restructured. If we are successful to localize underground criminal organizations, such concentrated decisive action to separate illegal global networks into pieces would help the government authorities ease the situation. Flushing out politics-related illegal criminal organizations should be the impending target because it might be connected to the terrorism in a long run wherever they have been sprout out. The low-density wars against non-institutional origin should be started from the basic remedial conduct from the resolution of menesysal connection.

Whatever the situation may be, investigating economic data of the late 1980's and early 1990's – the initial period when capitalism became dominant after the collapse of communism – will provide insight upon this phenomenon of the transition zone.

Vitalizing and Humanizing of the Menesys of Post-global Economy

The subject of post-globality in this book aims at the exit towards the solar system. We might have to administer the parts of the early development of the initial solar economy in a pattern of planned economy. The human race needs to prepare menesysal post-global software and at least seeds for further variation.

The creative peace will implement a time schedule of global participation, an individual and communal life patterns, the scope of works of the initial initiative, etc. Post-global economy should be prepared for the reorientation and repositioning of the unhealthy parts of enterprises in the present economic structure by remodeling the whole global economic continuum.

9-8. Essential "Peace Common Ground"

Common Ground for the Creative Peace
Multiplicity from Virtuality of Knowledge System
Continuum of Peace Knowledge System (PKM): Ecology, Personhood and Humanity
Integrity of Peace at the Center: PKM, PSM and PIM

It is the spirit of our time to establish the foundation of perpetual peace for the next millennium. That is the reason why some people frequently comments on World Government as Emmanuel Kant told about long time ago. It is our belief that perpetual peace is subject to its essential infrastructure; knowledge system, education, and multilateral collaboration in the global condition. In this study we are focusing at the creative peace which is a vitalized and stabilized condition of the peace status. Peace knowledge "menesys," not a system, will have to be one of the essential parts of the global common ground for the creative peace.

Some readers might raise questions about if knowledge system can be constructed centered at the peace purpose in reality. It is thought that a system cannot comprehensively embrace all the knowledge while a menesys can. System has its own closed chain(s) of logic. When the logic is not sufficed, the concerned constituents will be disregarded and cast out of the system or enfolded in functional networks in a system. Enfolding is an actual isolation or alienation.

Menesys works different from system. The immature formation of the global menesys is structured following the ramification of netsys (networked system). Netsys is developed by the networking of systems. These systems have grown up from nodes of network. Nodes are supposed to become systems when it attains a chain of enough number of logical progresses. Present globality is about to reach an early menesysal maturity because number of systems are ready to connect with other systems by various forms of connexion. The variety of connexion is based on coincidental logics, corresponding interests, and accidental chances in the cruxological development.

The frame of global knowledge system has been built in menesysal form since around the 25th century B.C. Blowout of the human intellectual power of that time was affected by mutual communication, even though the number of contacts was very rare compared to nowadays. Religious knowledge, legal knowledge and commercial knowledge had primitive mixture of systems under the local religious thought system. Religious knowledge system is mostly closed to the external intercommunication unless they are from the same root or oppressed to be comingled.

The academic knowledge system, which came after the early mixed up, is essentially systematic but open to the external input. Mathematics started from geometry and continued to add up new systems until it was transformed into the present extensive and complex system. Physics started from dynamics and later added thermodynamics, electromagnetism, relativity theory, and quantum theory.

Very recently in the human history, it is observed that mathematics and physics are about to start a significant exertion towards the integration of knowledge among various disciplines. The academic tools of science and technology has attained a vast and deep foundation in the realm of the human intellectuality.

It is certain that such logical tools in science and technology helped the global propagation of the present civilization. However, current menesysal development of knowledge seems to attach too much importance to the economic, linguistic and logical background.

The web of knowledge in internet is an important manifestation of menesysal development of the human knowledge. The semantic web has been developed for the systematic treatment of the human knowledge. The ontological approach in the artificial intelligence aims at the construction of the logical frame to manifest a detailed formal expression of communally agreed knowledge. These current developments of the human knowledge preponderate over the systemicity and logics.

The common ground for the creative peace should be founded on the knowledge menesys to take liberty of humanity in advancement towards the transcent of the human race.

Common Ground for the Creative Peace

Peace knowledge system has not been constructed at any time. Peace study of Emmanuel Kant is presented in “Perpetual Peace” in 1795. “Positive Peace” is suggested by Johan Galtung in the 1990s. Positive peace is the integration among mutually antagonistic societies without discordances.

Our study aims at “the creative peace.” The creative peace is a very delicate concept. It is a dynamic condition of the human existence. But it is manageable when the human race successfully keeps the tools in operation in accordance with the maintenance manual. However, it is not mechanical and automatic. The factor of humanity is the most important variable in the mechanism of maintenance of the creative peace.

The factor of humanity in the maintenance of peace is governed by cognition, communality and environment. Knowledge system is a kind of complex road map of how human reacts and progressively responds, because the intrinsic chain of logics formulates the paths of knowledge connection and builds the basic body of the knowledge continuum.

The pursuit of peace has to be problem-solving, not ideological. The pitfall of peace study is that the complete elimination of the logics threatening the peace status in the global power structure does not promise the desired prosperity. It means when we are successful in establishing a trouble-free peace status, we apt to excessively weaken the creative and productive attitudes which have flourished the human race. The ignorance about prosperity brings only a withering peace.

However, prosperity comes after the logic of competition, the necessary evil of the human history which has always led us to conflicts and demolitions. We need to partly modify the model of the competitive structure in order to pursue peace and prosperity through competition.

The range of peace knowledge menesys starts from the quantitative definition of the peace status and encompasses to the methodology of maintenance and administration of the creative peace.

The fundamental understanding of macro and micro state of peace and overall studies on the peace of different dimensional arrangement are required to generate a mutual correlation. The major context of peace study would be the reprogrammation of academic systems in accordance with the sincere peace pursuit. Humanity has to be altered in its way of thinking to induce comprehensive changes in general frames and methodology of academic efforts.

Multiplicity from Virtuality of Knowledge Systems

The knowledge system of the entire humanity is like a hydra in Greek myth. Nine heads are all alive respectively. One body brings the nine heads. Each head thinks differently. Global knowledge systems have respectively differentiated abstraction, symbols, and linguistic frame. Thus, multiplicity of the human knowledge system stems from various spatiotemporal environments, which we used to call "history and circumstances."

Multiplicity in the human knowledge system is to be cured by integrism. Integrism consists of nexusity and systemicity. Nexusity is the factor that heals multiplicity if it works well. Nexusity provides communication channels to be developed among hundreds of different religious, cultural, political, legal and other systems. Such entangled multiplicity is the reality of humanity of this day.

Present multiplicity in the global knowledge system in a great extent originally comes from different number of sources of unexamined virtual parts of humanity. We call those sources, guess, speculation, superstition, and self-seeking rationality, and so on. These sources of virtuality produce very wide loci of knowledge arc scattered in the system. They also crack humanity bringing discontinuity and inharmony of rationality on which knowledge system is usually based on.

Some virtuality of humanity guides the human cognition to transcend to a higher integration. Some others lead to a merely closed systemizing. Many forms of dictatorship or tyranny load heavy burdens of individual or oligarchic, simple and monotonous intentions and the scheme of limited rationality on people to draw them down to the uncivilized condition. It is also true in an intellectual society. Big wheel among us used to commit such faults.

Being rational and scientific means clarifying much of these irrational integrations of improper knowledge and in some sense eliminating much part of good provoking factors for creativity which otherwise have given a good extension of the realm of creativity and activation of daring humanity.

In order to discourage the virtuality, we had better give constraints on the free virtuality and connect the virtual parts with the core part of the continuum of knowledge by applying at an adequate level of scientific principles. Virtuality has been associated with superstitious activities in the past far from scientific rationality.

The most vicious source of multiplicity is the conflicts in political economics engaged with knowledge system. The fights in the domain of thoughts are supposed to call wars in the real world eventually. The conflict of thoughts develops to the psychological warfare and to the physical wars and battles in turn.

Multiplicity is cured by integrism; the first step of cure is done by nexusity and the second step is by systemization of certain nodes to establish necessary systems to construct a netsys (networked system). This kind of phenomenon is found in the stage of mixing up two different civilizations.

When the Greek culture met the Persian Empire, brilliant Hellenism prospered. Hellenism is actually known to have a deep influence in political aspects. The kings including Alexander the Great was contaminated by the absolute power of the Persian political system. It is known that even the emperors of Rome in subsequent history lost their spirit of republic ideals when they contacted oriental autocratic atmosphere.

Continuum of Peace Knowledge Menesys (PKM): Ecology, Personhood and Integrity

The continuum of knowledge systems has exceedingly more complicated problems than generally anticipated when number of different knowledge systems are intermixed. But we have to face the integration of knowledge systems. Present emerging trend of terrorism could be a result of the mixture between cultures.

When knowledge systems and cultures meet together to produce excessive stress conditions, the peace status could be easily lost. The achievement of the continuum among knowledge systems should be cautiously approached to avoid serious damages to humanity. Have we thought about the possibility to bear a bastard? What would you prefer if a baby is the violent terrorism? The peace theory has an unexplored frontier in the construction of the continuum of the global knowledge system.

Anyway let us think about a generally known example. Wikipedia is an emerging outfit of knowledge continuum in the internet society. But until now it is assessed to possess certain shortcomings of some undemocratic process, of less reliable edition of knowledge, of anonymity, of truthfulness. In my opinion, Wikipedia will be gaining a greater smoothness in continuity and dynamicity. When we regard the knowledge continuum as a dynamic process, multi-dimensional maps of reflections among relevant phenomena of humanity could be possible.

Ecology is our hope and prospect for the future but deemed insufficient as the tool and method to sustain the nature. It is generally said that ecology of the nature cures the planet by itself. It means humanity is just a superfluous stream of conscious entity attached to the nature with comparatively short history. However it is interesting that science have had to grasp an idea of anthropic theory in a cosmological expression.

Humanity in our viewpoint is the substantial body of history that is capable to fix the nature. Nothing can substitute humanity. Why? Humanity is the one which can correct oneself intentionally. Ecology is the constraints of humanity. Humanity is also contained in the realm of ecology. The thought about ecology is the evidence of a corrective process being undertaken. It is a current trend that ecological thought should govern our future civilization. However, the human determination to refresh the ecological environment cannot penetrate into the self-sacrifice of our needs for basic necessity and convenience.

The knowledge system upholding humanity is under reformation to find a way to a pattern of a more workable knowledge menesys. Such effort is observed in the active fusion and mixing between knowledge systems. We have to be careful if we have due principles in the integration of thought systems.

Personhood aims at integrity, at least to a certain sufficient degree. We can see it, ironically, by observing how much a person could be destructed by schizophrenia. It is thought that schizophrenia is one of representative symptoms that human personhood is disintegrated in its intrinsic condition of schizoid.

Transciousness without the aid of consciousness builds a fixed concentration of meaningless past residues. Sensciousness without consciousness leaves an animal kingdom.

The creative peace is essentially born from the ecological scheme of thought. The ecological thought means the way of thinking that takes into account all the cognitive process and environmental conditions associated with thinking. Ecological scheme is integrismic. Ecology is concerned with the relationship between living organisms and their environments. Thus, when we consider the thinking process as vivid living, which surpasses the level of mere rationality, it has to be called the ecological scheme of thinking.

We wrote above that the continuum of knowledge can be restored by taking a form of menesys, the combination of system and network. Recovery of the continuum of knowledge systems will be commented later in Ch. 10.

There could appear several centers but will be unified by the efforts of establishing a novel standard of weights and measures. Humanity has almost come to the point of defining a new standard system for a wider application in the solar system and in the universe as the measure of meter has been defined during the French Revolution.

Integrity of Peace at the Center: PKM, PSM and PIM

In this part of discussion we will use Peace Knowledge Menesys (PKM), Peace Sensing Menesys (PSM) and Peace Intelligence Menesys (PIM). Intelligence herewith means a wider concept than the intelligence of an individual and the intelligence of a nation. It is a menesysal body of the interpretation of global knowledge menesys. Therefore intelligence has a function of normalizing the jumping impact of knowledge upon a society. Knowledge is thought to be excessively governed by the impact of emotional basis of an individual. A part of knowledge system used to be exaggerated too much in history.

Anyhow, let us go back to the discussion about the integrity of peace. It is found by Claire Rampon et. al. in Princeton University that an animal has activated more than 15 genes in the brain within several hours when it was moved from a less stimulative environment to a highly stimulating condition. The chemicals produced by the activated genes are responsible for the growth of neurons, generation of synapses, and production of chemotransmitters. They are NGF (Nerve growth factor), BDNF (Brain-derived neurotrophic factor), CTNF (Ciliary neurotrophic factor), and so on. (Chapter 5, Joachim Bauer, 2002, Das Gedächtnis des Körpers, Chapter 5)

The comprehensive picture of the whole that is activated selectively in a specific situation is the major doctrine of dynamic modularity. Peace take variable integrity in accordance with the situational conditions. It is also noteworthy that peace knowledge has the format of different integrity in different situation.

In present status of the global political economy, we need to concentrate on the establishment of global institutional education and the civil service examination (科擧試驗制度, state-administrative wisdom of Far East designed in order to ease the stress of suppressed people giving fair and equitable opportunities for upper-orientating route). The global system of lawyers alone cannot take care of all the explosive increase of illegal and emotional repulsions of local societies.

The above mentioned systems of education and examination has to give gradual participation of local leaders in at least ten diciplines of academic and administrative knowledge and intelligence.

If we are really frank to practice communal politics in the future globalized world, we have to understand that both local and global communal politics must have accents on the due process itself, assisted by profound communication and many contact points that enable mutual adjustment in decision making.

It is the author's opinion that systems should be changed to menesys in order to vitalize global humanity. However we need the integrismic criteria of decision making, practical rules based on cruxological understanding, and managerial method as menesysal technique in administration.

Peace knowledge system founded on several specific cultures will have to be changed gradually into peace knowledge menesys (PKM). PKM will have to be steered under good collaboration with PSM and PIM.

PKM has to possess three kinds of centers spread over "the world and beyond." They are accumulation centers, modulation centers and creative centers. These centers will be managed on IT bases. The number of modulation centers may have smaller number than others at one lower level of magnitude. Creative centers and accumulation centers could be managed as a form of university at the first stage.

PSM, peace sensing menesys, is going to be managed for the detection of potential frustration. We need to attenuate most frustration of initial stage by effective by-pass installation. It will be implemented by reducing the over-pressurized atmosphere and the local complaints by solving local problems by the input of commodities, service and necessary institutionalization. It is observed in this day that NGOs are taking parts of the global burdens avoiding serious pitfalls of bureaucracy. PSM would be a weak link in the network theory comprising with background network preserving affection, goodwill and love.

Effective by-pass installment by PSM will be activated by the global collaboration menesys, initiated from the people of G-2, G-7, G-20 and whatever the hierarchical flexible Gs-number is, and it will be utilized to construct sensing menesys. Gs-countries are relatively creative and have been successful in leading the accumulation of capital power. They are responsible in the historical context of the human race to maintain volunteering to sense peace-disrupting atmosphere from the initial stage.

PIM, peace intelligence (maintenance) mechanism, will have to collaborate with PSM to preserve the initiative of transcendental intentionality of the contemporary humanity taken from the weak links of global menesys for peace keeping. These equalities between PSM and nexusity-prone weak linked menesys, and between PIM and intermediately nexal menesys, are a step-wise approach to suffice peace condition with fewer impacts on the global development.

In practice, it will be implemented by organizing elite menesys of global scale in a loose way to support PSM. We recollect that elite and commons are in the same dynamic modules in Section 10-1. This connection will help us to construct the global continuum of peace common ground.

EITech (Educative Information Technology) menesys will have to be established globally in order to distribute educative information. Global EITech menesys helps construct global continuum of PKM, PSM and PIM. EITech is a notion of information processing in peace knowledge menesys to efficiently and fluently distribute peace knowledge in a digestible pattern to the under-benefited people of the world.

9-9. Forfeiture of Motherhood in Current Humanity

Power of Motherhood in Humanity: Communal Continuum in Physicality
Preservation of Noise in Motherhood: Cultural Continuum in Maternal Emotion
Fundamental of Genetic Humanity: X-chromosome and Bisexual Reproduction
Half of Womanhood Endangered

Why do we have to leave a comment on the motherhood at the end of our status quo diagnosis? It is widely acknowledged that motherhood is endangered nowadays. We have to examine the secret hidden in motherhood and restore it. The loss of humanity comes from the forfeiture of motherhood in womanhood.

If well directed, meditation, the thinking process associated with emotional backup and physical instinct, can change the brain. It is proven by the brain test of deeply trained monks. Empathy and sympathy, especially the emotional activities concerned with maternal love and happiness, expelled the influence of negative feelings (Daniel Tammet, Embracing the Wider Sky, 2009, Ch. 1).

Power of Motherhood in Humanity

Empathizing and systemizing is deemed to have strong ties (Simon Baron-Cohen, the Essential Difference, 2003). Empathy is generally stronger in female than in male. Empathy is much stronger in motherhood than in most female-hood.

Motherhood should be regarded as the standard form of female. Certainly I agree that female is the basic genetic form of human as shown by the genetic studies. I believe the stance of motherhood in a genetic scheme is very special. Usually people think mother is a special presentation of being female. To find out that motherhood is the essential standard of femaleness is a very challenging subject.

It is suggested that the communal continuum in physicality is generated when we experience the fact that we co-own emotion and intent together. It is the essence of the social relation. If empathy hardwired, then the communal continuum takes the foundation in the abstractive form, and consequently on mentality.

We need to meet a fact about mirror neurons and reflectivity in order to attain another insight into the correspondent matching of humanity in the abstraction process; which would come first, reflective matching or neuronal mirroring?

Maternal imitation of the baby's facial expression activates an action plan of the baby. SMA (Super mirror neurons) which is found in greater dynamic modules of mirror neuron groups controls the on-off state of the entire mirror module. Neuronal mirroring is thought to be the essential brain activity of the human intellectuality and also of the maternal breeding which are two pillars of the human history.

Preservation of Noise in Motherhood

Preservation of the impeccable process of humanity is the purpose of the motherhood. Humanity could have been successful to preserve the main characteristics unaltered, which I would call the elasticity of humanity. It was largely due to the motherhood, which has sustained the connectivity of the temporal continuum of humanity.

Cultural continuum reserved in the maternal emotion is the foundation for the human mimetic. Mimetic is actually the downward mental influence from abstraction, which is hardwired symbolized, simplified, omitted and enfolded. Mimetic is not a sufficient terminology for what I think of motherhood. I would rather use "mometic" in order to contain a biological, personal and cultural successiveness and everything-inclusive mimetic.

Motherhood is the special carriage of personhood in a manner of white noise. White noise is a random signal of a flat power spectral density. White noise is a status of an adsorbing medium that we usually experience when we admit a specific signal of life holding our breath. We do not concentrate but we take our emotional status all flatly maintained to receive the external signal.

It happens like white and pink noise as described in Chapter 5 when we use our flat feeling with our eyesight 180 degrees wide to catch any special change of situation that appears with certain dash of a fast vehicle which approaches fast. It is like a spark moment of first love between a boy and a girl concentrating to each other but not doing any special action. A flat feeling with full of noise is what I believe happens in the reflective mirroring between a mother and a baby.

White noise or "flat noise (which I prefer)" is the neuronal expression of the entirety of personhood. The communication between people is most efficient when flat noise is formulated on the surface of the emotional sphere. Flat noise is the secret of motherhood by which mother transfer her personhood to her baby in a neuronal expression.

The most delicate maternal discernability in situations when she has to choose something, out of innumerous emotional signals in her noise set, is the fine modulation between two personhood of infant-maternal relationship. I would not worry if I am overstepping saying that this close relationship has to be studied in the fetomaternal situation too. A baby's experience of transmitting its intention to the surface of maternal flat feeling is the learning the basic scheme of the transcendental intentionality of humanity. The transcendental intentionality is in turn the base for the transcent of humanity.

Fundamental of Genetic Humanity

It is reported by the genetic science on the bisexuality of humankind that the fundamental system of humanity is preserved in woman. (David Bainbridge, the X in sex: How the X-chromosome controls our lives, 2003)

Male is produced by excessive aggressiveness attained by the Y-chromosome in order to connect the extroversive potentials, sometimes, by protection and, in other times, by introducing new potentiality for a better future. It is done by displaying the centrifugal function for the preservation of self.

The female gene does nothing and proceeds as programmed in the genetic sequence. Female-systemicity is the empathy-based sociality. Male-nexusity is for protection and preservation of the self from externality. In hostile situations, male is superior. When we enter a stabilized condition, female prevails. It could be uneasy to some readers to write X-sysy, Y-nexy, X-system, and Y-network, which is adopted here largely for emphasis and convenience.

I want to say that the X-chromosome is more ontological and existential in terms of nature. The X-chromosome and the ovarium are the essence and basis for human birth. It is usually considered that the Y-chromosome is manhood itself. The Y-chromosome is the additive to the X-chromosome. But that additive is the source of extension of the sphere of life. Bisexual reproduction is generated from the intersexual spectrum. The bisexual system of the human race is a tendency selectively intensified from intersexuality (間性) for bipolar population.

Half of Womanhood Endangered

Half of womanhood is motherhood. Womanhood is different from female-hood. Womanhood has the highest rank in humanity. It is actually a common ground of humanity.

If the chain of pain which has been reserved in motherhood is continually omitted from womanhood, as prevailing nowadays, it will cause the decadence process of humanity starting from womanhood. Flat noise of abundant gift in humanity implicated in motherhood fundamentally stems from the pain process consistently alive in the pattern of behavior of mothers.

The chain of pain and the chain of delight take good balance together in motherhood. Chain of delight usually does pursue systemization of behavioral pattern achieved when disregarding the need of pain.

Torn maternity severely distort womanhood; social abandoning of bisexuality has the beginning in the forfeiture of motherhood,

Abortion, shedding the innocent blood of the unborn, whose neural system is produced 14 days after the conception, can be alive with the support of incubator and has an independent thinking process from the 23rd week and on. We are afraid of killing humanity by abortion. But losing the motherhood is a real genocide and the committing suicide of the human kind.

Chapter 10. Integritium: Essential Strategy for the Creative Peace

The concept of menesys was introduced to peace study, especially focusing on the creative peace. Global *Pax Consortis*, the peace among multiple entities on Earth, is reinterpreted as the sets of glocal and permunal menesys.

Stephen Hawking commented recently that the human race will expand out to the universe if they successfully survive the next 200 years on the Earth. I believe more critical would be our survival in the first 50 years, because initiality management may not be as easy as anticipated. It looks definite to me that we, the human race, do not properly understand which problems we should tackle.

None of the terrestrial countries and governments would pass the omega path of today if they do not have a picture of Integritium, the scheme of successful humanity which will be given if and only if they concentrate on solving the problem with the sincere cooperation. The days of Imperium have passed away in the last two millennia. Forthcoming is Integritium based on the menesysal management of the creative peace. The management of menesys is the matter of connectivity and growth synergy, as notably seen in the human history, if interpreted right.

In this concluding chapter, two global menesysal projects are recommended to start in the next few decades. One is the Solarche Menesys and another, Keiros Menesys. Solarche menesysal schema will provide tractive force of the global continuum of the economy and Keiros project will be efficient to sustain the global continuity. Global Intelligence Modular Urban Planning (GIMUP) is a toolkit to tailor the intrinsic properties of global urban areas. The last section about Aasadal City implies how these global projects will be successful by grading up the menesys of global metropolitan areas with various characteristic technopolis.

10-1. Global Pax Consortis Based on the Creative Peace

Global Pax Consortis
Basics for the Achievement of the Creative Peace

The creative peace continuum is an entity, living as an integer, moving in the phase space, transcending to a higher level of hierarchical life, and extending to the infinity and the eternity. It might not be correct to say that the continuum is an integer because it is only a quasi-integer until an ideal state is reached. We are ready to embrace the idea of synchronization and syntalansis to realize the global continuum

Global Pax Consortis

Pax consortis is a peace status among multilateral parties. Authenticity and factualization of global pax consortis is a strong question to most people in the world. Prevailing diversity and plurality idealizes the global peace of today. Super power would not easily and thus practically use its military power for ultimatum due to the excessive development of the warfare technology. Inevitable restructuring of entire military forces is foreseen in the near future and will soon be around us within the next couple of decades. It will be a more advanced concept than current changes being taken because the technological alteration should be completely different from the present one.

Nations would not take the principle of “eye for an eye” but rather prefer “hound for a wolf.” They would decide to fight not for vengeance but for preventive means, and choose to reduce the scale of conflicts. But future warfare will post many small groups of warriors with stronger firearms and greater intelligence potential at widely spread locations at vaguely defined fronts.

Global pax consortis has already been introduced in international relations. The difference found in current diplomacy of global pax consortis is the generating innumerable civilian contacts in many peaceful objectives, which are the prerequisites for the world to flourish.

The contemporary civilian contacts are defined to be the networked individualism. It is a novel pattern of sociality proven, for instance, by the public digital culture in Amsterdam called “De Digitale Stad (DDS).” DDS became a perfect network community – digital techno polis, which provides information resources and free communicability. The city of Amsterdam is the first local administration that transparently connects internal networks and internet. DDS transcended to an internationally renowned "Communal Digital Culture of Amsterdam." (Manuel Castells, the Internet Galaxy, 2001) From then on, the world has been experiencing a greater leap and enhancement in technology. Constructors consider building up ubiquitous cities of a mega scale in the near future.

Nations are ready to suffer the diminution of their political power in order to be quickly admitted into the territory of digital culture. They have traded their power off with the substitutive civil power. In this situation, do we have anyone to trample down the global pie or to threat to destroy the banquet? The answer is yes and no. The world has always had its periphery in the history and that portion of humanity has reacted positively at sometimes and negatively at others with the main core of humanity. Dialectics of Heraclitus and Hegel emerged from such confronting situations.

Basics for the Achievement of the Creative Peace

There are certain basic requirements to materialize the creative peace, the three major factors are; first, abundance of technological and economic basics, second, activation of educational and creative basics, and third, reliability of legal and civilian basics.

Technology today is known to have almost an absolute impact on economy. This trend will be accelerated in the next several decades until a revolutionary innovation for eventual energy technology is introduced. Green energy industries are one of the intermediate steps toward the final solutions of energy technology. Food problems also have major constraints in energy problems if monopoly issues in business world are to be resolved. The creative peace stands on the technological enhancement.

The educational and creative basics ought to be activated in order to assimilate and mobilize global forces within the common people to the realization process of the creative peace. Education has to open the novel phases of humanity by provoking creativity in the growth of the next generations from their births. It means we have to design "Educative Information Technology" to improve the creative thinking and the lifestyle for their everyday life. Even the present level of creativeness in education is astonishingly advanced in comparison to that of a half century ago. One thing to note is that creativity is the key factor of humanity in the timely management and firm governance of the scientific revolution that is approaching ever faster.

The third requirement, the reliability of legal and civilian basics, has a slightly different implication from the other two. A genuine civilian legality had not been enjoyed enough, in my perceptive, until the latter half of the last century even in the Western European countries. Although the Hippies in the 1960s are prone to fall into an anti category, such dissidents are sometimes constructive in a long run for the whole society. When faced with unexpected course of events, people tend to look back and reexamine themselves. Any implications in the existing regime cannot take the role of deep ploughing even if the strong opponent political party were a shareholder in the power regime.

Civilian legality is still growing in power from the periphery of the society such as from the non-governmental organizations or the non-profitable organizations. It could be representing a force seeking for new dimensions of humanity. The real nature and the real identity of globalization is the growth of networks. NGO and NPO movements are the result of the increase in nexusity. Civilian legality is formulating the societal authenticity of the networking aspects. It might imply significant yet basic changes in the integrismic criteria.

Motives and incentives of today's common people have been growing to pursue the creative peace. The natural spontaneity is to be intensified by strengthening the individuality and mutual communications, and the consequent increase of nexusity. Nevertheless, what is noteworthy is that systemicity has always assumed the right of primogeniture.

10-2. Integritium

Integritium is a derivative terminology from integrism and integrator. Ancient Roman conception of the idea, the source of which is missing in my memory though, was that an integrator was a greater figure than an emperor. An integrator is a person who reforms, revives, and resurrects the social integrity. Integritium, on the other hand, concerns on all people, common or governing, disregarded or chosen, and rich or poor.

Imperium

In the western civilization, an imperium is the concentration and centralization of power in the regime, it started when King Alexander created and changed his own self-image into the absolute Persian Emperors, the desire to an absolute power disseminated into the Greek spirit and the succeeding Roman intellects, fossilizing the spirit of liberty inherited from the age of Polis and the times of Republics. In the civilizations of the Far East, human emperor imitated the Heavenly Emperor. In the Lexicon of Human Nature and Natural Law (性理字義), it is expressed that the Heavenly Emperor possesses his attributes of Li (理, reason, rationality). Such greedy dictatorships of both eastern and western empires are in fact far from their supposed ideals.

However, the definition of an empire, Antonio Negri and Michael Hardt has accentuated the four characters of Euro-American empires, explained as follows. First, the empire has no boundary. Second, the dominant power causes a historical discontinuity to make itself the utmost limits of regimes. Third, the infiltration of power encroaches into the human microcosm (human interactions). Fourth, is that the empire pursues the universal peace through oppressions and destructions.

Joseph S. Nye Jr. has a more complicated yet practical thinking in his Paradox of the American Power (2002). A simplified picture of the current world is shown in a three leveled chess game. The highest is military factors that prevail as a decision parameter. The middle is that of economy. These two levels are called a hybrid uni-multipolar world. The lowest level is the domain of denationalized relationships (I would prefer the word de-nationalized).

In the lowest board of the three dimensional chess, bankers of the highest rank, asymmetrical warriors of both sides of terrorism and antiterrorism, hackers and business persons of the underground economy are busy in their businesses. But volumes of multinational business have been increasing rapidly since the end of World War II. The interflow in the domain of the dissolved national boundaries is exposed by the fact that the overseas production of the U.S. is the double of export and the domestic sales of foreign corporations are twice of the U.S.'s imports. Such microeconomic links of the non-territorial regions are generating the space-of-flows instead of old space-of-places.

It is doubted that Europe will be a candidate for the potential global empire. Nye discussed that Europe would most likely be converted into a confederate with less power concentration, not a federal, in the future, since it acknowledged the United Kingdom and Scandinavian countries. He thinks that reemergence of China into a leading role is anticipated. But it is not so prospective as before, from the 5th century B.C. to the 15th century A.D. during the time when China had maintained the strongest leadership in the world of loosely connected conditions. Current conditions are different from the past because the world has become a conglomerate of competing countries.

The underneath dynamics penetrating such global state of present conglomeration has to be, in my opinion, interpreted freshly to show the characteristics of centralized colonia-netsys, judging from the menesysal frame of this book. China was capable of managing its ancient imperial situation that existed in a semi-isolated system that was connected with a dozen of its peripheral countries. Such regime is classified as a systema-netsys in menesysal frame as we discussed in Chapter 6 and 9.

The present world is too closely interrelated and conglomerated in some parts to simply centralize or decentralize as we did in the 20th century. It is not caused by the internet environment only, but it is due to the change of humanity itself. Nowadays we widely experience a hysteric symptom of the anxiety for human fates. Could imperium of ancient days, like Rome or China, be an answer to a current instability to get firm grips on human race? Should anyone of candidates for the center of future imperium practice their military or certain form of compulsory power on the tenants of the world?

Integritium

Currently the world is uni-multipolar. Unipolarity, as mostly agreed, is describing a global military aspect. It is well expressed in D. Kaplan's series of works and I believe his realistic idea on the military unipolarity is still effective even after the Russian's small success in invading the territory of Georgia in 2008.

However, global problems shall be solved only when it is viewed by the bird's eye. The world will be a vagabond planet if the impending objectives are not given instantly. It is a day of restructuring. The day of reformers and developers for new civilization is at the gate. The promoters of the global-administrative projects have to start their missions. It is not only the matter for elites but also the commons. Therefore, we must let everyone in the world to feel the same ambitions.

Upheavals in Wall Street and energy prices are not separate issues but issues that should be revealed even facing the complexity of problems. We are facing the overall inter-penetration between entities integratingly and pragmatically. Those are issues deeply entangled with values, and that have to do with the emergence of novel dimensions for the creative humanity. It is necessary to give right intentionality and right aim to the human race at this time. When we don't meet the challenge, the aftermath would be serious and immense relatively.

Integrator, who reorganizes the society in a whole, is valuable on this occasion. Integritium has its semantic origin from integrator as can be seen in the relationship between imperium and emperor. Integritium has a definition of "global administrative regime supported by menesysal concept, under management of integrismic factors and cruxological measures."

We are responsible to ask if there have been certain reference models found in the past history for our conversion to integritium; it might be the integrated reflection of complex factors associated with number of references such as Hanseatic League, Mediterranean colonization of Greeks, any factors from the anthropological development to extended family, any attributes of national states, the immigrant expansion of overseas Chinese, the westward flow of wagon streams in America not so long ago, Persian imperial administration and the chain of the cities of Alexandria. An establishment of the creative peace is the right answer to this vague problem that none of the single country, the candidate for imperium, can make a right approach.

The human history has entered into a transition zone. A transition, started from the 1990s, will be able in our proper response to change all the phases of our civilization into an integritium. Overall phaseal change into a new imperial political economy has usually taken 50 years, equivalent to one and half generation. Phaseal change into an integritium would also take a little more than that, about two and half generations, because it is a little bit more complicated condition to suffice than the centralized imperium. The two and half generation should be around 100 years. Every generation starts their governance in their society later than previous generations and lasts long, for instance the 40-40 year's model.

The change can be imagined if we think of the embryological development of the brain and the neurophysiologic growth. I believe the role of humanity on nature is more than a simple metaphor. The menesysal establishment of integritium centered on the planet of Earth would take the whole 21st century. We have to activate "dynamic modularity" in the human civilization. We are not dreaming, we are mobilizing the strongest and *maximus* imagination of our thoughts and our hearts.

The supports of the technological system, the law-abiding intellectual system, economic abundance, the interchanging nexus, the establishment of functional hierarchy, the non-stratification of people and the increase of the degree of freedom, are all waiting for the fruit of integritium. Real humanizing is a long way to go but we, already, have plenty to enjoy in our lives. If we have received the words of the Almighty and the Divine attributes of thousands of years ago, it is in the truthful humanization of the creative peace in realistic meaning in this world.

Integritium Politics and Economics

Thus, we have absolutely no idealism in our way of thinking, but only the firm connection between the practical approach and our transcendental intentionality. We need practical assessment methodology and confidence in our pragmatism to access the target. The human race ought to take the next global objective in a tight grip and lift the banner into the high wind and let it fly and flutter powerfully. We need to publicly announce this matter and focus on it.

To attain our goals, our realism will not loosen on us until we develop and utilize our tools of policy administration. This book was written to suggest and recommend the basic frame for the whole idea. It has absolutely nothing to do with the idea of universal governments of past generations. It is also of a greater scale than that which the United Nations can pursue and achieve alone. It should not be a banquet for just the cartel of powerful states. But the pies distributed for powerful cartel, if any nations intend to become, would be bigger than that when they have done it alone. And the other countries will be able to take the rest which will satisfy them greater than before. In order to realize the intangible dynamic elasticity of the global humanity, on informal implementation of the global consensus process is desirable. Representatives with hearts to discuss and collaborate for the common objective are required.

Recommended here is the concept of "governance without power", which means temporary and voluntary governance. In my personal view, perhaps G-8, G-20 and/or G-50 would be eligible and responsible to organize such movement for this historical mission the initialization of the marathon toward integritium.

The key to opening the gate to integritium is the one-time administration of a non-formal governance in order to give birth to necessary menesysal movement and stop to hand the baton over to the next nexus of people, not the Davos of elite leaders like always, or the Hainan in the process to hold the power, which is not directed by any power netsys but by varying sysnet governance.

The list of minimum practical tools for policy administration at the initialization process of the first integritium is as follows;

1) Mappings of the Basic Menesys of Global Humanity: Various menesysal mappings of global humanity are made based on the existing geographical maps. Dynamic patterns should be recognizable in a digital pattern of data collected on time series maps. The maps in the initial stage are not deemed to be many.

2) Analysis of the Mechanism of Values by Triradixism: Humanity will cause the creativity value to accumulate quickly if the bias in past value theories comes out of the single radix, which has been mainly the labor value theory since the modern age. The major advantage of Triradixism is that it deals with the phenomenon of the human creativity based on the land price and/or rent cost, which quantify the psychological effects in the consideration of economics. Values are only measurable by the determination of price in monetary units in a market. The transference of values between respective value domains are organized to understand how the whole economy built up by the human race has been distorted and isolated and sometimes enabled to interflow in accordance with our permunal intentions. It very much resembles to human cognition and psychology. The relation of rent vs. creativity is a good indication of the vitality of the society.

3) Mappings of the Hierarchy of Complexity: The topographic mapping of hierarchical complexities in the menesys of global humanity and its associated iso-potential mapping, which means the readiness of next possible outcomes, and the measure of complexity and its hierarchical classification of various complexities, are the precedent requirement of the mapping procedure. The factors of integrity and multiplicity are important in the classification of complexities. The above iso-potential mapping will be done by the proper correlation of time series parameters in a continuous manner. The noticeable thing with the complexities of humanity is the separation by boundaries due to antagonism among nations and societies. We need to find out some measure to quantify the trends of emotionally associated complexities as well.

4) The Assessment Method of the Paths of Events: Concerned data will have to be accumulated for the paths of plexus of accidental events and affairs, until sufficient research can be made based on them. The effects of the cancellation of symmetrical and fluctuating factors in the data collection would leave significant asymmetrical moves to efficiently influence the paths of accidental events in the long run. These paths of events are very meaningful when boundary obstacles are too high to get past; and thus bypassing is needed or sometimes even a long way around is considered.

5) The Advanced Technique of Dynamic Modularity: Dynamic modularity is at the center of the whole concept. The entire phenomenon of humanity is explicable by dynamic modules including integrismic conditions that decide the balance of the cruxological process. Dynamic modularity changes the boundary conditions of entities and clusters. It means the menesysal properties would be changed by the intrinsic conditions of concerned dynamic modules. Dynamic modularity is also devoted to the highest capabilities of humanity, which are usually called intuition, insights, integrated feelings of personal affection, and full volitional devotion and voluntariness accompanied by the emotional "flow." (Flow in creativity means full concentration of the free mind.) Thus, the advanced technique of dynamic modularity will be able to find out how to initiate and maintain the best possible outcome in specific situations without disturbing the whole menesys and the neighborhoods.

6) The Advanced Technique of Value Administration: Value administration in a menesys is essential in keeping the degree of generation of creativity. If it were an individual case, the value would not do much to the level of creativity. Individuals have a greater variance than communal entities in generating creativity. However, if our concern is the macro phase of global situations or any complex conditions of menesysal entities that include nations, societies, and divisions of profit pursuing systems and networks, we need to think of enhanced techniques to avoid unnecessary conflicts. It is suggested in GIMUP in this chapter that such higher technology be considered to cope with the present discontinuity problems that have confronted our global society.

7) Advanced Technique of Integrismic Administration: The science of integrism, in chapter 5, presents that integrism is the matter of ratio between systemicity and nexusity of menesys. The problem arises when we do not have enough set of reliable data and information to practically apply the integrismic measure into a real social circumference. As a rule of thumb, it was at first suggested to apply the "Proportio Divina", which is simply useful for the beginning of trial and error methods when an order has to be found in a natural chaos. However, if it is intended to understand the scientific bases for the integrismic phenomenon, systematic data collection should be implemented for a variety of situations, at least at the minimum level.

8) Advanced Technique of Cruxological Administration: The science of cruxology matters with the growth procedure of matured menesys and other cruxological development of lower stages. With cruxological view, the growth or fall of humanity can be observed analytically in connection with integrismic idea. Nations and communities are specific cases of cruxological existence. Cruxological administrations for networks and systems are different from each other and the composite models of sysnet and netsys must be managed by differentiated administrative techniques. An advanced technique will be developed with its focuses on the complex administration of businesses and national bureaucratic organizations, but the real target is the refined micro treatment of humanity in various cruxological mechanisms.

9) Integrated Management of Global Creativity Menesys: the creative peace has coherence with the global creativity menesys. In sizable societies with a lower

integrity due to the variety of constituents, it is observed that a rent has a significant effect on preserving the creativity of the society. Many political revolutions through the human history have burst out because of the rent problem. In these days, the rent for environment has become so critical that the poor classes in developing countries may pay increasing attentions to the destiny of their children. We need to take care of these on-going problems. However, it is also considered that the creativity menesys ought to be extended up to the global scale to upgrade the total level of creativity. In formulating the creative menesys, we will find practical strategies to solve the retarded conditions of nations and provoke incentives to activate the overall creativity of humanity toward the outer space and deeper inner phases of ourselves.

10) Continuum Political Economics: It is about time to write a new political economics in an integrated manner with novel scientific findings and previous experiences. Our observation on the political economy proves that it has got to be a socio-psychological political economics. Nobody wants to deal with psychology that is associated with economy. Very recently we found some activities to make a research on marketing and organizational problems. However, in a continuum economy, we should cope with psychological factors to view the wholeness of the political economy and integrate the efficacy of multiple divisions into the continuum.

11) Integrated Technology of Menesysal Management: All menesys need menesysal management. The integration of the technology comes from the understanding of the existential foundation, which I call here persontological basics, of concerned menesys. Menesysal management shall reveal the cardinality (not a mathematical terminology) in the management of humanity from the point of time when the human race sends communities, small or big, to the outer space. It is a technology of software that combines communities together with other communities and strongly adheres to their mother menesys. Another aspect of menesysal management will enlighten the synergy of mechanisms in various patterns of menesys. However, it is our recommendation that the central and initiative menesys should be identified as the global trans-space projects to motivate the world and concentrate the global resources. The towing trans-space project should be balanced with another connexion project between advanced markets and retarded markets, which would permit developing countries to be efficiently integrated into the global market continuum so that they can enjoy the wealth of the future technology and the space exploitation.

12) Theoretical Development of Pure Science of Humanity: Humanity cannot be dealt as an object for cultural and philosophical consideration any further. Humanity, nowadays, has turned out to be a matter of science. Without this progress in our standpoint, the human race will not be admitted to the space. Are we merely fungus that spreads its hypha from its mycellium over all the surface of the earth, or the real children of the Eternal? It is to be determined by the forthcoming response of humanity in realizing the essence of the human race.

The methods mentioned above would yield a good approach to administer integritium in accordance with the quantified data of integrismic governance.

To Build the Menesys of Integritium

Integritium might be an idealistic idea to certain people. It might look idealistic at the beginning but will be improved steadily and gradually manifesting its hidden glory for the next one hundred years. It is a matter of consensus as we have reached in many global agreements on number of agenda. Desires for integritium have been surging in the hearts of humankind for millennia. Simply we did not have pierced concerned pearly ideas with an adequate language thread. To get integritium actualized, the menesys designed for integritium must be initiated at the first stage.

The first stage of the menesys of integritium would take a form of "randomly systemized network" (a sysnet randomly manned from the global manpower out of reliable and virtuous organizations), which are composed of formal organizations to blow consecutive bubbles of temporary volunteering organizations around it. Many of these bubbles will remain and be settled as formal institutions.

These volunteering random sysnet would look like citizens' movement in appearance. Actually, currently prevailing citizens' movements have essentially an identical character with random sysnet.

The network of these random and temporary organizations, to be called T-1-sysnet, will execute preparatory planning steps for ten times, which require ten temporary "sysnet" to last, consecutively, the first 50 years. T-1-sysnet will not be left in the hands of limited number of influential people. If properly funded and with their authority transferred to new successor, the old T-1s can be altered into peripherals of new T-1 sysnet. (The peripherals of T-1 are to be called T-0 in our discussion.) This is in order to handle widely recognized and well defined global problems, so that it only produces internal disciplines and rules of those specific agenda for the formation of the prospective menesys of integritium.

The temporary organizations, with the longevity of five years each, will also be examined for its pertinence to the existing global conditions and remodeled each time. T-1-sysnet would take a form of sysnet which has the smallest body of organization equipped with function to plan, a little bit of organizing power and with enough fund. They have to publicize all the results of their works.

The second stage to perform projects of planning and preparatory execution for minor global issues will start with the sixth T-1-sysnet. The T-2-sysnet, ten temporary sysnet of the second stage, is going to execute small to medium sized global projects and will last 50 years, gradually increasing until seventy fifth years is reached to hand over their baton to the sixth T-3-sysnet of the next third stage.

The third stage, as now we can easily imagine, will start from the fiftieth year counted from the start of the first T-1-sysnet. This will be activated for the next 50 years executing ten times, expanding from medium to a full scale, As a result, it will enable major issues to induce their solutions and implement their own role under a close collaboration with formal governments, United Nations, etc.

As presented in Figure 10-2-1, T-1 sysnet will probably induce 100 peripheral organizations if well adapted to the global situation. T-2 might provoke the rise of 1,000 peripheral organizations and T-3, hopefully, 10,000. In order to make such global trend effectively materialized, a traction force originated from a globally recognized and acknowledged objective will have to be given. It will be discussed in a later section about Solarche schema.

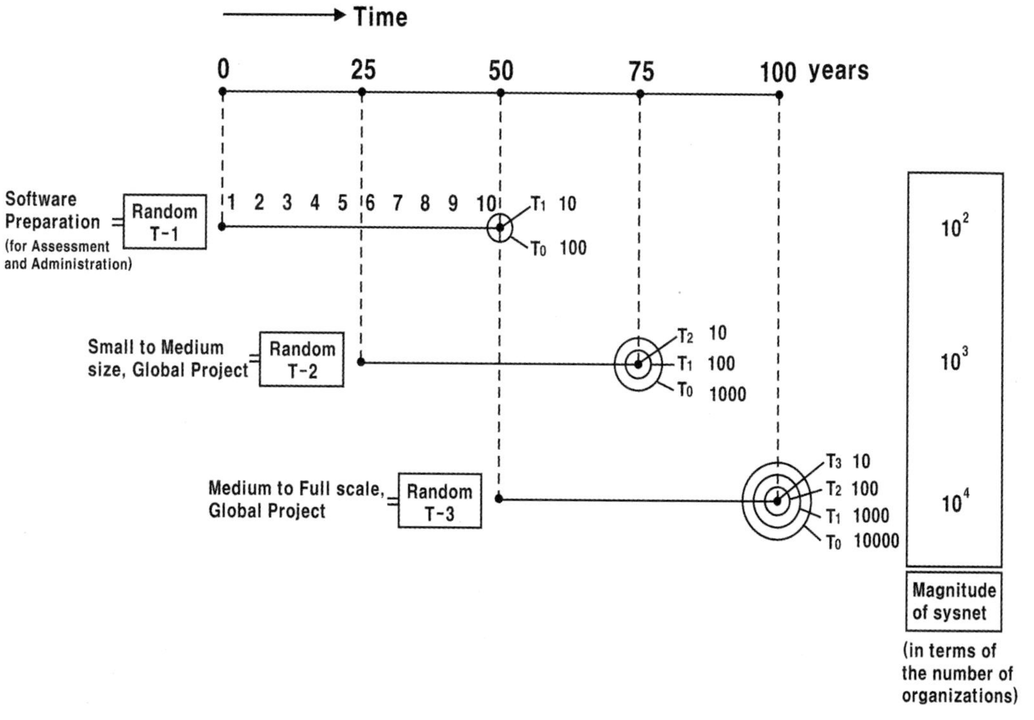

Figure 10-2-1. Schemes and Time Table for the Randomly Systemized Network

Periodic reports are to be submitted to formal organizations that are globally interwoven in the whole pattern of "netsys," a networked system at the last stage. This whole pattern of netsys will become the seed of the globally matured menesys of integritium. However it is noteworthy that the local activities (contrary to global activities) from regional and national level will have to be encouraged to add up their own scent and life styles to the world trend.

The sponsoring guardianship of the United Nations and G-8 and/or G-20 will be very important in performing this whole project in constructing the framework for the menesys of global integritium. Existing organizations and networks have their own scent and too much, maybe not, of systemicity in their intrinsic attributes, the organizations may be either public or private.

We need more nexusity. Nexusity has been jeopardized due to the negative development of the illegal side of underground markets. Nexusity has been strengthened by NGOs but they have been also exposed to public more than its necessity. That is why, in my personal opinion, nexusity cannot grow further into a fully grown sysnet scale. Most of the capability of nexusity furnished to the global society has eventually turned out to be organizational and provided new job opportunities, which were positive to the economy but not exactly to the future of the world.

During 100 years, (I believe it to be most likely accomplished within the first quarter of a century) most of the agenda for global problems will be extracted and organized to help to interweave "the peace knowledge menesys." The peace knowledge menesys will be the skeleton to generate "the educative information technology."

Global Pax Consortis by the Creative Peace of Integritium

Aforementioned menesys for Integritium is the initial sample for developing external mechanisms of the creative peace. In the process of developing the patterns, various methods will be designed to provoke the participation of stratified systems and networks of the world, including national states, confederal and federal states, the United Nations, global NGO/NPOs, etc.

We have defined the creative peace to be the "the continuum of activated humanity in stabilized condition." The creative peace will be materialized by Global Pax Consortis founded on the menesys of Integritium and these three concepts are mutually interconnected. The world has always been a sort of immature menesys of Integritium. Even an imperium is a special kind of generalized menesys, which is too much solidified and strengthened by the excessive systemicity.

The strategy-G for Global Pax Consortis would be more material, political and diplomatic. The strategy-I for integritium would give more emphasis on the cruxological measures. The strategy-C for the creative peace will be focusing on the peace knowledge menesys and the educative information technology. This kind of multi-orientated essential set of strategies would give firmer promise for peace in the humanity's future. However, these strategies presumes the constraints of the military stability of the world, for which summits of global leadership has to talk from the basic, hand in hand. This comment also implies that we ought to have a certain measure to safely treat the trends of low density warfare too.

10-3. Peace Knowledge Menesys

Peace Knowledge Menesys
Alpha Path Finding
Menesysal Education Process
Educative Information Technology: Floating Education Menesys

It was mentioned above that during the first quarter of next 100 years the agenda for global problems will be extracted and organized to help the interweaving of "the peace knowledge menesys." The peace knowledge menesys (PK menesys, hereafter) will be the skeleton to generate "the educative information technology (EIT)." The strategy for the creative peace will be focusing on establishing the peace knowledge menesys and the educative information technology derived thereon. However, it is recommended to start the EIT to be applied for the retarded economies first and proceed to higher level of economic entities successively.

Peace Knowledge Menesys

Knowledge systems in humanity are mostly diversified according to the geographical, ethnical, linguistic, religious and national factors. If the knowledge contents to be present in a dispersed condition, the peace on earth is yet to be accomplished. Knowledge systems have to be integrated into a single menesys and the subframes should be concentric to the theme of the creative peace.

The first rule to formulate a menesys is the acknowledgment and conservation of all the knowledge systems. It will enable our future humanity to trace the developmental process of past humanity. Even the thoughts about deception, violence, and decadence (DVD factors) are not the objects to be cancelled. These are the three main sources of the failure of humanity as discussed in triradixism.

However, it was thought that, if any closed circuits of these three factors exist, either combined or alone, "those closed circuits" should be most strictly taken care of. These closed circuits can be named "the axles of evil" which tend to generate a fluctuation mode in the global humanity. We would call it axles instead of axis because these are dynamic sources of sinuous fluctuations. These dynamic sources should assume and behave as necessary evils. But it is important that infinite expansion and infinite contraction of humanity should be properly treated to avoid bursting of bubbles. The bubbling effect sometimes appears in the form of a violent and devastating war.

PK menesys also has to start from three alpha entrances and finish at the three omega exits. These concepts are commented in Chapter 3 explaining the agnostical realism of science. Three alpha entrances into PK menesys are in their past terminology, nihilism, agnosticism and anarchism.

We have to expand the meaning of these negative thinking to make tools for the powerful initialization of creative thoughts. Reconstruction of thought frames are made possible when we find a basal stratum (agnosticism), with nothing left on hand (nihilism), by cleaning from the bottom (anarchism). Thus we can absorb the good advantages of these negative thoughts even though those were neglected in the past because of their untouchable character in our everyday lives.

In fact, the three brethren of extinction are inevitable prerequisites for those new constructors of ancient civilizations, who built their prerequisites in spite of the ruins of the past. We need them in a mental pattern, since we are going to tame those destructive prerequisite factors without actual destruction by adopting and domesticating into our knowledge system.

Three omega exits of our thoughts would be the way of thinking that human being has not dared to access because they were all energy-exhaustive concepts for the human mentality. But for the humanity, it is noteworthy that the three attitudes furnished from infinities-matrix have led us to overcome all the difficulties on our shoulders.

They do not even have proper names, for which suggested are ultimatism, integritism and continuitism; ultimatism in order for humankind to mobilize all the capability, integritism to activate humanity based on personhood to attain the integrity of humanity, and continuitism in which discontinuities are discouraged to globally achieve effective *koinonia* (ancient Greek terminology adopted to represent mutual communication, interchange and association) in external conditions of the earth and internally in the structure of humanity.

Three entrances and three exits make the widest structure of phase spaces of the human attitudes in their way of positive thinking. The attitudes give a difference to the menesys generated thereon.

The two main compatible pillars that we need to stand in front of the PK menesys are: the philosophies of Dyon (Dynamic ontology) and Trep (Transcendental epistemology). Dyon is the way of erecting the whole scheme of thought of humanity, while Trep is powerful in constructing and managing the substantial body of thoughts. Dyon and Trep are both dynamic. Dynamic phenomena associated with personage are the carrier of the PK menesys. The foundation of the permunal dynamicity stands on the four cornerstones of humanity - concrete sincerity, overviewing Intellect, dare execution, and exchange of goods.

Persontology is the main skeleton of PK menesys. Persontology can be illustrated as the integration of the two ontological frames as shown in Figure 10-2-b. The two ontological frames include static ontology and dynamic ontology. They have a long history of conflicts against one another; nominalism and realism. These contradicting and complementing fields of philosophy have collided hard several times in history. William of Ockahm from the thirteenth century has attained his fame throughout such conflicts. In the process of debates for a long time, some of the contents were mutually interchanged. But here, we will discuss in a simpler way to disregard it.

Existing notion of "ontology" is relatively close to static ontology. Static ontology starts from naming things through nominalism. For instance, when Helen Keller, the renowned triple handicapped, was told for the first time the name of "wa~te~r" as she felt the cold water running through her bare hands, her new world was opened. Through learning and acknowledging objects and facts she never knew before, she is said to have restored her affection toward the environment and the others.

Dynamic ontology is definitely close to realism, if properly interpreted. Dynamic ontology is a matter of realistic cognitive process. The human cognitive process is well interpreted and analyzed when the persontological view is applied. In most of the current interpretations about the human cognition, intellectualism supported by epistemological approach has been prevailing. It means our imagination cannot comprehend the extended understanding of the ultimate picture. It is suggested here to take the emotional aspects, beyond intellectualism, into account.

In this context, a diagram is provided to illustrate the persontological cognition-map, which contains and is correlated with an ontological K-map (with some changes to Takashi Saito, Ontology Algorithm, vol. 1, 2004).

Here is a very important implication of the Peace Knowledge Menesys. The interconnected structure of infinities-concepts in the human thought and knowledge has essential impacts on the health, lifetime and adaptability of the knowledge system or knowledge menesys.

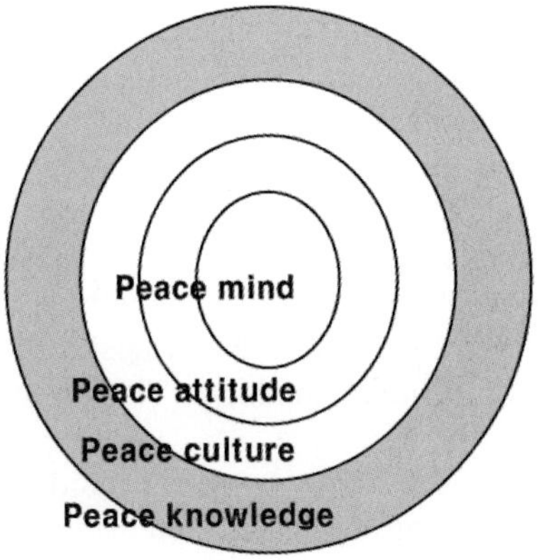

The Onion of the creative peace

Figure 10-3-1. The Onion of the Creative Peace

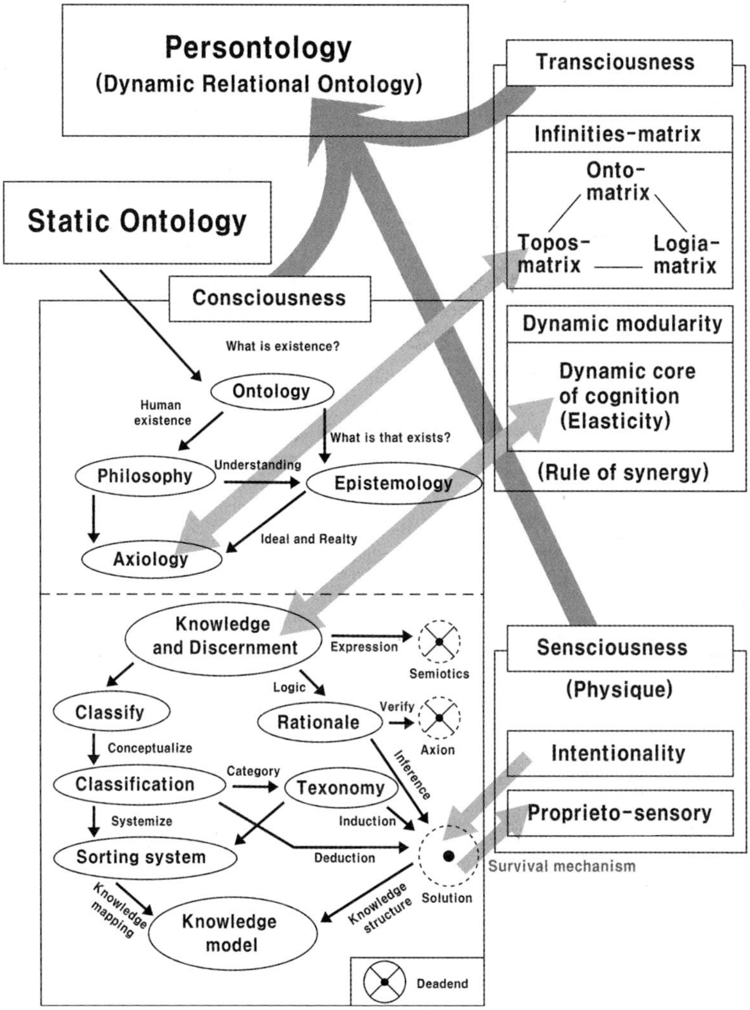

Figure 10-3-2. Persontology and Peace Knowledge Menesys.

The diagrams shown on Figure 10-3-1 and 10-3-2, give us an idea of the connectivity among three cognitive levels. K-map, as a part of static ontology, is based on the logic and linguistic system, but it has to be realized that *veritas* (or axiology) cannot be communicated with the others, because it is self-complete. This happens for the systems of semiotic and axiom too. A system of semiotics and axiology should prepare its communicative tools to get together with other systems of semiotics and axiology. Then, it is not an isolated system anymore and should be regarded as a systema in a comprehensive and a-level-higher menesys. The epistemology and problem solution are also interesting, in that they integrate all the information then supply the conclusion to the next item in the course of the survival mechanism as shown in the Figure 10-3-2.

It is noteworthy that dynamic modularity has a firm connection with knowledge and discernment, and that it is interrelated with epistemology, semiotics and axiology. In particular, axiology must have a connection with the infinities-matrix. Problem solution has a deep relevance to the survival mechanism in the realm of sensciousness.

The basic idea of the diagram has been explained in depth in Chapter 3 and 4. The key points of this idea are how to settle down a knowledge menesys to enhance the human creativity and the peace-oriented knowledge. The third kind of creativity to build the foundations of new civilization would be the answer for the whole query. In my opinion, the third kind of creativity that alters most of the topography of an existing menesys happens to emerge from the least compressed environmental condition in which a completely nucleic seed can survive and preserve its whole genetic peculiarity.

Nevertheless, this section is the tee off spot for the menesysal education process. It means that we need to clarify, at the least, which major innovative process needs to be added, as well as which negative portions need to be weakened or excluded, concerned with menesysal education.

Peace cannot be sustained with just the peaceful thinking. It is indeed true that certain social constraints need to suppress the negative factors; yet these same social constraints can also impose compulsory means and undesirable burdens on individuals and communities. Our first issue is to find a way to induce positive creativity from those negative burdens. Some of the individuals escape from the stressful environment, to take liberty in finding new solutions. Can we manage to restructure the knowledge menesys and furnish it as a general solution for the entire society?

Secondly, we need to consider the gyroscopic maintenance of orientation toward a creative society. Gyroscopic orientation to creativity is subjected to the abundance of resources for the creative thinking and the way of thinking to creativity.

The "flow" status of an individual person has, as suggested by Mihaly Csikszentmihalyi, a significant efficacy on the productivity of the value of creativity when he is positioned "between boredom and anxiety." Then we have another axis for communal creativity, which is "stress and relaxation." Are these effective for all the global situations? It is frequently observed that in a society indulgence does not always promise healthy creativity but rather bring decadence in ethical aspects.

Our desire to mend this through education has come up in the current edutainment solutions. But such tactical and technical approach is not enough for all global situations. It is almost the last chance for the human race to reduce the undesirable excessive influences of sex and violence. We will see that any further significant development in this direction would cause a collapse of humanity to the bottom. The problem is of a deep psychological origin that humanity stands on.

In this context, we must eliminate the path of humanity to Titanism in the peace knowledge menesys of the future. Titanism is named in this book for a tendency of humanity to exaggerate its self-image to the position of Titanic deities of the Greek mythology. It is a kind of psychic phenomenon when human beings usually encounter summit experiences by which they identify themselves with the figures of infinities. Infinities-encounters sometimes induce us to dehumanization such as mental gigantism, psychological upscale of human identity, and self-recognition at an excessive eye-level, instead of factual self-image. The addicts to violent internet games, in which the players can behave as a semi-deity, are the victims of Titanism.

Peace knowledge menesys rejects Titanism, sexual perversion, and deceptive spirit because they seriously damage the integrative process of human knowledge. Integration of human mentality is subjected to the emotional equilibrium of expressive patterns of human knowledge. The disciplines of human knowledge have respectively specific patterns of expression. Expressions for music usually have rhythmic expression. It is from time to time found in reality that even painting has rhythmic expression.

We do not yet have good methods to analyze the fundamental expressions of human mentality. But it is conceived that reversal integration of human knowledge is possible and desirable to get back to the peace knowledge. Meditation is an individual way back to the peace status. However we want to recover the peace status of the whole society. The answer is the peace knowledge menesys for the whole human race.

Alpha Path Finding

Finding the third kind of creativity will be rewarded if we position ourselves at the entrance of the alpha path. Alpha path is located where people do not feel excessive stress and can integrate their abilities as easy as possible but away from the three negative factors of deception, violence and decadence (DVD factors). Alpha path is not to be discovered along with any strict logical constraints but with the relaxed reactions of ordinary persons and families in usual circumferences. It will be found in most frequent emotional reactions revealed naturally, which we would call the communal "flow". The attenuated and hidden elements of common responses of the communal flow ought to be deciphered too.

For the contemporaneous human race, there are good focusing points that can concentrate our efforts to the very productive orientation. Recommended herewith are three categories; the first is our actual dreams and endeavors toward the entrance of outer space, the second, global collaboration projects for developing countries, and the third, software enhancement projects in human civilization in everyday life.

Space programs are ongoing globally under the global collaboration but it needs to speed up to make it a great concern for all the global people even in Madagascar, Guyana and Tahiti. The earlier the commencement of global collaboration projects occurs, the better chance for the world peace. It will be explained in the following sections. Software enhancement has to do with our study of PK menesys. Afghanistan, for instance, is a good place for a new challenge that we can introduce a new approach in software of global PK menesys.

Menesysal Education Process

Menesysal education ought to be differentiated from the past in a sense that motivation is the main frame and thus education process will be associated therewith. Motivation process supposedly sprung out of the learner himself, in other words, the educator should provide a deeper care to him and his motif. It is not likely that the usual global learning environment is as effective as it is in technologically advanced conditions. Menesysal education pays attention to the needs of the learners and reconstructs educative information to fit their practical situation concerned in peaceful, pragmatic and ethical ways of life, eliminating violence, decadence, and deception.

For instance, the global goal for the human race, visions for the development of local economy and means for the increase of income would be good contents in the first stage. Thus we need to improve the "educative information technology" to make people adapt to their own life more easily and to provoke their own wisdom and insights for their immediate future. It is recommended that menesysal education ought to be synchronized with the implementation of PK menesys project and the continuum economy.

Educative Information Technology: Floating Curricular Education

Educative Information Technology (EIT or EITech) includes kinds of software to encourage people to participate in the global education system in a menesysal pattern. Forthcoming educative software technologies would be quickly altered to admit "menesys for floating curricular education," since the hardware part of the educative system can support the floating mechanism of curricula at a far less expensive way than in the past.

The major idea to cope with, in the floating curricula system, will be actually the concept of "dynamic modularity." A quickly changing dynamic status of local people wherever they live would change the environmental condition of the educative system easily and frequently. Educative information technology will be an important aspect of the answer for the creative peace of the world. Such trend should be seen more evident in developing countries, since the introduction of new hardware technology will make rapid and distinctive differences.

The methodology to apply the technology to a specific cultural module should be based on a reasonable analysis of the NPSP political economy of the concerned module. The most probable developmental routes will be set forth in accordance with the application of EIT strategy. EITech will follow futuristic principles with which the most probable strategies could be found.

Futuristic principles of our interest that govern the EITech procedure would be; reshaping pattern of imagination, rearranging fundamental sets of rationale, and reconstructing conduit of social reaction.

"Puzzle of symbolism" is the name of the game of EITech. It starts from simple matching problems to construct a more complicated structure. Solving puzzles would spontaneously guide most of the least educated people on this planet toward the inquisitiveness for their own future, in a peaceful state not in a harsh struggle for their lives. The fundamental culture of EIT recipients will generally coincide altogether in a sense that communal relations are maintained based on their social humaneness.

In the structure of their knowledge system, the core people have a strongest adhesion to the power maintenance. In pursuing their common objective they try to mobilize and assimilate their people around the core. Means to mobilize and assimilate their people, even since the ancient days in which extended family was prevalent, have been always dependent upon their own peculiar symbolism. Symbolism has been usually strengthened by repetition and concentration.

Basically, it is recommended to adopt the frame of the educational idea of John Amos Comenius, who initiated the thought on essential global education with the ideals of *Pansophia* (entire wisdom) and *Pampaedia* (entire education). It is deemed that his ideas should be revived to fit our present situation, especially in today's globalization. He was a serious Bohemian idealist who knew the suffering and tear of losing one's home country.

Comenius (1592 ~ 1670) had his own clear vision of the millennium society very far ahead of his time and that closely fits with the ideals of today. His vision for the millennium was based on the socio-biological evolution, illumination on humanity, establishment of the global peace, and thus a healthy viewpoint of the spontaneous emergence of an ideal society reached, solidly found on an ethical and political affinity. (J. D. Bernal, Comenius Visit to England. And the Rise of Scientific Societies in the 17th Century, Cambridge 1942)

Our futuristic learning procedure needs to clarify numbers of our intrinsic learning mechanism generated throughout centuries hidden in our genetic *a priori*. Maybe they are located in our ensconced multi-personages. It would have a greater importance, especially dealing with the educative information system of developing countries.

The keys to this technique are in digital information and multi media education through a low cost mobile communication system. It has to do with physical expressions such as native melody, rhyme, dance, etc, those that are familiar to the native people.

Such educative information will have to follow the paths from the phonetic and the semantic to the syntactic frame. This procedure will help the learners to become emotionally stable.

10-4. Continuum Economy

Origin of Continuum Economy
Generation of Global Continuum Economy
Admission of Global Countries into the Boundary of Continuum Economy
Socio-psychological Econometrics of Continuum Economy
Global Currencies to Support Economic Continuum

At this moment, we have been ambushed by Sub-prime Mortgage Crises. Concurrent economy of the world can be hardly analyzed. Our remedy is merely achieved by instinct and many are wandering around the modifications of the Keynesian emergency measures. The panic and global quake's origin is the lack of novel philosophy to diagnose the current complex situation. But the fluctuation after the panic status actually comes from the absence of worldwide objectives to mobilize the mental energy and all the resources toward one great consensus.

We do not have to worry about whether or not a collective measure will ever happen in the world if we concentrate on certain teleological aims. There are measures to avoid the collectivism of "1984." The well spread global political economy is a good placenta for a new flexible systems. The failure of financing and investment industries today is due to their lack of responsibility in their position within our global structure. They did not even understand why the mechanism worked with their alert readiness to recognize in mapping and directing the concerned industries could have prevented our economy from fallibility.

The phenomenon of multiple personages, which visualizes the essence of complexity associated with dynamic modularity, would enforce us to confront the problem and see the real substance of the crisis. The viewpoint can be logically pierced by "the continuum political economy." Multiple manifestations in collaboration with complex dynamic modularity will appear to materialize Pax Consortis.

Origin of Continuum Economy

What everybody knows is that "eating is the main core of life." Economy is a means to get food. Politics is taking and distributing power to use. Sometimes it is done by an outright monopoly of the power-holder, and other times by balancing in order to achieve a smooth relationship within the community. It depends whether or not humanity has a definite idea of the identity of the apparent market and monetary menesys. Without a firm monetary and financial background for the exchange of commodities based on credit system, market would not behave as efficient as it does now.

Market mechanism originates from an internal, psychological, and cognitive mechanism of humanity which has been manifested apparently by being associated with the material foundation such as monetary currency system. In other word, "manifested apparently" means the rest of unrevealed portions of humanity are omitted in appearance by the enfolding mechanism because the omitted portions do not have proper coincidental communication mechanism with the environmental infrastructure such as the currency system.

Mind, usually in our everyday conversations, substituted for an internal mechanism for humanity, has a symmetrical reflectivity, which is activated, intervenes with the real world and tries to achieve a balance every moment. It is an intrinsic regulating mechanism. Such regulative mechanism is the origin of the continuum economy and the condition to be used for the establishment of the global continuum economy.

It is well understood by most people that the apparent governance to effectively release the internal energy of a market relies on the political governments. Thus, the origin of the continuum of economy is the market that is a part of nexusity, but the incubator of market is the system that is either the government or governing organizations. And we have a third factor, emerging later in the history compared with the previous two, called science and technology.

Generation of Global Continuum Economy

At present, the global continuum economy is building its formation but is in need of directives guided by a concrete overarching philosophy. First of all, the global consensus and objective have to be secured for the human race to advance to the outer space. Psychological consensus in the orientation throughout the whole world is evidently necessary.

We usually think that the international market aroused after the emergence of the Industrial Revolution for the last three centuries due to the imperialism and multinational corporations. However, if we extend our perspective, we can see that the market, manifested in various morphological menesys, has always been present around us since Mediterranean was occupied by the early Grecian colonization. The centralized pattern of governance has also been organized as early as the Babylonian Empire at latest.

In terms of completing the internal structure and mending the apparent mechanism of the global continuum of economy, we are at the stage where we can set up common objectives, understand what is going on in the real economy in the long term perspective, and distribute resources throughout multi-level. The understanding of overall basic mapping of psychological mechanism of the global market is crucial to complete the process.

The emergence or the birth of the global continuum economy depends on the securing the center of the continuum to be formed. Until now at the center of the existing global continuum there has been formed a nexus of logics supported by legal and contractual transactions, which is partly associated with the psychological ties originated from the fashion and popularity of superficial trends. Psychological ties have assumed the intangible central role. It is not necessarily recommended that tangible ties, such as the centralized body of institutions, are compulsory to build up and sustain a healthy body of continuum. It is suggested that the minimum requirement of systemized institutions are needed unless worldwide political economy stays safe with present groups of national systems.

We need to settle down and adhere to a major execution mechanism based on complex and integrismic measures. For the new approach, it is essential to understand the world as an entity of multi-personage with diverse intentionality based on emotion and psychology. It means to acknowledge the multiplicity of global economy and to find out a new path from there on. We need to eliminate psychological partitions from the center of continuum. It is quite likely that nations would have a hard time reaching a mutual agreement. Therefore it is very important to provide incentives to every nation to reach an agreement.

I have been impressed by the respectful work of David Held et al., "Global Transformations," published in 1999 after 10 years of in-depth study. According to them this study could be categorized as a transformationalist approach. However, we have a very distinguishable difference from other transformation theories. We are devoted to the practical prospect of the real world transformation within a foreseeable future. It means, in our terminology, we have transcendental intentionality.

The work of Joseph Stiglitz, "Making Globalization Work (2006)" also gave an insight for this study. He has dealt with problems of development, trade, intellectual property rights, natural resources, global warming, multinational enterprises, debt, and global currency. We have more concentrated on the solution of the continuum economy of a global scale to give a definite route of escape from this dead end situation. We are sure we have a strategic solution.

World industries deserve a sincere consideration for restructuring and adjustment for competition within tolerable range of impacts. An econometric simulation model will be necessary. For the global continuum economy, a prediction made by a simulation of both flowing mechanism and change of medium should be proposed. The reasonable econometric treatment of the flowing mediums, in which the flow of economic utility in dollars and the concurrent flow of intangible values take place, have to be well designed to simulate the reality.

It is suggested that such simulation would resemble the flow mechanics in porous and permeable media as what is usually practiced in the underground reservoir engineering of oil and gas fields in which multiphase (water, oil and gas) flow together at the same time in a varying situation of stratified and local mediums. It is thought that the physical condition of the underground reservoir is basically a good metaphoric model for the market menesys which can cope with and simulate the complexity of the highest degree we meet in the realms of humanity.

Admission of Global Countries into the Boundary of Continuum Economy

The process of affiliation of new countries into the continuum is one of the major concerns of the global economy. When capital transfer occurred from Europe to the new continent and later from America to East Asia, enormous changes swept over the global economy. The gradual and stepwise progress had affected all aspects of human life. The whole process ignited new dynamic modularity and multi-personage issues in our terminology.

It is a prerequisite of dynamic modularity in the continuum economy to establish the centric part of continuum with excellent connectivity at the initial stage. At the lower hierarchy, where the dynamic core of modularity is actively operated, the connectivity from the core to the individual partitions (of the dynamic modules of an economy) will be vitalized. Currently, the global connectivity has to be achieved in any circumstances. The connectivity will have to be mended to take gentle and smooth transitional contacts. The connection from the center of the continuum will reach to the periphery, where many partitions of multitudes are developed, through the medium in between.

The present status of the world economy is assumed to be in a relatively good interconnected condition due to the past efforts of international collaboration. However, poor countries yet have to be fully admitted to the main flow of economy. The introduction of those countries into the continuum might bring unexpected loss of the global economic output or the efficiency of economy. We have to believe that we can overcome such scenario by rationally planning the 'value exchange' that promises the mutual benefits of all nations.

An adequate integrismic management is required to take good ratio between sysy and nexy and maintain the ratio or adapt it to the circumferential changes as the reality responses. My opinion is that the world could be restructured in a way that it gives the most favorable considerations to the existing countries and to the four layers (hierarchy) of sysnet. They consist of the first layer of seven to fifteen countries, the second of forty to fifty, the third of two hundred to three hundred, and the fourth of five hundred to one thousand. The connections between the four layers of global sysnet will be made and then the center of it will be kept vacant. It will not be good to compose a group of only two countries on the top layer, G2, if we want to avoid probable conflicts.

On this planet, we need the expansion of the global continuum boundary and fill it up with the circulation of wealth. On the other hand, we are at the stage where we can touch the flash of lightning of Trans-space-vision. Trans-space vision is the only way out that the human race can concentrate to solve the future problems in this century. Environmental vision would solve only a part of the global problems and would not be able to mobilize the global energy to the extent that subsidiary problems would be all adsorbed into. It is recommended to solve global problems in a package.

In the next section, we recommend the development of Educative Information Technology and the materialization of GIMUP in the next 30 to 50 years through the Keiros Project in solving such situation.

Socio-psychological Econometrics of Continuum Economy

It was more than twenty years ago that the author reviewed a book written by a Japanese economist that commented that economics has to be changed to the socio-psychological political economics. But it has been observed that there have been no more profound and significant approaches to revise economics by the socio-psychological viewpoint. It is true that we have seen significant pragmatic change in the financial engineering and in other tools of mathematical approach in order to bypass the psychological factors by synthesizing ideas based on the science of complexity.

The continuum economy should be based on socio-psychological factors, due to the fact that the continuum of the global economy is neither measurable nor manageable by existing economics. It was aforementioned that the global economy should be dealt with fluid mechanics that describe the flowing mechanism within permeable media in which inertial terms in movement of commodities has to be considered. The commodity flows are associated with flows of various currencies and relevant expectations to describe the influence of socio-psychological factors.

Establishing an econometric model associated with the continuum economy will become a serious matter when the human race goes out to the space or other extreme environments and builds isolated residential colonies of considerable sizes.

The way we can move forward is by grasping the general idea of defining boundary conditions of present economic modules in the global continuum. Modules exist in a dynamic state in which they are subject to changes in their long-sustained core parts as in their easily changing peripheral parts. They are comparable to computer simulations of a porous and permeable reservoir under ever changing physical conditions, containing and allowing various mobile fluids flow inside. The medium consists of plural strata and clusters to which the basic principles of dynamic modularity can be assigned.

It should be relatively more complicated and of a greater size in many magnitudes to simulate the global econometric model than in modeling the underground reservoir situation. But it looks still manageable if the models are well designed to simplify the various parameters and their concerned extent.

The discussions made in the earliest part of Chapter 9, regarding the elite and commons, and enclosure and connexion (cap-con relation) are helpful to understand the factors and parameters which enable the simulation process. Such basic philosophical works, which correlate with practical economy, should be accumulated before entering the simulation of the continuum econometric model.

Global Menesys of Currencies to Support Economic Continuum

Currencies which would support the global continuum would become profoundly diversified in their design stage in the kinds and the levels of currencies in the economic continuum. Currency movement will be an epidemic like the citizens movement of the last century. Institutional organizations will have to be prepared within several decades for the impact of real economic movements of a high participation from citizens movements, in how to introduce such flows into the global economy and efficiently organize them in a healthy status.

It is natural to have such movements in the economy during the period of financial crisis, which hits citizens who have no protection. Citizens in such a circumstance have always been helped by themselves. Such voluntary development of monetary movement will help the global economy become healthier if it is well prepared and admitted by official institutions in local and global administrations.

As we discussed in Section 9-3, before-burst-bubble (BBB) currency and capitation-grant for less-capitalized group countries would be in need of the structural sustainability of the global economic continuum. We will soon recognize that the content of the global continuum has to be filled up in order for continuance to be guaranteed.

Capitation-grant for slowly developing countries would be applicable initially for three groups of countries at isolated locality in the currency space: some countries in Africa, the Caribbean and the South Pacific Groups. In our view, the Keiros project presented in Section 10-7 gives us certain ideas about how to approach capitation-grant.

Before-burst-bubble (BBB) currency is designed to be a counter-flow from the continuum center towards peripheral localities to heal the over-pressurized (thus, subject to collapse) continuum centers by transforming them into the foam status. We recommended starting from the level of an annual 3% infra investment of intangible and tangible. Such an amount can be used to practice Educative-information-technology in badly retarded developing countries.

The above ideas would help the global economy to be structured, rectified and strengthened. There could also be other remedies for a country which is too slowed down or isolated. For instance, socialist countries used a dual system of currencies at their earliest stage of introduction to the global economic continuum.

China had a differentiated domestic currency, Lunminpe, and a foreign exchange currency in order to split their economy from global capitalism and keep isolation with the foreign market economy. Such policy looks to have helped the Chinese economy avoid excessive exposure to external pressure.

For a badly retarded economy, it could be helpful to use a modified dual currency systems for a certain period of time. It is deemed that communal value of the society manifested as land price could be well managed by a special domestic currency by slowing down outlet flow, which occurs when such special currency is managed in connection with the construction cost of social overhead capital. This could be good for North Korea for instance. I would call it Lock-Canal method.

10-5. Solarche Menesysal Trans-Space Schema

Now it has become our major concern how to succeed in the menesysal management of the global economic continuum. As we mentioned, it is necessary to initiate two menesysal bodies within a few decades. One is tractor mechanism and another, connexion mechanism to be explained in the next section. Tractor mechanism is necessary in the global economy in order to advance to the future of greater-than-earth, stronger-than-system, and tolerable-than-science. The resource of the mankind that has to be utilized at this point is the trans-space global menesys. We need to touch the flash of lightning in trans-space vision to gain the traction of motivation.

Trans-space vision is essentially destined to pursue the infinity. Human beings will be able to earnestly digest the infinity problem when they face the factual affairs of the outer space at every corner of their lives.

These two global projects will have to do with T-1-sysnet, the randomly systemized network project in Section 10-2.

Current Structure of Space Projects in Global Economic Continuum

It is observed in the collaboration and competition among EU, the USA, Russia, China, and Japan that space projects are not well integrated for the best efficiency of the global resources. GPS revealed the definite crevice fractured between the US and EU. Japan and China has different history of birth in the space industry. A Japanese writer stated that China has military-oriented and national-centric objectives in the space project. However, it is not right if anyone asserts that their country does not have nation-centric targets, because every country is established in its own national pursuit. Some differences in national orientation might exist in their respective international politics.

Our major concern is how this discontinuous structure can be converted into the global continuum. Recently, international collaboration in the space development has made a great advancement, which is not yet sufficient to provide directives to the global economy. For a great leap in the expansion of the global continuum, we have to conceive a novel frame in order to make a breakthrough.

The reason we have to launch a steering body of global trans-space project is in order to provide the on-time supply of energy and mineral resources for human civilization within one century. As far as resources issues are concerned, the human race shall face an impending situation supposedly within fifty years. I believe environmental issues will not be choking us as tight as resources problems.

The resources issues have to be tackled based on profit-seeking enterprises in order to yield high efficiency. I believe that the human race has to concentrate on resources issue in the next two centuries. The highest potential of existing intelligence and military systems and forces (except for public security) need to be allocated to trans-space project, which will analyze, plan, and implement in order to acquire energy and mineral resources.

Above profit-seeking enterprise is a mammoth integration of worldwide network of space technology and knowhow. The mammoth enterprise will be positioned at the center of the Solarche Menesysal Trans-Space Schema. The financial crisis of the latest several years happened partly because financing institutes could not found sufficiently profitable ventures that fulfill the appetite of current potential of global financial institutions. Seeking peace is identical with seeking due profit. Seeking due profit is the same with seeking bread for the next breakfast.

However, we have to consider what the due profit is. It is why we search for a new profit-distribution method. The profit from the Solarche Schema should be redirected to global people even to all the non-participating countries. It is the wisdom for the creative peace.

Providential Opportunity in Trans-Space Project

The human race has to face the fact that the day is coming, the day in which those that do not become a body of the continuum of the *Homo sapiens* species will be extinct. The reason the author mentioned *Homo sapiens* is the preservation of survival knowledge, which is exposed and grown up as the peace knowledge menesys. The peace knowledge menesys of the human race has been structured unbalanced in a sense that the dimensions of infinities-matrix of humanity are not well correlated. The dimensions of human infinities-matrix will have to be studied in depth.

A foreseer like Freeman Dyson has made a significant extension of the human imagination into the solar sphere (The Sun, the Genome, and the Internet). The human race has had desires for a long time to utilize Asteroids belt, Kuyper belt, Moon and Mars. Dyson preferred to establish isolated and creative small societies to initiate novel and advanced civilizations (Disturbing the Universe, 1979).

Our view is different from his idea in that isolation is not likely to induce real creations for the extension of humanity. The continuity of character is the essence of human anthropomorphosis. We know that Dyson's isolation is not a complete isolation but freedom to seek a completely different future. However, the constituents of the new society should be a part of the human continuum in terms of knowledge and desire.

In the macroscopic point of view, we believe that the accumulation of knowledge, the practical objectives and the change in the global situation has moved toward the Trans-space schema of the universal continuum. We believe that global Trans-space project toward the solar sphere is on the line of providential opportunity for the species of *Homo sapiens* at the turning point of the human civilization.

Recently, it is widely acknowledged that Phobos landing is preferentially anticipated. With its close proximity to Mars and stunning vistas, it is easy to picture the Red Planet's biggest moon as our next outpost in space. It is cheap to proceed for the next step than simply landing to the Moon first. (Stuart Clark, Newscientists, 2010, Feb.)

Trans-Space Project and Global Continuum Economy

Trans-space project will bring spontaneous downward propagation of economic growth if well planned and properly managed for establishing the global continuum economy.

Trans-space project is most likely to disturb the continuum of the global economy if it is not carefully designed. This book recommends that the explosive amplification of future economic values should be treated beforehand. The industries which will accompany such imprudent value amplification will have to be identified.

We need to provide certain policies to adjust the developmental progress of such industries and the excessive value amplification. The policies will curtail and reduce the development and in some cases mitigate the return flow of value in value circulation.

To preserve the driving energy of the global continuum essential for the concerned industries are the adjustments mentioned above; they are categorized into three, which are physical space development, the exploitation of cyber space, and the extensions of human abstractivity.

The abstractivity in humanity is most likely going to be soon extended very fast and wide in scale in the overall economic continuum adsorbing excessive monetary flow. We can find such signs in the visualization of design industry such as the I-phone, financial presentation in graphic analysis and scientific illustration of higher dimensions into 3-D environment. However, the real explosion of the core of human abstractivity has not yet come. The core of human abstractivity is hard to define at this moment.

Anyway, it is the author's view that current financial crisis was caused by the insufficient and indiscreet administration of the economic growing points. Financial industry is a specific tool for the management of future values. Human being is the only distinctive entity on Earth who has invented methods to utilize un-conceived future and lives at present time on the future value, draining the future value so quickly.

Future value can be converted to present value by the application of interest rate. Interest rate is the rate of avarice in bad expression and the rate of wish in good expression. Without future value, we will not enjoy the fruit of creativity. The value of creativity is connected to the future in our psychological phase space.

Therefore, we need to design a methodology of the classification of future value. In utilizing one of those classifications, we will be able to induce a successive return flow of future value to the people of past-oriented class from the people of future-oriented class which preferentially consumes, accumulates, and reserve future value.

The classification of creativity value was explained in Chapter 8 to have three categories from the first kind to the third kind. We can consider some other categories for the classification of creativity value based on the social grouping of the producers of creativity and the temporal changes of the creative process. Such categorization has yet to be clearly made in a whole picture.

The global creative class (such as the social power class of digital information nomads which directs the development of future society, or the preferential class which governs financial liquidity like banker groups) are able to pop up the early accumulation of the value of creativity which is transferred to the economic utility value. They enjoy the privilege of their governance on the introduction of new dimensions.

Trans-space project has the application problem of value system for which we need to be careful to manage in order to strengthen the continuum economy. Figure 10-5-1 shows how a globally integrated Trans-space project can enhance the continuum of global economy. The project is named Solarche Menesysal Trans-space Schema.

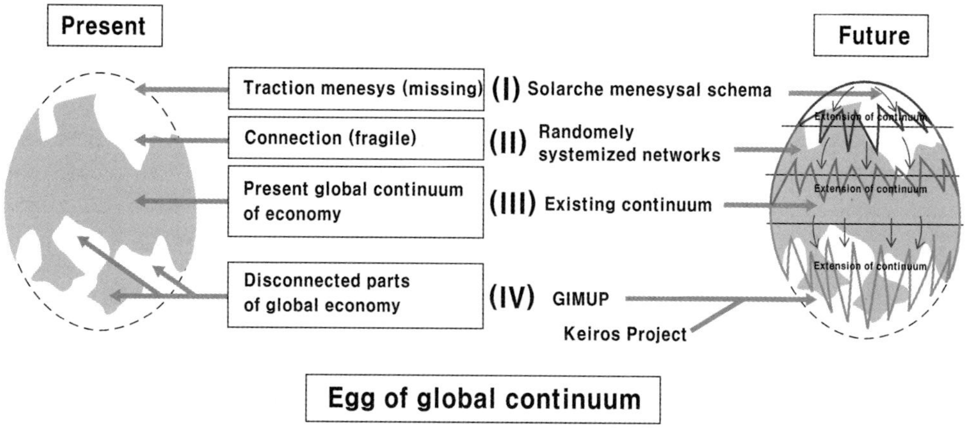

Figure 10-5-1. Healing of the Egg of Global Continuum

Solarche Menesysal Trans-space Schema

The compound of solar and *arche* is "Solarche," which means the beginning of the solar sphere. It is to suggest for the global human race to come to a conclusion that there is not much time to bypass the most probable alternative of the human race directed to the upcoming quick and drastic diminishment of our civilization.

The reasons that I don't agree with the statement that we might be at the edge of a cascade in our imminent future are; that contemporary human civilization has been bound to the complex boundary problems of the global continuum, that handled without overall strategy is the lack of leading frontal movement and its energy, and that observed is most likely the failure of insufficient current frontal ends in the global collaboration to pursue the whole solar space projects.

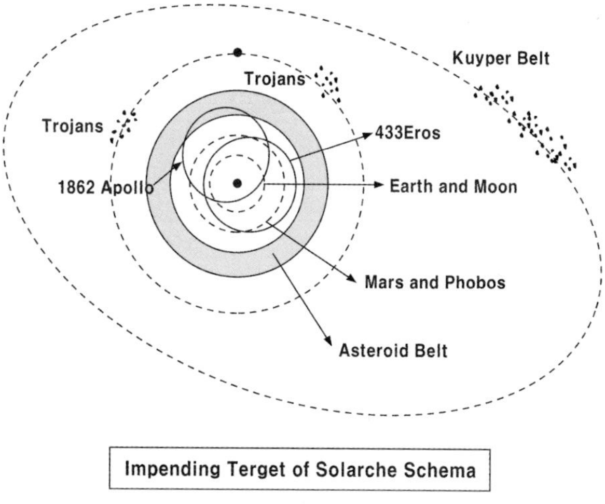

Figure 10-5-2. Solarche Menesysal Trans-space Schema

We have to see our integrated future in the Solarche schema. Solarche is a seed. The seed contains every possibility of the human race in the era of going out of this small Earth. We did not have shaped up even a zygote. We are about just after the conceptus, the start of cell division.

The most important issue we are facing would be the complex boundary problems, because the human race does not yet understand the essence of boundary problems nor recognize its importance. The boundary of the continuum of empires has been caused by the conquest of hard military forces, by the governance of moderately hard economic forces, or by the infiltration of soft cultural trends, against which posed are most of the passive partners of such connecting activities. We can see the seriousness of current boundary issues because most boundaries have abrupt discretion between opposing parties situated at contradictory stances, which generally causes mutual hostility and antagonism. It is also noteworthy to comment that such antagonisms are usually generated from the excessive self-respect of concerned parties.

Table 10-5-1. Frame of Solarche Menesysal Schema

(1) Solarche Menesysal Schema is the Top-notch Tractive Menesys of Global Creativity-intensive Menesys
(2) Chartered and Authorized by Global Treaty for Outer-space Initiative
(3) Mainly Profit-orientated based on Economic Feasibility
(4) Institutionalized Complex Menesys of Science and Technology
(5) Umbrella Menesys to Direct and Concentrate Global Resources
(6) Interconnected with Randomly Systemized Network (Figure 10-1) Random T-sysnets are Designed for the Assessment and Administration of Solarche Schema
(7) Extended to the Continuum Depth to the Bottom to Contribute to Connectivity Connection to Global Intelligence Modular Urban Planning and Keiros Project
(8) Preference of Direct Utility and Declination of Indirect Utility in Resources Usage
(9) Suffice Urgent Business Needs of Global Continuum of Economy
(10) Three Major Global Problems are to be Tackled by Urgency: Resources, Environment and Population, First Stage: Energy and Mineral Resources Second Stage: 1) Enhancing Environmental Condition of the Earth 2) Use of Physical Property of the Universe Third Stage: Space Immigration(?) or Advanced Base

Solarche Menesysal Schema first aims at the establishment of the healthy continuum of the global economy. Only the healthy and peaceful continuum of the global economy can feed the human race. Therefore the top-notch frontal menesys of the Solarche Menesysal Schema, which stands on the tractive position of the global economy, has a role of tentacle.

Solarche Schema will have to be authorized based on the initiative of global treaty, which needs to be established with and governed by a certain level of practical constraints and enforcement as a menesys. We have to make studies on the integration of dimensional introduction into our global continuum to fit the stable orientation of anthropomorphosis. Otherwise, the current human civilization will be evaporated according to the singularity problem.

Solarche Schema will be administered under the charter and authorization by global treaty for outer-space initiative. It is believed that the current scheme of the proliferation treatise of nuclear weapons can become a good starting point for outer-space initiative, provided that the initiative widens its scope of concern. Let's call it Solarche Initiative.

The expansion of the above scope of concern of Solarche Initiative is of no doubt connected to the huge menesys of current destructive innovations. Destructive innovations constitute a huge manifold of the human intellect. The strange thing is that such destructive menesys has a positive influence to a certain significant extent. It is probably due to the fact that most destructive ideas originally stemmed from the instinct of survival.

Anyway we know that the new manifold of Solarche Schema reorganizing destructive ideas will induce a lot of new creativity of the second kind if well directed. If we want to establish a new entity chartered and authorized by Solarche Initiative, we have to reach to the point of attaining the 3rd kind of creativity. The author believes that it will not take long once we agree to establish a loose umbrella organization in a global scale starting from aforementioned T-1 sysnet and so forth.

We have frequently said that corporations are new entity succeeding the concept of nations. But corporation was considered as a lower entity in the social hierarchy than nations and other legalized institutions. However, the concept of corporation will light greater in the coming ages, because of its liberty and openness, if we take a proper remedy of how to deal with the profit of corporation, which is allowed legally to be accumulated and preserved in the corporation. Present capitalism is centered at the concept of corporation and can be almost called as corporation capitalism.

However, multinational corporations are a pattern of typical dynamic cores in the global continuum and should be identified as a stitching node pattern which connects nations across national boundaries. For the last two centuries, multinational corporations have assumed the role of networking seams which enabled the development of full-fledged institutions of global business environment; financial institutions, legal environment of modern marketing, global information system and so forth.

Corporations are not only meaningful in terms of profit conservation but as a connexion by their corporative aims and functions. Therefore, for a considerable span of time in the future, corporations of current pattern will assume the essence of continuum capitalism which guarantees the connectivity in the continuum by the flow of profit. Profit brings desire and psychological flow together. Flow of profit is one of the keywords of continuum capitalism, and as commented before the essence of forthcoming value capitalism or pretium capitalism.

Thus, the chartered corporation of Solarche Schema has to consider the connectivity of the global continuum. It is why we need to design a complementary project for the reinforcing of the global connectivity of economic continuum, which is Keiros project. Keiros project will be dealt later in this chapter.

J. Micklethwait and A. Wooldridge wrote that the future of corporations is in three probable alternative paths; a small number of super corporations taking over the entire business world, corporations with vague substantial body, the calm encroaching of networks into the realm of corporations (The Company, 2004, Conclusion). But the author believes that the change will bring more complicated spectrum in a broader scope.

We are facing a new boundless expanse of menesysal world in our economic continuum. Niche industries and interdisciplinary niche disciplines come up on the new menesysal horizon. Davos is at one end of sysnet and try to steer its connexions in global menesys. WSF (World Social Forum) is at another end of sysnet to be confronted with Davos.

In our view, Davos and WSF are just the most relaxed sysnet (MRSN) which do not possess a realistic steering ability. The opposite concept of MRSN is the most centralized netsys (MCNS). Solarche Schema will become a menesys with strong sysnet character.

Chartered company received the most significant systematic support of the power authorization and military enforcement from the nation when East India Company of Netherland launched and existed for 274 years. Solarche Schema is recommended to take the form of chartered corporation with a broader spectrum.

Solarche Menesysal Schema will eventually be retailized step by step to the common mass of the world. Retailization is a term to denote the spread of stocks to the entire society so that the governance of Solar Schema is in the hand of people, although the administration must be given to the expertise.

Therefore Solarche Schema will be managed under check and balance by properly specialized systems of technology, management and auditing. The multinational penetration of the Schema will provide strong fundamentals in order to sustain balance with beyond-globe extension.

Our objective is to establish a corporation seeking the maximum profit which resembles the East India Company, to design a koinonia-flow of the profit into the overall global economic continuum, and to direct the corporation of Solar Schema in order to tract the growth of the global economy at the front and keep eyes on not to harm the connectivity of the global continuum.

The chartered authority of the East India Company had the complete assignment of developmental decision (but with mechanisms regulated by quasi-congressional organization). It was the differentiated character of the company in the aspect of governance from general limited corporate and made the company a new institution in its age. We live in an age comparable to the 17th century which we have to struggle to open up unidentified conditions of survival. The East India Company had to have a defense procedure of public hearing every score years to justify the necessity of the existence of the company and its chartered right before the government authority.

The new menesysal institution, Solarche menesysal Schema, will have to keep tightly knit connection with the global continuum. GIMUP in the next section is designed to bring forth the sound flows of koinonia. The emergence and adaptation of new rational menesys is necessary to achieve the urgent need of the human race. GIMUP is the device of planning to suffice the needs mentioned above.

10-6. GIMUP (Global Intelligence Modular Urban Planning)

GIMUP
Complementarity between Major Carrying Values of Countries
Comparison, Exchange, Resonance, Covalence of Value
Tri-Taiji Principle in a So-called Nomadic Era

Anyhow, our present concern is at first with our planetary situation, which has already too many nuisances. We have to concentrate our direct eye on the present issues and tackle the root of those problems. It is our life line. That also contains the solutions to resolve the international discrepancy between political economics of South and North. GIMUP is recommended to cope with this set of problems. Soft touches will not hold a thistle tight.

GIMUP

GIMUP means the global intelligence modular urban planning. In this age we acknowledge that the scope of urban planning has been widened to the end of the world. GIMUP is a design method of connecting all urban areas spread over the planet into an inter-communicable and inter-connected intrinsic environment.

Intelligence modules are at work everywhere the human intellectuality effectively governs human life. Intelligence is not a word exclusively monopolized by the intelligent institutions for the national security. We have to understand that intelligence has become a general social phenomenon. It is recommended for contemporary intellectuals to regard intelligence as the body of information flow which enables the dynamic core of a dynamic module to sustain its integration. Therefore intelligence module is analogous to the network of neurons which are fired together. Intelligence module is at the center of the governance of dynamic modularity, but it has social functions greater than national intelligence. The social functions encompass all causes and factors that effectively converge and synchronize the energy and resources of society for the same objective under concurrently effective logics.

However, this intelligence module has to be open to all the other systems of culture, knowledge and political economics when we deal with it in global integrative purposes. GIMUP adopts a concept of tendons in the intelligence module in order to connect plural intelligence modules. Various tendons have existed between intelligence modules in history as we can find in empires, economic alliances, leagues of polis and cultural communities. The discovery and connection of major corresponding or resonating tendons between intelligence modules are the major technique of GIMUP. Tendons have greater influence on the globalization than those blue jeans and Coca Cola. If several tendons are well matched between two adjacent intelligence modules, the two modules will formulate a good integrative process after a mutual contact.

The economic continuum in a country is generally entangled and bound together very densely and intensively. It is not yet so dense in an international environment. GIMUP is a method to weave the world into a huge effective network. The world has many retarded areas in economy and politics, which are obstacles to achieving the extensive global continuum.

GIMUP will accelerate in weaving the world into a huge network of systems and later into a menesysal continuum. GIMUP is a methodology in terms of both hardware and software to achieve the objective. Keiros (Κείρος, hand in Greek) project to be discussed is the earliest pattern of the fruit of GIMUP. The core concept of GIMUP is the intelligence of humanity. By giving a new definition to intelligence and extending present intelligence institutions to softer functions than before, we will be able to successive, stable and creative development of the human race. Signs of such novel trends are observed when intelligence institutions have started to work for the collection of non-military information. The future notion of intelligence is far broader and deeper than just being national, industrial, educational and military. Intelligence menesys is different from knowledge menesys.

As discussed in Chapter 7 and as illustrated further in Figure 10-6-1, value-utility menesysal structure is manifested by the onion-like strata until people recognize the market price formulation.

In Section 10-3, we explained two main compatible (but mutually interrelated) pillars that stand at the front of the PK menesys, which are the philosophies of Dyon (Dynamic ontology) and Trep (Transcendental epistemology). Dyon erects the whole scheme of thought of humanity, while Trep constructs and manages the substantial body of thoughts. Dyon and Trep are both dynamic. Dynamic phenomena associated with personage are the carrier of the PK menesys. Dyon pillar has three-fold menesysal constituents of value, intelligence and pretium, while another Trep pillar has three-fold menesysal constituents of utility, knowledge and market price.

They have value-intelligence-pretium (VIP) pillar at one side and utility-knowledge-price (UKP) pillar at another side. When the human cognition shifts from VIP to UKP pillar, diffusion and diversification occurs, and UKP-to-VIP shift generally induces integration and unitization in the cognitive process. The change of cognition from diversification to integration (which is identical to UKP-to-VIP) gives a relatively smooth stability, because the mental manifolds in cognition are mended to discard the segmental isolation of logical and emotional effects.

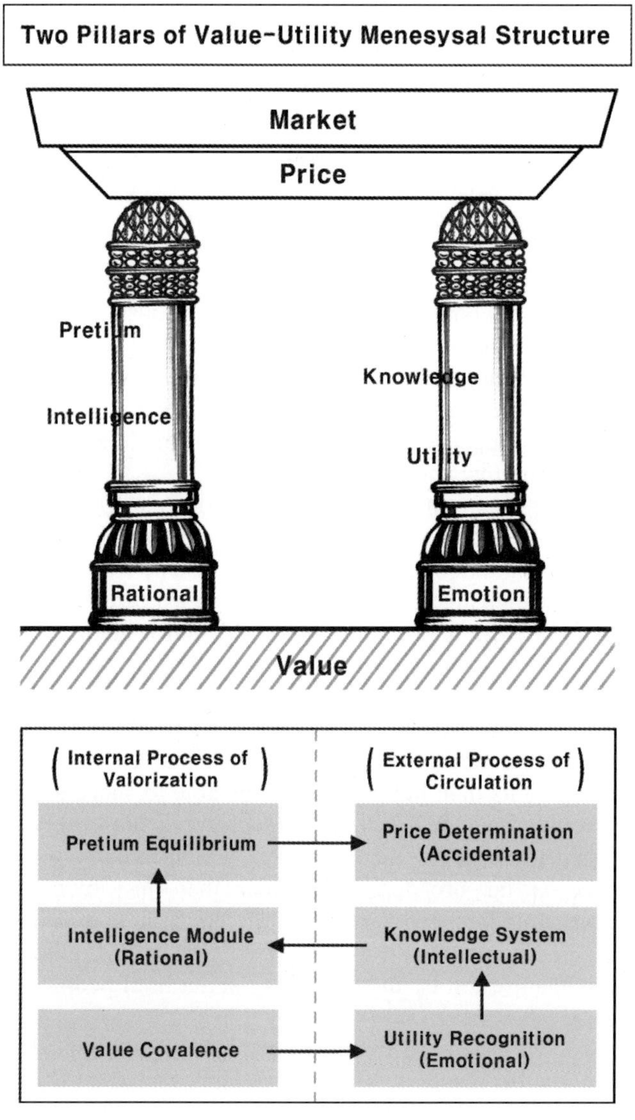

Figure 10-6-1. Two Pillars of Value-Utility Menesysal Structure

Table 10-6-1. Frame of GIMUP

1. **Value covalence** will become the major concern rather than wealth distribution. Global metropolitan areas must be converted from the place of exploitation and dominant accumulation of wealth to the place of value covalence among profit seeking nations and institutions.
2. **Early Neural Embryology** shows similar process of integration and diversification. Sync: comparable to the unitization of frequency at around 40Hz(30-80) active in integration and attention : **anthropomorphism of humanity** de-Sync: diversification of neurons in function and shape by different gradient and locality due to differing accidental situation: **Menesysal Development by cruxological articulation**
3. **Intelligence modularity** of Humanity The concept of Intelligence has been evolved to mean the integrated principles for the interpretation of a dynamic core. Intelligence is comparatively less creative than the highly volatile creativity. Thus intelligent information is processed and concentrated by Educative Information Technology (EIT). Modularity(分格性) is governed and directed by Information Technology, which is respondent (resonant) to peace knowledge menesys EIT is a part of Information technology.
4. **Commerce-University(C-U, 商-學) Compound** C-U compound is the key of Global-urbanization. Global-Urbanization is the key of matured globalization.
5. **Reverse-Capitation Grant**: the prerequisite of GIMUP The granted amount of allocation to respective countries needs to be calculated. Calculation method should be discussed globally. Minimum investment for living, education, and basic social overhead capital. Minimum reinvestment for R&D and creative education.
6. **Global-Urbanization Development Funds (GUDF)** should be established. The global consensus of fund management agreement will be necessary.
7. **Keiros (Κείρος) Project** will be launched, before the full settlement of GUDF, in order to show the effect of EIT to the economy of beneficiary countries. It will be implemented by voluntary basis for suppliers and recipients.

The main concern of GIMUP is the generation of the intelligence module in the global urbanization. But it is also another compatible concern that the productivity of creativity value is to be enhanced. Commerce-University compound will be evolved in a sense that commerce is deemed as a dynamic capital flow and the university is seen as a core of dynamic intelligence modularity.

In order to attain the value covalence, the communal affinities in macro and micro scales and in global and regional scales have to be pursued. The sustainability of environmental terrenity value is another key to stabilize the global, social and communal affinity.

Complementarity between Major Carrying Values of Countries

Axiology, usually called value system, is largely relevant to the indigenous culture of a country. The culture stemmed from its history and the ethnic character of a nation. The major carrying values of a country are restrained by these factors of constraints, which are a part of complex plexus of the global history. Such entanglement and mutual relevance are supposed to produce complementarity among values of countries.

These interrelations among nations and countries can open a way to collaboration if we take it in a positive, progressive and comprehensive manner. Complementarity exists between countries in their major carrying values.

Triradix axiology provides the foundations for a novel methodology on how to start thinking in connection with infinite matrixes of the human transciousness. In Chapter 3, we discussed that the human transciousness has three basic elements of submatrixes originated from the experimental results of brain science; onto-submatrix, topos-submatrix and logia-submatrix.

The major carrying values of countries or regions have to be analyzed depending on the positioning of the concerned object in triradix axiological topography. The analysis method of axiological positioning will be developed adopting a material balance mechanism if the interflow of values across the mutual boundaries of value domains can be defined based on the measurement of socio-psychological change and land price change.

Land price change can be measured and correlated in the international environment as it has been done in the past land market. The measurement method of socio-psychological change in a global scale has to be designed increasing the complexity based on the thinking method of triradix axiology. The triradix axiological topography will be refined as empirical data is accumulated from our past and in the future. But the simplest method we can start with is to systematically investigate the swollen coefficient of ill-psychological expectation in the local land market price.

Certain modifications of land price versus GDP due to the swollen effect should be made if we proceed further to accommodate a greater complexity in our model. However, this earlier part of theoretical trial will yet have to be refined to fit the interflow of values across the boundary of different value domains. Per Espen Stoknes, a Norwegian psychologist wrote Money and Soul (2009) about the psychological aspects of multi-strata monetary system which reflects the axiological layers: social capital, natural capital and so forth.

We have discussed in Chapter 8 that the land price and its change comprehensively represent the intangible factors of socio-psychological factors.

Comparison, Exchange, Resonance, Covalence of Values

The earliest step of GIMUP is the Keiros Project all over the planet. Keiros Project focuses on the value covalence of contradictory parties in profit seeking.

Values are mutually compared, exchanged, resonated, and covalenated. Covalence of values is decreasingly covalenated until the tie is dissolved, or increasingly covalenated until mutual parties are bonded chemically unalterable.

The comparison and exchange of values are related to the wealth distribution. The resonance and covalence of values are related indeed to the common wealth of our planet and the creative peace.

The next era of our planet would not be characterized by the wealth distribution but by the value covalence. The resources-rich developing countries want to establish the global economic developmental model which has accents on the covalent function of economic values. The value covalence by power intervention is not necessary and nor desirable as had been done by Global Powers in the last few centuries. Therefore, advanced countries have to consider establishing a conglomeration of upper middle size countries to guide the line of global development in value covalence. They are in and around the group of G-20 to G-50.

The international covalence of values is founded on the complementarity of bilateral or multilateral value systems.

Tri-Taiji Principle in a So-called Nomadic Era

In order to deal with the aforementioned complex situation comprehensively under sufficient stability, we have to adopt a new method of thinking which can organize unpredictable factors in an integrated predictable pattern. This is to be done by the science of complexity in a special approach by grouping new initial factors into the third fractal. Maybe Bandung Conference in 1960 was one of such possible approaches. However, present globalization has made much greater progress than ever that our future plan may avoid the probable conflicts between profit-seeking powers.

It is suggested to rearrange the thinking methods concerning I-ching pattern of prediction. There are many approaches to overcome the lack of scientific foundation of I-ching (The History of I-ching Philosophy, Liao Ming-Chun et al., 1991). It is not proper to utilize the thinking pattern of past methods in order to forecast the developmental cases of our present days. For instance, I-ching does not present a definite reason how the Eight-Signs (of divination, 八卦) were generated from the Four-Symbols (四象) (The Comprehension of Zhou-Yi Philosophy, Gao Huai-Min, 1978, Chapter 1, Section 2-(1)).

However, we have discussed in Chapter 7 that the internal and external management of cruxological condition will guide us to the best efficient solution in any situations. The best menesysal output will be initiated from the internal cruxological management of a menesys and later manifested as a healthy menesys that soundly copes with the confronting situation.

We have seen that, as Jacques Attali stated, current civilization is transferring to the nomadic era. Nomadic life is so dynamic that the symmetry of the society may be easily damaged. It means dynamic symmetrical balancing of Bi-Taiji (二太極, Yi-Taekeuk in Korean) will not work stably in our era. The thinking method of Bi-Taiji is said to be developed by the agricultural culture and Tri-Taiji by the nomadic people. In fluidal, or fluiditive condition of the nomadic life, it is suggested that Tri-Taiji (三太極, Sam-Taekeuk in Korean) Principles are more applicable as shown in Figure 10-6-2.

Forthcoming years will have to add up dynamicity in knowledge and culture which will eventually lead people's life in a dynamic Tri-Taiji balance. The added parts represent the fractal third which is the undefined portions of new trends of anthropomorphosis and usually called the source of creativeness.

Actually the fractal third is the compound of creative part and the debris or the remnants of the past non-mainstreams. New initiality of next development of history usually begins out of these remnants of non-mainstreams. But they have to possess the adaptive insights over the significance of the entire past. They are called the adaptive node of complex adaptive systems in the science of complexity. However, the author would like to name them as the node of creativity of a society.

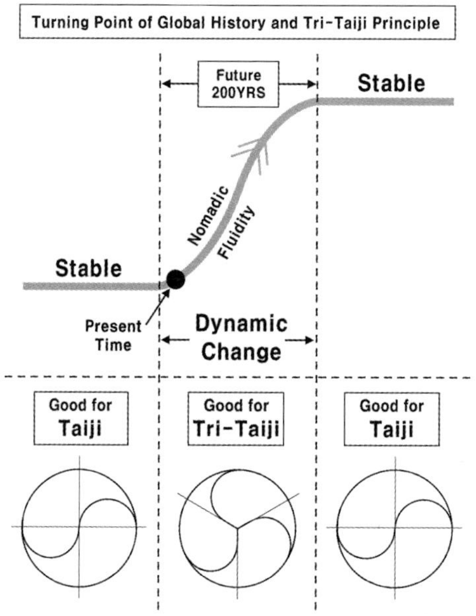

Figure 10-6-2. Turning Point of Global History and Tri-Taiji Principle

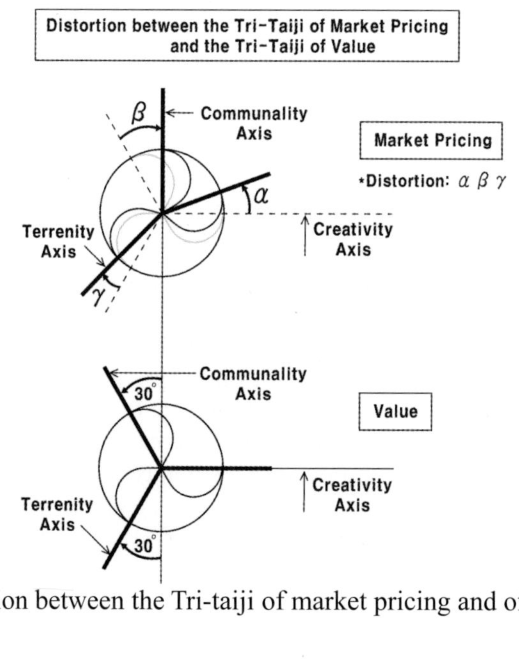

Figure 10-6-3. Distortion between the Tri-taiji of market pricing and of value

When a society is over-stabilized and has started the hidden process of inner deterioration, it is generally governed by the static Bi-Taiji. The creativity of such social condition is degraded to lower kinds of creativity. On the contrary, dynamic Tri-Taiji allows sometimes excessive instability. But as far as the whole scheme successfully keeps continuous peaceful adjustments among the constituents, creativity will be enhanced to bring adequate changes to renew the whole continuum.

10-7. Keiros Project of Healing the Planet Earth

Keiros Project and Educative Information Technology (EIT or EITech)
Value Covalence in Reality among Countries
Pax Consortis and Pretium Consortis
Additional Global Currency based on Contra-Capitation Grant

The earliest step of GIMUP is the Keiros Project to be materialized all over the planet. Keiros Project focuses on the value covalence of contradictory parties in profit seeking. Keiros (Κείρος, hand) is the healing hand of our planet Earth. It is the most urgent global program that humankind has to discuss together for the global implementation. The core idea of Keiros Project is the installation of the hardware and software of EITech to the core nodes of isolated parts of the global economy.

The Keiros is the hand to pull up those hands that are below, within the boundary of the global economic continuum. The furnishing of educative information to dynamic modules of the retarded political economy will draw our deficient brethren to participate and build the conduit of wealth into their economy. For example, refugee camps, destructed cities by interracial battles, the core cities of isolated countries and so forth would be identified as the object of our plan.

Keiros Project and Educative Information Technology (EIT or EITech)

In Section 10-3, we discussed about the thought of Comenius, *Pansophia* (the entire wisdom) and *Pampaedia* (the entire education). To achieve the goal, we need to follow an easier way to implement. As described in Table 10-6-1, modern city and metropolitan areas are the offspring of commerce-university compound. The heritage of commerce in current metropolitan areas can be observed in the connectivity of transportation, communication and shopping mall plaza. The legacy of university can be seen in the pragmatic diversification and segmentation of current knowledge system.

The beginning of universities is originated from the pursuing systematic knowledge which generated the essence of modern rationality. The systemicity of rational knowledge paved a way of the cores of modern cities which had to support the production of various knowhow in many branches of business and public administration. If we disregard the public sector and the supply of public assets, cities and metropolitan areas are the integrated adult organism of the commerce-university compound. In origin, public sector is deemed to be the eventual extension of commercial activities which was successful to construct the core of the background infrastructure.

We said before that the nucleus of globalization is the urbanized metropolitan areas stemmed from the commerce-university compound for the past several centuries. The basis of the global urbanization is reliant to the fundamental education for citizens which was originated from the national primary education since the 19th century. Current nations have a converted trend that voluntariness is more important than assimilation and assimilation than mobilization. Voluntariness stands on the citizen's discernment and participating spirit.

If we take a glance at the flow of current citizen's movement, the voluntariness of common people stands on the factors such as extra economic allowance, the stable legal foundation for the citizen's life and the network of common people. These conditions which contribute to current voluntary movements change the intangible infrastructures of society to become easy, fast, essential, indigenous, less expensive and mutually connective. Such changes of a society imply the alteration of intrinsic social properties in forming the social continuum.

Educative information technology (EITech) aims at supporting the development of the global network of commerce-university compound. Educative information focuses on the nexusity of the continuum and adopts the most essential extendable part of the simple, clear and articulatory systematic information. The extendibility is well designed to behave as the original embryo for the diversification of knowledge and intelligence.

In connection with the viewpoint of this book, the extendibility of EITech is endowed with the creative growing points and the essential continuum environment. In order to satisfy the premise, the central ideas of EITech are balanced by the two major orientations: the **resonant mesh method** and the **floating reception method**. It is shown in Figure 10-7-1.

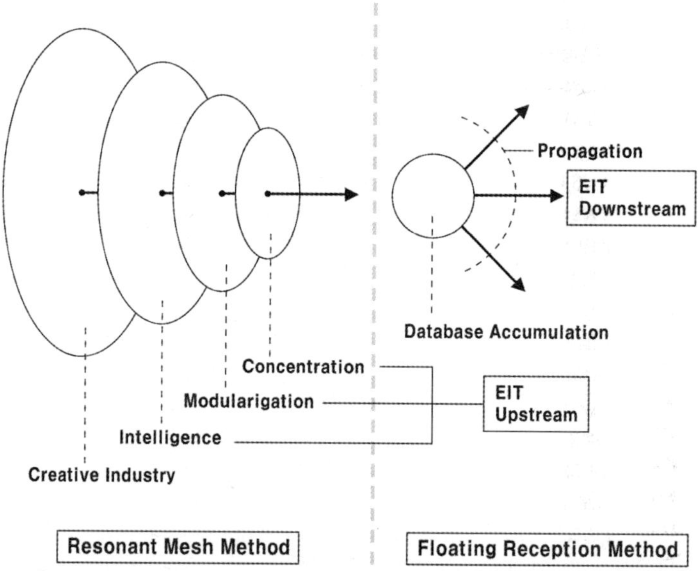

Figure 10-7-1. Up and Down Stream of EITech and Global Urban Planning.

The resonant mesh method is how we are going to treat the upstream production of educative information. The upstream of educative information has to reflect the basic components of human knowledge and intelligence. The basic components of upstream have to be sorted and rearranged by the formulation of mesh structures expressed by the least number of axial definitions.

The axial definition of educative information for a specific community or society is relevant to the native rhythms, dancing gestures, rhymes, and expressions which are engraved in their emotional system. The essential peaceful emotions are also inducible from the systematic research. The software of EITech should be closely attached to the basic ideas on the local nativity. The author believes that the *Pansophia* of Comenius can be materialized by the resonant mesh method.

The basic philosophy of the resonant mesh method is to start the *Pansophia* from the *mimesis* (mimesis) of *apate* (illusion, or deceptive virtual reality in the author's concept) to realize *katharsis* (catharsis). It is suggested to correlate the these three important Greek notions for aesthetics (Wladyslaw Tatarkiewicz, A History of Six Ideas, 1980) in order to construct many profound sets of virtual educative information to initiate EITech. Catharsis caused by the educational mimesis of virtual reality will enhance the educative process to materialize GIMUP in the historical reality.

The floating reception method is concerned with how we design the downstream usage of educative information. Floating reception has to do with floating curricular system by which learners can design his educative process by himself most effectively. Floating reception means the reception of educative knowledge depending on the prior achievement of recipients.

The elementary beginning of floating reception provides the peace value and fundamentals of individual emotional stability. The intermediate stage deals with the cultivation of social courtesy in the course of promotion. The final stage opens a new field of pretium, which provides a room for *apatheia* (the mind or nous of non-pathos from Eastern Orthodox Theology, that is 不受動心 after Rev. San-Choon Kim, psychologically close to the transcendental summit of virtues).

The transcendental aesthetics is the backbone philosophy of EITech. The transcendental intentionality as discussed in Chapter 4 brings a novel understanding to the transcent of humanity. The truth-goodness-beauty (*verum-bonum-pulchrum*) is the transcendental characters of existence. They, the three forward faces of transcent, listen to each other mutually communicating. They are actually the unity beyond their backward interpretation. *Pansophia* and *Pampaedia* aims at the understanding of summital virtue.

In this context, the floating reception method will take a learning strategy from the concept of *aptum* of Augustin. *Aptum* is what Cicero described *decorum* as the synonym of *prepon* in Greek. The author thinks that the ancient definitions of aesthetic possess the initial simplicity and are good to apply to the plain commons in their search for the fitness (aptness, *aptum*) of self-designing of floating curricular activity. Aptness is actually the spontaneous articulation that is manifested out of the circumferential conditions. Beauty is not alone when *modus* (measure), *species* (appearance) and *ordo* (order, arrangement) are considered to fit specific situation. Articulation suffices the condition of fitness which is most probably admitted and agreed by communal recognition in certain situations. Human being in general does not react only by the motives of truth and goodness. Beauty is the compensation that human reacts most delightfully.

If we accept the idea that human transcent is the matter of existential teleology, the transcendental intentionality is the flame of ignition. The floating reception method has to be given with the rank of measure (*modus*), the kind (or appearance, *species*) of learning and the mapping (or arrangement, *ordo*) that indicates a learner where to head forward. Fulfillment at a specific level of learning activates a learner the transcendental intentionality for a broader and a higher target of "aboutness". The floating reception is a liberated education process that can cope with the current process of eventual diversification and vitalized interdisciplinary activities.

Table 10-7-1. Framework of Educative Information Technology

Cost reduction	Hardware and software costs are reduced due to the accommodation of essential functions only (linked with native emotional and rational system)
Simple design	Applicable from the primary intelligence level (the simplest design of resonant mesh and floating reception)
Educative purpose	For the integration of menesys of utility, knowledge, intelligence, and pretium (the process of social cultivation of intangible values to materialize reasonable and healthy menesys of market prices)
Extendibility	Connection guided to a wider coverage and a higher hierarchy by assessment (Upward development to the summit of the tree of knowledge and intelligence)
Resonance and covalence of global values	Resonance of values stands on anthropomorphic fundamentals (both physical and mental, null to infinite, parts and the whole) Covalence of values achieved by the settlement of criteria for value exchanges
Peace doctrine	Peace is the backbone axiom of the intelligence menesys in interpretation
Keiros Project	Vessel project of the EITech Global currency based on Contra-capitation (逆人頭稅) Grant will be used for the project for the first time.

Value Covalence in Reality on Earth

Recently reported is that Afganistan has the probable reserves of lithium greater than Bolivia in a single dried lake on the highland. Natural resources can be a good medium to invite matchmaking for covalence.

The main issue of value covalence is how to satisfy each other when concerned parties cannot estimate the fair values of mutual exchange.

The fair value can be converted into the fair market price. In practice, it is not very easy to mutually agree upon the right price since either party may install a structural obstacle, sometimes legally and at other times politically, to protect its interest against the other. Such obstacles have prevented us from establishing the economic continuum. The process of globalization is to discard those blocking.

Globalization has several representative issues of mineral resources, capitals (assets, technology, and knowledge and so forth), and political power. As we discussed, the process of urbanization is at the center of the globalization. Urbanization is not only limited to the construction of material environment, but also extended to the vitalized continuance of the human living. The emergence of telecommunication in the 20th century and internet environment in the 21st century have revealed the social depth of human connectivity.

We can say that cities and metropolitan areas of our planet will become a global web of material and mental intercommunication, finally formulating a unified global plexus of lumps and clusters. The urgent issue of urbanization is how we can create the socio-political-economic devices to handle the morbid phenomenon of sharp and sudden value concentration. Value concentration or the density of economy can be cured by value covalence not by the fight but by the light. The light is from the infinite sight in humanity unlike the fight which is characterized by competition and survival.

The path to light has been blocked by the fight of mighty powers, which enforced their brethren the model of compulsory value covalence for their own profit. The covalence model of power intervention is not effective and applicable anymore in this era. Therefore we come to the spontaneous conclusion that the world, in the course of achieving the global economic continuum, needs intermediary entities to mediate between the advanced and the retarded, between the strong and the weak, and between the big and the small.

The lost memory of the advanced is in fact the link of the economic thought that connects from bottom to top. The sympathy for the small and weak will have to find the proper seat in the heart and thought of the humankind if the strong countries really desire to survive together on this unsubstitutable small planet. The strong generally does not allow such sympathy in pursuing the power.

However, the methodology based on value covalence does not provide all means of breeding our global continuum to achieve mutual survival and mutual profit. Long term pax consortis will have to be governed by the fair value distribution. The start of the fair value distribution is to find the measure toward the value covalence.

Pax Consortis and Pretium Consortis

Following the discussion mentioned above, we acknowledge that pax consortis will be governed by the fair value distribution. Value distribution is proven to be stable and fair in terms of structure when it is initially tested by value covalence. Value covalence decides the distribution of utility value or utility. Utility is influenced and determined by the emotional topography in neurophysiological aspects. Triradixism is founded on the emotionally balanced response of human beings.

Therefore, we came to the conclusion that triradixism is the main criteria of value covalence in GIMUP. GIMUP is the urbanization process of the global economic continuum. M of GIMUP stands for "modular." Modular urbanization means that dynamic modules of many hierarchical orders should be considered valuable as far as it does not hurt the other entities and general dynamic equilibrium. In general, dynamic equilibrium of the global continuum, pax consortis goes together with pretium consortis. Pretium is specially adopted from the equivalent Latin terminology of value in order to express the unquantifiable value, which is normalized and quantified by the interpretation processes through utility-knowledge-intelligence-pretium menesysal hierarchy. Thus pretium menesys is in a gentle dynamic equilibrium as far as intelligence menesys works sound. When pretium menesys (or partial pretium consortis) acts healthy, we would enjoy reasonable pricing in the global market.

In fact, GIMUP should assume the core role of the systematic administration of value covalence among plural parties. GIMUP would be a practical tool for the long-term peaceful governance of global value conflicts. Especially, in the space entrance era, GIMUP will take an important role to ease the discrepancies between classes and groups of distinctively different value mixes.

In the course of the establishment of the global continuum economy, we will confront a comparatively short period in which implosive burst of creativity value prevails. The period will grasp the whole global economy if not sufficiently prepared.

Industries, like space industry, cyber industry and the other creative industries, will be generating such excessive energy and values that the global economy will not be able to hold firmly long enough with the current systems. It is like a flood of values or should be considered as a burst of values in the human cognition. In the last half century, some of the global inhabitants experienced such bursts, for instance, in Dubai, where most of people have experienced vertical shift of cultural and economic status from desert living to the most modern residence within one generation.

It had better be prudently considered to avoid localized downpour of creativity value produced by creative industries. Creative industries are concentrated in advanced countries. If it happens, the global economy will be badly distorted to come back to the normal continuum for years. When the property of global economic continuum is violated too deeply, the time to get to the creative peace will be postponed and lagged.

The implosive burst of creativity value is different from the explosive burst of creativity value. In the explosive burst, creativity value should disappear abruptly due to the lack of container of the created value or the want of centripetal force such as the system core that holds the produced creativity value. In Dubai we can see that the financial capability and modern management technique have been holding the creativity value in the economy by the physical construction of metropolitan area.

The implosive burst of creativity value can be observed when a society is equipped with good enough containability or sustainability such as the socializing process of produced creativity value. In the implosive burst, the produced creativity value can be conserved in the continuum safely in comparison with the explosive burst. The implosive burst, if it does not adequately circulate the creative value to the global continuum, should affect the concerned society to decay as observed in the ancient Rome in the ethical point of view. Thus humanity will lose the balance.

Future management of the value of humanity would for some time be concentrated in the administration of the creativity value. It means pax consortis will be governed by pretium consortis principally. Pretium consortis is essentially value consortis, but pretium is the superficial part of value as we have feelings at our skin. As we discussed in the menesys of value-utility-knowledge-intelligence-pretium-price, the global flow of pretium is the theoretical object we have to concentrate to manage the creative peace. The best way of doing this is to stabilize the domain of intelligence-pretium by introducing a healthy structure to be connected to the educational information flow.

Parallel Global Currency Based on Contra-Capitation Grant

As observed in the agro-socialist revolution in Cambodia and the massacre in the Killing Field, the simple remedy to increase just the industrial production in developing countries is not the proper answer to solve the global problem.

Recommended is the Keiros (Κείρος) Project, which will utilize global or local currencies based on Contra-capitation (逆人頭稅) Grant for the first time in history. Capitation is the tax levied on every head of person. Contra-capitation Grant is on the contrary the allowance of subsidy granted to every person from the government. The Keiros project needs the capital allowance within the global monetary "capacitance" to invest in the construction of the Keiros project.

(Capacitance is used in electronics to denote the property of a system that enables the store of electronic charge, which is analogically adopted here for the socio-psychological store of monetary charge in the economy. It is the task of economics how to calculate the amount of the grant.)

The purport of hand (Keiros, Κείρος) is to install the relational ties and covalent links in globalization. KEIROS is the initial pattern of GIMUP to mend the connectivity between the economies within the global economic continuum and isolated or alienated economies. The earlier part of Keiros can be implemented without the settlement of the parallel global currency based on the contra-capitation grant.

For the global implementation of GIMUP in a full scale, we have to solve aforementioned prerequisite tasks, which are the establishment of the Global-Urbanization Development Fund (GUDF) and the parallel contra-capitation currency that will take the major part of the GUDF. The calculation method of allocated amount to respective country needs to be determined. The calculation should be based on the figures of the minimum requirement of living, education, basic social overhead cost, etc. and the minimum requirement for reinvestment, R&D, creative education, etc.

However, seriously to be discussed are the maximum dispersion of the distribution of the global wealth to sustain the economic continuum in the creative peace and the requirement for the global economic continuum (eventually, for the creative peace continuum).

Resources problem should be straightened out. Resources values and the in-and-out flows need to be calculated by mutually agreeable methods. Those countries which sell their value rights for resources calculated for a certain period of time will be listed in the participants of social development funds.

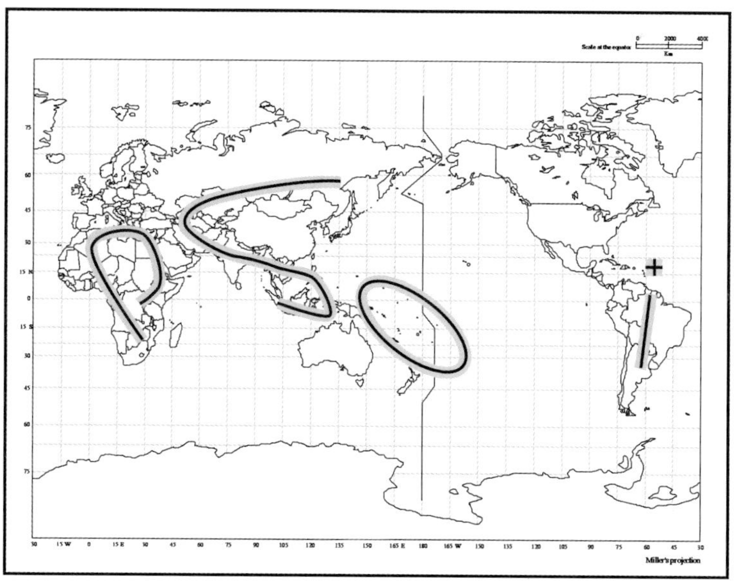

Figure 10-7-1. Global Map of Keiros Project

Keiros will be launched before the full settlement of GUDF to show the effect of EITech to the economy of beneficiary countries. It will be implemented by the voluntary basis for suppliers and recipients. The locations of Keiros cities will have to be decided first by the financial entities at the first stage but will be gradually authorized to concerned global institutions. In Figure 10-7-1, shown is a tentative and insignificant expression of Κείρος.

10-8. AASADAL Creative Pax Consortis in DMZ of Korean Peninsula

The ideas about Aasadal City in DMZ of the Korean peninsula were introduced to concerned people in 1994 for the formulation of the global Creative Pax Consortis. The idea has been formed in the author's mind and writings since 1990. Please acknowledge that this section is regarding the 1990s' situation in the description of the international political economy.

Aasadal means "morning shining." Aasadal City will provide good standards for the head basis for the global GIMUP and the Keiros projects. When the author was about to start this section, the Korean Navy corvette Cheonan has been hit and sunken by the attack of an unidentified torpedo (May, 2010), most probably of North Korea as acknowledged by the United Nations. It does not look proper to consider the construction of Aasadal City in this peculiar condition that North Korea, a hostile militant country, is threatening peace in the Korean peninsula. However, the purpose of this writing is to present a grand scheme and planning philosophy of constructing the global intelligence modules and how to establish the head cities.

Most Significant Cruxological Spot of Human Race

The Korean peninsula is considered as a place to where the plain peace and the common harmony are to be introduced, however, neither the great peace nor the great harmony. "Great peace and great harmony" asserted by big countries is written TaiPing-Daiwa (Chinese + Japanese pronunciation of 太平大和), while "plain peace and common harmony" really wished by usual nations of this globalized world is ChungPyeong-ChungWha (Korean Pronunciation of 中平中和), which translates "middle flatness and middle mixture."

Between those humble and non-ambitious countries which have decent ideas of middle flatness and mixed-up, to be raised is only the mutual peace. We had a song in our heart for DMZ in the Korean peninsula as follows;

Invite gentle winds to our valley of the misty morning.
Lo the vivid rainbow on the grey sky of mourning.
Bring the birth of continuum, to the place of vacuum.
Ring the land of our abandoned DMZ,
Sing the anthem of the unified Corea.
Born here will be a propitious nation.

Korean DMZ is one of the most significant places for the establishment of global integritium. All the spots of conflict in the global map are where peace should be invited.

When they become the peace centers, the world will be peaceful. The last DMZ of the world remains in the Korean peninsula. The last consciences of the human kind should respond to the ringing from the last DMZ.

DMZ Peace City: Creative Pax Consortis

The most stabilized form of the creative peace is found in Tri-Taiji format. The least number of Pax Consortis is operated generally based on three axes. Even in the days of the confronting two-blocs, we had the third world. It is the general trend of humanity.

World triple cores we usually acknowledge are the US, EU, and AU-to-be. AU stands for Asian Union. I am afraid too much simplification has been done in my above comment. But in the next half century, the creative Pax Consortis in AU will gain a greater drive as the global peace structure gets settled down.

As seen in Chapter 1, the dimension II in Figure 1-2-1 shows that the best policy to achieve the peace is the balance of power. Due to the inertia of potential opposition, increase in costs, in geometrical figures, the negotiation and acquisition of information is inevitable as the state of confrontation is prolonged. Whereas in dimension I we can see a rise in knowledge and produce in geometrical figures, it is induced through the positive peace, promoting active interchange among nations as well as between local governments, and thus resulting in the elimination of political-economic conflicts and barriers.

For instance, the task of peace settlement in DMZ of the Korean peninsula is not to be solved by the mere liquidation of existing hostile relations, but through activating the physical distribution and the industrial flow for the establishment of political and economic harmony in North-east Asia. We would like to say that the peaceful and effective utilization of DMZ can be the new paradigm for the peace settlement in North-east Asia. This kind of essential challenge will be leading us to some substantial result, which the world history will remember.

The Heart of Creative Pax Consortis

Throughout her history, Korea had experienced invasions by both the continental and oceanic forces. For the past few decades, Korea came under the influence of Pax Americana focused on Korea-USA international security. It has been customary for the Japanese to set foot on the Korean peninsula whenever influences of the past dynasties of Sina weakened. In the late 19th century, Korea saw its role as a buffer zone and so gave ground to foreign continental forces and maritime nations, who collided in war on its own territory.

Regarding the latest movements of China, who are gaining more and more influence; it is most unlikely for the U.S. to hold its position as the superpower in the Northeast-Asian region for good. Japan is growing into a political Big Power, China is emerging out of the curtains of communism, and North-east Asia needs a policy of order that can pull it through the gradual and complex diplomatic process. We suggest a Pax Consortis. However, shall the role of the Pax Consortis be limited to balance of power, it would be an inefficient model to dissolve the global hostility lying underneath. What is needed is a nation to take the initiative in creating an international link for the flow of information and logistics among nations instead of supporting the supremacy of one nation such as is the case of China and Japan in the past.

Global Creative Peace and Universal Perpetual Peace are somewhat identical in essence. Northeast Asia will be a central issue in the next few decades for peace purposes. For this purpose, all nations concerned with peace in Northeast Asia should come together in Northeast Asian Peace Committee where political interchange and mutual cooperation is possible down to the local government level for the maximization of productivity and true purchasing power and multi-national harmony—the creation of a Creative Pax Consortis.

In the process of healing its past wounds, the Korean peninsula, once the center of anguish, needs to contribute to the world peace and the prosperity. If the use of the DMZ can be converted constructively into an area for the creative peace in this world where, though in times of post-cold war, peace has yet to settle down so that the flow of the economy and political relations in North-east Asia revives, so that the global village experiences a time of harmony and prosperity never witnessed before. Korean DMZ located in the center of North-east Asia is the optimal place to build the menesys of the creative knowledge for the world peace and prosperity.

Profile of DMZ Peace Project

The profiles of DMZ Peace Project are presented in the Figures from 10-8-1-a to 10-8-1-i. The figures illustrate the contents, institutions, and structures of DMZ Peace Project.

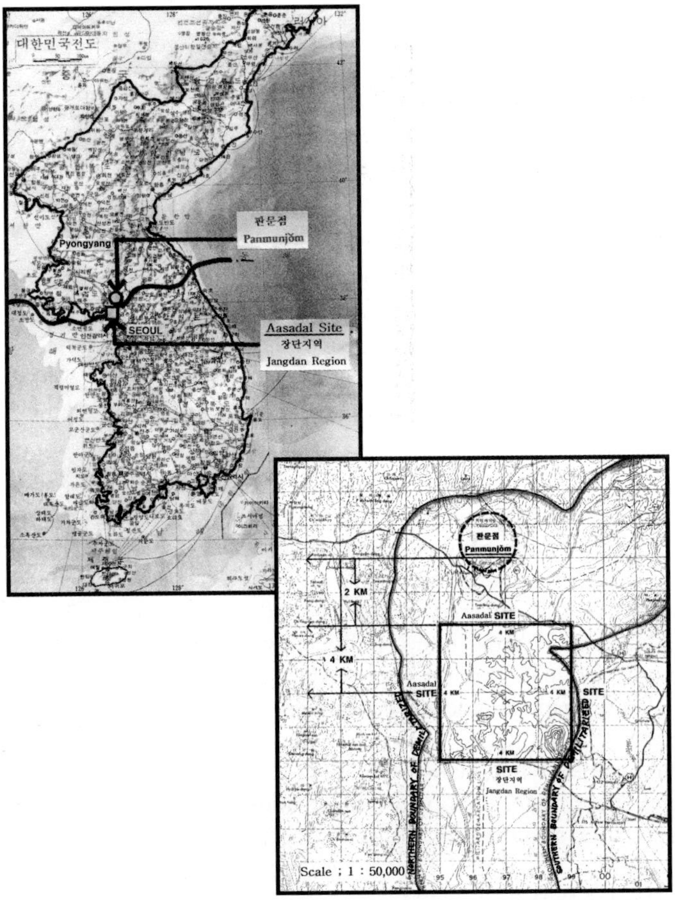

Figure 10-8-1-a. The Proposed Location of the City of Aasadal (Changdan Area around Panmunjom)

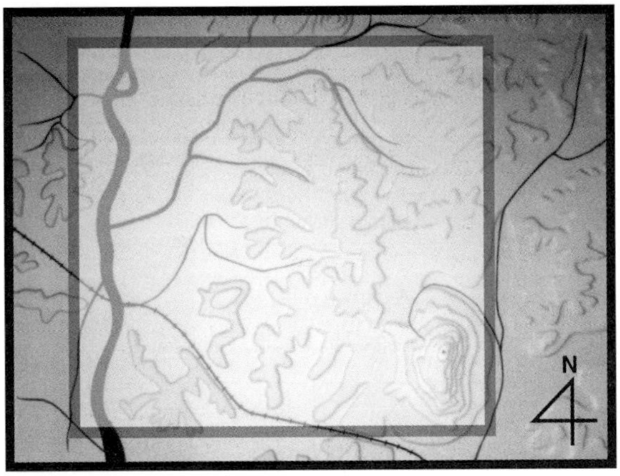

Figure 10-8-1-b. Topological View of Changdan Area

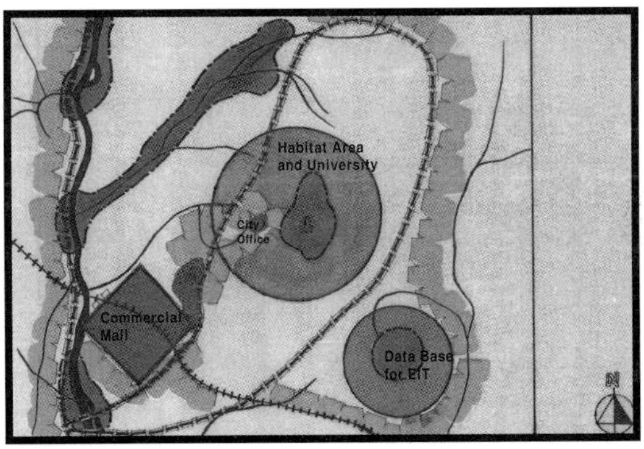

Figure 10-8-1-c. General Arrangement of Major Districts

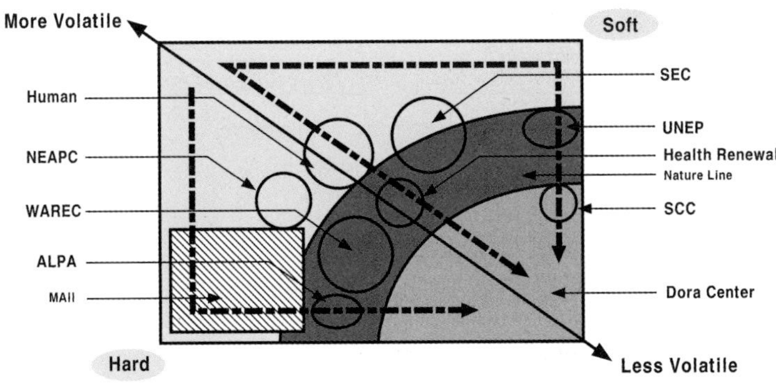

Figure 10-8-1-d. General Schema of the Functional Flow of Creative Industries

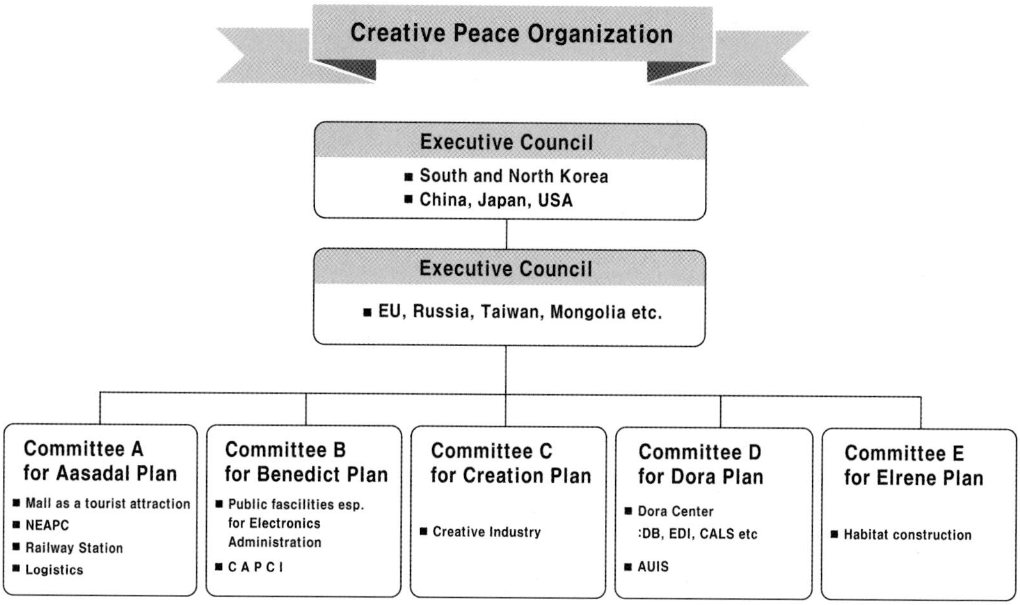

Figure 10-8-1-e. Creative Peace Organization and Relevant Committees.

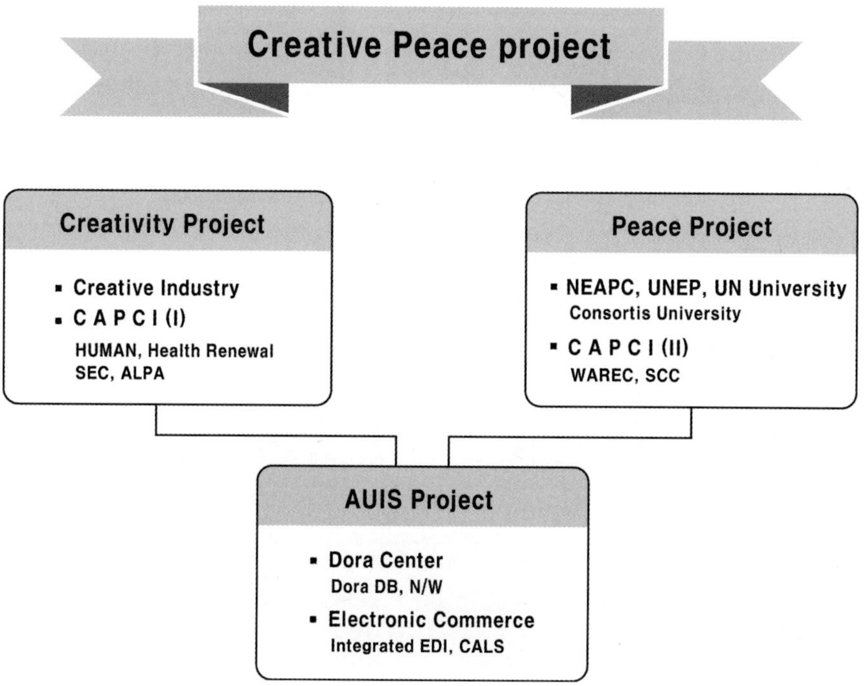

Figure 10-8-1-f. Profile of Overall Creative Peace Project

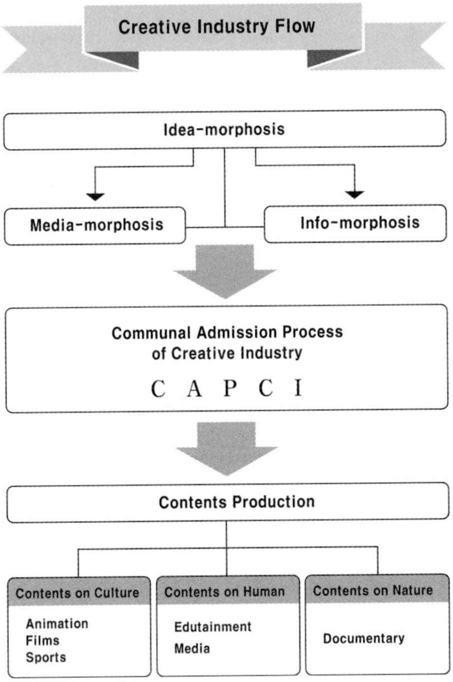

Figure 10-8-1-g. Contents Production and Business Flow of Creative Peace Project

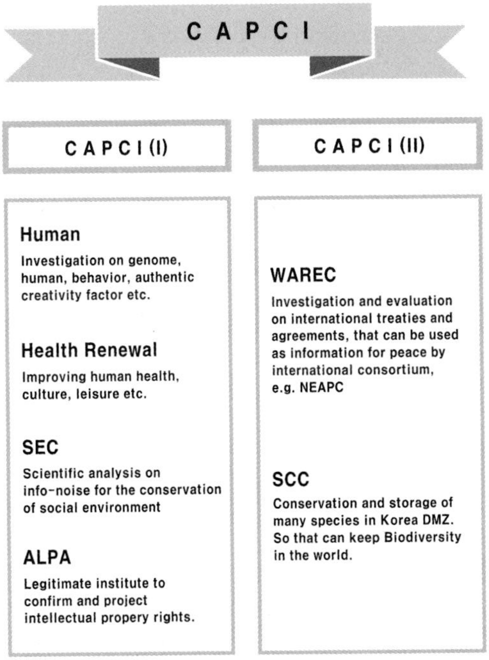

Figure 10-8-1-h. Flow Lines of Communal Admission Process of Creative Industy (CAPSI)

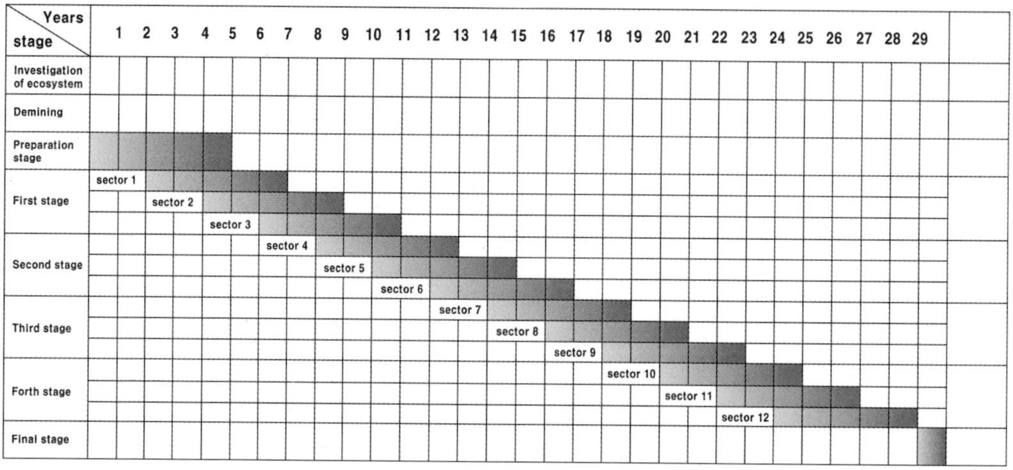

Figure 10-8-1-i. Long Term Schedule of DMZ Peace Project for Third Years

Menesysal Point of View in Aasadal Project

Aasadal city is planned to be located at the disconnection point of the two opposing trends of the world. It is the author's observation that the world has been shifting from bi-Taiji pattern toward tri-Taiji pattern, because our global civilization has attained a greater mobility and dynamicity.

The increase in mobility and dynamicity of the human cognition and lifestyle will accelerate the need of educative information and the connectivity between the advanced societies and other retarded societies to formulate a healthy economic and cultural continuum in the global situation.

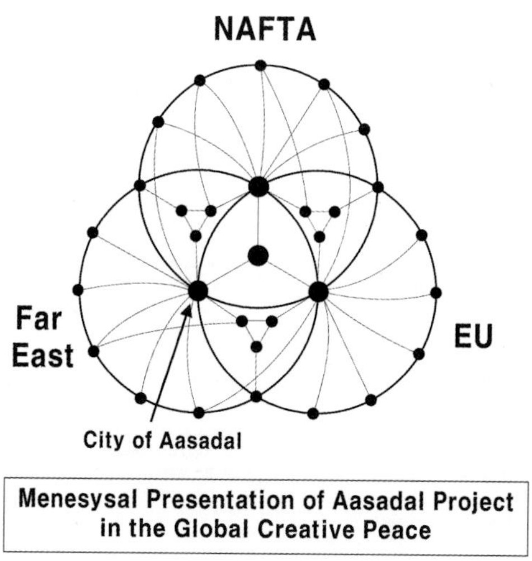

Figure 10-8-2. Menesysal Presentation of Aasadal Project in the Global Creative Peace.

The city of Aasadal will provide good effective links among the enclosures of societies of different value and pretium through adequate interpretation to the distinctive historical development of respective societies. As we will explain in the next Section 10-9, Aasadal project is the link between EITech and global development of metropolitan areas. The global dynamic modularity will get gradually matured by the menesysal effect of head cities that will be built on our planet at the major spots of conflict areas. Those head cities have to be planned and managed at the top notch of creative industries.

"Aasadal" Techno-Metropolitan Development Plan

The development of Aasadal city is the project pursuing for peace and reunification of Korea. Aasadal city is to be located in the DMZ where the peace of the Korean peninsula and North-east Asia is threatened. This project planned to develop the city as an environment-friendly ecopolis, a city that pursues peace in international politics, as well as a creativity-centric technopolis.

This project aims to achieve the following three goals.

First, a stronger means to peace is necessary. At the point of the project development, peace is pursed rather passively. Peace in the Korean peninsula is to be achieved by rather actively than settling for a simple power balance. Thus it is essential to connect the BeSeTo line by developing a city of reconciliation in the DMZ.

Second, the location of Aasadal can be harnessed for economic prosperity in North-east Asia because it connects China and Japan, ocean and continent, South and North Korea. If this location were effectively utilized, especially in the cutting-edge creativity industries, the city could bring an economic output that is comparable to that of NAFTA and EU.

Third, the Creative Pax Consortis can arise from this city. This city is designed to pursue reconciliation and creativity. Because the Creative Pax Consortis pursue peace by concentrating creativity and productivity, it applies nicely to the purpose of Aasadal city.

The project were to be operated by the administration of the 'Creative Peace,' an organization equivalent to KEDO, in Changdan area (4km x 4 km), located four kilometers south of the Panmunjom.

This project prospected 40,000 populations in the area, among which 20,000 would be inhabitants, 15,000 would commuters, and 5,000 would be visitors per day. It was expected that the 70 percent of the residences would be foreigners and the rest Koreans. The city was designed to attract foreigners with the cutting-edge industries. The city was to be utilized to systematically cultivate creative industries in order to develop the creative industrial structure interconnected to the overall peninsula. Moreover, the city was designed based on the communal and environment-friendly philosophy, in accordance with the requirement of the UNEP.

The industries to enter the city were those in the field of image visualization, media, information, edutainment, and pharmaceutics. The institutions to branch out within the city were HUMAN, Health Renewal, SEC (Socio-environmental Center), ALPA (Aasadal Legitimate Process Center), NEAPC (North-east Asia Peace Center), UNEP, UN University, Consortis University, WAREC (World Agreement Research and Evaluation Center), and SCC (Species Conservation Center).

The overall investment contained infrastructures and five plans for the industrial development (i.e., Plan A, B, C, D, and E).

Table 10-8-1. Guestimation of Investment Plan and Project Contents (as of 1994)

Plans	Expected Investment (KRW/USD)		Contents
Demining	6.8 Bil. KRW	4.5 Mil. USD	Implemented stepwise
Infrastructure	15 Tril. KRW	10 Bil. USD	Civil engineering, Environment, Urban infrastructure
Aasadal Plan	200 Bil. KRW	0.14 Bil. USD	Commercial logistics, Convention
Benedict Plan	500 Bil. KRW	0.34 Bil. USD	Public institutions, Censor creative industries
Cresyion Plan	1 Tril. KRW	0.66 Bil. USD	Creative industries
Dora Plan	300 Bil. KRW	0.20 Bil. USD	Urban information systems, DB
Eirene Plan	700 Bil. KRW	0.46 Bil. USD	Residential area
Total ('98.3.15)	17.7 Tril. KRW	11.8 Bil. USD	30 years of investment

(The construction schedule is designed into six stages of the 12 sectors as shown in Figure 10-8-1-i.)

Several diplomatic agreements were expected to be necessary to execute this project. The agreement on the peaceful utilization of DMZ and the gradual demilitarization was necessary through the North Korea-United States Talk.

Additionally, the agreement on the improving ceasefire, the usage of land in DMZ, and the Aasadal city construction was necessary between the two Koreas at their summit. Lastly, the 4-Way Talk would have provided to agree on the initiation of civilian North-east Asia peace conference and the dispatch of representatives from local governments.

The project planned to have three years of an initial implementation. The plans for these first three years included introducing the Consortis University that contains the overall image of Aasadal city as the Future Society Campus of the UN University, initiating meetings between North-east Asia nations through the North-east Asia Peace Conference, and establishing Cyber University and Peace Forum at the Panmunjom area.

More specifically, in the first year, part of Panmunjom would be utilized and the Consortis University would be commenced. In the second year, 100 Reconciliation Habitat Houses in the Changdan area would be built and the Future Society Campus of the Consortis University would be initiated. In the third year, Cyber University and Peace Forum would be constructed at the Panmunjom and Changdan area.

This project were to be financed by the both the South and North Korean government, the international funds, and the international consortium funds. Private capital invested through fund raisings was also considered. The risk of the whole investments was to be reduced by applying the project financing.

The Aasadal city development has important expected effects. The development would not only promote the unification of two Koreas, but also with the unification, jointly lead the peace in North-east Asia. Aasadal has the potential to be the hub of the creative industries and politics in North-east Asia.

10-9. Administration of the Global Dynamic Modularity

The essence of our recommendation up to this point is that there are three major managerial strategies in the current developmental stage of the global dynamic modularity. The strategies are founded on the undiversified (undifferentiated) embryonic approach that is the developmental process of the physical life on our planet. It is thought that the intangible anthropomorphic circumstances to be developed on our planet should follow the orientation of the tangible process of physical forms of life on Earth.

Dynamic modularity in a single set of integrated global condition has to be equipped with three factors of dynamic movement; the frontal advancing head with sufficient energy, the centers of local integration, and the connectivity of the continuum.

The Solarche Scheme aims to induce the whole global civilization towards an entirely transformed status within the next half century. Solarche in short is the frontal advancing head with sufficient energy. As NASA project has assumed the tracting role of top notch technology in the past, Solarche will be positioned so.

The head cities at the strategic locations all over the world as the city of Aasadal will enable to alter the metropolitan areas into the productive container of the creativity value. The head cities will also behave as the center of local integration around global metropolitan areas.

There will be some early types of global head cities in some designated retarded countries, which will admit the EITech (Educative Information Technology) as the basic philosophy of the global urban planning. The global urban planning will connect the head cities of Keiros Project.

GIMUP (Global Intelligence Modular Urban Planning) is a method to take the advanced EITech as the backbone of philosophy and to enhance the role of intelligence menesys in order to produce a more stabilized but vivid pretium for the inhabitants on the Earth.

GIMUP is the initiative to develop the future picture of the pearly Earth embracing all nations and people on the planet. The connectivity of the tenants of our planet will be duly improved by adopting EITech to orientate them in order to assimilate the global trend in a unitary way but with sufficient diversity.

The global dynamic modularity has three directives to achieve the administrative objectives. The first directive is Solarche Scheme, the second directive is the Head City Strategy, and the third directive is EITech accompanied with GIMUP.

10-10. Closing Summary

It is the end of our remarks on the creative peace on our small planet which has changed the global village in the last fifty years as most contemporary intellectuals agree. It is the pleasure and privilege of the author to summarize the content of the interwoven complex ideas as following.

The creative peace is the solution to all struggles occurring on this planet.

Integritium (instead of ancient imperium) will flourish on the earth as the creative peace will get matured.

The science of humanity is recommended to organize a new way of thinking
by interweaving **the science of integrism**, **the science of cruxology** and **the science of creativity** as dealt in this book.

Dynamic Modularity is the answer to the old query of human race
to solve the problems between the big and the small, the advanced and the retarded, and the strong and the weak.

We cannot solve our problem if we don't have **the vision beyond our planet**, which is the solar system and the universe.

We will not be able to achieve the global peace
if **the global economic continuum** is not to be generated.

Epilogue

Being a Presbyterian, I consider the Bible to be the Word of God and the world to be the manifestation of the good will of God. It is my firm belief that the sincere thoughts and honest activities of every human being are worth being preserved.

In my upper teenage years, I went through the life-changing summit experiences of being overwhelmed by the charismatic blessings of the Holy Spirit. What had been grasped in my mind was that "Christ is all and in all (*alla panta christu*, Colossian 3:11)." Reading that passage from a cosmological point of view, Christ became a powerful concept that yielded to me an adequate answer on the problem I had been struggling with for a long time ever since the reading teachings of eastern civilizations, Buddhism, Confucianism, Taoism, and other works.

There is a trend in the society that criticizes on Christianity regarding its exterior elements at the expense of overloading its intrinsic attributes. It is true that a large part of Christianity is in some sense conservative, being loyal to the past values and methodologies. However, what is not true is that humanity needs conservatism to evaluate and select the appropriate offspring of its own to inherit. Science and technology have also had similar kinds of misunderstandings made against them.

Despite Galileo Galilei being accused wrongfully and the honorable Darwin mistreated, the changes they insisted to be correct had to be introduced. We now know that their ideas are also mended in accordance with new concrete evidences. Similarly, as humanity and science have developed or evolved, the external manifestation of Christianity can also be altered into a better shape for a better understanding by contemporary people. Christianity is a profound treasure of the human race. Science and technology should not exert to form stubborn doctrines to regulate the human intellectuality from any directions.

Now in this era of entrance into the universe, all human assets must be concentrated to open the gate. Such collaboration can only be based on agreement with the answers of the queries of what a human is, what a person is, what a communality is, what the phenomenon of humanity is, what creativity is, what integrity is, and so forth. These questions have been answered in this book to the best of my capability. I started from my personal limitations but later liberated myself opening to what may be prevalent in the future. This was done not only because of the necessity but also because of my conviction that everything manifested is due to the Divine Providence for the betterment of humanity.

My search for the true identity of Christ Jesus in its full meaning succeeded in and out of Christianity throughout all my life. There was another ingredient added to the search of Christ. It was a ceaseless search for the method of how the global peace could be achieved and maintained. In pursuing this answer, I have tried to keep myself far away from the doctrinal obstacles that stem from religion, race, emotion, politics, locality and so forth. Rational and intellectual thoughts led me to keep my stance firmly on the ground of my position in science and technology.

I have always left myself in a transition zone in between the Christ who was intuitively found to be ontological, and the humanity in myself that was incrementally increased in a manner of epistemology. This book is an outcome of the combination of the centrifugal and centripetal attitudes together in parallel, centrifugal meaning epistemology and centripetal meaning ontology.

I have a deep trust in the Divine Providence. It does not intersect with my belief in reliance on scientific approaches. As a number of intellectuals maintain their decisions withheld in the transition zone, I will have to keep my position out of the spheres of conventional creationism and evolutionism (not necessarily exact Darwinism or Neo-Darwinism) until a horizon of new certainty appears in my field of vision. The vocabulary, "crevolution," is the compound word of creation and revolution, not of creation and evolution. Crevolution is an expression of the leaping mechanism hidden in the phenomenon of humanity, which has certain strong implications on evolution.

It is my frank opinion that currently existing conflicts between interpretations of creationism and evolutionism and also debates between monism and dualism will cease when an integrating explanation arises in human minds. It is my belief that the discord between monism and dualism might be a play directed by the human brain, based on a scenario written by the same director. Creative integration pops out of the vague but energetic plasmic status of reasons and emotions, which I would call an "awkward conformity in creativeness," especially when concerned with the third kind of creativity.

In both of realistic and religious lives, pretext and practice compete mutually with one another: pretext, equivalent to the personal justification of preference, usually overburdens humanity, while practice usually cannot help escaping from mannerism due to the fatigue plasticity of the human mentality. Then comes niche from which is generated the transitional phenomenon of the human mind. The human mind feels uneasiness in the ambiguity of a mental transition zone and thus creates numerous probable sets of explanations, which are the contents of humanity and ends up as the well-known human soul. However there is another aspect of humanity which is still inexplicable, which is usually called the human spirit.

Humankind will, I am convinced, overcome forthcoming obstacles and labyrinths within the next 50 to 100 years. However, we must realize that we humbly confess that we are just at the starting point of 100 years of struggle in order to stabilize sustainable life styles on the earth. After that, we expect the next 1000 years to be years of silence bearing the fact that the human race may not be successful in the introductory process to the outer universe if we are not prudent enough.

This book is dedicated to my parents, the late Mr. and Mrs. Sung-Hoon and Ju-Shim (Kang) Koh, the two intellectual figures I respect the most, who have lived their lives pursuing their ideals and faith. They implanted the seeds of this adamantine challenge into my life for the peace of the human race and my nation.

My appreciation goes out to my better half, Jung-Hee Lee. I consulted with her when I was afraid that specialists of every discipline would condemn me on their review of my study for not being receptive towards current trends, to what she responded with encouragement, challenging me to write what I have desired to write. I have done just that, with many thanks.

There is one more person I would like to thank from my memory: Mr. Yong Jip Kim, my geometry teacher from my high school days. Despite having difficulties with mathematics, his personal recommendation for me was to go on to take the mathematics major in the undergraduate course of university. He encouraged me because he thought that I had a special gift in integrating and framing capability from abstract works.

Even to this day, I am still grateful for his comments even through my career as an engineer and onto different stages of my life: as one of the founding leaders in the earliest citizen's movement in Korea, as the founder of social work in the Habitat for Humanity of Korea, as a mentor reviving several religious movements for young people, as one of initiator of land and natural resources economics, as a publisher of books and a periodical, as a counselor guiding Christian community movements, a leader of a movement for the reunification of Korea, a researcher in the science of complexity, a manager or managing director of governmental institutions such as mineral resources research institute and state oil company, a lecturer in universities, director in petroleum related business in private industry, Hyundai Group, and a long-time devoted researcher on the national and international strategy and global peace study.

I believe the unusual diversity of my past career has contributed to the contents of my present writing. I am confident that the faith and the intellect can go forward together in the realm of humanity. The realm of humanity is a good sphere, able to tolerate all sorts of mix-ups in a continuum.

My sincere longing is that thinkers and researchers would seek to implement my integrating theories, and find their own horizons thereon. It is my only wish that humanity move forward and pass quickly over this stage of violent change. The last thing I want readers to acknowledge is that the main line of thinking presented here has been kept unaltered since my nineteenth year's prototype.

The First Day of July in 2010,

Dr. Wong In Koh, the author.

List of Tables and Figures

TABLES

FIGURES

Acknowledgement

It is the author's acknowledgement that following collegues, fellows, and assistants have done significant contributions to the preparation and publication of this book:

Administration: Jung Hee (Lee) Koh, my better half, President of Cedar Publishing Co.
Ryu Ju Esther Koh, Ryu Oak Jade Koh, my daughters, who are partners of the above company.
Proof readings: Matthew Ferrick, Choong Jung Raimond Chae, Sara Han,
Choong Nam Royce Chae, Sung Soo Hyun
Indexation: Yun Hee Cecile Choi
Drawings of cover and illustrations: Young Eun Sim
Data Base: Aram Kang, Il Lee, Chong Chan Lee
Management: Min Ki Kang
Edition: Kyung Hee Kim
Publication: Eui Seong Lim, President of Geul Planner,
Yong Taek Kim, Design Team Leader of the above company

Great appreciation is also extended to my friends, who are mostly in academic positions, for their encouragement and counseling. However, all the responsibility as to the content of this book should be levied only to the author.

Wong In Koh, Ph. D.

Glossary

- **Absolute Other, the** a potential infinity; the total ontological frame that creates the notion of centeredness in human cognition, such as the Divinity and the self
- **actual infinity** the real infinity that surpasses all infinite series
- **alpha path** a lowest stress condition with innate complexity in general region of initiation; a path found when new stressless dimensions are detected by the tentacles of human's transcendental intentionality
- **animotus** composite word of *anima* and *motus*
- **anthropomorphosis** a process of manifestation of the intrinsic attributes of humanity
- **apoptosis** a death mechanism or the controlled death of an organism's cells as part of its natural growth and development
- **articulus** a knotting mechanism between nodus and modus that is formed due to the influence of the environment or other internal condition
- **axiology** the theory of values, moral or aresthetic
- **Axis Tempus** time axis intruded into human physicality in a manner of rhythms of organic pulse
- **background radiation** radiation trace of cosmological initial energy level
- **cap-con relation** an vitality indicator in a society that is formed between capacity and connectivity of values in a society
- **centrifugality** a trend of human mentality related to individual subsciousness, communal nexusity and epistemological aspects
- **centripetality** a trend of human mentality related to individual self and consciousness, communal systemicity, and ontological aspects of human phenomenon
- **che (體, ti)** the literal meaning of body; an established substantial order
- **communal intelligence** intelligence that contains lots of different systems and nexus, is beyond that of individual, and wiser than swarm in terms of rationality
- **communalism** a system or theory of government in which the state is seen as a loose federation of self-governing communities
- **communal value** the value created from the economic efficiency from cooperation in community
- **Complex Adaptive System(CAS)** a semi-stabilized strange attractor in between diffused networks and condensed systems; usually dealt in present complexity science
- **continuum** a continuous series or whole, no part of which is perceptibly different from the adjacent parts
- **continuum economy** an internal story of the market growth by admission

- **creative pax consortis** multilateral creative peace that pursues peace by concentrating creativity and productivity
- **creative peace, the** the continuum of activated humanity in stabilized condition
- **creativity value** the individual cognitive value manifested by creativity
- **crevolution** the way in which human creativity creates revolutionary changes; a compound word of creation and revolution, not of creation and evolution; a leaping mechanism hidden in the phenomenon of humanity, with strong implications on evolution
- **cruxology** the framework of the most natural way of human initiatives and responses to spatiotemporal environments
- **Das Ganz Andereheit** *see* **the Absolute Other**
- **dualism** the doctrine, as opposed to idealism and materialism, that reality consists of two basic types of substance usually taken to be mind and matter or two basic types of entity, mental and physical
- **dynamic modularity** the physical neuronal cognitions that balances between unity and diversity to correctly interpret the phenomenon within a reasonable range of accuracy and semantic validity
- **dynamic ontology** a type of ontology distinct from previously known static ontology; the ontology focuses on plexus of affairs
- **dynamic relational ontology** *see* **dynamic ontology**
- **epistemology** the theory of knowledge, especially the critical study of its validity, methods, and scope
- **exaptation** a non-adaptive evolution that has taken place in the transcent of humanity
- **figura** a pattern of diversification; characteristic pattern of changing alignment due to meeting and parting of herds and modules
- **figurarity** an expression of integrismic composition
- **flaring tetrahedron matrix** a stabilized geometry which is not easily changed by interrelationships of topos, onto, and logia matrix, with animotus flaring on the zenith
- **gestalt** a perceptual pattern or structure possessing qualities as a whole that cannot be described merely as a sum of its parts
- **GHZ entanglement** the quantum states of infinite particles
- **Grand Chi, the** a geographical belt that crosses the Grand Slant and forms a line of densely populated area in Eurasia
- **Hsing-Li philosophy** a philosophy applicable to the wholistic view of humanity and to the fundamental idea of the problems of a person
- **Homeochaos** the maintenance of dynamic stability of nature which is free from substantial constraints such as the embedment to materiality

- **Homeomenesys** the maintenance of equilibrium within a certain stabilized range achieved by social actions in the human society that resemble the patterns of brain-neuronal actions
- **Homeooikonomia** the general paradigm that includes the patterns of homeochaos, homeomenesys and homeostasis
- **Homeostasis** the maintenance of metabolic equilibrium within an animal by a tendency to compensate for disrupting changes
- **humanification** a new vocabulary to denote an excessive systemization toward human-centric humanity
- **humanization** a proper way that leads to humane transformation of humanity
- **Hyper-Topological Conversion** a higher step in abstractive metamorphism in human cognition
- **imperium** a part of immature integritium or a core part of integritium
- **infinities-encounter** the individual experiences that highly surpass the intensities of past experiences, for instance, the experience of the initial conceptus, the impact of the sperm penetrating into the surface of an ovum, or child labor
- **infinities-exile** a blank for each summit experience or equivalently for single infinity-encounters and single infinity-transfer; these blanks successively connect to the episodic edition of the brain to form an understanding of time frame
- **infinities-matrix** a set of ITRings of an individual person that is constructed based on onto-, topos-, and logia-submatrix; yielded by time-series episodic memory
- **infinities-transfer** the process of human cognition that the impact of infinities-encounter eventually changes scope and content of cognitive phase space of an individual
- **infinities-Transfer Ring (ITRings)** a ring placed over the surface of blank time tunnel to make a strand of episodic memory with an emphasis of the self-interpreted summit experiences in a row
- **integrism** a measure of integrity of humanity; the mixed dynamic frame of systemicity and nexusity
- **integritism** the pursuit of the integrity of a person; the attitude that represents the human effort to overcome present constraints and obstacles in order to accomplish human integrity
- **intentionality** the aboutness of the human consciousness without any scalar value and immunity
- **knotting** an achievement of connectivity
- **koinonia** a Greek biblical terminology adopted to represent mutual communication, interchange and association
- **logia-submatrix** a submatrix of infinities-matrix that represents relational attributes of locality
- **menesys** MEgalo-NEtwork-SYStem; the outcome of cruxology in accordance with the hysteresis, or path-dependence of cruxological development

- **mimesis** a process of mixing; picking up the location of peak value and the offset discrepancy of active location in the cognition
- **minima** an "initial infinitesimal" move of dynamic incidences, specifically designed to describe the least emotional move; the least string:
- **nexusity** the networking characteristic for understanding the ontological relation; a parameter to connect a person with external factors and environments
- **nexy** *see* **nexusity**
- **nullity** a creative and productive vacuity when a nihil becomes a part of a wholeness of the substantial universe; it exists as a hole, a vacancy or a vacuity in human consciousness
- **omega path** a highest stress condition generating apparent complexity in near-completion region of a crevolution process
- **penumbra** a transition zone between consciousness and transciousness; the fringe region of a half shadow resulting from the partial obstruction of light by an opaque object
- **positive peace** a kind of peace that promotes the active exchange of knowledge, information and goods and creates a continual synergy effect in knowledge and production
- **persona** derived from the Latin word "persona," which is an individual substance of a rational nature
- **personage** a socio-psychological entity with a centripetal "self" who has an independent interpretability of the environment and an autobiographic memory
- **personalism** the usual philosophical tendency of dealing with the human personality
- **personhood** the outcome of a person which includes personage and homimmunage
- **personism** a perspective which deals with a person embedded in physicality and focuses on the substantial individual person and the relationship-based communality
- **persontology** the ontology based on the universal phenomenon of person
- **phenotype** the opposite of genotype, thus the physical manifestation of an organism, determined by its genetic constituent
- **physicalism** the phylosophical doctrine that is supported by epistemological evidences, and thus states that all phenomena can be described in terms of space and time
- **Planck condition** a condition that is expressed in terms of the Planck constant, which equals to the energy of any quantum of radiation divided by its frequency; it has a value of 6.62×10^{-34} joule seconds
- **plasticity** the ability of living things or humanity to change in order to adapt and survive
- **Plexion** the whole universe deducted by its single observer

- **postefference** the process of concentratred attention to assimilate objects to the cognitive self
- **pragma** the growing substantial environment of the existential humanity that indicates the range of the human cognition, in which a human can perceive or feel a definitely limited individual entity, and can be influenced by externality
- **preafference** the process of concentrated attention to assimilate oneself to an object
- **pretium menesys** pretium means value in Latin, pretium menesys is the result of interpretation of utility done by the intelligence menesys
- **Proportio Divina** the ratio between the long line segment and the short line segment, obtained by the equal division of a line (the ratio has an approximate numerical value of 1.618)
- **psycon** a mental unit, a unitary conscious experience
- **qualia** *Phylosophy;* essential properties or qualities
- **quantia** a kind of qualia that responds from our neuronal sensation for mathematical measure, logic, and complex notions
- **quantum** the smallest quantity of some physical property, such as energy, that a system can possess according to the quantum theory.
- **quasi-universal value system** a kind of value system that enables us to provide the value foundations for public law, which is deemed universal
- **regeneracy** similar to survival; the act of undergoing moral, spiritual, or physical renewal or invigoration
- **rumination** the act of meditating or pondering upon, thus the source of being and becoming human through introspection
- **sadism** the gaining of pleasure or sexual gratification from the infliction of pain and mental suffering on another person
- **science of complexity** the science of humanity that includes all human history and even the dimensions of the spirituality
- **science of creativity** the science which organizes the overall phenomenon of creativity
- **science of cruxology** the study of cruxology that includes the methodology of the creative peace and the conceptual foundation for the technology of menesys
- **science of humanity** the science that aims at the scientization of humanity (Scientizing is the restructuring of the internal attributes of humanity to achieve the real scientific conversion of human life)
- **science of integrism** the study of integrism that aims to achieve the integrity of both internality and externality of an integer person; in Asia, it is called the Tong-Ghi science
- **semiotics** the study of signs and symbols, esp. the relations between written or spoken signs and their referents in the physical world or the world of ideas
- **sensciousness** the ability of humans to feel by sensing organs

- **Seven Sensations** the doctrine of Neo-Confucianism that indicates the practical emotions distorted by the need of an individual, and these seven sensations are joy, anger, sorrow, fear, love, hate, and lust
- **singularity** discontinuity in mathematics (a point at which a function is not differentiable although it is differentiable in a neighborhood of that point); singularity is used in this book to present the discontinuity in civilizations
- **strange attractor** a pattern existing in abstract mathematical space, representing the path traced by a point that expresses the essential parameters of a chaotic system
- **sciousness** "Scious" comes from Latin word "scire," which means "know," thus sciousness indicates the condition, state or quality of knowing and it consists of sensciousness, consciousness and transciousness
- **summit experience** a different name for peak experience
- **superposition** in this book, superposition is of a similar usage as that used in the quantum systems describing multiple overlaps, not as in mathematics
- **syntalansis** the extensive synchronization of the oscillators of network elements
- **sysnet** it stands for SYStemized NETwork, which means the systemization of the intrinsic attributes of a network in nature
- **system** a group of combination of interrelated, interdependent, or interacting elements forming a collective entity; a methodical or coordinated assemblage of parts, facts, concepts, etc
- **systema** a system that is not isolated by boundaries
- **systemicity** the ontological attributes of an integer that indicates the centripetal effects of composing the integrism of an integer
- **sysy** *see* **systemicity**
- **Taiji** *see* **yin and yang**
- **tensorial phase space** human brain is adaptive to the infinite number of dimensions in tensorial manner
- **terrenity value** land value, or terrestrial value that can admit the existence of human being
- **topos-submatrix** a submatrix of infinities-matrix that represents the position of a specific locality
- **transcendental epistemology** *see* **Trep**
- **transcendental intentionality** it is different from the intentionality suggested by Brentano, in a sense that humans have teleological tendencies to trascend to a higher hierarchical level of humanity
- **transcent of humanity** the shift and the entrance of humanity into a novel phase
- **transciousness** the highest domain of the human cognition that guides humanity to the transcent
- **Trep** it stands for TRascendental EPistemology, which aims at the epistemological examination that leads us to the trascent of humanity

- **Trinitarianism** the Christian theology that believes in the "One God in Three Persons" doctrine, in which the Three Persons are the Father, Son, and Holy Ghost
- **Triradixism** the thought that states the three roots of value; the creativity (heaven-root), the value of labor (human-root), and the value of rent (land-root)
- **Tsim-tsum** the Divine Self-contraction suggested by Naimonides in the medieval jewish thought of Kabbalah
- **Ultimatism** one of the attitudes of humankind that aims to mobilize all capability in an earnest and thorough manner to surpass the present omega path
- **Unicrux** UNItary original CRUX, the general continuum of the original universe composed of the real infinite numbers of original dimensions
- **utilities menesys** the composition of systems and networks that hold the complex relations of the human society interweaving pains and joys, which are emotional phenomena
- **vacuity** the vacant building blocks of the emerging point of a quantum matter (vacuity is different from nullity)
- **value** in this book, value indicates the individual cognitive value which covers the sensciousness, consciousness and transciousness of a person, thus it corresponds to the humanity of each person
- **values menesys** from the persontological viewpoint, it is a tree diagram which develops in the following sequence; from onto (heaven), topos (terrenity), and then to logia (human)
- **virtual reality** a computer-generated environment that, to the person experiencing it, closely resembles reality
- **vocalization** articulation of a speech sound with voice
- **yin and yang** two complementary principles of Chinese philosophy: Yin is negative, dark, and feminine, Yang positive, bright and masculine. Their interaction is thought to maintain the harmony of the universe and to influence everything within it
- **Yong (用)** literally means the usage of body and/or indicates the movement in the body of a person
- **zygote** the totipotent cell resulting from the union of an ovum and a sperm

Bibliography

Abe, Y. 2009. Kinyuu No Shikumi Wa Zenbu Rothchild Ga Tsukutta. Eldorado.
Abo, T. 2005. Meneki Kakumet (The Immune System Revolution). Bukwang Publishing Co.
Abraham, R. 1997. Chaos, Geia, Eros. Doosan Dong-A Publishing Co.,Ltd.
Aczel, A. D. 2002. The Mystery of the Aleph: Mkathemathics, the Kabbalah, and the search for Infinity. Seung San Publishers.
Aczel, A. D. 2007. Entanglement: The Greatest Mystery in Physics. Vista Books.
Adkison, L. R., and Brown, M. K. 2007. Elsevier's Intergrated Genetics. Mosby, Inc., an affiliate of Elsevier, Inc.
Alon, U. 2007. An Introduction to Systems Biology: Design Principles of Biological Circuits. CRC Press, Taylor & Francis Group.
Alper, M. 2006. The God Part of the Brain: A Scientific Interpretation of Human Spirituality and God. Source Book, Inc.
Altmann, G. T. M. 2005. The Ascent of Babel. Youkrak.
Amabile, T. M. 1996. Creativity in Context: Update to the Social Psychology of Creativity. Westview Press, A Member of the Perseus Books Group.
Amyx, J. W., et al. 1960. Petroleum Reservior Engineering: Physical Properties. McGraw-Hill Book Company.
Anheier, H., et al. 2002. Global Civil Society. Arche Publishing House.
Arbib, M. A. (eds.). 2006. Action to Language: via the Mirror Neuron System. Cambridge University Press.
Archer, M., et al. 2005. Critical Realism : Essential Reading 1, Transcendental Realism and Science. Hanul Academy.
Arthur, B. W., et al. 1998. Economics of Complexity. Pyungbeomsa.
Arthur, B. W., et al. 1998. Lectures in Complex Systems. Pyungbeomsa.
Ascott, R. 2002. Technoetic Arts. Yonsei University Press.
Asia Research Instiue of Economy. 1975. The Community, Collective Farm, and General Federation of Labour in Israel. Daehakseorim.
Atkins, P. 2007. Four Laws: That Drive the Universe. Oxford University Press.
Attali, J. 2009. La Crise et Aprés?. Wisdomhouse Publishing Co. Ltd.
Audretsch, J. (ed.). 2006.Entangled World: The Fascination of Quantum Information and Computation. Viley-VCH Verlag GmbH & Co. KGaA
Avila, C. 1983. Ownership: Early Christian Teaching. Orbis Books.
Bailey, K. D. 1996. Social Entropy Theory. Shin-Yu Publishing Co.
Bainbridge, D. 2006. The X in Sex: How the X chromosome controls our lives. God's Win Publishers, Inc.
Banich, M. T. 2008. Cognitive Neuroscience and Neuropsychology. Sigma Press, Inc.
Barabási, Á. L. 2002. Linked: The New Science of Networks. Perseus Publishing.
Barabási. Á. L. 2002. Linked: The New Science of Networks. EAST-ASIA Publishing Co.
Barbalet, J. 2009. Emotions and Sociology. Ehak Publishing.
Barker, D. K. 1987. Jubilee on Wall Street: An Optimistic Look at the Coming financial Crash. Prescott Press.
Barraclough, G. 1982, 1992. The Times Concise Atlas of World History. Times Books, A division of harperCollinsPublishers.
Barrow, J. D. 2003. The Book of Nothing. Henamu Publishing Co.
Bauer. J. 2006. Das Gedächtnis des Körpers. Ez-book.

Bauman, Z. 2000. Liquid Modernity. Polity Press.
Bear, M. F., et al. 2007. Neuroscience: Exploring the Brain (3rd ed.) Lippincott Williams & Wiliins, a Wolters Kluwer business.
Begley, S. 2007. Train Your Mind Change Your Brain:How a new science reveals our extraordinary potential to transform ourselves. Ballantine Book.
Behe, M. J. 2001. Darwin's Black Box. Pulbit Publishing Co.
Behe, M. J. 2007. The Edge of Evolution: The Search for the Limits of Darwinism. Free Press, A Division of Simon & Schuster, Inc.
Beinhocker, E. D. 2007. The Origin of Wealth: Evolution, Complexity, and the Radical Remaking of Economics. Randomhouse Korea, Inc.
Bengtsson, I., and Zyczkowski,K.2006. Geometry of Quantum States: An Introduction to Quantum Entanglement. Cambridge University Press.
Bergson, H. 2008. Matière et Mémoire; essai sur la relation du corps à l'esprit. Salimbooks.
Berman, M. 2009. Hedge Funds and Prime Brokers. BMS, Inc.
Beutelspacher, A., and Petri, B. 2007. Der Goldene Schnitt. Ichi Science.
Binnendijk, H. (ed.). 2004. Transforming America's Military. Korea Institute for Defense Analyses.
Bloom, H. 2003. Global Brain: The Evolution of Mass Mind from the Big Bang to the 21st Century. Pascal Books Co.
Bodanis, D. 2001. $E=mc^2$. Thinking Tree Publishing Co.
Boden, M. A. (ed.). 1994. Dimensions of Creativity.A Bradford Book The MIT Press.
Bohm, D. 1994. Thought as a System
Bohm, D., and Hiley, B. J. 1993. The Undivided Universe. Routledge, Taylor & Prancis Group.
Bolz, N. 2000. Das Kontrolierte Choas: Vom Humanismus zur Medienwirklichkeit. Moonye Publishing Co.
Bonabeau, E., et al. 1999. Swarn Intelligence: From Natural to Artificial Systems. Oxford University Press.
bonner, J. T. 2008. Why Size Matters: From Bacteria to Blue Whales. Eclio Publishing Co.
Börner, G. 2009. Schöpfung ohne Schöpfer? Book House Publishers.
Boyle, F. A. 2004. Foundations of World Order. Pakyoungsa.
Brenner, R. 2002. The Boom and the Bubble. Ahchimyisul.
Brockman, J. (eds.).2002. The Next Fifty Years: Science in the First Half of the Twenty-First Centry. Vintage Books, A Division of Random House Inc.
Brockman, J. (eds.).2003.The New Humanists: Science at the Edge. A Barnes and Noble Book.
Brockman, J. 2002. The Next Fifty Years: Science in the First Half of the Twenty-First Century. Thinking Tree Publishing Co., Ltd.
Brockman, J. 2006. The New Humanists. Sosoh Publishing Co.
Brown, P. R. L. 2004. The Rise of West Christendom. Saemulgyul Publishing House.
Bublath, J. 2003. Chaos im Universum. Thinking Tree Publishing Co., Ltd.
Buchanan, M. 2003. Nexus. Sejong Institution Press.
Buchanan, M. Nexus: 2002. Small Worlds and the Groundbreaking Science of Networks. W. W. Norton & Company, Inc.
Bulkeley, K.(eds.). 2005. Soul, Psyche, Brain: New Directions in the Study of Religion and Brain-Mind Science. New York: Palgrave Macmillan.
Buller, D. J. 2005. Adapting Minds: Evolutionary Psychology and the Persistent Quest for Human Nature. A Bradford Book The MIT Press.
Burgin, L. 2008. Der Urzeit-Code. Dosol Publishing Co.
Burrow, R. Jr. 1999. Personalism: A Critical Introduction. Chalice Press.

Burt, R. S. 1992. Structural Holes: The Social Structure of Competition. Harvard University Press.
Buss, D. B. 2005. Evolutionary Psychology: The New Science of the Mind. Nanomedia Publisher.
Buzsáki, G. 2006. Rhythms of the Brain. Oxford University Press.
Byrne. J. H., and Roberts J. L.(eds.). 2004. From Molecules to Networks: An Introduction to Cellular and Molecular Neuroscience. Elsevier Academic Press.
Cacioppo, J. T., et al. (eds.). 2006. Social Neuroscience: people thinking about thinking people. MIT Press Books.
Canton, J. 2007. The Extreme Future. Gimm-Young Publishers, Inc.
Capra, F. 2003. The Hidden Connetions. Whisler Publishing Co.
Carroll, J. M. 1987. The Story of the Laser. Hundaikwahakshinseo.
Carroll, S. B. 2007. Endless Forms Most Beautiful: The New Science of Evo Devo and The Making of the Animal Kingom. Chiho Publishing House.
Carroll, S. B. 2008. Making of the Fittest: DNA and the Ultimate Forersic Record of Evolution. Chiho Publishing House.
Carter, R. 2002. Exploring Consciousness. University of California Press.
Castells, M. 2000. The Information Age Economy, Society and Culture Vol.3: End of Millennium(2nd ed.). Blackwell Publishing.
Castells, M. 2003. The Information Age Economy, Society and Culture Vol.1: The Rise of the Network Society (2nd ed.). Blackwell Publishing.
Castells, M. 2004. The Information Age Economy, Society and Culture Vol.2: The Power of Identity (2nd ed.). Blackwell Publishing.
Castells, M., and Hall, P. Technopoles of the World: The making of 21st Century Industrial Complexes.
Chardin, P. T. 2001. Teilhard de Chardin's Theology of the Christian in the World. Benedict Press.
Chardin, P. T. 2003. Le Coeur de la Matiére. Benedict Press.
Cheng, H. C., et al. (Trans.). 1985. Shan Hai Ching: Legendary Geography and Wonders of Ancient China. The Committee for Compilation and Examination of the Series of Chinese Classics, National Institute for Compilation and Translation.
Chio, C. H., and Park C. H. 2007. Complex System and Orientalism: Selforganization and Organization Management. Jisaem Publishing.
Chio, D. K. 2006. Introduction in Analysis Tensor. Pan Korea Book Co.
Chiu, L. S. 2006. When a Gene Makes You Smell Like a Fish. Hanerl Media.
Cho, K. H. 2002. Der Mensch als Da-Sein zwischen Zeit und Ewigkeit: Einführung in Plotin's Denken von Trinität. St Pauls.
Cho, Y. H. 1992. The Scientific Philosophy of Karl Popper. Seokwangsa.
Choi, H. C. (ed.). 1982. Hymns From the Rig-Veda. Imun Publishing Co.
Choi, W. H. 2000. The Central Government of Organizations of the People's Republic of China. The Korea Institute of Public Administration.
Chomsky, N. 2003. Problems of Knowledge and Freedom. Meeto Publishers.
Chomsky, N. 2003. The Spectacular Achievements of Propaganda. Mosek Publishing Co.
Chomsky, N. 2004. Noam Chomsky deux heures de Incidité. Window of Times.
Christian, D. 2009. This Fleeting World: A Short History of Humanity. Seohaemunjip Publishing Co.
Christie, D. 2002. The Unbroken Wholeness Power of God: Bible Screts Revealed. Virginia: Xulon Press.
Church, D. 2007. The Genie in your Genes: Epigenetic Medicine and the New Biology of Intention. Energy Psychology Press.
Churchland, P. 1986. Neurophilosophy: Toward a Unified Science of the Mind/Brain. A

Bradford Book The MIT Press.
Churchland, P. 2007. Neurophilosophy at Work. Cambridge University Press.
Churchland, P. M. 1995. The Engine of Reason, the Seat of the Soul: A Philosophical Journey into the Brain. A Bradford Book The MIT press.
Churchland, P. S. 2002. Brain-Wise: Studies in Neurophilosophy. A Bradford Book The MIT press.
Churchland, P. S. 2006. Neurophilosophy: Toward a Unified Science of the Mind-Brain. Chulhak kwa Hyunsil Sa Publishing Co.
Churchland, P. S., and Sejnowski, T. J. 1999. The Computational Brain. A Bradford Book the MIT Press.
Clegg, B. 2006. The God Effect: Quantum entanglement, science's strangest phenomenon. ST.Martin's Press.
Cohen, B. J. 1999. The Geography of Money. Shi Yu Shi Publishing & Edition.
Cohn, N. 1993. The Pursuit of the Millennium: Revolutionary Millenarianism and Mystical Anarchists of the Middle Ages. Korea Institute of Theology.
Cole, K. C. 2004. The Hole in the Universe. Hainaim Publishing Co., Ltd.
Coligan. J. E., et al. (eds.). 2005. Short Protocols In Immunology: A Compenduim of Methods From Current Protocols In Immulology. John Wiley & Sons, Inc.
Combs, A. The Radiance of Being: Complexity, Chaos and the Evolution of Consciousness.
Complex System Network. 2006. Complexity Sciences Work Shop. Samsung Economic Research Institute.
Cook, V. J. 1995. Chomsky's Universa Gremmar. Hanshin Publishing Co.
Cooper, G. 2008. The Origin of Financial Crises: Central Banks, Credit Bubbles and the Efficient Market Fallacy. Vintage Book, A Division of Random House, Inc.
Cooper, G. 2009. The Origin of Financial Crises: Central Banks, Credit Bubbles and the Efficient Market Fallacy. Readers House.
Cover, T. M., and Thomas, J. A. 2006. Elements of Information Theory (2nd ed.). Wiley-Interscience, A John Wiley & Sons, Inc., Publication.
Cowan, G. A., et al. 1994. Complexity: Metaphors, Models, and Reality. Perseus Book.
Craft, B. C., et al. 1991. Applied Petroleum Reservoir Engineering (2nd ed.). Prentice Hall.
Crichlow, H. B. 1977. Modern Reservior Engineering-A Simulation Approach. Prentice Hall.
Cristol. S. M., and Sealey, P. 2000. Simplicity Marketing: End Brand Complexity, Clutter, and Confusion. The Free Press.
Csermely, P. 2006. Weak Links: Stabilizers of Complex Systems from Proteins to Social Networks. Springer.
Csikszentmihalyi, M. 1996. Creativity: Flow and the Psychology of Discovery and Invention . Harper Perennial, A Division of Harper Collins Publishers Inc.
Csikszentmihalyi, M. 1999. Finding Flow. Hainaim Publishing Co.
Csikszentmihalyi, M. 2004. Creativity: Flow and the psychology of discovery and invention. Bookload Publishing.
Csikszentmihalyi, M. 2004. Flow: The Psychology of Optimal Experience. Hanulim Publishing Co.
Dacey, J. S., and Lennon, K. H. 1998. Understanding Creativity: The Interplay of Biological, Psychological, and Social Factors. Jossey-Bass, A Wiley Company.
Dake, L. P. 2001. The Practice of Reservoir Engneering (Revised ed.). Elsevier.
Damasio, A. 1994. Descartes' Error: Emotion, Reason, And the Human Brain. Penguin Books.
Damasio, A. 1999. The Feeling of What Happens: Body and Emotion in the Making of Consciousness. A Harvest Book Harcourt, Inc.
Damasio, A. 2007. Looking for Spinoza: Joy, Sorrow and the Feeling Brain. ScienceBooks Co., Ltd.

Davidson, E. H. 2006. The Regulatory Genome: Gene Regulatory Networks In Development and Evolution. Elsevier Academic Press.
Davis, P. 1995. Superforce: The Search for a Grand Unified Theory of Nature. Pumyang Co.,Ltd.
Dembski, W. A. 1999. Intelligent Design: The Bridge Between Science & Theory. IVP.
Denda, M. 2009. Daisa no nou. Openscience Publishing Co.
Dennett, D. C. 2006. Kinds of Minds: Toward an Understanding of Consciousness. ScienceBooks Co., Ltd.
Dent, H. 2005. The Next Great Bubble Boom. The Chungrim Publishing.
Derbyshire, J. 2009. Unknown Puantity: A Real and Imaginary History of Algebra. Seung San Publishers.
Devin, K. J. 2002. The Math Gene. Kachi Publishing Co. Ltd.
Doidge, N. 2007. The Brain That Changes Itself: Stories of Personal Triumph from the Frontiers of Brain Science. Penguin Books.
Doidge, Norman. 2008. The Brain that Chages Itself. Chiho Publishing House.
Doran, K. P. 1996. Solidarity : A Synthesis of Personalism and Communalism in the Thought of Karol Wojtyla/ Pope John Paul II. Peter Lang.
Draaisma, D. 2006. Metaphors of Memory. ECO-LIVRES Publishing Co.
Drescher, G. L. 2006. Good and Real: Demystifying Paradoxes from Physics to Ethics. A Bradford Book The MIT Press.
Düweke, P. 2005. Kleine Geschichte der Hirnforschung. Motive Book Publishing House.
Dworkin, R. 2008. Law's Empire. Acanet.
Dyson, F. 1979. Disturbing the Universe. Basic Books.
Dyson, F. 1998. Imagined Worlds. Harvard University Press.
Dyson, F. 1999 Origins of Life (2nd ed.). Cambridge University Press.
Dyson, F. 1999. The Sun, The Genome And The Internet: Tools of Scientific Revolutions. Oxford University Press.
Dyson, F. 2000. Imagined Worlds. ScienceBooks Co., Ltd.
Dyson, F. 2001. The Sun, the Genome, & the Internet: Tools for scientific Revolutions. Sagoonja.
Dyson, F. 2007 A Many-Colored Glass: Reflections on the Place of Life in the Universe. University of Virginia Press.
Dyson, F. 2009. Disturbing the Universe. ScienceBooks Co.,Ltd.
Earle, S. A., and Glover, L. K. 2008. Ocean: An Illustrated Atlas. National Geographic.
Eddy, M. B. 1875, 1986. Science and Health with Key to the Scriptures. The Mary Baker Eddy Foundation.
Edelman, G. M. 1987. Neural Darwinism: The Theory of Neuronal Group Selection. Basic Books Inc.
Edelman, G. M. 1988. topobiology: An Introduction to Molecular Embryology. Basic Books.
Edelman, G. M. 1992. Bright Air, Brilliant Fire: On the Matter of the Mind. Basic Books.
Edelman, G. M. 1998. Bright Air, Brilliant Fire: On the Matter of the Mind (Korean ver.). Pumyang Co., Ltd.
Edelman, G. M. 2004. Wider the the Sky: the phenomenal gift of consciousness (Korean ver.). Pumyang Co., Ltd. Henamu.
Edelman, G. M. 2005. Wider than the Sky: A Revolutionary View of Consciousness. Penguin Books.
Edelman, G. M. 2006. Second Nature: Brain Science and Human Knowledge. Yale University Press.
Edelman, G. M. 2009. Second Nature: Brain Science and Human Knowledge. Eum Publishing Co.
Edelman, G. M., and Mountcastle, V. B. 1982. The Mindful Brain: Cortical Oranization and the Group-Selective Theory of Higher Brain Function. The MIT Press.
Edelman, G. M., and Thiery J. (eds.). 1985. The Cell in Contact: Adhesions and Junctions as

Morphogenetic Determinants. A Neurosciences Institute Publication.
Edelman, G. M., and Tononi, G. 2000. A Universe of Consciousness: How Matter Becomes Imagination. Basic Books.
Edelman, G. M., and Tononi, G. 2000. Consciousness: How Matter Becomes Imagination. Penguin Books.
Edelman, G. M., el al. (eds.). 1985. Molecular Bases of Neutal Development. A Neurosciences Institute Publication.
Edelman, G. M., et al. (eds.) 1990. Signal & Sense: Local and Global Order in Perceptual Maps. A Neurosciences Institute Publication.
Edelman, G. M., et al. (eds.). 1987. Synaptic Function. A Neurosciences Institute Publication, John Wiley & Sons.
Edelman, G. M., et al. 1985. How We Know. Harper & Row, Publishers, San Francisco.
Edelman, G. M., et al. 1990. (eds.). Morphoregulatory Molecules. A Neurosciences Institute Publication.
Eichengreen, B. 2008. Global Imbalances and the Lessons of Bretton Woods. Mizibooks.
Elman, B. A. 1990. Classicism, Politics, and Kinship: The Chang-Chou School of New Text Confucianism in Late Imperial China. SMC Publishing Inc.
Evans, D. 2002. Emotion: The Science of Sentiment. Yeesaw Publishers.
Faber, M. 2003. Tomorrow's Gold. Philmac Publishing Co.
Falk, D. 2003. Universe on a T-Shirt. Whisler Publishing Co.
Fanchi, J. R. 2001. Principles of Applied Reservoir Simulation (2nd ed.). Gulf Professional Publishing.
Fauconnier, G., and Turner, M. 2009. The Way We Think: Conceptual Blending and the Mind's Hidden Complexities. Chiho Publishing House.
Feldman, J. A. 2006. From Molecule to Metaphor: A Neural Theory of Language. A Bradford Book The MIT Press.
Ferguson, N. 2002. The Cash Nexus. Gimm-Young Publishers, Inc.
Ferguson, N. 2003. Empire: How Britain Made the Modern World. Penguin Books.
Ferguson, N. 2008. The Ascent of Moneyt: A Financial History of the World. Penguin Books.
Figeys, D.(eds.). 2005. Industrial Proteomics: Applications for Biotechnology and Pharmaceuticals. John Wiley & Sons, Inc.
Fischl, J. 1987. Geschichte der Philosophie: von den Griechen bis zur Gegenwart. Seokwangsa.
Fleckenstein, W., and Sheehan, F. 2008. Greenspan's Bubbles. Hans Media.
Foray, D. 2003. Economics of the Knowledge (L'économie de la connaissance). Hanul Books.
Foster, J. B., et al. 2009. Monthly Review 2. Philmac Publishing Co.
Frankl, V. E. 2005. The Unheard Cry for Meaning: Psychotherapy and Humanism. Chung-A Publishing Co.
Freeman, W. J. 1995. Societies of Brains: A Study in the Neuroscience of Love and Hate. Lawrence Erlbaum Associates, Publishers.
Freeman, W. J. 2000. How Brains Make Up Their Minds. Columbia University Press.
Freeman, W. J. 2007. How Brains Make Up Their Minds. Boogle Books.
Freeman, W. J., and Núñez, R. 1999. Reclaiming Cognition: the primacy of action intention and emotion. Imprint Academic.
Fuller, S. 2007. Kuhn vs. Popper. Thinking Tree Publishing Co.
Galtung, J. 2000. Peace by Peaceful Means. Dulnyuk Publishing Co.
Gardner, H. 2001. Intelligence Reframed. Gimm-Young Publishers, Inc.
Gardner, J. N. 2006. Biocosm: The New Scientific Theory of Evolution: Intelligent Life is the Architect of the Universe. Kachi Publishing Co.

Garreau, J. 2007. Radical Evolution. Forest of Knowledge.
Gazzniga, M. S., et al. (eds.). 2004. The Cognitive Neurosciences (3rd ed.). A Bradford Book the MIT Press.
George, H. 1979. Progress and Poverty: An inquiry into the cause of industrial depressions and of increase of want with increase of wealth... The Remedy. Rober Schalkenbach Foundation.
Gilbert, M.1969, 1993. Atlas of Jewish History. JM Dent Ltd.
Gleick, J. 1993. Chaos: Making a New Science. Dongmoon Publishing Co.
Glen, J., and Gordon, T. 2007. Future Research Methodolog. Kyobobook.
Goffi, J. 2003. La Philosophie de la Technique. Hangilsa Publishing Co., Ltd.
Goldstein, R. 2007. The Great Discoveries Series, Incompleteness: The Proof and Paradox of Kurt Gödel. Seunng San Publishers.
Gordon, W., and Adams, J. 1995. Economics as Social Science: An Evolutionary Approach. Bee Bong Publishing Co.
Gorßmann, K., and Schröer, 1991. H. Auf den Spuren des comenius: Texte zu Leben, Werk und Wirkung. Vandenhoeck & Ruprecht in Göttingen.
Gould, S. J. 2007. Punctuated Equilibrium, The Belknap Press of Harvard University Press.
Graeber, D. 2009. Toward an Anthropological Theory of Value. Greenbee Publishing Co.
Greefield. S. 2004. Brain Story. Chiho Publishing House.
Green, S. 2009. Good Value: Reflections on money, morality and an uncertain world. Allen Lane an imprint of Penguin Books.
Greene, B. R. 2002. The Elegant Universe. Seunsan Publishers.
Greenwald, B. C. N., and Kahn, J. 2009. Globalization: n. the irrational fear that someone in China will take your job. Segyesa Publishing Co., Ltd.
Griffin, E. 2006. A First Look at Communication Theory (6th ed.). McGraw Hilll.
Grousset, R. 2002. L'Empire des Steppes . Sakyejul Publishing.
Guha-Khasnobis, B., Kanbur, R., and Ostrom, E.(eds.). 2006. Linking the Formal and Informal Economy: Concept and Policies. Oxford University Press.
Ha, Y. S. (ed.). 2002. The Peace Study in 21st Century. Pulbit Publishing Co.
Halpern, P. 2006. The Great Beyond: Higher Dimensions, Parallel Universes, and the Extraordinary Search for a Theory of Everything. Chiho Publishing House.
Hamer. D. 2004. The God Gene: How Faith is Hardwired into our Genes. New York: Anchor Books, A Division of Random House, Inc.
Han, H. J. 2002. From Xu Shi to Jung Yak Yong: Philosopical Paradigmatic Study on Confusionism of Yi'Dynasty. Segyesa Publishing.
Harrison, N. E. (eds.). 2006. Complexity in World Politics: Concepts and Methods of a New Paradigm. State University of New York Press.
Hartmann, N. 1997. Neue Wehe der Ontologie. Seokwangsa Publishing Co.
Hasegawa, K. (長谷慶太郎). 2008. Chou Kakusa Kakudai No Jidai. Woogok Publishing.
Havil, J. 2008. Gamma: Exploring Euler's Constant. Seung San Publishers.
Hawking, S., et al. 2002. The future of Spacetime. Henamu Publishing Co.
Hazelrigg, G. A. 1993. Systems Engineering. Institute for Advanced Engineering.
Hazen, R. M. 2008. Genesis: The Scientific Quest for Life's Origin. Hanseung Publishing Co.
Held, D., et al. 2002. Global Transformations. Changbi Publishers Inc.
Heller, M. 2009. The Gridlock Economy. Woongjin Think Big Co., Ltd.
Hilborn, R. C. 2000. Chaos and Nonlinear Dynamics: An Introduction for Scientists and Engineers (2nd ed.). Oxford Univesity.

Hirose, T. 1996. Shitsuryou No Kigen. Chonpakwahaksa.
Hoffmann, R. 2009. The Same and not the Same. Kachi Publishing Co.
Hofstadter, D. 2007. I am a Strange Loop. Basic Books, A Member of the Perseus Books Group.
Holland, J. H. 1995. Hidden Order: How Adaptation Builds Complexity. Helix Books.
Holland, J. H. 1998. Emergence: From Chaos to Order. Perseus Books.
Hong, D. S. (ed.). 1996. Applied Social Psychology. Pakyoungsa.
Hoshi, K. 2000. Stress and Immunity. Blue Backs.
Huang, R. 2001. The History of Capitalism and Chinese 21st Century. Yeesan Publishing Co.
Hüther, G. 2001. The Compassionate Brain: How Empathy Creates Intellingence.Trumpeter.
Hüther, G. 2007. Biologie der Angst: Wie aus Streß Gefühle werden. Kungree Press.
Ikegaya, Y. 2008. No Wa Nanikato Itwake Suru: Hito wa Shiawase ni Naruyoni Dekiteita!?. Leader's Book Publishers.
Ikegaya, Y. Shigesato itoi. 2006. Kaiba Nou Ha Tsukarenai. Ehbook.
Ivancevic, V. G., and Ivancevic, T. T. 2008. Complex Nonlinearity: Chaos, Phase Transitions, Topology Change and Path Integrals. Springer-Verlag.
Izhikevich, E. M. 2007. Dynamical Systems in Neuroscience: The Geometry of Excitability and Bursting. The MIT Press.
Jablonka, E., and Zeligowski, 2005. A. Evolution in Four Dimensions: Genetic, Epigenetic, Behavioral, and Symbolic Variation in the History of Life. A Bradford Bood the MIT Press.
Jacobs, C. S. 2009. Management Rewired: Why Feedback Doesn't Work and Other Surprising Lessons from the latest Brain Science. Portfolio.
Jang, H. I. 2008. Science and Metascience. Jisiksanupsa.
Jantsch, E. 1995. The Self-Organizing Universe: Scientific and Human Implications of the Emerging Paradigm of Evolution. Bumyang Publishing.
Je, S. H. 1997. The Study of Demilitarized Zone in the Korean Peninsula. Seoul Press.
Jensen, H. J. 1998. Self-Organized Criticality: Emergent Complex Behavior in Physical and Biological Systems. Cambridge University Press.
Jeong, B. M. 2001. Creativity: The Psychology, Person, and Society. Kyoukkwahaksa.
Jeong, I. S. 2003. Transpersonal Psychology. Daewangsa.
Jeong, W. C., and Kim, H. B. 2008. Money & Financial Markets (3rd ed.). Yulgok Publishing.
Jin, K. H. 2007. The Person: The Meaning of Person from Ancient to Modern Times. Seoul National Univesity Press.
Jin, K. H. 2008. Economics of Free International City. Gak Publishing.
Joe, J. H. 2003. Autonomia. Galmuri.
Johnson, S. 2004. Emergence. Gimm-Young Publishers, Inc.
Johnson, S. 2006. Mind Wide Open: Your Brain and the Neuroscience of everyday Life. Woongjinthinkbig Co., Ltd.
Joseph, R.(eds.). 2002. Neuro Theology: Brain, Science, Spirituality, Religious Experience. California: University Press.
Kaku, M. 1997. Hyperspace. Gimm-Young Co., Publishers.
Kanaya, O. 1999. Eki No Hanashi. Hanul Publishing Co.
Kandel, E. R. 2009. In Search of Memory. Random House Korea Inc.
Kaneko, K., and Tsuda, I. 1996, 2000. Complex Systems: Chaos and Beyond: A Constructive Approach with Applications in Life Sciences. Springer-Verlag.
Kaneko, K., and Tsuda, I. 2006. Life: An Introduction to Complex Systems Biology. Springer.
Kaneko, M., and DeWit, A. 2008. World-wide financial crisis. Jisangsa.
Kang, E. J., et al. 2004. Brain, Affecting Mind and Mind, Affecting Brain. Henamu

Publishing Co., Ltd.
Kaplan, R. D. 1993, 1996, 2005. Balkan Ghosts: A Journey Through History. Picador.
Kaplan, R. D. 1996. The Ends of the Earth: A Journey to Frontiers of Anarchy. Vintage Book, A Division of Random House, Inc.
Kaplan, R. D. 1999. An Empire Wilderness: Travels into America's Future. Vintage Book, A Division of Random House, Inc.
Kaplan, R. D. 2000. Eastward to Tartary: Travels in the Balkans, the Middle East, and the Caucasus. Vintage Departures, Vintage Book, A Division of Random House, Inc.
Kaplan, R. D. 2000. The Coming Anarchy: Shattering the Dreams of the Post Cold War. Vintage Book, A Division of Random House, Inc.
Kaplan, R. D. 2001. The Coming Anarchy: Shattering the Dreams of the Post Cold War. Dulnyouk Publishing Co.
Kaplan, R. D. 2002. Warrior Politics: Why Leadership Demands a Pagan Ethos. Vintage Book, A Division of Random House, Inc.
Kaplan, R. D. 2004. Mediterranean Winter: The Pleasures of History and Landscape in Tunisia, Sicily, Dalmatia, and the Peloponnese. Vintage Departures, Vintage Book, A Division of Random House, Inc.
Kaplan, R. D. 2005. Imperial Grunts: On the Ground with the American Military, from Mongolia to the Philippines to Iraq and Beyond. Vintage Book, A Division of Random House, Inc.
Kaplan, R. D. 2007. Warrior Politics. Thinking Tree Publishing Co.
Kaplan. R. D. 1988, 2003. Surrender or Starve: Travels in Ethiopia, Sudan, Somalia, and Eritrea. Vintage Book. A Division of Random House, Inc.
Kaplan. R. D. 2007. Hog Pilots, Blue Water Grunts: The American Military in the Air, at Sea, and on the Ground. Random House.
Kaser, M. 1998. Römische Rechtsgeschichte. Beopwonsa.
Katsutoshi, K. 2009. Chi No Bunruishi. Korea Publishing Marketing Research Institute.
Kauffman, L. H. 1991. Knots and Physics (3rd ed.). World Scientific.
Kauffman, S. 1995. At Home in the Universe: The Search for the Laws of Self-Organization and Complexity. Oxford University Press.
Kauffmann, S. 2002. At Home in the Universe: The Search for Laws of Self-Organization and Complexity. Science Books Co.,Ltd.
Kaufman, J. C., and Sternberg, R. J. (eds.). 2006. The International Handbook of Creativity. Cambridge University Press.
Kelso, J. A. S. 1995. Dynamic Patterns: The self-organization of Brain and Behavior. A Bradford Book The MIT Press.
Khanna, P. 2008. The Second World: Empires and Influence in the New Global Order. Random House.
Kiernan, J. A. 2006. Barr's The Human Nervous System: An Anatomical Viewpoint, 8th Edition. E*Public.
Kim, J. 2004. Whitehead und Huayen-Metaphysik. UUP.
Kim, J. 2005. Phänomenologie und Hermeneutik. UUP.
Kim, J. B. 2001. I Ching and Socialogy. Yemoonseowon.
Kim, J. H. 2004. The Federal Reserve System: Secret of Decision of Interest Rate. Whiz Press.
Kim, J. K. 1996. Philosophy of Mind. Chul Hak Kwa Hyun Sil Sa Publshing Co.
Kim, J. K. 1999. Mind in a Physical World. Chul Hak kwa Hyun Sil Sa.
Kim, J. K. 2007. Physicalism. Acanet.

Kim, K. H. 2002. Gestalt Psychology. Hakjisa.
Kim, N. K. 1988. Heaven Thought of Confucianism. Soongsil University Press.
Kim, S. C. 2003. Senses and Transcendence. Benedict Press.
Kim, S. I. 1993. Whitehead and Orientalism. Seokwangsa.
Kim, S. I. 1999. Hyper'Space and Korean Culture: Hyper'Spatial Structure of Korean Consciousness in the light of Möbius, Escher, Russel. Kyohak Yunkusa.
Kim, S. I., and Lee, S. E. 1995. Fuzzy Beauty, Fuzzy Philosophy, Fuzzy Management. Joenjashinmoonsa.
Kim, S. S. 1989. Fata All Guide of Acupuncture. Sung-Han Publishing Co.
Kim, T. M. 1998. 中國土地經濟史硏究 (The Study of Chinese Econmic History of Land) Korea University Press.
Kim, Y. J. 2005. In Between Science and Religion. Dolbegae.
Kim, Y. K. 2008. 萬國公法. 小花
Kim, Y. W., and Kim, Y. K. 1995. 數學序說. (An Introduction to Mathematics) Woosung Publishing.
Kim, Y. W., and Kim, Y. K. 2002. From Figures To Spaces. Woosung Publishing.
King, B. J. 2007. Evolving God: A Provocative View on the Origins of Religion. Doubleday.
Kitano, H. (eds.). 2001. Foundations of Systems Biology. The MIT Press.
Klein, S. 2006. Alles Zufall. Woongjinthinkbig Co., Ltd.
Kleindorfer, P. R., and Wind, Y. J. 2009. The Network Challenge: Strategy, Profit, and Risk in an Interlinked World. Wharton School Publishing.
Koch, C. 2006. The Quest for Consciousness: A Neurobiological Approach. Sigmapress.
Kock, W. E. 1993. Lasers and Holography. Hyundaikwahakshinseo.
Koh, B. S. 1992. A Study of Values. Nanam Publishing House.
Kolko, G. 2009. World in Crisis: The End of the American Century. Via Publisher.
Koonin, E. V., Wolf, Y. I., and Karev, G. P. 2006. Power Laws, Scale-Free Networks and Genome Biology. Springer Science+Business Media, Inc.
Kriwaczek, P. 2005. Yiddish Civilisation: The Rise and fall of a Forgotten Nation. Phoenix.
Krugman, P. 2002. The Self-Organizing Economy. Bookie Publishing House.
Kuratani, S., and Osumi, N. 1997. Up Biology (Japanese Version). University of Tokyo Press.
Kuratsu, Y. 2006. Sekai Ga Wakaru Gendai Money Muttsu No Shiten. Chikumashobo Ltd.
Kuratsu, Y. 2009. Toshiginko Bubble No Syuen. Shinwonmunhwasa.
Kuroda, A. 2005. Kahei Shisurtemu No Sekaishi. Nonhyung.
Kurzweil, R. 2005. The Singularity is Near: When Humans transcend Biology. Gimm-Young Publishers, Inc.
Kwak, S. H. 1990. Comprehension of I Ching: the View of Nature and Human in I Ching. Seokwang Publishing.
Lafon, C. 2006. Idées Recues En Biologie. Woongjin Think Big Co. Ltd.
Lamfalussy, A. 2001. Financial Crises in Emerging Markets: An Essay on Financial Globalisation and Frgility. SungWoo Publishing Co.
Landry, C. 2004. The Creative City: A Toolkit for Urban Innovators. Haenam Publishing Co.
Lane, N. 2009. Power, Sex, Suicide-Mitochondria and the Meaning of Life. Puriwa Ipari Publishing Co.
Larsen, W. J. 2001. Human Embryology (3rd ed.). Churchill Livingstone, A Harcourt Health Sciences Company.
Ledoux, J. 1996. The Emotional Brain: The Mysterious Underpinnings of Emotional Life. Simon & Schuster Paperbacks.
Ledoux, J. 2002. Synaptic Self : How Our Brain Become Who We Are. Penguin Books.

LeDoux, J. 2005. Synaptic Self: How Our Brains Become Who We Are. Source Publishing.
LeDoux, J. 2006. The Emotional Brain (Korean ver.). Hakjisa publisher.
Lee, C. 1990. History of Development in Siberia. Minumsa.
Lee, C. K. 1998. Speculation Capital and Hegemony of USA. Yeongoosa.
Lee, D. H. 2008. Network Economy: East Asia Books.
Lee, H. S. 2006. Knowledge Administration System for the Prescription of Oriental Medicine based on the Compound Ontology. Korea Studies Institute Co.,Ltd.
Lee, J. H. 1975. The Ideas of Yulgok. Hyunamsa.
Lee, J. J. 2005. Economics of Land. Pakyoungsa.
Lee, J. J. 2005. Mathematical Design of Life. Books Hill Publishing Co.
Lee, J. P. 2009. The Gpd Particle. Mati Publishing Co.
Lee, K. H., et al. 2001. Transcendence of Establishing Personality. Chungkey Publishing.
Lee, N. (eds.). 2003. The Essential David Bohm. Routledge, Taylor & Prancis Group.
Lee, S. E. 1999. The Liftime and Study of Toegye. Yemoonseowon.
Lee, S. J. 1996. Educational Ideas of John Amos Comenius. The Kyoyuk Kwahak Sa.
Lee, S. S. 1987. Laser Speckles and Holography. Kyohak Yunkusa.
Lee, Y. 2004. Physics of Space-Time. Hongreung Science Publishing.
Lee, Y. P. 1996. The Complex System Model of Unification of South and North Korea. Shin-Yu Publishing Co.
Lee, Y. P., et al. 1997. System Science and National Polices. Shinyu Publishing.
Lee. S. W. 2004. Music Phychology. Shimsuldang Publishing Co.
Legge, J. (trans.). 2000. The Chinese Classics 3: The Shoo King. SMC Publishing Inc.
Legge, J. (trans.). 2000. The Chinese Classics 4: The She King. SMC Publishing Inc.
Legge, J. (trans.). 2000. The Chinese Classics 5: The Ch'un Ts'ew with The Tso Chuen. SMC Publishing Inc.
Legge, J. (trans.). 2001. The Chinese Classics 1 & 2 : Confucian Analects, The Great learning, The doctrine of the Mean, The Works of Mencius. SMC Publishing Inc.
Lesk, A. M. 2007. Introduction to Genomics. Oxford Univesity Press.
Levin, J. How the Universe Got Its Spots. HanSeung publishers.
Levin, S. 1999. Fraglie Dominion: Complexity and the Commons. Helix Books.
Levitin, D. J. 2008. This is your brain on Music. Mati Publishing.
Levitin, D. J. 2009. The World in Six Songs. Mati Publishing.
Lévy, P. 2000. Technology of Intelligence. Chu Hak Kwa Hyun Sil Sa Publishing Co.
Lévy, P. 2000. Technology of Intelligence. Chul Hak Kwa Hyn Sil Sa.
Lévy, P. 2003. World Philosophie. Thinking Tree Publishing Co.
Lévy, P. 2007. L'intelligence Collective: Pour une Anthropologie de Cyberspace. Moonji Publishing Co.
Lewin, R., and Regine, B. 2000. The Soul at Work: Listen…respond…let go: Embracing Complexity Science for Business Success. Simon & Schuster.
Lewin, R., and Regine, B. 2000. Weaving Complexity and Business: Engaging the Soul at Work. Texere.
Lindblom, C. E. 2009. The Market System: What It Is, How It Works, and What to make of It.
Lindley, D. 2009. Uncertainty: Einstein, Heisenberg, Bohr, and the Struggle for the Soul of Science. Marubol Publications.
Llinás, R. 2007. I of the Vortex. Booksense.
Lloyd, D. 2009. Radiant Cool. Wisdomhouse Publishing Co.
Lloyd, S. 2007. Programming the Universe. Chiho Publishing House.
Lowental, M. M. 2008. Intelligence: From Secrets To policy. Myung In Publishers.

Luecke, R. 2003. Managing Creativity and Innovation. Harvard Business School Publishing.
Luhmann, N. 2007. Soziale Systeme. Hangilsa Publishing Co., Ltd.
Luminet, J. & Lachiéze-Rey, M. 2007. De L'Infini… Mystéres et limites de l'Univers. Henamu Publishing Co.
Magueijo, J. 2005. Faster than the Speed of Light. Kachi Publishing Co.
Mainzer, K. 2005. Zeit Von der Urzeit zur Computerzeit. Dulnyouk Publishing Co.
Mankiewicz, R. 2002. The Story of Mathematics. Kyngmoon Publishing Co.
Maor, E. 2008. e: The Story of a Number. Kyung Moon Publishers.
Marcus, G. 2005. The Birth of the Mind. Henamu Publishing Co.
Matsufuji, T. 2007. Sekai Bburu Keizai Owari No Hajimari. ONE & ONE BOOKS Publishing Co.
Mazlish, B. 2001. The Fourth Discontinuity: The Co-evolution of Humans and Machines. Science Books Co., Ltd.
McArthur, J. V. 2006. Microbial Ecology: An Evolutionary Approach. Elsevier Academic Press Publications.
McMillan, J. 2007. Reinventing the Bazaar: a natural history of markets. Minumsa.
McMillan, J. 2007. Reinventing the Bazaar: a natural history of markets. Minumsa.
Mehlman, M. J. 2006. Wondergenes: Genetic Enhancement and the Future of Society. Chonpa Kwa Hak Sa.
Michel-Kerjan, E. and Slovic, P. (eds.).2008. The Irrational Economist: Making Decisions in a Dangerous World. Public Affairs
Micklethwait, J. 2004. The Company. Eulyoo Publishing Co.
Minsky, M. 1985. The Society of Mind. Simon & Schuster Paperbacks.
Mitchell, W. J. 1999. City of Bits: Space, Place, and the Infobahn. Gimm-Young Publishers, Inc.
Miura, K. 2003. Shushi To Ki To Shintai. Yemoonseowon.
Miyazaki, Y. 1992. 複合不況. The Korea Economic Daily.
Mizuho Research Institute Ltd. 2008. Subprime Kinyu Kiki. Joenryak Kwa Munhwa Publishing.
Moffett, S. 2007. The Three-Pound Enigma: The Human Brain and the Quest to Unlock its Mysteries by Shannon Moffett.
Mogi, K. 2004. No To Kaso. Yangmoon Publishing.
Mogi, K. 2005. Nou To Souzousei 'Kono Watashi' To Iu Kuoria E. PHP Institute.
Mogi, K. 2007. Hirameki Nou by Mogi Kenichiro. Brain World.
Monahan, D. 2001. World Atlas of the Oceans: More than 200 Maps and Charts of the Ocean Floor. A Firefly Book.
Monsma, S. V. 2000. Responsible Technology. Christian University Press.
Morange, M. 2000. A History of Molecular Biology. (Cobb, M. trans.) Harvard University Press.
Morowitz, H. J. The Emergence of Everything: How the World Became Complex.
Morrison, R. 2003. The Spririt int the Gene: Humanity's Proud Illusion and the Laws of nature.(Korean Ver.). Chon Pa Kwa Hak Sa.
Motoyama, Y. 2008. Kinyu Kenryoku. Strategy & Culture Publishing Co.
Mounier, E. 1952. Personalism. University of Notre Dame Press.
Murakami, K. 2003. Inochi no Baka-Jikara. Yang Moon.
Murphy, M. P., and O'Neill, L. A. J. (eds.). 1995. What is Life? The Next Fifty Years: Speculations on the future of biology. Cambridge University Press.
Murphy, P., and O'Neill, L. A. J. 2003. What is Life? The Next Fifty Years. Chiho Publishing House.
Mushegian, A. R. 2007. Foundations of Comparative Genomics, Elsevier Academic Press.
Nakagome, Y. 2004. An Illustrated Guide to the Human Genoce and Genes. Dong Hwa Technology.
Nakano, F., and Godai, T. 2009. 日中宇宙戰争. Nbook.

Narby, J. 2002. The Cosmic Serpent, DNA and the Origins of Knowledge. Dulnyouk Publishing Co.
National Geographic Society. 1998. National Geographic Satellite Atlas Of The World. National Geographic.
National Intelligence Council. 2009. Global Trends 2025: A Transformed World. Yemun Publishing.
Newberg, A., and Waldman, M. R. 2006. Why We Believe What We Believe: Uncovering Our Biological Need for Meaning, Spirituality, and Truth. Free Press.
Newberg, A., D'Aquili, E., and Rause, V. 2001. Why God Won't Go Away. Ballantine Book.
Newberg, A., et al. 2001. Why God Won't Go Away. Hanulim Publishing Co.
Newman, M., Barabási. Á. L., and Watts, D. J. 2006. The Structure and Dynamics of Networks. Princeton University Press.
Noaa. 2007. Hidden Depths: Atlas of the Oceans. Smithsonian.
Noble, D. 2006. The Music of Life: Biology Beyond the Genome. Oxford University Press.
Noble, D. 2009. The Music of Life: Biology Beyond the Genome. Yolinkwahak Publishing.
Nolte, J. 2007. Elsevier's Integrated Neuroscience. Mosby Elsevier.
Norgate, S. 2009. Beyond 9 To 5: Your Life in Time. Woongjin Think Big Co., Ltd.
Nowak, M. A. 2006. Evolutionary Dynamics: Exoploring the Equations of Life. The Berknap Press of Haravard University Press.
Núñez, R., and Freeman, W. J. (eds.). 1999. Reclaiming Cognition: the primacy of action intention and emtion. Imprint Academy
Nüsslein-Volhard, C. 2006. Das Werden des Lebens. Ichi Publishers.
Nye. J. S. Jr. 2002. The Paradox of American Power. Sejong Institute Press.
Ogoshi, T. 1987. Opto-Electronics and Opto-Communication. Hundaikwahakshinseo.
Ogura T. , et al. 2006. The Lifetime of Protein. World Science Publishing.
Oh, S, K. 2003. Basic Concepts of Supersymmetry. Chembum.
Pagels, H. R. 1991. The Dreams of Reason. Pumyang Co.,Ltd.
Palladino, M. A. 2006. Understanding the Human Genome Project (2nd ed.). Pearson Education, Inc.
Pannenberg, W. 1996. Anthropologie in theologischer Perspektive 1. Benedict Press.
Pannenberg, W. 1996. Anthropologie in theologischer Perspektive 2. Benedict Press.
Pannenberg, W. 1996. Anthropologie in theologischer Perspektive 3. Benedict Press.
Pannenberg, W. 2002. Toward a Theology of Nature: Essays on Science and Faith. Korea Theolohical Study Institute.
Parham, P. 2005. The Immune System (2nd ed.). Garland Science, Taylor & Francis Group.
Park, S. K. 2006. Searching for the Meaning of Time. Science Culture Publishing.
Park, W. H. 2005. Basic Principle of Economy: Discovery of Science, Philosophy, art and Economic Principle. SNU Press.
Parker. A. 2007. In the Blink of an Eye: The Cause of the Most Dramatic Event in the History of Life. Puriwa Ipari Publishing.
Pater, R. 2009. The Value of Nothing: How to Reshape Market Society and Redefine Democracy. Porto Bello Books Ltd.
Peacocke, A. 1993. Theology for a Scientific Age: Being and Becoming-Natural, Divine, and Human. Theology and the Sciences.
Penrose, R. 2004. The Emperor's New Mind 1: Concerning Computers, Minds, and The Laws of Physics (Korean ver.). Ewha Womans University Press.
Penrose, R. 2005. The Emperor's New Mind 2: Concerning Computers, Minds, and The Laws of Physics (Korean ver.). Ewha Womans University Press.
Perkowitz, S. 2008. Universal Foam. Science Books Co., Ltd.

Pieper, J. 2005. Wahrheit der Dinge: Eine Untersuchung zur Anthropologie des Hochmittelalters. The Catholic University of Korea Press.
Pinker, S. Language Instinct: How the Mind Creates Language. Sosoh Publishing.
Pinker, S.2007. How the Mind Works. Sosoh Publishing.
Playfair, J. H. L., and Chain, B. M. 2005. Immunology at a Glance (8th ed.). Blackwell Publishing.
Popper, K. 2008. Unended Quest: An Intellectual Autobiography. Galapagos.
Powelson, J. P. 1998. The Story of Land: A World History of Land Tenure and Agrarian Reform. The Korea Economic Daily.
Prigogine, I., and Stengers, I. 1989. Order out of Chaos: Man's New Dialogue with Nature. Jeongeumsa Publishing.
Ptashne, M., and Gann, A. 2002. Genes & Signals. Cold Spring Harbor Laboratory Press.
Ramachandran, V. S. 2006. The Emerging Mind. Bada Publishing Co.
Ramachandran, V. S. 2007. Phantoms in the Brain. Bada Publishing Co.
Randall, L. 2005. Warped Passages: Unraveling the Mysteries of the Universe's Hidden Dimensions. Harper Perennial.
Randall, L. 2008. Warped Passages. ScienceBooks Co., Ltd.
Rees, M. 2003. Before the Beginning. Henamu.
Regev, O. 2006. Chaos and Complexity in Astrophysics. Cambridge University Press.
Restak, R. 2006. The Naked Brain: How the Emerging Neurosociety is Changing how We Live, Work, and Love,Three Rivers Press.
Restak, R. M. 2006. The Brain. Hakjisa.
Rho, E. K. 2001. Turbulence. Sigma Press.
Rho, O. H. 2007. Theory of Viscosty Flow. Pakyoungsa.
Richerson, P. J., and Boyd, R. 2009. Not by Genes Alone: How Culture Transformed Human Evolution. E-UM Publishing House.
Ridley, M. 2000. Genome: The Autobiography of a species in 23 chapters. Gimm-Young Publishers, Inc.
Ridley, M. 2004. Nature Via Nurture: Genes, Experience, and What Makes Us Human. Gimm-Young Publishers, Inc.
Riemann, B. 2008. Prime Obsession: Bernard Riemann and The Greatest Unsolved Problem in Mathematics. Seung San Publishers.
Roberts, S. 2009. King of infinite space: Donald Coxeter, the man who saved geometry.. Seung San Publishers.
Rogers, T. T., and McClelland J. L. 2004. Semantic Cognition : A Parallel Distributed Processing Approach. A Bradford Book The MIT Press.
Rötzer, F. 2000. Megamaschine Wissen. Thinking Tree Publishing Co., Ltd.
Ruelle, D. Chance and Chaos(Koean ver.). 2000. Ewha Womans University Press.
Russell, P. 2007. From Science To God. Henamu Publishing Co., Ltd.
Russell, R. J., et al. (eds.) 2002. Neuroscience and the Person: Scientific perspectives on Divene Action. Vatican Observatory Publications.
Russell, R. J., Murphy, N., and Peacocke A. R. (eds.). Chaos and Complexity: Scientific Perspectives on Divine Action (2nd ed.). Vatican Observatory Publications.
Ryuichi. K. 1996. Ameba Organization of Kyocera. Changhyun Publishing.
Sachs. J. D. 2008. Common Wealth: Economics for a Crowded Planet. The Penguin Press.
Sacks, O. 2008. Musicophilia: Tales of Music and the Brain. Alma.
Samsung Economic Research Institute (ed.). 1997. Understanding and Application of Complexity Science. Samsung Economic Research Institute.

Sanders, 1998. T. I. Strategic Thinking and the New Science: Planning in the Midst of Chaos, Complexity, and Change. The Free Press.
Sarkar, S. 2005. Molecular Models of Life: Philosophical Papers on Molecular Biology. A Bradford Book the MIT Press
Schneider, E. D., and Sagan, D. 2005. Into the Cool: Energy Flow, Thermodynamics, and Life. The University of Chicago Press.
Schrödinger, E. 2000. What is Life?: The Physical Aspect of the Living Cell. Hanwool Publishing.
Schröeder, G. 2000. The Science of God. Pumyang Publishing Co.
Schroeder, G. L. 1997. The Science of God: The Convergence of Scientific and Biblical Wisdom. Broadway Books.
Schroeder, G. L. 2007. The Hidden Face of God: Science Reveals the Ultimate Truth. Heaven Treasury.
Schroeder, G.L. 2001. The Hidden Face of God: Science Reveals the Ultimate Truth. New York: A Touckstone Book.
Scott, A. 2001. Stairway to the Mind: The Controversial New Science of Consciousness. Ewha Woman University Press.
Seife, C. 2008. Decoding the Universe. Forest of Knowledge, a division of Nexus Press, Ltd.
Seki, O. 2009. Subete No Keizai Wa Baburu Ni Tsujiru. Iaso Publishing.
Seo, K. 2002. The Korean-English Buddhist Dictionary. Bulkwang Press.
Shanor, K. N. 2003. The Emerging Mind. Tree Planting People Publishing Co.
Shimizu, H. 1994. Life and Place. Joenpakwahaksa.
Shin, S. J., et al. 2000. Quantum Field Theoretical Consideration of Phase Transition. Hanyang University Press.
shin, S. J., et al. 2000. Quantum Field Theoretical Consideration of Phase Transition. Hanyang University Press.
Shinohara, Y. 2006. Choukakunou Mimi Wo Kaereba Jinsei Ga Kawaru. Joongang Life Publishing Co.
Siefer, W., and Weber, C. 2007. ICH, Wie wir uns selbst erfindedn . Dulnyouk Publishing Co.
Siegfried, T. 2003. The Bit and the Pendulum (Korean ver.), Gimm-Young Publishers, Inc.
Simonton, K. 2004. Creativity in Science: Chance, Logic, Genius, and Zeitgeist. Cambridge University Press.
Simpson, C. 2009. Science of Coercion. Sunin Book.
Singer, P. 2006. The Expanding Circle: Ethinics and Sociobiology. Ingansarang Publishing Co.
Smith, A. 1984. The Geopolitics of Information: How Western Culture Dominates the World. Sungkyunkwan University Press.
Smolin, L. 2007. Three Roads to Quantum Gravity. ScienceBooks Co.,Ltd.
Sober, Elliott. 2004. Philosophy of Biology. Chul Hak Kwa Hyun Sil Sa Publishing Co.
Soh, K. S. 2001. Physics and 'Awakening of Faith in the Mahayana'. Seoul National University Press.
Sóle, R., and Goodwin, B. 2000. Signs of Life: How Complexity Pervades Biology. Basic Books, A Member of the Perseus Books Group.
Solms, M., and Turbull, O. 2005. The Brain and the Inner World: An Introduction to the neuroscience of subjective experience (Korean ver.). Hana Medical Publishing Co.
Song, H. 2008. Currency Wars. Random House Korea, Inc.
Song, S.G., Bae, M. I., Kwon, K. J., Shim, J. R., Jung, B. J., Cho, S. T., and Choi, M. J. 2003. The Teaching of Buddha. Korean Buddhism Promotion Foundation Press.
Soros, G. 2008. The New Paradigm for Financial Markets. Wisdomhouse Publishing Co., Ltd.
Soros, G. 2008. The New Paradigm for Financial Markets: The Credit Crisis of 2008 and

What It Means. Public Affairs.
Stacey, R. D. 1992. Managing the Unknowable: Strategic Boundaries Between Order and Chaos in Organizations. Jossey-Bass, a Wiley Company.
Stacey, R. D. 2001. Complex Responsive Processes In Organizations: Learning and knowledge creation. Routledge, Taylor & Francis Group.
Stacey, R. D., et al. 2002. Complexity and Management: Fad or Radical Challenge to Systems Thinking? Routledge, Taylor & Francis Group.
Stein, M. 2008. Solar Conscience Lunar Conscience: An Essay on the Psychological Foundations of Morality, Lawfulness, and the Sense of Justice. Chulhak kwa Hyunsil Sa Publishing Co.
Sternberg, R. J. (ed.). 1999. Handbook of Creativity. Cambridge University Press.
Steven, S. 2003. The Emerging Science Sync of Spontaneous Order. Hyperion Books.
Stiglitz, J. E. 2008. Making Globalization Work. Book21 Publishing Group.
Stirling, J. 2003. Introducing Neuropsychology. Sigmapress.
Strobel, L. 2005. The Case fo a Creator. Tyrannus Press.
Strogatz, S. H. 1994. Nonlinear Dynamics and Chaos: With Applications to Physics, Biology, Chemistry, and Engineering. Westview Press, a member of Perseus Books Group.
Strogatz. 2005. Sync. Gimm-Youngsa Publishers, Inc.
Styer, D. F. 2004. The Strange World of Quantum Mechanics. Book's-Hill Publishing Co., Inc.
Sun, T. (Trans.). 2001.Tao Te Ching: The greatest wisdom of Lao-Tze. CHI CHU Moral Culture and Education Foundation.
Swedenborg, E. 1996. True Christian Religion: Containing the universal theology of the new church foretold by the Lord in Daniel 7:13-14 and Revelation 21:1-2. Pennsylvania: Swedenborg Foundation.
Synnestvedt, S(eds.). 1977. The Essential Swedenborg: Basic Religious Teachings of Emanuel Swedenborg. Swedenborg Foundation.
Szpiro, G. G. 2008. The Secret Life of Numbers: Fifty Easy Pieces on How Mathematicians Work and Think. Kachi Publishing Co.
Tada, T. 2006. Semantics of Immunity. Hanulbooks.
Takashi, S. 2008. Ontology Algorithm 1: The World of Recordation, Information, and Knowledge. Hanul Publishing Co.
Talbot, M. 1999. Holographic Universe. Junshinsekyesa.
Tammet, D. 2009. Embracing the Wide Sky: A Tour across the Horizons of the Mind by Daniel Tammet. HongIk Publishing Co.,Ltd.
Tancredi, L. 2005. Hardwired Behavior: What Neuroscience Reveals about Morality. Cambridge University Press.
Tapscott, D. 2000. Creating Value in the Network Economy. Mulpure Publishing Co.
Tatarkiewicz, W. 2001. A History of Six Ideas: An Essay in Aesthetics. Misulunhwa Publishing Co.
Tattersall, I. 2006. The Monkey in the Mirror. Essays on the Science of What Makes us Human. (Korean Ver.). Henamu.
Tattersall, I. 2007. Be-Coming Human: Evolution and Human Uniqueness. (Korean Ver.). Henamu.
Taylor, E. L. H. 1978. Economics Money and banking: Christian Principles. The Craig Press.
Taylor, J. B. 2008. My Stroke of Insight: A Brain Scientist's Personal Journey. Hodder & Stoughton.
Taylor, M. C. 2001. The moment of complexity: emerging network culture. The University of Chicago Press.
Terrien, S. 2000. The Elusive Presence: Toward a New Biblical Theology. Wipf and Stock Publishers.

The Center for Strategic and International Studies. 2009. A Smarter, More Secure America: Report of the CSIS Commission on Smart Power. Samin Books.
The Research Group of Socialogy of SNU. 1988. Contemporary Capitalism and Community Theory. Hangilsa Publishing Co., Ltd.
The Whitehead Society of Korea. 1999. Metaphysics of Creativity. DongkwaSeo.
Thelen, E., and Smith, L. B. 1994. A Dynamic Systems Approach to the Development of Cognition and Action. A Bradford Book the MIT Press.
Tjeng, E. C. 2000. Fundamental Problems of Being. Yolin.
Todaro, M. P. 1988. Economic Development in the Third World. Bee Bong Publishing Co.
Toffler, A., and Toffler, H. 2009. Beyond Depression: Yesterday, Today, and Tomorrow. Chungrim Publishing Co., Ltd.
Trocmé, A. 1986. Jésus et la révolution non violente. Korea Theological Study Institute.
Tsuzuki, M. 2002. Shinsouban Yojigen No Sekai. Hongbooks Publishing Co.
Tullock, G. 2007. Rent Seeking. Kyungsung University Press.
Turner's B. S. 2006. The Body & Society: Explorations in Social Theory. Body and Mind Books.
Varela, F., et al. 1991. The Embodied Mind: Cognitive, Science, and Human Experience. The MIT Press.
Varela, F., et al. 2000. The View From Within: First-person approaches to the study of consciousness. mprint Academy
von Baeyer, H. C. 2007. Information: the New Language of Science. Seung San Publishers.
von Rad, G. 1972. Genesis: A Commentary (Revised Edition). The Westminster Press.
Wada, H. 2004. Tajuujinkaku. Hakjisa.
Waldrop, M. M. 1992. Complexity: The Emerging Science at the Edge of Order and Chaos. A Touchstone Book, published by Simon & Schuster.
Wallerstein, I. 2005. World-Systems Analysis: An Introduction. Dangdae Publishing Co.
Walton, J. H. 2009. The Lost World of Genesis One: Ancient Cosmology and the Origins Debate. IVP.
Warfield, J. N. 2006. An Introduction to Systems Science. World Scientific.
Watson, L. 2002. Jacobson's Organ: And the Remarkable Nature of Smell. Jeonshinsekyesa Publishing.
Watts, D. J. 2003. Six Degrees: The Science of a Connected Age. W. W. Norton & Company, Inc.
Weber, A. Alles fühlt. Woongjin Think Big Co., Ltd.
Weinberg, S. 2005. The First Three Minutes. Yangmoon Publishing Co.,Ltd.
Weinberg, S. 2007. Dreams of a Final Theory. Science Books Co.,Ltd.
Werkmeister, W. H. 1999. Historical Spectrum of Value Theries 1. Seokwangsa Publishing Co.
Whitehead, A. N. 2005. Process and Reality: An Essay in Cosmology.
Wilczek, F. 2008. The Lightness of Being: Mass, Ether, and the Unification of Forces. Basic Books.
Witkowski, J.(eds.). 2005. The Inside Story: DNA to RNA to Protein. Cold Spring Harbor Laboratory Press.
Wuketits, F. M. 2006. Gene, Kultur und Moral: Soziobiologie-Pro und Contra. Science Books. Co., Ltd.
Yamadori, A. 2003. Neuropsychology of Memory. Koonja Publishing.
Yan, J. F. 2002. DNA and the I Ching: The Tao of Life. Body and Mind Books.
Yang, K. S. 1994. The Philosophy of Peace of Zamenhof. Jayumoonko.
Yoshida, S. 2008. Hito To Saru No Aida. Senior Communication Inc.
Yoshinaha, Y. 1997. 'Fukuzatsu-Kei' Towa Nani Ka. Korea Economic Daily.
Yuasa, T. 2003. Imangination of World History. Hundaimihaksa.
Yun, Y. S., and Chae, S. B. 2005. An Introduction to Complex System. Samsung Economic Research Institute.
Zakaria, F. 2008. The Post-American World. Vega Books.
Zankl, H. 2007. Das verflixte X. Woongjin Think Big Co., Ltd.

Zee, A. 1995. Fearful Symmetry. Pumyang Co., Ltd.
Zimmer, C. 2004. Soul Made Flesh: The Discovery of the Brain and How It Changed the World. Free Press.
Zizizek, S. 2009. The Parallax View. Mati Publishing.
加藤恭義, et al. 2006. Cellular Automata. Intervision Publishing.
康學偉, et al. 2009. 周易研究史. Yemoonseowon.
孔子, 2009. Confucius: The Analects (Chinese-English Edition, 楊伯峻, et al. trans.). 聯經.
老子, 2009. Lin Yutang Chinese-English Bilingual Edition, The Wisdom of Laotse Ⅰ (Lin Yutang, trans.). Cheng Chung Book Co.,Ltd.
大塚久雄. 1982. A Baisic Theory of Community. Dolbegae.
馬鳴菩薩. 2000. Awakening of Faith in the Mahayana. 玄音社
孟子, et al. 2009. Lin Yutang Chinese-English Bilingual Edition, Harvest Moon on West Lake. (Lin Yutang, trans.). Cheng Chung Book Co.,Ltd.
名東孝二. 1990. Political Economy of Underground Economy. Joenchaek.
武者利光. 1995. World of Noises: The Mystery of 1/f Noises existing in the Natural World. Jeonpakwahak Publishing.
民族과思想 研究會 編. 2003. 四端七情論. Seokwangsa.
本間龍雄. 1995. The Way to Topological Phase Space (Korean ver.). Chon Pa Kwa Hak Sa.
北溪 陳淳. 2005. 性理字義. Yeogang.
司馬遷, et al. 2009. Lin Yutang Chinese-English Bilingual Edition, The Wisdom of Confucius Ⅱ. (Lin Yutang, trans.). Cheng Chung Book Co.,Ltd.
三石善吉 (Jenkichi, M.). 1993. The Millennium of China. Koreawon Publishing.
蘇東坡, 2008. Lin Yutang Chinese-English Bilingual Edition, Selected Poems and Prose of Su Tungpo. (Lin Yutang, trans.). Cheng Chung Book Co.,Ltd.
矢野暢. 世界單位論. 講座現代の 地域研究 2. Jeonyewon Publishing.
岩村 充. 2003. Cyber Economy. Shinronsa.
龍岡靜夫. 2007. Laser and Image. Gijeonyoengusa.
元曉. 1999. 大乘起信論 疎,別記. 一志社.
李申. 1991. 周易之河說解. 中國社會科學院 世界宗教研究所.
日本平和學會 編輯委員會. 1987. Peace Studies(平和學): Theory and Practice. 文佑社.
莊子, et al. 2009. Lin Yutang Chinese-English Bilingual Edition, The Concubine Market of Yangchow. (Lin Yutang, trans.). Cheng Chung Book Co.,Ltd.
張潮, 2008. Lin Yutang Chinese-English Bilingual Edition, Dream Shadows. (Lin Yutang, trans.). Cheng Chung Book Co.,Ltd.
長晋藩 編. 2006. 中國法制史. Sonamoo Publishing.
錢穆. 1987. A New Comprehension of Chinese History. Jipmoondang Publishing.
鄭板橋, et al. 2009. Lin Yutang Chinese-English Bilingual Edition, Family Letters of a Chinese Poet. (Lin Yutang, trans.). Cheng Chung Book Co.,Ltd.
周敦頤. 2004. 通書解. Chungkye Publishing.
陳來. 2004. 宋明理學. Yemoonseowon.
陣寶良. 2001. 中國流氓史. Acanet Publishing.
韓國東洋哲學會 編. 1996. Ontology and Personalism of Eastern philosophy. Yonsei University Press.

Index

A

B

C

D

E

F

G

H

I

J

Q

R

S

T

U

V

W

Y